P9-CJT-096

RENEWALS: 691-4574

DATE DUE

4995848			
7-2-93			
DEC 0 1			
NOV 0 9			
MAY 0 1			
NOV -			
APR 1 6			
NOV 2 9			
APR 1 2			

Demco, Inc. 38-293

WITHDRAWN

Microbial Technology

SECOND EDITION/VOLUME I

WITHDRAWN
UTSA LIBRARIES

Second Edition/Volume I

Microbial Technology

Microbial Processes

Edited by

H. J. PEPPLER
Universal Foods Corporation
Milwaukee, Wisconsin

D. PERLMAN
School of Pharmacy
University of Wisconsin
Madison, Wisconsin

ACADEMIC PRESS
New York   San Francisco   London   1979
A Subsidiary of Harcourt Brace Jovanovich, Publishers

LIBRARY
The University of Texas
At San Antonio

COPYRIGHT © 1979, BY ACADEMIC PRESS, INC.
ALL RIGHTS RESERVED.
NO PART OF THIS PUBLICATION MAY BE REPRODUCED OR
TRANSMITTED IN ANY FORM OR BY ANY MEANS, ELECTRONIC
OR MECHANICAL, INCLUDING PHOTOCOPY, RECORDING, OR ANY
INFORMATION STORAGE AND RETRIEVAL SYSTEM, WITHOUT
PERMISSION IN WRITING FROM THE PUBLISHER.

ACADEMIC PRESS, INC.
111 Fifth Avenue, New York, New York 10003

United Kingdom Edition published by
ACADEMIC PRESS, INC. (LONDON) LTD.
24/28 Oval Road, London NW1 7DX

Library of Congress Cataloging in Publication Data

Peppler, Henry J
    Microbial technology.

    Includes bibliographies and index.
    CONTENTS: v. 1. Microbial processes.
1. Industrial microbiology. 2. Fermentation.
I. Title.
QR53.P45 1979   660'.62   78–67883
ISBN 0–12–551501–4

PRINTED IN THE UNITED STATES OF AMERICA

79 80 81 82   9 8 7 6 5 4 3 2 1

Contents

Chapter 17 Carotenoids
L. Ninet and J. Renaut

Subject Index

List of Contributors

Numbers in parentheses indicate the pages on which the authors' contributions begin.

Ralph F. Anderson* (1), Velsicol Chemical Corporation, Chicago, Illinois 60611

Otto Andresen (281), NOVO Industri A/S, Bagsvaerd, Denmark

Knud Aunstrup (281), NOVO Industri A/S, Bagsvaerd, Denmark

Joe C. Burton (29), Research Department, The Nitragin Company, Milwaukee, Wisconsin 53209

F. Constabel (389), Prairie Regional Laboratory, National Research Council of Canada, Saskatoon, Saskatchewan S7N OW9, Canada

I. W. Cottrell (417), Research and Development Department, Kelco Division, Merck & Company, Inc., San Diego, California 92123

Edvard A. Falch (281), NOVO Industri A/S, Bagsvaerd, Denmark

J. Florent (497), Centre Nicolas Grillet, Rhône-Poulenc Industries, Vitry-sur-Seine, France

Yoshio Hirose (211), Biochemical Department, Central Research Laboratories of Ajinomoto Company, Ltd., Kawasaki-ku, Kawasaki 210, Japan

Carlo M. Ignoffo (1), U.S. Department of Agriculture, Science and Education Administration, Federal Research, Columbia, Missouri 65205

K. S. Kang (417), Biochemical Development, Kelco Division, Merck & Company, Inc., San Diego, California 92123

W. G. W. Kurz (389), Prairie Regional Laboratory, National Research Council of Canada, Saskatoon, Saskatchewan S7N OW9, Canada

John H. Litchfield (93), Battelle, Columbus Laboratories, Columbus, Ohio 43201

Lewis B. Lockwood (355), Biology Department, Western Kentucky University, Bowling Green, Kentucky 42101

* Deceased

J. L. Martin* (187), International Minerals & Chemical Corporation, Terre Haute, Indiana 47808

Yoshio Nakao (311), Microbiological Research Laboratories, Central Research Division, Takeda Chemical Industries, Ltd., Jusohonmachi, Yodogawa-ku, Osaka, Japan

Tage Kjaer Nielsen (281), NOVO Industri A/S, Bagsvaerd, Denmark

L. Ninet (497, 529), Centre Nicolas Grillet, Rhône-Poulenc Industries, Vitry-sur-Seine, France

Hiroshi Okada (211), Biochemical Department, Central Research Laboratories of Ajinomoto Company, Ltd., Kawasaki-ku, Kawasaki 210, Japan

H. J. Peppler† (157), Universal Foods Corporation, Milwaukee, Wisconsin 53201

D. Perlman (241, 483, 521), School of Pharmacy, University of Wisconsin, Madison, Wisconsin 53706

Randolph S. Porubcan (59), Chr. Hansen's Laboratory, Inc., Milwaukee, Wisconsin 53214

J. Renaut (529), Centre Nicolas Grillet, Rhône-Poulenc Industries, Vitry-sur-Seine, France

O. K. Sebek (483), Infectious Diseases Research, The Upjohn Company, Kalamazoo, Michigan 49001

Robert L. Sellars (59), Chr. Hansen's Laboratory, Inc., Milwaukee, Wisconsin 53214

M. T. Walton** (187), International Minerals & Chemical Corporation, Terre Haute, Indiana 47808

* Present address: 1334 S. Center Street, Terre Haute, Indiana 47802.
† Present address: 5157 N. Shoreland Avenue, Whitefish Bay, Wisconsin 53217.|
** Present address: RFD No. 3, Shoals, Indiana 47581.

Preface

In the decade since the first edition of "Microbial Technology" appeared, applied microbiology has changed, expanded, and diversified. As new products were introduced in this period, and greater demand for some of the old ones developed, the total fermentation capacity increased at about the same rate noted in the previous 25 years. The number of fermentation products, it is estimated, has quadrupled, while the volume of products manufactured has increased tenfold. This growth has prompted publication of a second edition, completely revised and enlarged.

To accomplish a worldwide survey of industrial microbiology and to describe its contributions to agriculture, industry, medicine, and environmental control, the editors are indebted to 57 willing and expert contributors. Their comprehensive reviews of traditional fermentations and propagations, as well as newly developed microbe-dependent processes and products, are presented in a two-volume set.

Volume I, subtitled "Microbial Processes," describes the production and uses of economic bacteria, yeast, molds, and viruses, and reviews the technologies associated with products of microbial metabolism.

Volume II, subtitled "Fermentation Technology," deals principally with fermentations and modifications of plant and animal products for foods, beverages, and feeds, while reviewing salient aspects of microbial technology: general principles, culture selection, laboratory methods, instrumentation, computer control, product isolation, immobilized cell usage, economics, and microbial patents.

H. J. Peppler
D. Perlman

Contents of Volume II

FERMENTATION TECHNOLOGY

Microbial Technology

SECOND EDITION/VOLUME I

Chapter 1

Bioinsecticides

CARLO M. IGNOFFO
RALPH F. ANDERSON*

I. INTRODUCTION

Control of insect pests costs more than one-half billion dollars annually in the United States. Yet less than 5% of this is spent for microbial insecticides, despite the fact that entomopathogens (pathogens of insects) have been suggested as controlling agents of insect pests for over a century. Synthetic organic chemical insecticides in vogue today were not even available until about 50 years after control of an insect pest was demonstrated by using an entomopathogen. Obviously, the development of microbial insecticides has been pitifully slow. However, within

* In Memorium: Ralph F. Anderson (1924–1978).

1

the last decade there has been a significant revival. This was largely brought about by recent successes in developing entomopathogenic viruses and bacteria and was further fortified by the public's increased awareness of the impact of toxic, broad-spectrum chemical insecticides.

Of the nearly 1 million species of known insects only about 15,000 species are considered pests and only about 300 are sufficiently destructive to require some control. Fortunately, most insect pests have pathogenic microorganisms associated with them. About 1500 entomopathogens belonging to the fungi, viruses, protozoa, or bacteria are known. Of these, bacteria and viruses, because of their known effectiveness and relative lack of toxicity or pathogenicity to nontarget animals and plants, have been developed into commercial products in the United States.

A candidate entomopathogen, whether it is a bacterium, virus, fungus, or protozoan, must meet certain technical prerequisites before it can be developed into a microbial insecticide. Three of the most important technical prerequisites are: (1) availability or feasibility of a systematic, continuous production technology; (2) minimal or no toxicity or pathogenicity to man, nontarget animals, and plants; and (3) proved effectiveness against the intended target pest. However, from a commercial, nontechnical viewpoint the ultimate and most important question is whether this development will result in a reasonable profit to the industrial developer. Obviously, then, the decision whether or not to develop a candidate entomopathogen into a microbial insecticide is based upon favorable answers to both technical and nontechnical questions.

II. HISTORICAL BACKGROUND

Aristotle's description of diseases of the honey bee in *Historia animalium* was probably the first documentation of an insect disease. Observations of insect maladies, particularly those of bees and silkworms, were reported in early Greek and Roman literature as well as in the works of various poets and naturalists of the sixteenth to nineteenth centuries. For example, Kirby and Spence (1826) included a chapter called "Diseases of Insects" in their classical text "*Introduction to Entomology*."

Insect pathology per se probably had its beginning in the nineteenth century under the stimulus of Bassi (1835) and Pasteur (1870). Bassi was the first to demonstrate that a microorganism, the fungus *Beauveria bassiana,* could cause an infectious disease in an animal (*the silkworm*).

Bassi not only contributed to man's understanding of infectious diseases but also implied that infectious diseases could be used to control insects. Pasteur's investigations were concerned not with insect control but with control of diseases in populations of the beneficial silkworm and honey bee. These studies drew attention to the impact of diseases on insect populations and their feasibility for use as microbial insecticides.

A significant contribution to microbial control of insects was made by Metchnikoff (1879) and Krassilstschik (1888). They were the first to document that an entomopathogen, a muscardine fungus, *Metarrhizium anisopliae,* could be mass produced and applied as a microbial insecticide to control the grain weevil and the sugar beet curculio. These early efforts inspired Giard (1892), Forbes (1898), and Snow (1891) to widely use another muscardine fungus (*B. bassiana*) in an unsuccessful attempt to control chinch bugs.

Control of insect pests with bacteria was probably first attempted by d'Herelle (1914) approximately 35 years after Pasteur's description of silkworm diseases. Apparently, control was not consistent and therefore interest in bacterial pathogens was muted. However, after a lag of nearly 30 years, White and Dutky (1940) succeeded in demonstrating control of the Japanese beetle by distributing spores of the milky disease bacterium, *Bacillus popilliae.* Undoubtedly, this success stimulated other investigators to reinvestigate bacteria, and literature began appearing on the effectiveness of *Bacillus thuringiensis.* Issuance of eight patents between the years 1960 and 1963 for *B. thuringiensis* further attested to the revived interest in bacterial insecticides (Briggs, 1964). A more detailed history of the use of bacteria to control insects is presented by Steinhaus (1964), Heimpel (1965), and DeBach (1964).

Use of viruses to control insect pests was stimulated by the studies of Balch and Bird (1944) and Steinhaus and Thompson (1949) during the years immediately following World War II. This initial interest, which lagged after the first successful demonstration, is presently having a rebirth, as evidenced by the recent registration of the first viral pesticide (Ignoffo, 1973a, b,c) in the United States by the Environmental Protection Agency (EPA).

Documentation of the historical development and use of entomopathogens as microbial insecticides is an empty exercise without noting the contributions of an inspiring, enthusiastic giant, Edward A. Steinhaus, who died in 1969 (Linsley and Smith, 1970). Dr. Steinhaus not only developed the first university curriculum of insect pathology and laboratory of pathology in the United States but also wrote texts, reference sources, and numerous articles on invertebrate pathology. He was responsible for organizing the Society of Invertebrate Pathology and starting the *Journal of Invertebrate Pathology*. His guiding principles and vision still inspire and influence invertebrate pathologists. Through his efforts there are laboratories throughout the world devoted to both basic and applied research in invertebrate pathology and a flourishing society and a journal.

III. CANDIDATE MICROBIAL INSECTICIDES

A. Entomopathogenic Bacteria

From a stockpile of nearly a hundred species associated with insects, only three bacteria (*B. popilliae, B. thuringiensis,* and *B. moritai*) were developed into commercial microbial insecticides (Table I). *In vitro* and *in vivo* production processes both are used (cf. Section V,A). The bacterium *B. popilliae*, produced only in larvae of Japanese beetles, is formulated into dust (Dutky, 1963). In contrast, *B. thuringiensis* and *B. moritai* are produced by conventional fermentation techniques. It is significant that the bacterial species that have been most useful for insect control have been sporeformers, perhaps because they can be

TABLE I.   Trade Names of Commercial or Experimental Preparations of Microbial Insecticides Formulated from Bacteria

Bacterial species	Trade name	Producer
Bacillus moritai	Rabirusu	Sumitomo, Japan
Bacillus popilliae	Doom, Japidemic	Fairfax Biological Labs. (U.S.A.)
Bacillus thuringiensis		
β-Exotoxin	Biotoxksybacillin	All Union Inst. Agr. Microbiol. (U.S.S.R.)
	Eksotoksin	Glavmikrobioprom (U.S.S.R.)
	Toxobakterin	Glavmikrobioprom (U.S.S.R.)
δ-Endotoxin	Agritrol	Merck & Co. (U.S.A.)
	Bakthane	Rohm & Haas (U.S.A.)
	Bactospeine	Roger Bellon (France)
	Bathurin	Chemapol-Biokrma (Czechoslovakia)
	Biospor	Farbwerke Hoechst (Germany)
	Biotrol BTB	Nutrilite Prod (U.S.A.)
	Dendrobacillin	Glavmikrobioprom (U.S.S.R.)
	Dipel	Abbott Labs (U.S.A.)
	Entobacterin	Glavmikrobioprom (U.S.S.R.)
	Insektin	Glavmikrobioprom (U.S.S.R.)
	Parasporin	Grain Proc. Lab (U.S.A.)
	Sporeine	LIBEC Laboratoire (France)
	Thuricide	Sandoz-Inc. (U.S.A.)

easily mass produced and are stable enough to be handled in commerce. A relatively large number of entomopathogenic spore-forming bacteria were listed by Steinhaus (1959). However, none of the nonsporeformers, (which also can be mass cultured) have been commercialized, probably because they are not easily stabilized for field use. Safety to humans, other vertebrates, and plants was demonstrated for *B. popilliae, B. thuringiensis,* and *B. moritai* but has not been established for other candidate bacterial pathogens of insects (cf. Section V,B). Surprisingly, 19 proprietary names are reported for *B. thuringiensis* preparations (Ignoffo, 1967, 1975). Thirteen formulations are based upon the δ-endotoxin and three on the β-exotoxin. Two trade-name products are formulated from *B. popilliae* and one from *B. moritai.*

B. Entomopathogenic Viruses

Many insect pathologists believe viruses have the greatest potential for development into microbial insecticides. Entomopathogenic viruses are specific and are active against many economically important insects (Ignoffo, 1967, 1968, 1975). About 650 insect viruses are described (Martignoni and Iwai, 1977), and more are being discovered each year. In fact, there has been a phenomenal increase (over 250%) in the number of new viruses reported within the decade covered by this revision (Anderson and Ignoffo, 1967).

Just as the more useful candidate bacteria are protected by spores, the most useful insect viruses are protected by a proteinaceous inclusion body. Those without this protective body are called noninclusion viruses. The inclusion body of insect viruses varies in shape from the irregular or regular polyhedrons (PIB) characteristic of the polyhedrosis and entomopoxviruses (EPV) to ovoid bodies representative of the granulosis viruses (GIV). The site of viral replication also is commonly used to describe insect viruses, e.g., nuclear polyhedrosis (NPV) or cytoplasmic polyhedrosis (CPV). Entomopathogenic viruses are currently placed in seven genera based upon the morphology and physiobiochemical characteristics of the virion (Ignoffo, 1973a; Vago *et al.*, 1974). Latinized binomials are used to name the viruses and the suffix-*virus* is used for all the generic names, e.g., *Baculovirus, Entomopoxvirus, Iridovirus.*

The NPV and GIV, because of their specificity, safety, virulence, and stability, are probably the most promising candidates for consideration as viral insecticides. The CPV and EPV are less virulent, but they have the other characteristics of a good microbial pesticide. Noninclusion viruses, although virulent, are not currently considered likely candidates because less is known about their safety, production feasibility, and efficacy. Considerable information has been obtained on the safety and specificity of entomopathogenic viruses within the last decade (cf. Section V,B,3).

Because of their specificity, insect viruses are more difficult to produce on an industrial scale than bacteria or fungi. They are obligate parasites and consequently must be mass produced *in vivo,* i.e., in living hosts (cf. Section V,A,3). In spite of this inherent difficulty, insect viruses representing every major group have been produced and used in the field as viral insecticides (Ignoffo, 1967, 1975). In addition, more than a dozen commercial or experimental preparations of entomopathogenic viruses have been evaluated sufficiently enough to warrant a trade name designation (Table II). As of 1977, two insect viruses have been commercialized: the *Heliothis* NPV, trade-named Elcar (USA)*; and a CPV of *Dendrolimus* called Matsukemin in Japan. The U.S. Forest Service recently registered an NPV of the tussock moth, TM-BioControl-T (Martignoni and Iwai, 1978), and submitted a registration request to EPA for the gypsy moth NPV. Data are also being collated for submission of a registration of another NPV, i.e., *Autographa,* by the Agricultural Research Service. Insect viruses continue to be attractive candidates for microbial insecticides. Within the last decade production feasibility has been established, safety and specificity to nontarget organisms has been confirmed, and effectiveness against target pests has been demonstrated.

C. Entomopathogenic Fungi

No fungi are presently available as commercial microbial insecticides in the United States, but several were produced and used on a large scale in the Soviet Union (Table III). All classes are represented among the more than 500 entomopathogenic fungi (Ignoffo, 1967). Most of the promising fungi are in the class Phycomycetes and Deuteromycetes. Fungi associated with insects are considered effective pathogens if and when field conditions are optimal for spore germination, invasion, and growth. Because these conditions are not predictable, the usefulness of fungi for insect control is often questioned (Yendol and Roberts, 1970). Some mycologists (Bucher, 1964; Latch, 1965) contend that entomogenous fungi are ubiquitous and that field application for insect control is

---

* This paper reports the results of research only. Mention of a pesticide in this paper does not constitute a recommendation for use by the USDA, nor does it imply registration under the Federal Insecticide, Fungicide and Rodenticide Act (FIFRA) as amended. Also, mention of a commercial product in this paper does not constitute an endorsement of this product by the USDA.

TABLE II. Trade Names of Commercial or Experimental Preparations of Microbial Insecticides Formulated from Viruses

Viral type and host	Trade name	Producer
Cytoplasmic polyhedrosis		
Dendrolimus	Matsukemin	Chugai Pharmaceutical Co., Ltd. (Japan)
		Kumiai Chemical Industry Co., Ltd. (Japan)
Nuclear polyhedrosis		
Heliothis	Biotrol VHZ	Nutrilite Prod. (U.S.A.)
	Elcar	Sandoz, Inc. (U.S.A.)
	Virex	Hays-Sammons (U.S.A.)
	Viron/H	Int. Minerals & Chem. Corp. (U.S.A.)
Lymantria	Virin-ENSH	Glavmikrobioprom (U.S.S.R.)
Mamestra	Virin-EX	Glavmikrobioprom (U.S.S.R.)
Neodiprion	Polyvirocide	Indiana Farm Bureau (U.S.A.)
Orgyia	TM BioControl-1	Forest Service (U.S.A.)
Pieris	Virin GKB	Latvian Agr. Acad. (U.S.S.R.)
Prodenia	Biotrol VPO	Nutrilite Prod. (U.S.A.)
	Viron/P	Int. Minerals & Chem. Corp. (U.S.A.)
Spodoptera	Viron/S	Int. Minerals & Chem. Corp. (U.S.A.)
Trichoplusia	Biotrol VTN	Nutrilite Prod. (U.S.A.)
	Viron/T	Int. Minerals & Chem. Corp. (U.S.A.)

unnecessary. Others (Dunn and Mechalas, 1963; Müller-Kögler, 1965; Roberts and Yendol, 1971; McCoy *et al.*, 1976) are convinced that fungal insecticides are effective if properly used and that their usefulness can be increased when combined with chemical insecticides (Pristavko, 1963).

Fungal mycelium, spores, and toxins can be mass produced, either *in vitro* or *in vivo* (cf. Section V,A). Obviously, the *in vitro* method, either via submerged or surface fermentation, is preferred (Ignoffo, 1967; Dulmage and Rhodes, 1971; Roberts and Yendol, 1971). Some of the fungi that have been successful in controlling insects are *Aschersonia aleyrodis,*

TABLE III. Trade Names of Commercial or Experimental Preparations of Microbial Insecticides Formulated from Fungi

Fungal species	Trade name	Producer
Aschersonia aleyrodis	Aseronija	All Union Inst. Agr. Microbiol. (U.S.S.R.)
Beauveria bassiana	Biotrol FBB	Nutrilite Products (U.S.A.)
	Boverin	Glavmikrobioprom (U.S.S.R.)
Hirsutella thompsonii	ABG-6065	Abbott Labs (U.S.A.)
Metarrhizium anisopliae	Biotrol FMA	Nutrilite Products (U.S.A.)

B. bassiana, M. anisopliae, Entomophthora thaxteriana, and *Nomuraea rileyi* (Ignoffo, 1975).

One advantage of entomopathogenic fungi is that they are generally less specific than other groups of entomopathogens and consequently can be used against different types of insect pests. For example, species of fungi were successful in controlling scale insects, whitefly, plant hoppers, aphids, chinch bugs, phytophagous beetles, flies, caterpillars, and stored-products pests. More extensive development of fungal microbial insecticides, however, must await additional evidence of field efficacy (cf. Section V,C, 2) and safety (cf. Section V,B,3). Fortunately, there has been considerable effort in both these areas within the last decade, which increases the probability that entomopathogenic fungi can be commercially available within the next decade.

D. Entomopathogenic Protozoa

Protozoa of insects never have been industrially produced or sold as microbial insecticides, largely because they are low in virulence, unstable, and difficult to produce. Currently, production can be accomplished only *in vivo* (cf. Section V,A,3). In spite of this difficulty, at least 15 different protozoa were propagated for experimental use against insects (Ignoffo, 1967). Protozoa are promising candidates because many pest insects that are not attacked by other entomopathogens are susceptible to at least one of the 300 described species.

Only five species of protozoa are currently being seriously investigated for development into microbial insecticides. The protozoan *Nosema locustae* has been evaluated for production feasibility and safety and is currently being evaluated for efficacy against rangeland grasshoppers (Henry, 1976); *Mattesia trogodermae* is being evaluated for use against stored-products pests (Schwalbe *et al.*, 1973); *Nosema algerae* is a possible controlling agent of mosquitoes (Undeen and Maddox, 1973); *Vairimorpha necatrix* was suggested for control of caterpillar pests of soybeans (W. M. Brooks, North Carolina State University, personal communication) and *Nosema (Perezia)* pyrausta is being considered for control of corn borers (Frye and Olson, 1974; Lewis, 1975). Largely because of problems discussed earlier, none of these have been widely applied (Decker, 1960; Ignoffo, 1967; McLaughlin, 1973); however, successful uses were recorded against caterpillars (Weiser and Veber, 1956; Weiser, 1957; Wilson and Kaupp, 1975), grasshoppers (Taylor and King, 1937; Henry, 1976), and beetles (McLaughlin, 1966; Schwalbe *et al.*, 1973). More recently, favorable information on safety (cf. Section V,B,3), ease of production, environmental stability, and efficacy (cf. Section V,C,2) have encouraged the reconsideration of protozoa as microbial insecticides.

IV. DEVELOPMENTAL PHASES OF A MICROBIAL INSECTICIDE

The initial impetus for development of a candidate microbial insecticide or any other pesticide is the demonstration of its activity against a target pest(s). Fortunately, most of the screening of candidate microbial insecticides has been done by nature, which has relieved developers of the inherent high cost of finding one potential candidate among thousands that would have to be screened. Approximately 10% of the cost of developing a chemical insecticide is that of synthesis and screening (Anonymous, 1969). Actual development toward commercialization of a microbial insecticide usually begins with a concept in the laboratory that progresses through a pilot-plant phase and attains technical realization as a registered product (Ignoffo, 1973a, c). Three technical parameters, i.e., production feasibility, safety (man, vertebrates, beneficial insects, and plants), and effectiveness against important pests (cf. Section I), are evaluated during each phase of development (Fig. 1). These technical parameters, broad generalizations during the laboratory phase, are more closely examined at each successive step as more data are accumulated. The questions asked at each phase are: Can technical success be attained? Can technical success be translated into commercial success? Technical success is attained when an entomopathogen can be continuously produced, is safe, and will effectively control a pest.

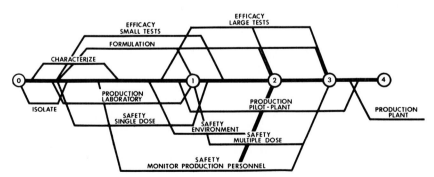

FIGURE 1. A critical pathway scheme for development of entomopathogens into microbial insecticides. (Reprinted from *Environmental Letters*, p. 27, by courtesy of Marcel Dekker, Inc.)

Commercial success is achieved when the product is produced, is effective, and can be used at costs that will return a reasonable profit to the producer. Generally, it takes 5–7 years to develop a known entomopathogen into a microbial insecticide at a cost two to five times lower than for a chemical insecticide. However, if the entomopathogen is similar to those previously investigated (*B. thuringiensis, B. heliothis*) and no problems of safety are encountered, costs may be less, and the entomopathogen may be available within 5 years. A previous success facilitates development of new agents because a framework has been established for constructive dialogue and exchange of information between persons responsible for development and registration.

The time required for development will undoubtedly vary for specific cases and for a variety of reasons, but the sequence (Fig. 1) is applicable to many candidate entomopathogens. Estimated dollar cost and research team years required to develop a candidate microbial insecticide during the first 3 years is about $1–$3 million (Ignoffo, 1975). Most technical questions concerning characteristics of the entomopathogen (biological identity, specificity, mode of action, stability, compatibility, etc.) can be resolved during the first year. Questions of production feasibility, safety, and effectiveness are resolved within the first 2 years. By the end of the third year, all major problems, both technical and nontechnical, should be defined or resolved. Therefore, shortly after the start of the fourth year, sufficient information should be available to decide whether to proceed toward commercial production, marketing, and sales. Optimistically, a product then could be available during the field season of the fourth year of development.

V. TECHNICAL PARAMETERS: MICROBIAL INSECTICIDES

A. Production Feasibility

1. General Comments

Entomopathogens are produced either by using fermentation (*in vitro* process) or in living insects (*in vivo* process). Fermentation technology (submerged or surface) has been used with facultative entomopathogens; *in vivo* technology with obligate entomopathogens; both processes have been employed to successfully produce commercial products. For example, submerged fermentation is generally the technology of choice for commercial production of the bacteria *B. thuringiensis* and *B. moritai;* surface fermentation is used for fungi, e.g., *B. bassiana* and

Hirsutella thompsonii (McCoy and Kanavel, 1969). The *in vivo* approach is used exclusively to produce insect viruses and protozoa. There are, of course, some notable exceptions; e.g., Doom, a formulation of spores of the bacterium *B. popilliae* can only be produced in living beetle larvae.

Regardless of the process, a large quantity of the desired isolate is first produced and stabilized with a suspending medium. Then, small subsamples of this mix are dispensed into glass vials. The samples are lyophilized, sealed under vacuum, and stored under refrigeration until needed for production. In the production of bacterial entomopathogens test tube slants prepared from these vials are used as the initial inoculum for a stepwise increase in volume of inoculum through a series of two shake flasks and a seed fermentor. This procedure provides a large population of clean cells of the same stage and thereby increases the probability of synchronous growth in the production fermentor. The sequential steps from initial inoculum to seed tank for surface fermentation are similar to those for the submerged fermentation process, but production and recovery are significantly different.

Various sources of carbon (molasses, dextrose, grain mash, grain by-products), nitrogen (cottonseed meal, fish meal, soybean meal, whey, yeast, or casein hydrolysates) are used for fermentation. In submerged fermentation, the total volume of solids is generally limited to about 5%. Essential minerals and growth factors are also included, when necessary.

Production costs of surface fermentation were estimated at $0.50 per lb of technical product and submerged fermentation at $1.69 per lb (Dulmage, 1971). Formulated, wettable powder products of *B. thuringiensis* retail for about $9.00 per lb; use rates, depending upon the pest species, range from ca. ⅛ to 1 lb/acre.

Patents for production of *B. thuringiensis* with submerged fermentation were issued to Bonnefoi (1960), Megna (1963), Drake and Smythe (1963), and Shell International (1969). Production of *B. thuringiensis* using surface fermentation was patented by Mechalas (1963). Production of *B. thuringiensis* by submerged or surface fermentation, based upon the patents of Megna (1963) and Mechalas (1963), respectively, are used here as examples of both fermentation processes (Fig. 2). Further details of production are described by Briggs (1963), Fisher (1965) and Dulmage and Rhodes (1971).

2. *In Vitro Process*

a. Submerged Fermentation. In submerged fermentation, the inoculum from slants is incubated in nutrient broth of the first flask for 24 hours.

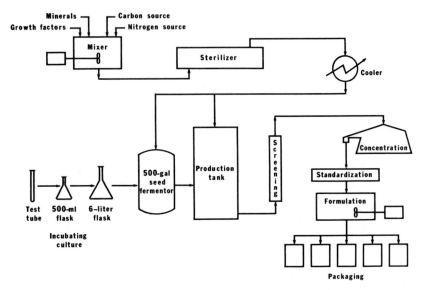

FIGURE 2. Submerged fermentation pathway used to produce the entomopathogenic bacterium *Bacillus thuringiensis.*

Medium for the second flask and the seed fermentor contains beet molasses (1%), corn steep solids (0.85%), and calcium carbonate (0.1%). The production media contain beet molasses (1.86%), corn steep solids (1.7%), cottonseed flour (1.4%), and calcium carbonate (0.1%). Development to sporulation and crystal formation with a viable spore count of 2–5×10^9 spores/ml is attained within 28–32 hours. Drake and Smythe (1963) obtained spore concentrations five times greater than Megna (1963) by using a production medium containing corn starch (6.8%), corn steep (4.7%), casein (1.94%), sucrose (0.64%), yeast (0.6%), and phosphate buffer (0.6%).

Safety of each fermentation batch is confirmed by intraperitoneal administration to white mice (Harvey, 1960; Anonymous, 1973). The insect-active ingredients of *B. thuringiensis* are recovered by centrifugation, filtration, precipitation, spray-drying, or combinations of these methods and processed into a flowable liquid, wettable powder, dust, or granular products, depending upon the type of fermentation, economics of the process, and the need for a specific formulation. Commercial products of *B. thuringiensis* are available in all four formulations (Fig. 3). Products of fermentation prior to and after blending with adjuvants or concentrated active ingredients are characterized by a bioassay of insecticidal activity and a viable spore count.

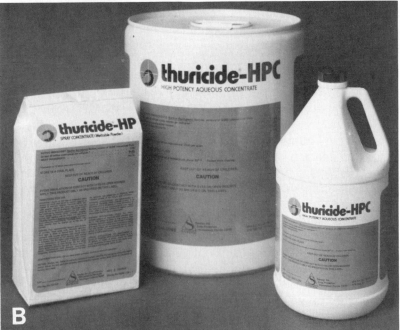

FIGURE 3. (A) Dipel, a wettable powder formulation of *Bacillus thuringiensis* (courtesy of Abbott Laboratories, Inc.). (B) Thuricide, flowable liquid and wettable powder formulations of *Bacillus thuringiensis* (courtesy of Sandoz, Inc.). (C) Elcar, the commercial wettable powder formulation of *Baculovirus heliothis* (courtesy of Sandoz, Inc.).

Only the viable spore count was used as an index of activity during early phases of commercialization of B. *thuringiensis*. However, it soon was apparent that, although spores were easy to count and spores may contribute to the total insecticidal activity, the primary source of caterpillar activity was the parasporal, proteinaceous crystal, i.e., the δ-endotoxin. Therefore, insect bioassays were developed to measure insecticidal activity of commercial products. Unfortunately, at first, producers used different insects and bioassay procedures. This permitted adequate comparisons for in-house standardization of activity but made it very difficult to compare directly the insecticidal activity of products of different producers. Largely through the efforts of H. T. Dulmage (Dulmage *et al.*, 1971), a standardized insect bioassay was developed and adopted for the United States. Similar bioassays in which indigenous insects were used were developed in other countries (Burges and Thomson, 1971; Aizawa *et al.*, 1975).

b. Surface Fermentation. In the surface fermentation process, the inoculum from the seed tank is absorbed on a carrier, usually a mixture of grain or grain by-products and insoluble inert substances (e.g., vermiculite, diatomaceous earth), to obtain a high surface area to volume ratio (Mechalas, 1963). The growth medium in the stepup flasks (adjusted to pH 7.0–7.2 with sodium hydroxide and incubated for 8 hours at 30°C) contains dextrose (1.5%), corn steep (0.5%), yeast hydrolysates (0.5%), and potassium phosphate (0.4%). The contents of the seed fermentor (incubated for 16 hours at 30°C) are dextrose (1.0%), yeast hydrolysates (0.6%), corn steep (0.45%), potassium phosphate (0.35%), sodium hydroxide (0.04%), and calcium chloride (0.01%). An inoculum ratio of two parts of the seed fermentor to five parts of the semisolid medium is used for production.

Large, perforated-bottom metal bins containing about 500 lb of semisolids are used instead of fermentation tanks. The semisolid production medium contains wheat bran (1800 parts), perlite (1200 parts), water (500 parts), soy meal (200 parts), dextrose (120 parts), lime (12 parts), sodium chloride (3 parts), and calcium chloride (1 part). During incubation, air at 30°–34°C and 95–100% relative humidity is forced through the medium via the perforated bottom of each bin for 36 hours. The moisture content during incubation drops from about 60 to 50% while the pH rises from 6.9 to 7.5. Moisture accumulated during incubation is reduced by forcing hot (50°–55°C) dry air through the bins for another 36 hours to achieve a moisture content of about 4%. The final product is then harvested, milled, passed through screens, characterized, blended with various adjuvants to achieve a standard insecticidal activity and spore count, and then packaged as either a dust or a granular formulation.

3. In Vivo Process

Techniques for producing entomopathogens in living hosts, primarily caterpillars, were developed over the last decade and are currently operational (Ignoffo and Hink, 1971). For example, annual production of 1–2 million European pine sawfly larvae, 2–10 million gypsy moth larvae, 15–45 million bollworm larvae, 35–110 million cabbage looper larvae, and 350 million mosquito larvae can be achieved (Ignoffo, 1975). Alternate hosts that are more easily reared and produce more spores than the primary hosts can also be used, e.g., caterpillars for production of a boll weevil protozoan or a mosquito protozoan (Ignoffo and Garcia, 1965; Undeen and Maddox, 1973).

A process to produce insect viruses in yeast and bacteria by using submerged fermentation was patented by Wells (1970) but never commercialized, and laboratory-scale production of insect viruses and protozoa in lines of insect cells both are possible, Virus produced in cell lines was as effective under field conditions as virus obtained from caterpillars (Ignoffo et al., 1974; Jaques, 1977). Although these latter methods have potential as production techniques, the living insect is still the only feasible substrate for commercial production of an obligate entomopathogen.

The production process of the bollworm–budworm NPV, *Baculovirus heliothis* (Ignoffo, 1966a, 1973a) is diagrammed in Fig. 4 as a typical example of an *in vivo* process. As with production of other microorganisms, a large quantity of the selective isolate is first produced. For viruses, the quantity should be sufficient to provide inoculum for several years of production. Subsamples from the first large-scale passage, stabilized, lyophilized, sealed in individual glass vials under vacuum, and stored at −20°C, are maintained as reference standards.

Production of entomopathogenic viruses is often done in several isolated facilities. One facility is devoted to preparation of the diet used to rear the living insect which is used as the substrate for virus production. Another facility, commonly called the insectary, is used to rear the insects, generally caterpillars of 0.1–0.5 ml capacity. The insectary may be further divided so that one section is used solely for maintaining a standard, disease-free, thriving culture of the living insect and the other section is used for mass producing insects to be used in production of virus. A third facility, the actual virus production unit, is used to infect the living insect. Another facility is used for final recovery and formulation of the product. Other functions associated with production, such as quality

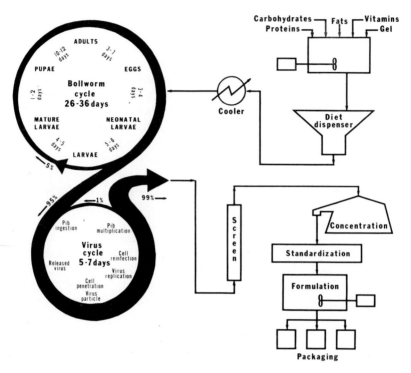

FIGURE 4. The *in vivo* process used to produce the nucleopolyhedrosis virus, *Baculovirus heliothis,* in larvae of species of *Heliothis.*

control bioassay, and standardizations are routinely done in an analytical laboratory.

The semisynthetic diet used to rear bollworm larvae for production of *B. heliothis* is a water-based mixture of casein, sucrose, wheat germ, yeast, Wesson salts, and growth factors (Ignoffo, 1966a). Alphacel is used as a filler, and agar is used to solidify the diet. Antimicrobial substances (e.g., formalin, sorbic acid, methyl parahydroxybenzoate, aureomycin) are often added to inhibit growth of secondary contaminants, such as bacteria, yeast, or fungi. Technology of the food industry was used to automate the virus production process.

Hot (70°–80°C) liquid diet, prepared in a large tank (Fig. 5A), is piped

FIGURE 5. (A) Mixing tank for preparing the semisynthetic diet used to rear *Heliothis* *larvae.* (B) Tray-filling machine that automatically dispenses a uniform volume of semisynthetic diet into individual cubicles. (C) Cages used to obtain eggs from *Heliothis* moths to be reared into larvae for virus production. (D) Mobile racks used to move trays containing diet and larvae. (E) Spray tower used to infect larvae and diet with virus. (F) Recovery of the technical virus product by acetone precipitation and filtration.

to a filling machine that automatically dispenses a uniform volume of diet (Fig. 5B). Plastic trays of individual cells containing the semisynthetic diet are used as the basic larval-rearing and virus-propagation unit (Ignoffo and Boening, 1970). Trays of cells are formed from plastic sheets, and the cells are filled with diet and sealed automatically on a continuous production line. Diet dispensing and cell sealing may be done at the rate of up to 200 trays per hour.

Neonatal caterpillars obtained from eggs laid in the insectary (Fig. 5C) are individually placed in the cell of each tray. Hundreds of these trays are thus semiautomatically infested with larvae and stacked on mobile racks for incubation in controlled environmental rooms (30°C) for 5–7 days (Fig. 5D). Approximately 95% of all larvae are used for virus production, while the other 5% are used to perpetuate the insect culture.

A standard volume of virus is sprayed on the diet of each cell (Fig. 5E). During the next 5–7 days (26°C) (Fig. 4), the virus replicates within the caterpillars and produces 5000–10,000 times more virus than that originally used. Bollworms weigh about 400–600 mg when they die. One caterpillar can produce as much as 36 billion PIB, ca. 30% of its dry weight.

The recovery of virus begins with collection of virus-killed larvae from individual cells by use of a suction tube. The larvae are then slurried with water, and the suspension is filtered through a screen to remove large particles. Further processing to concentrate PIB is done by centrifugation, precipitation, filtration (Fig. 5F), or spray-drying. As with other microbial insecticides, safety and quality control of each batch of virus is confirmed by a mouse injection test, evaluation of the types and levels of secondary microbial contaminants (Duggan, 1970), and bioassays.

A bioassay (mortality of neonatal bollworms) is used to determine the insecticidal activity of each virus batch (Ignoffo, 1966b). Several modes of administration, i.e., surface treatment, diet incorporation, and feeding may be used. The easiest large-scale, analytical method, however, is to incorporate the virus into the diet and feed the diet to neonatal or 1-day-old larvae. All preparations are compared against a reference standard, and their activity is related to this standard. The standard is prepared according to a designated protocol and stored under conditions of optimum stability. Insecticidal activity is measured as an LD_{50} or LC_{50} with 95% fiducial limits are recorded as the number of insecticidal units per unit weight of virus product. The number of PIB per unit weight of virus product (counted under phase microscopy) is also used to characterize viral preparations.

Significant reductions in the cost of producing viruses and other obligate entomopathogens are anticipated since present techniques are still new. Current cost of producing caterpillars is about $0.1 to $0.4 per insect. The product from 1 to 100 insects is generally used to treat 1 acre. The current commercial formulation of B. heliothis (ELCAR) is a wettable powder mixed with several adjuvants to increase stability and efficacy (Fig. 3C). The cost of ELCAR to the grower is about $1.55 per oz. Recommended rates for control of Heliothis on cotton are from 2 to 4 oz/acre.

The major problem in virus production has been the time required to infest each cell of a tray with a newly hatched larva. If the process from tray forming to larval infestation were completely automated, production of 10,000 larvae per hour would be possible. Pilot-plant production capacity is about 1 million larvae per month, an increase over laboratory production of approximately 20-fold with no significant increase in production personnel. Maximum capacity for one semiautomatic production line operating continuously is estimated at about 7 million larvae per month.

B. Safety of Entomopathogens

1. General Comments

Because entomopathogens are infectious, replicating, living entities, there has always been a justifiable concern about their specificity, that is, their possible infectivity to nontarget organisms. Most previous evidence of specificity was indirect, such as the lack of reported cases of toxicity or pathogenicity in laboratory and field personnel, the ubiquitous presence of entomopathogens on food products, and the fact that entomopathogens are a natural part of our environment. This indirect evidence that microbial insecticides pose no serious hazard to vertebrates is generally supported by direct evidence (Heimpel, 1971; Ignoffo, 1973b). Nevertheless, safety, as discussed previously (Ignoffo, 1967), cannot be absolutely guaranteed for all entomopathogens in all living systems for all time. Therefore, it is necessary that inferences of lack of safety be corroborated by tests specifically designed to uncover possible risks to man and the environment. Within the last decade, guidelines have been developed to evaluate this impact (Ignoffo, 1973b; Engler, 1974). The establishment of these guidelines is a highly significant achievement not only because they are available but because they result in a regulatory procedure for a systematic review of the safety of many potential microbial insecticides.

2. Types of Tests

Many types of tests and test organisms have been used to evaluate the safety of entomopathogens (Table IV) but not all of those listed are presently required (Engler, 1974). Initially, only acute studies were conducted to evaluate possible toxicity–pathogenicity to nontarget species. More extensive testing began after each entomopathogen was seriously considered a viable candidate. For example, susceptibility was evaluated at all levels, such as in cells, tissues, organs, and organisms. Test organisms also were challenged with both purified entomopathogens (spores, cells, mycelium, crystals, virions PIB) and with formulations of the entomopathogen. Microbial insecticides known to be relatively nontoxic or nonpathogenic were generally tested in animals at 10–100 times the average rate per acre by using a conversion ratio of weight of test animal to the body weight of man (Ignoffo, 1973b). Animals are generally sacrificed at the end of the test and carefully examined. Criteria used to evaluate the effects of entomopathogens *in vivo* included clinical examinations, histology, serology, and necroscopy. The following paragraphs give a brief summary of the kinds of tests that have been used. Detailed discussion of the protocols and results are presented elsewhere (Ignoffo, 1973b).

a. Acute Toxicity–Pathogenicity. Feeding, inhalation exposure, and dermal toxicity are used to evaluate acute toxicity or pathogenicity of candidate entomopathogens for both invertebrates and vertebrates. In these tests, the entomopathogen is generally administered once and the animal is observed for a short period (generally less than 3 weeks). However, observations may extend for 8 weeks when serological examinations are conducted. Vertebrates, both laboratory-reared and wildlife, are used in these studies.

b. Subacute Toxicity–Pathogenicity. Subacute or chronic studies are initiated after toxicity–pathogenicity has been established. The objective is to detect subtle abnormalities in the test species that are not apparent in acute studies. Since most entomopathogens are nontoxic, large doses are generally administered in the diet or by inhalation, subcutaneous injection, or skin contact throughout the test period, which may extend from 2 to 13 weeks. Two species of animals, one of which is a nonrodent, usually are used. Serological evaluations are done at regular intervals, generally every 4–5 weeks.

c. Primary Irritation. The skin and mucous membrane of the eye frequently are exposed to pesticides; therefore, both systems are used to

TABLE IV. Types of Tests and Organisms Used To Evaluate Safety to Invertebrates, Vertebrates, and Plants[a]

Test type	Organisms used to test:	
	Baculovirus heliothis	*Bacillus thuringiensis*
Acute toxicity–pathogenicity		
Per os, diet	Rat, mouse, birds, fishes, oyster, shrimps, human	Rat, mouse, guinea pig, birds, fishes, swine, human
Inhalation	Rat, guinea pig	Mouse, human
Dermal, skin	Rat, rabbit, guinea pig, human	Mouse, guinea pig, rabbit
Intraperitoneal	Rat, mouse	Chicken, mouse, rat, guinea pig, rabbit
Intravenous	Rat, mouse	
Intracerebral	Mouse	
Eye	Rabbit	Rabbit
Subacute toxicity–pathogenicity		
Diet	Monkey, dog, rat, mouse	Chicken, rat, human
Inhalation	Monkey, dog, rat, mouse	Rat, human
Subcutaneous	Monkey, dog, rat, mouse	Guinea pig, human
Teratogenicity	Rat, mouse	
Carcinogenicity	Rat, mouse	
In vitro replication	Man, primate	
Phytotoxicity	Agricultural crops	Agricultural crops
Invertebrate specificity	Beneficial and other arthropods	Beneficial arthropods

[a] From Ignoffo (1973b).

evaluate the potential of a candidate material to induce irritation. Rabbits are the animals of choice, but rats, mice, guinea pigs, and man have also been used to assess entomopathogens. Tests generally are continued for less than 1 week (eye) or 3 weeks (skin).

d. Teratogenicity. Teratological tests have been used to evaluate several NPV but probably will not be required in the future unless toxicity is observed. The NPV was repeatedly administered (per os) to pregnant dams to ensure maximum effect during organogenesis. Rats and mice are usually the animals of choice, and sufficient levels are used to obtain responses that vary from no effect to a definite effect.

e. Carcinogenicity. Carcinogenic potential has been studied in only a few entomopathogenic viruses. As with the teratological studies, it probably only will be required if toxicity is observed. Strains of rats or mice bred for susceptibility to carcinogens are used as experimental animals. Tests are initiated when the animals are young and are terminated 18 months to 2 years later.

f. *In Vitro* Development. Tests are made to evaluate the potential for entomopathogens, primarily viruses, to replicate in a nonhomologous host (established vertebrate diploid cell lines, including human). Cytopathic effects, inclusion body formation, hemadsorption, transformation, and viral interference are used as criteria to evaluate this potential.

g. Invertebrate Specificity. Acute toxicity studies against beneficial arthropods and invertebrates are also included in an evaluation of the safety and host spectrum of potential microbial insecticides. The scope of the evaluation is based on a knowledge of the organisms that may be present in the ecosystem where the entomopathogen will be used. Representative members of major orders of insects and species of freshwater, estuarine, or marine invertebrates, such as daphnia, dragonfly nymph, oyster, crab, and shrimp, are included. Entomopathogens are commonly tested by topical and per os administration of up to 100 times the anticipated exposure rate.

h. Phytotoxicity–Pathogenicity. Phytotoxicity of formulations of microbial insecticides are included as part of the overall evaluation of safety. Usually, greenhouse-grown plants of species that represent different families of economic plants are selected. Tests are also performed on field-grown species of the crop to be treated as well as on other economic crops.

3. Results of Tests

Extensive tests in the United States have demonstrated that at least four entomopathogenic viruses (*Heliothis, Orgyia, Lymantria, Autographa*); two bacteria (*B. popilliae, B. thuringiensis*); two fungi (*B. bassiana, H. thompsonii*); and three protozoa (*N. locustae, N. algerae, M. trogodermae*) are not toxic or pathogenic to man, other animals, or plants (Heimpel, 1971; Ignoffo, 1973b). Also, sufficient testing has been conducted to demonstrate the safeness of a fungal preparation (Boverin, *B. bassiana*) produced in the Soviet Union and a virus (Matsukemin, *Dendrolimus* CPV) and a bacterium (Rabirusu, *B. moritai*) produced in Japan (Ignoffo, 1973b).

Not one of about 50 entomopathogenic viruses tested *in vivo* was toxic or pathogenic to vertebrates. In fact, all but two of a dozen insect viruses were unable to replicate in vertebrate embryos or cell lines derived from birds, fishes, amphibians, or mammals and the possible replication of virus or development of cytopathic effects reported in the two tests has not been confirmed by other investigators. No deleterious effects at normal rates were reported in tests with *B. thuringiensis* endotoxin, *B. popilliae,* and *B. moritai* (Ignoffo, 1973b).

Presence of allergens may be encountered among fungi (*Beauveria, Entomophthora, Hirsutella, Metarrhizium, Nomuraea*) currently under consideration. Results of recent studies, however, indicate that *Beauveria, Entomophthora, Nomuraea,* and *Hirsutella,* are nontoxic or noninfectious to vertebrates (Ignoffo, 1973b). Only three entomopathogenic protozoa, one from grasshoppers (*N. locustae*), one from mosquitoes (*N. algerae*), and one from beetles (*M. trogodermae*), have been evaluated for safety to nontarget organisms. Initial *in vivo* tests with all three indicated no apparent risk to vertebrates.

Entomopathogens under consideration as microbial insecticides must be evaluated for possible deleterious effects to man, other animals, and plants. Risks cannot be assumed from studies of closely related groups. Undoubtedly, all tests described (Ignoffo, 1973b) will not be required for all entomopathogens, but each candidate should be broadly evaluated prior to large-scale field use. Because infectious microorganisms are being used, studies of risks should be carefully designed to detect presence of the microorganisms in addition to any possible effects induced by the microorganisms. In the final analysis, the risks of field use must be balanced against the benefits obtained. Prudent evaluations against vertebrates and other organisms will, in the long run, hasten development and ensure successful use of safe, effective entomopathogens as microbial insecticides.

C. Effectiveness of Entomopathogens

1. General Comments

Effectiveness of an insecticide is generally interpreted as its ability to kill pests as quickly as possible. However it is better also to consider protection of yield or quality balanced against risks to man and the environment. Initially, a novel microbial insecticide may not provide immediate control equal to that of a highly toxic, broad-spectrum chemical insecticide. Continual use of selective microbial insecticides may, however, preserve beneficial parasites and predators and ultimately result in yields comparable to those obtained with chemical insecticides.

2. Examples of Effectiveness

Entomopathogens belonging to each of the major groups have been successfully used to control at least 100 serious arthropod pests (Ignoffo, 1967; Anderson and Ignoffo, 1967). For example, about 60 insect viruses have been used to control caterpillar pests of agricultural crops, forests, and stored products, phytophagous mites, and beetles. Species of fungi have successfully controlled scale insects, whitefly, plant hoppers, aphids, chinch bugs, phytophagous beetles, flies, caterpillars, and stored-products pests. Bacteria other than the commercially available *Bacillus* preparations have been used to control grasshoppers, termites, caterpillar pests of forests and agricultural crops, and some species of beetles. Applications of entomopathogenic protozoa have suppressed pest populations of boll weevils, mosquitoes, grasshoppers, and about a half-dozen species of caterpillars.

3. Use Patterns and Impact

Microbial insecticides are most often used like chemical insecticides. Need we be so restrictive? There is no question that the use of spray technology is an important approach, but is it the only one available (Ignoffo, 1978)? For example, entomopathogens may be successfully introduced and established to provide long-term seasonal control of pest populations (White and Dutky, 1940; Balch and Bird, 1944; Hall and Dunn, 1958; Henry, 1971). Insects themselves can be used to disseminate entomopathogens (Elmore and Howland, 1964).

An epizootic of virus and fungi might be induced to occur earlier than normal in order to control a target pest (Morris, 1965; Ignoffo *et al.*, 1976). It also may be possible to manipulate the environment to create conditions wherein naturally occurring diseases would exert their greatest effect (Burleigh, 1975). Some of these approaches should provide levels of control equal to or better than that currently obtained with spray technology. However, they should be further explored and additional research should be devoted to evaluating their full potential. At least some of these approaches can be immediately applied, and more may be utilized in the near future.

The potential environmental impact of the substitution of microbial insecticides for some chemical insecticides may be visualized from the following examples. Commercial utilization of the virus, *B. heliothis,* as a control for bollworms and budworms on cotton alone, could decrease the use of at least 17 million lb of chemical insecticides annually. Use of the fungus *H. thompsonii* in Florida against the citrus rust mite and of the bacterium *B. thuringiensis* for cabbage looper control on vegetables could annually eliminate the need for about 4–16 and 1–3 million lb of chemical insecticides, respectively. Development of *N. locustae* for use against western-range grasshoppers and *Entomophthora thaxteriana* against the green peach aphid on potatoes in Maine could annually replace 1–2 or 0.2–0.7 million lb of chemical insecticides, respectively.

VI. CONCLUSIONS

There have been some notable developments in the area of microbial insecticides since publication of our last edition (Anderson and Ignoffo, 1967). Guidelines for the evaluation of the infectivity of entomopathogens to nontarget organisms have been formulated by the Environmental Protection Agency (EPA). The *Heliothis* NPV has been registered in December 1975 by the EPA as the first viral pesticide. Shortly thereafter (September 1976), another insect virus (*Orgyia NPV*) was approved by the EPA after development by the U.S. Forest Service. Two other viruses, the *Lymantria* NPV and the *Autographa* NPV, have been granted a temporary permit for further testing leading to possible registration. A bacterium (*B. moritai*) and a virus (*Dendrolimus CPV*) were registered in Japan for control of flies in feces of livestock pests and a forest caterpillar pest, respectively. A new more active strain of *B. thuringiensis* was produced which has increased the performance and acceptance of commercial products and broadened its use against other insect pests. Data on a fungus (*H. thompsonii*) are being collated by industry in anticipation of an experimental temporary permit for control of mites of citrus, and two other fungi (*E. thaxteriana* and *N. rileyi*) are being intensively investigated for control of aphids and soybean caterpillars, respectively. The protozoan (*N. locustae*) is being extensively field tested against grasshoppers on range crops under a temporary permit issued by the EPA. Currently, about four dozen trade-named products formulated from about 20 entomopathogens have been proposed as either experimental or commercial microbial insecticides (Tables I, II, and III).

Development of entomopathogens or their by-products into microbial insecticides is a reality. Safe, effective entomopathogens formulated as insecticides are being produced by industry and are being effectively used by growers. The technical successes reported herein undoubtedly will continue to be translated into commercial successes and will in turn stimulate additional research on use of entomopathogens as safe, selective insecticides.

REFERENCES

Aizawa, K., Fujiyoshi, N., Ohba, M., and Yoshikawa, N. (1975). *Proc. Int. Congr. International Association of Microbiological Societies* 1st, 1972 Vol. 2, pp. 577–606.

Anderson, R. F., and Ignoffo, C. M. (1967). *In* "Microbial Technology" (H. J. Peppler, ed.), pp. 172–181. Van Nostrand-Reinhold, Princeton, New Jersey.

Anonymous (1969). *Chem. Week* **104,** 36–38.

Anonymous (1973). *Code Fed. Regul.* **38,** 10643.

Balch, R. E., and Bird, F. T. (1944). *Sci. Agric.* **25,** 65–80.

Bassi, A. (1835). *Orcesi Lodi.*

Bonnefoi, A. (1960). French Patent 1,225,179.

Briggs, J. D. (1963). *In* "Insect Pathology" (E. A. Steinhaus, ed.), Vol. 2, pp. 519–548. Academic Press, New York.

Briggs, J. D. (1964). *Bull. W. H. O.* **31,** 495.

Bucher, G. E. (1964). *Ann. Soc. Entomol. Que.* **9,** 30.

Burges, H. D., and Thomson, E. M. (1971). *In* "Microbial Control of Insects and Mites" (H. D. Burges and N. W. Hussey, eds.), pp. 591–622. Academic Press, New York.

Burleigh, T. G. (1975). *Environ. Entomol.* **4,** 574–576.

DeBach, P. (1964). "Biological Control of Insect Pests and Weeds." Van Nostrand-Reinhold, Princeton, New Jersey.

Decker, G. C. (1960). *Agric. Chem.* **93,** 30–33.

d'Herelle, F. (1914). *Ann. Inst. Pasteur,* Paris **28,** 280.

Drake, B. B., and Smythe, C. V. (1963). U.S. Patent 3,087,865.

Duggan, R. E. (1970). *Fed. Regist.* **35,** 18690.

Dulmage, H. T. (1971). *In* "Microbial Control of Insects and Mites" (H. D. Burges and N. W. Hussey, eds.), pp. 581–590. Academic Press, New York.

Dulmage, H. T., and Rhodes, R. A. (1971). *In* "Microbial Control of Insects and Mites" (H. D. Burges and N. W. Hussey, eds.), pp. 507–540. Academic Press, New York.

Dulmage, H. T., Boening, O. P., Rehnborg, C. S., and Hansen, G. D. (1971). *J. Invertebr. Pathol.* **18,** 240–245.

Dunn, P. H., and Mechalas, B. J. (1963). *J. Insect Pathol.* **5,** 451.
Dutky, S. R. (1963). *In* "Insect Pathology" (E. A. Steinhaus, ed.), Vol. 2, pp. 75–115. Academic Press, New York.
Elmore, J. C., and Howland, A. F. (1964). *J. Insect Pathol.* **6,** 430–438.
Engler, R. (1974). *Dev. Ind. Microbiol.* **15,** 200–207.
Fisher, R. A. (1965). *Int. Pest Control* **7,** 4.
Forbes, S. A., II (1898). *16th Rep. State Entomol. Noxious Beneficial Insects State Ill.*
Frye, R. D., and Olson, L. C. (1974). *Farm Res.* September.
Giard, A. (1892). *Bull. Sci. Fr. Belg.* **24,** 1.
Hall, I. M., and Dunn, P. H. (1958). *J. Econ. Entomol.* **51,** 341–344.
Harvey, A. M. (1960). *Fed. Regist.* **25,** 3207–08.
Heimpel, A. M. (1965). *World Rev. Pest Control* **4,** 150.
Heimpel, A. M. (1971). *In* "Microbial Control of Insects and Mites" (H. D. Burges and N. W. Hussey, eds.), pp. 469–489. Academic Press, New York.
Henry, J. E. (1971). *J. Invertebr. Pathol.* **18,** 389–394.
Henry, J. E. (1976). *Agric. Res.* **24,** 5.
Ignoffo, C. M. (1966a). *In* "Insect Colonization and Mass Production" (C. N. Smith, ed.), pp. 501–530, Academic Press, New York.
Ignoffo, C. M. (1966b). *J. Invertebr. Pathol.* **8,** 547–548.
Ignoffo, C. M. (1967). *In* "Insect Pathology and Microbial Control" (P. A. van der Laan, ed.), pp. 91–117. North-Holland Publ., Amsterdam.
Ignoffo, C. M. (1968). *Curr. Top. Microbiol. Immunol.* **42,** 129–167.
Ignoffo, C. M. (1973a). *Exp. Parasitol.* **33,** 380–406.
Ignoffo, C. M. (1973b). *Ann. N. Y. Acad. Sci.* **217,** 141–172.
Ignoffo, C. M. (1973c). *Misc. Publ. Entomol. Soc. Am.* **9,** 57–61.
Ignoffo, C. M. (1975). *Environ. Lett.* **8,** 23–40.
Ignoffo, C. M. (1978). *J. Invertebr. Pathol.* **31,** 1–3.
Ignoffo, C. M., and Boening, O. P. (1970). *J. Econ. Entomol.* **63,** 1696–1697.
Ignoffo, C. M., and Garcia, C. (1965). *J. Invertebr. Pathol.* **7,** 260–262.
Ignoffo, C. M., and Hink, W. F. (1971). *In* "Microbial Control of Insects and Mites" (H. D. Burges and N. W. Hussey, eds.), pp. 541–580. Academic Press, New York.
Ignoffo, C. M., Hostetter, D. L., and Shapiro, M. (1974). *J. Invertebr. Pathol.* **24,** 184–187.
Ignoffo, C. M., Marston, N. L., Hostetter, D. L., Puttler, B., and Bell, J. V. (1976). *J. Invertebr. Pathol.* **27,** 191–198.
Jaques, R. P. (1977). *J. Econ. Entomol.* **70,** 111–118.
Kirby, W., and Spence, W. (1826). "An Introduction to Entomology: or Elements of the Natural History of Insects," p. 197, Longman, Rees, Orme, Brown, & Green, London.
Krassilstschik, I. M. (1888). *Bull. Sci. Fr. Belg.* **19,** 461.
Latch, G. C. M. (1965). *N. Z. J. Agric. Res.* **8,** 384.
Lewis, L. C. (1975). *Iowa State J. Res.* **49,** 435–445.
Linsley, E. G., and Smith, R. F. (1970). *J. Econ. Entomol.* **63,** 689–691.
McCoy, C. W., and Kanavel, R. F. (1969). *J. Invertebr. Pathol.* **14,** 386–390.
McCoy, C. W., Brooks, R. F., Allen, J. C., and Selhime, A. G. (1976). *Proc. Tall Timbers Conf.* **6,** 1–17.
McLaughlin, R. E. (1966). *J. Econ. Entomol.* **59,** 909–910.
McLaughlin, R. E. (1973). *Misc. Publ. Entomol. Soc. Am.* **9,** 95–98.
Martignoni, M. E., and Iwai, P. J. (1977). *U.S.D.A., For. Serv. Gen. Tech. Rep.* PNW 40-28.
Martignoni, M. E., and Iwai, P. J. (1978). *U.S.D.A. For. Serv., Tech. Rep.* (in press).
Mechalas, B. J. (1963). *U.S. Patent* 3,076,992.
Megna, J. C. (1963). *U.S. Patent* 3,073,749.

Metchnikoff, E. (1879). "Issue III. The Grain Weevil" *Commission attached to the Odessa Zemstvo Office*, Odessa.

Morris, O. N. (1965). *J. Insect Pathol.* **5,** 401–414.

Müller-Kögler, E. (1965). "Pilzkrankheiten bei insekten." Parey, Berlin.

Pasteur, L. (1870). "Etudes sur la maladie des vers à soie," Parts I and II. Gauthier-Villars, Paris.

Pristavko, V. P. (1963). *Sb.: 5, Soveshck. Vses. Entomol. Obshck.* p. 110.

Roberts, D. W., and Yendol, W. G. (1971). *In* "Microbial Control of Insects and Mites" (H. D. Burges and N. W. Hussey, eds.), pp. 125–149. Academic Press, New York.

Schwalbe, C. P., Boush, G. M., and Burkholder, W. E. (1973). *J. Invertebr. Pathol.* **21,** 176–182.

Shell International Research Maatschappij, N. V. (1969). Netherlands Patent 68,163,097.

Snow, F. H. (1891). *Annu. Rep. Entomol. Soc. Ont.* **21,** 93.

Steinhaus, E. A. (1959). *Trans. Int. Conf. Insect Pathol. Biol. Control, 1st, 1958* p. 37.

Steinhaus, E. A. (1964). *Proc. Int. Conf. Insect Pathol. Biol. Control, 2nd, 1962* p. 7.

Steinhaus, E. A., and Thompson, C. G. (1949). *J. Econ. Entomol.* **42,** 301.

Taylor, A. B., and King, R. L. (1937). *Trans. Am. Microsc. Soc.* **56,** 172.

Undeen, A. H., and Maddox, J. V. (1973). *J. Invertebr. Pathol.* **22,** 258.

Vago, C., Aizawa, K., Ignoffo, C., Martignoni, M. E., Tarasevitch, L., and Tinsley, T. W. (1974). *J. Invertebr. Pathol.* **23,** 133–134.

Weiser, J., III (1957). *Cesk. Parazitol.* **4,** 359.

Weiser, J., and Veber, J. (1956). *Z. Angew. Entomol.* **40,** 55.

Wells, F. E. (1970). Republic of South Africa Patent 700,831.

White, R. T., and Dutky, S. R. (1940). *J. Econ. Entomol.* **33,** 306–309.

Wilson, G. G., and Kaupp, W. J. (1975). *Can. For. Serv. Inf. Rep.* **IP-X-11,** 28.

Yendol, W. G., and Roberts, D. W. (1970). *Proc. Int. Colloq. Insect Pathol., 4th, 1970* pp. 28–42.

Chapter 2

Rhizobium Species

JOE C. BURTON

INTRODUCTION

Man lives in an atmosphere dominated by nitrogen gas, but he is unable to use this element in its molecular form. Air nitrogen (N_2) must first be "fixed," that is, incorporated into compounds which animals and plants can utilize as food.

N_2 can be fixed chemically or biologically. However, a large amount of fossil fuel energy is required to manufacture nitrogen fertilizers chemically. The recent shortage of fossil fuel and resulting high cost of N fertilizers has focused greater attention to biological processes.

Certain bacteria, *Rhizobium* spp., have the ability to infect roots of leguminous plants, form nodules, and work symbiotically with their host in fixing molecular N. It has been estimated (Hardy and Holsten, 1972)

29

MICROBIAL TECHNOLOGY, 2nd ed., Vol I
Copyright © 1979 by Academic Press, Inc.
All rights of reproduction in any form reserved ISBN 0-12-551501-4

that nodule bacteria (rhizobia), in association with leguminous hosts, fix at least 90×10^6 metric tons of N annually. This is more than twice the amount of N in chemical fertilizers and more than one-half the total amount of this element fixed biologically each year. Nonetheless, with the rapidly increasing world population and the acute shortage of fossil fuel to manufacture nitrogen fertilizer, large increases in biological N_2 fixation are mandatory. The *Rhizobium*–leguminous plant association offers the greatest promise of all systems for providing the nutritious protein food which will be needed in the years ahead.

Culture of *Rhizobium* species (rhizobia) in the laboratory for use in agriculture began shortly after Hellriegel (1886) revealed his discovery of the magic of nodulated leguminous plants in 1886. Beijernick, a Dutch microbiologist, isolated the causal organisms, rhizobia, from the leguminous nodules in 1888. Culture of *Rhizobium* species in the laboratory to increase legume crop production began soon afterward, but early attempts to use laboratory-produced cultures of the organism for improvement of leguminous crops met with little success. This should not be too surprising considering the many factors involved. The perplexing problems encountered during this early period are described in detail by Fred *et al.* (1932).

It was recognized early that leguminous plants have preferences for rhizobia and vice versa, but the full implication of this was not understood. The selection of effective strains of rhizobia for specific leguminous species was one of the key developments and continues to be one of the major responsibilities of the inoculant manufacturer today. Other key developments in understanding the *Rhizobium*–legume association and in development of the inoculant industry are described by Burton (1967).

II. *Rhizobium*: LEGUMINOUS PLANT ASSOCIATIONS

A. Plant Groups and *Rhizobium* Species

The family, Leguminosae, comprises more than 12,000 species of leguminous plants although fewer than 100 of these are currently used for food production. These are chiefly in the subfamily, Papilionatae, which consists predominantly of nodulating species with the ability to work symbiotically with rhizobia to utilize atmospheric nitrogen.

The nodule bacteria (*Rhizobium* spp.) are one of two genera in the family Rhizobiaceae and are characterized by their ability to infect and induce nodule formation on roots of leguminous plants (Buchanan and Gibbons, 1974). *Rhizobium* species are differentiated by the kinds of plants they nodulate. Plants mutually susceptible to nodulation by a particular kind of rhizobia constitute a cross-inoculation group. The bacteria capable of nodulating the plants in one of these groups are considered a *Rhizobium* species.

The six recognized *Rhizobium* species and their respective leguminous hosts are shown in the tabulation below.

Rhizobium meliloti	*Medicago, Melilotus, and Trigonella* spp.
Rhizobium trifolii	*Trifolium* spp.
Rhizobium leguminosarum	*Pisum, Vicia, Lathyrus*, and *Lens* spp.
Rhizobium phaseoli	*Phaseolus vulgaris* and *P. coccineus*
Rhizobium japonicum	*Glycine max*
Rhizobium lupini	*Lupinus* and *Ornithopus* spp.

The designation of new species of *Rhizobium* on the basis of cross-inoculation groups was discontinued about 1939 because the many unexplainable, incongruous plant and bacteria reactions cast doubt on the validity of the concept. Rhizobial taxonomists have now focused major attention on speed of growth, acid production, serology, DNA base ratios, numerical taxonomy, DNA hybridization, and phage susceptibilities. Graham (1976) and Vincent (1974) discuss these systems in detail and point out the inconsistencies. The difficulty is evidenced by the fact that no new *Rhizobium* species have been added for almost four decades. The practice today is to identify the nodule bacteria strain with its parent host plant, e.g., lotus rhizobia from *Lotus* species or baptisia rhizobia from *Baptisia* species, when there is no *Rhizobium* species name.

The grouping of leguminous plants on the basis of susceptibility to nodulation by a particular species or kind of rhizobia has been useful to inoculant manufacturers. With leguminous species, it is generally accepted that nodulation is essential for symbiotic N_2 fixation. The tests can be limited to strains capable of nodulating the particular plant and the best N_2-fixing bacteria can be selected from these.

B. *Rhizobium* Strain Selection

Rhizobium strain selection is one of the most important responsibilities of the inoculant manufacturer. *Rhizobium* strains are selected for the following qualities: (a) They must form N_2-fixing nodules on the particular plant and provide an adequate supply of nitrogen; (b) they must be competitive as evidenced by colonization of the root hairs and nodulation of the host in the presence of other highly infective rhizobia;

(c) nodulation must occur over a range of temperatures under which the particular plant will grow; (d) the *Rhizobium* strain must be able to grow well in culture medium, in the carrier medium, and in the soil after the seed is planted; and (e) the organism should be able to survive in the soil from one season to another.

1. Nitrogen-Fixing Ability

A prime consideration in selecting *Rhizobium* strains is nitrogen-fixing ability. The initial tests are generally made in a greenhouse or growth chamber (Vincent, 1970; Wacek and Brill, 1976; Date, 1976). Surface-sterilized seeds are planted and inoculated with the various strains of rhizobia. *Rhizobium* strains vary widely in ability to fix N_2 and stimulate growth (Fig. 1).

Through the years, investigators have sought simple, quick tests for measuring N_2-fixing efficiency, but the only way of knowing with certainty that a strain of rhizobia will fix N_2 with a particular host is to put them together under a favorable environment for growth and measure the N_2 fixed. More recently the discovery that nitrogenase, the enzyme in the legume nodule responsible for reducing N_2 to NH_3, also reduces

FIGURE 1.   Growth of white clover, *Trifolium repens,* inoculated with various strains of *Rhizobium trifolii.* (A) Noninoculated; (B) and (D) inoculated with effective strains; (C) and (E) inoculated with ineffective strains.

acetylene to ethylene has provided a quicker method of evaluating *Rhizobium*–plant associations and weeding out inferior strains. Hardy and Holsten (1977) discuss the merits of acetylene reduction and other measurements of dinitrogen fixation in legumes.

Rhizobium strains which are effective on a wide range of host plants are preferred over those which are effective on only a few. The responses of three *Vicia* spp. to seven strains of rhizobia are shown in Fig. 2. Strains 175G9, 175G10, and 128C2 were highly effective on *V. villosa* but very ineffective on *V. grandiflora*. In contrast, strains 128C53 and 128C56 were very effective on all three host plants and would thus be more desirable in an inoculum for *Vicia* species.

Certain leguminous hosts are very promiscuous and have the ability to work effectively with a wide range of rhizobia. Siratro (*Macroptilium purpureum*) and the cowpea, *Vigna unguiculata*, are examples. Other plants are highly specific and work only with a special *Rhizobium* strain. The Texas bluebonnet, *Lupinus subcarnosus*, and Kura clover, *Trifolium ambiguum*, exemplify this group.

In studying N_2 fixation by numerous species of legumes in association with many strains of rhizobia, it has been noted that within cross-inoculation groups, certain plant species tend to respond similarly to inoculation with strains of rhizobia. This information has led to a subdivision of plants on the basis of N_2 fixation predictability. These subdivisions are referred to as "effectiveness groupings" and constitute the basis for multiple host inocula (Table I).

It should be remembered that all plants within an effectiveness group may not respond similarly to all strains of rhizobia. However, there is

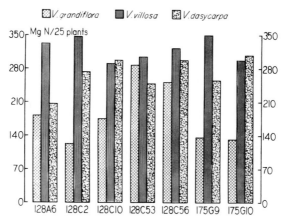

FIGURE 2. Nitrogen fixation by three species of *Vicia* inoculated with seven strains of rhizobia in growth chamber studies; plants were harvested at 6 weeks.

TABLE I. Effectiveness Groupings of Leguminous Species within Cross-Inoculation Groups

Cross-inoculation group	Reference
Alfalfa subgroups/*Rhizobium meliloti*	
A. *Medicago sativa, M. falcata, M. minima,*	Burton and Erdman, 1940
M. tribuloides, Melilotus denticulata	Burton and Wilson, 1939
M. alba, M. officinalis, M. indica	Brockwell and Hely, 1961
B. *Medicago arabica, M. hispida, M. lupulina,*	Brockwell and Hely, 1961
M. orbicularis, M. praecox, M. truncatula	
M. scutellata, M. polymorpha, M. rotata,	Kassim, 1976
M. rigidula, Trigonella foenum-graecum	
C. *Medicago laciniata*	Brockwell and Hely, 1961
D. *Medicago rugosa*	Brockwell, 1971
Clover subgroups/*Rhizobium trifolii*	
A. *Trifolium incarnatum, T. subterraneum,*	Burton and Briggeman, 1948
T. alexandrinum, T. hirtum, T. glomeratum,	Vincent, 1945;
T. arvense, T. angustifolium	Purchase and Vincent, 1949
	Strong, 1937
B. *Trifolium pratense, T. repens, T. hybridum*	Erdman and Means, 1946
T. fragiferum, T. procumbens, T. nigrescens	
T. glomeratum	Burton *et al.*, 1977
C. *Trifolium rueppellianum, T. tembense,*	Norris, 1959, 1965; Norris and
T. usambarense, T. baccarinii, T. steudneri,	't Mannetje, 1964
T. burchellianum var. *burchellianum*	
T. burchellianum var. *johnstonii*	
T. africanum, T. pseudostriatum	
D. *Trifolium semipilosum* var. *kilimanjaricum*	Norris, 1959; Norris and
T. masaiense, T. cheranganiense,	't Mannetje, 1964
T. rueppellianum var. *lanceolatum*	
E. *Trifolium semipilosum*	
F. *T. masiense*	Norris, 1959; Norris and
	't Mannetje, 1964
G. *Trifolium reflexum*	Burton *et al.*, 1977
H. *Trifolium ambiguum*	Parker and Allen, 1949
I. *Trifolium vesiculosum, T. berytheum, T. bocconi*	Burton *et al.*, 1977
T. boissiere, T. compactum, T. dasyurum,	
T. leucanthum, T. mutabile, T. vernum,	
T. physodes	
J. *Trifolium heldreichianum*	Burton *et al.*, 1977
K. *Trifolium medium, T. sarosience, T. alpestre*	
L. *Trifolium rubens*	
Pea and vetch subgroups/*Rhizobium leguminosarum*	
A. *Pisum sativum, Vicia villosa, V. hirsuta,*	Burton *et al.*, 1977
V. tenuifolia, V. tetrasperma, Lens esculenta,	
Lathyrus aphaca, L. cicera, L. hirsutus, L. odoratus	
B. *Vicia faba, V. narbonensis*	Burton *et al.*, 1977
C. *Vicia sativa, V. amphicarpa*	Burton *et al.*, 1977

(continued)

TABLE I   *(Continued)*

Cross-inoculation group	Reference
D. *Lathyrus sativus, L. clymenum, L. tingitanus*	Burton *et al.*, 1977
E. *Lathyrus sylvestris*	Burton *et al.*,1977
F. *Lathyrus ochrus, L. tuberosus, L. szenitzii*	Burton *et al.*, 1977
Cowpea subgroups/*Rhizobium* sp.	
A. *Vigna unguiculata, V. sesquipedalis,*	Burton, 1952
V. luteola, V. cylindrica, V. angularis,	Burton *et al.*, 1977
V. radiata, V. mungo, Desmodium sp.,	
Alysicarpus vaginalis, Crotalaria sp.,	
Macroptilium lathyroides, M. atropurpureus,	
Psophocarpus sp.*, Lespedeza striata, L. stipulaceae*	
B. *Phaseolus limensis, P. lunatus, P. aconitifolius*	Burton *et al.*, 1977
Canavalia ensiformis, Canavalia lineata	
C. *Arachis hypogaea, A. glabrata,*	Burton *et al.*, 1977
Cyamopsis tetragonoloba, Lespedeza sericea,	
L. japonica	
D. *Centrosema pubescens*	Bowen, 1959
E. *Lotononis bainesii*	Norris, 1958
Lotus subgroups/*Rhizobium* sp.	
A. *Lotus corniculatus, L. tenuis, L. angustissimus*	Erdman and Means, 1949
L. tetragonolobus, Dorycnium hirsutum	Burton, 1964
L. caucasieus, L. crassifolius, L. creticus,	Burton *et al.*, 1977
L. edulus, L. frondosus, L. subpinnatus, L. weilleri,	
Anthyllis vulneria, A. latoides	
B. *Lotus uliginosus, L. americanus, L. scoparius,*	Burton *et al.*, 1977
L. angustissimus, L. pedunculatus, L. strictus,	
L. strigosus	
Ornithopus sativum	Jensen, 1967
Lupine subgroups/*Rhizobium lupini*	
A. *Lupinus albicaulis, L. albifrons, L. albus,*	Burton *et al.*, 1977
L. angustifolius, L. arboreus, L. argenteus,	
L. benthami, L. formosus, L. luteus, L. micranthus,	
L. perrenis, L. sericeus	
B. *Lupinus densiflorus, L. vallicola*	Burton *et al.*, 1977
C. *Lupinus*	Burton *et al.*, 1977
D. *Lupinus polyphyllus*	Burton *et al.*, 1977
E. *Lupinus subcarnosus*	Burton *et al.*, 1977
F. *Lupinus succulentus*	Burton *et al.*, 1977
Coronilla subgroups/*Rhizobium* sp.	
A. *Coronilla varia, Onobrychis viciaefolia,*	Burton *et al.*, 1977
Petalostemum purpureum, P. candidum,	
P. microphyllum, P. multiflorum, P. villosum	
Leucaena leucocephala, Albizia julibrissin	

sufficient congruity within a group to justify this categorization until the precise *Rhizobium* requirements for each host genotype can be determined.

2. Competitiveness

Competitiveness implies aggressiveness and the ability to produce nodules on the roots of a particular leguminous host growing in a soil which contains other highly infective rhizobia. A *Rhizobium* strain is considered competitive when it produces a large proportion of the nodules on a particular leguminous plant growing in a soil or substrate heavily infested with other infective strains. The competitive nature of a strain of *Rhizobium* is difficult to assess but its importance is not questioned. Date (1976) asserts that controlled-environment facilities are not usually considered suitable for measuring competitiveness or other characteristics, such as pH and pesticide tolerance or longevity in culture or on seed. He considers field tests imperative to measure these qualities.

Johnson *et al.* (1965) have used serological techniques very successfully in studying competitiveness of strains of *R. japonicum* under greenhouse conditions and in field soils free of *R. japonicum*. However, this technique is too laborious and time-consuming to permit screening of large numbers of *R. japonicum* strains for competitiveness in the field, particularly where soybeans have been grown for several years. Only a few strains could possibly be studied.

The writer (unpublished results) has used a growth chamber technique in screening strains of *R. trifolii* for use on soils infested with numerous highly infective non-N_2-fixing rhizobia. A peat-base inoculum containing the native strains is added to the sterile sand in Leonard jars. Each strain of *R. trifolii* being tested is prepared as a peat-base inoculum which is applied to the seeds. The inoculated seeds are planted in Leonard jars containing sterile sand as a control. In a parallel series, the inoculated seeds are planted in the Leonard jars containing sand inoculated with the native ineffective rhizobia. Growth and nitrogen fixation obtained in the latter series is a good measure of strain competitiveness (Fig. 3).

Often effective strains of rhizobia, as judged by growth in the sterile sand, are ineffective in the presence of native rhizobia. These are considered noncompetitive strains. Results from growth chamber tests have been tested and verified under field conditions.

3. *Other Qualities of Rhizobia*

Rhizobium strains equally effective in N_2 fixation in the greenhouse often show marked differences when tested in the field. The reasons for

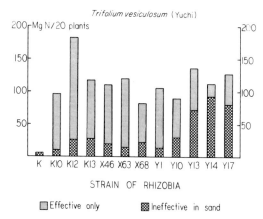

FIGURE 3. Competitiveness of effective rhizobia on arrowleaf clover (*Trifolium vesiculosum* Yuchi) as influenced by the presence of highly infective, non-nitrogen-fixing rhizobia in the substrate. Cross-hatched portion of bars show competitiveness of *Rhizobium* strains.

this are not completely understood, but certain factors have been suggested: (a) adaptability of the strain of rhizobia to the fertility level or the pH of the soil, (b) inherent variation in N_2-fixing potential of the *Rhizobium* strain which is manifested only under the nitrogen stress of a mature fruiting plant, or (c) variation in ability to tolerate an antagonistic soil microflora.

The question of whether rhizobial strains respond differently to levels of mineral nutrients and soil pH has not been answered adequately. One reason for this is that it has not been possible to separate effects on the bacteria from those on the plant. Also, information on acid tolerances of many of our important legumes is lacking.

Soil acidity influences nutrient solubility and growth of the host plant as well as growth and effectiveness of the rhizobia. With a low pH, nodulation failure may result from poor survival or inability of the rhizobia to colonize the root. Even if nodulation occurs at a low pH, this does not assure N_2 fixation. Optimal nodule function may require higher calcium and pH than is needed for nodulation (Munns, 1976, 1977).

Under acid conditions, aluminum and manganese are more soluble and may inhibit N_2 fixation simply by reducing growth of the host plant. *Rhizobium* strains tolerant to high concentrations of aluminum and manganese would be of little benefit unless the host plant were also tolerant to aluminum and manganese. Acid-tolerant strains of *R. meliloti* have been reported by Rice (1976) and by L. M. Bordeleau (personal communication), but these have been selected by screening under field conditions and perhaps for several other factors as well as pH. There is no substitute for field testing of rhizobial strains and this is particularly true in problem areas where conditions cannot be duplicated in the glasshouse.

III. CULTURE OF RHIZOBIA

A. Nutritional Requirements

Rhizobia are aerobic gram-negative rods ($0.5–0.9 \times 1.2–3.0 \mu m$), occurring singly or in pairs and motile while young. Older cells stain unevenly in bands because of granules or polymeric β-hydroxybutyrate (PHB) which does not stain.

Rhizobia are generally divided into what is called "fast growers" and "slow growers." The term, "fast grower" is assigned to cultures of *R. meliloti, R. trifolii, R. leguminosarum, R. phaseoli, Lotus* (birdsfoot type), *Astragalus, Sesbania, Caragana,* and various other varieties which produce turbidity in broth or visible colonies on solid media in 3–5 days at 28°C. The slow growers, *R. japonicum, R. lupini,* cowpea, *Lotus* (big trefoil type), and other varieties, require 8–10 days to attain an equal amount of growth. This division is not absolute. Differences in rates or amounts of growth among strains may be reduced, but not eliminated by supplementing the media with different forms of carbohydrate, nitrogenous compounds, and organic extracts.

The extensive literature on nutrition of rhizobia is reviewed by Wilson (1940), Allen and Allen (1950), and Vincent (1974, 1977). Rhizobia utilize monosaccharides and disaccharides readily, and to a lesser extent trisaccharides, alcohols, and acids. Pentoses, such as arabinose and xylose, are the preferred source of carbon for *R. japonicum* and other slow growers (Wilson and Umbreit, 1940; Graham and Parker, 1964). Graham and Parker reported that strains of rhizobia differ in their abilities to utilize carbohydrates. None of 16 different carbohydrates proved completely satisfactory for all strains of particular species of rhizobia. This should not be surprising, however, when consideration is given to the host preferences of the rhizobia. Also, the ability of rhizobia to utilize carbohydrates depends to some extent upon the basal medium, oxidation–reduction potential, size of inoculum, method of culture, and possibly other factors.

The composition of culture media used in industrial propagation of rhizobia is given in Table II. Sucrose is the most commonly used carbon

TABLE II. Media Used in Culturing Rhizobia

Ingredient (gm/liter)	Wright[a] (1925)	Bond (1940)	Van Schreven (1953)	Burton (1967)	Date (1976)
Sucrose	—	10.00	15.00	10.00	—
Mannitol	10.00	—	—	2.00	10.00
K_3PO_4	—	—	—	0.10	—
K_2HPO_4	0.50	9.50	0.50	—	0.50
KH_2PO_4	—	—	—	0.37	—
$MgSO_4 \cdot 7 \ H_2O$	0.20	0.20	0.20	0.18	0.80
NaCl	0.10	0.10	—	0.06	0.20
$(NH_4)_2HPO_4$	—	—	—	0.10	—
$CaSO_4 \cdot 2 \ H_2O$	—	—	—	0.04	—
$Ca(NO_3)_2 \cdot 4 \ H_2O$	—	—	—	—	—
$CaCO_3$	3.00	1.00	2.00	0.25	—
Ca gluconate	—	1.50	—	—	—
$FeCl_3 \cdot 6 \ H_2O$	—	—	—	—	0.10
Yeast water (ml)	100	50	100.0	—	100.0
Autolysed yeast	—	—	—	1.00	—
Water (ml)	900	950	900	1000	900

[a] Commonly referred to as Fred and Waksman Medium No. 79.

source, but mannitol and glycerol are sometimes used for the slow growers.

Yeast extract or autolyzate is used almost universally as a source of growth factors and inorganic nitrogen. While some strains of rhizobia can use nitrate or ammonium as a nitrogen source, better growth is usually obtained when low molecular weight amino acids are added. Yeast extract, hydrolyzed casein, and corn steep solids serve as sources of growth factors and amino acids.

The balance of amino acids appears to be important also (Holding *et al.*, 1960; Hamdi, 1965, 1968a, b, 1969; Badawy, 1965; Strijdöm, 1963; Strijdöm and Allen, 1965, 1967, 1969). Rhizobia cultured through several transfers on media enriched with glycine, alanine, and certain D-forms of amino acids lost their ability to fix nitrogen. Rhizobial strain efficiency was maintained when the organisms were cultured on media containing only the regular complement of yeast extract consisting predominantly of the naturally occurring L-isomers of amino acids. The composition of various organic autolyzates used in culturing rhizobia is given in Table III.

The growth factor requirements of *Rhizobium* spp. are simple and vary with species and strain. Strains of *R. leguminosarum, R. trifolii,* and *R.*

TABLE III. Amino Acid and Vitamin Content of Various Medium Supplements

Component	Difco yeast extract[a]	Hydrolyzed protein				Corn steep solids[c]
		Yeast autolysate Amberex 1003[b]	Casein[b] Amber EHC	Soy-Peptone Sheffield	Cottonseed[b] Pharmamedia	
Nitrogen (%)						
Total N	9.5	8.0	13.5	9.2	9.0	7.3
Amino N	73.0	50.0	30.0	20.0	50.0	37.0
Amino acid (%)						
Alanine	—	—	2.6	2.3	—	5.6
Arginine	1.0	3.7	3.7	4.6	7.1	5.6
Aspartic acid	5.0	—	5.7	5.3	7.5	0.2
Cystine	—	0.8	0.3	0.6	1.4	—
Glutamic acid	6.5	8.5	20.1	12.1	17.4	0.7
Glycine	2.5	5.0	1.0	2.8	3.7	1.6
Histidine	1.0	2.3	2.0	1.8	2.2	1.2
Isoleucine	3.0	4.2	4.8	0.7	2.9	1.3
Leucine	3.5	6.0	9.4	3.8	5.5	4.7
Lysine	4.0	6.0	6.8	3.7	3.0	3.3
Methionine	1.0	2.0	2.8	0.8	1.7	1.0
Phenylalanine	2.0	4.0	5.5	3.3	4.9	1.5
Threonine	3.5	3.2	4.3	1.9	2.9	1.2
Tyrosine	0.5	3.0	4.4	1.9	3.1	—
Valine	3.5	4.0	6.2	2.0	4.0	2.4
Vitamins (µg/gm)						
Biotin	1.0	2.0	—	—	0.6	—
Niacin	280.0	550.0	—	—	74.8	—
Pyridoxine	20.0	25.0	—	—	15.0	—
Riboflavin	20.0	35.0	—	—	13.6	—
Thiamin	3.0	50.0	—	—	20.0	—
Calcium pantothenate	—	100.0	—	—	58.0	—

[a] Analysis by Difco.
[b] Analysis by Amber Laboratories, Milwaukee, Wisconsin.
[c] D. D. Christianson et al. (1965).

phaseoli may require one, all, or any combination of biotin, thiamin, and calcium pantothenate (Wilson, 1940; Allen and Allen, 1950; Graham, 1963). Only certain strains of *R. meliloti, R. lupini, R. japonicum,* and the cowpea rhizobia require any vitamins and biotin is the only one needed. The vitamin needs of rhizobia from the many legumes have not been determined, but the majority grow well in yeast extract–mannitol—mineral salts medium.

Our knowledge of the role of inorganic salts in nutrition of rhizobia is scanty. Plant extracts have generally been used in culturing rhizobia. The mineral content of these have probably been adequate for most rhizobia. A synthetic medium is generally necessary to demonstrate mineral deficiencies. Yeast extract contains nutritionally significant concentrations of iron, calcium, magnesium, strontium, sodium, potassium, barium, manganese, copper, lead, aluminum, and vanadium (Steinberg, 1938). It is generally conceded that rhizobia need small amounts of iron, magnesium, calcium, potassium, manganese, zinc, and cobalt (Vincent, 1974, 1977). It appears that fast growers need more calcium than slow growers (Badawy and Allen, 1963). Since rhizobia in the nodules are nourished by the leguminous host, it is reasonable to assume that their mineral nutrient requirements are somewhat similar.

B. Growing Rhizobia

Unlike most microorganisms, *Rhizobium* spp. are not grown for their by-products or protein. The product envisaged is an inoculum, a live culture of rhizobia for application to leguminous seeds or to the soil where seeds are to be planted—a culture teeming with billions of effective rhizobia capable of producing nodules and fixing N_2 on the plant or plants for which it is prepared. Cell yield or biomass is not important; the main concern in formulating a medium for growing rhizobia is retention of the nitrogen-fixing capabilities of the organisms. It is desirable, of course, to have high viable counts in the broth culture and in the inoculant to prepare a strong inoculum and to assure early nodulation. The steps in large-scale growth of *Rhizobium* species are given in Fig. 4.

The fermentors used in growing rhizobia vary from small bottles which can be sterilized in an autoclave to large, jacketed stainless steel vessels built for high-pressure steam and equipped with mechanical agitators and various control devices. Regardless of the size of the operation, small and intermediate size fermentors which can be used either for production or starter cultures are desirable.

Rhizobium species are aerobic. The fermentors must be equipped to supply sterile air through spargers. Large vessels must have mechanical

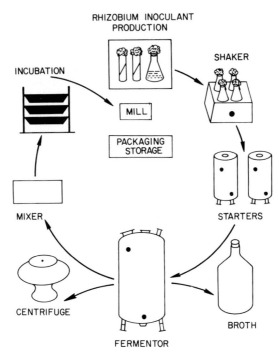

FIGURE 4. Steps in production of *Rhizobium* inoculants.

agitation to facilitate heat exchange and avoid overcooking of the medium and to increase aeration efficiency. The essentials are described by Date and Roughley (1977).

While rhizobia are considered obligate aerobes, excellent growth is obtained at oxygen partial pressures as low as 0.01 atm (Wilson, 1940). An oxygen partial pressure of 0.15 atm is optimum for respiration. As a rule of thumb, 1 liter of air per minute should be supplied for every 20 liters of medium.

Rhizobia grow best in the range of 28°–30°C, with the exception of *R. meliloti* which prefers a temperature of 35°C. With a good medium and adequate aeration, a cell population of 4 or 5×10^9 can be attained in 96 hours with a 1.0% inoculum. The incubation time can be reduced substantially by increasing the size of the inoculum.

Norris (1965) claims that the fast growing rhizobia produce acids when mannitol is broken down. However, Laird and West (1938), working with strains selected especially for their ability to produce acid, obtained a maximum of 2–3 ml of $0.1N$ acid per 100 ml of medium in 48 hours.

Even with a weakly buffered medium, *R. meliloti, R. trifolii, R. phaseoli,* and *R. leguminosarum* cultured in mass in fermentors produce no appreciable change in pH in 96 hours. When incubated for 144 hours, *R. meliloti* may lower the pH from 6.8 to 6.4; other fast growers produce no change in pH. An acid reaction is usually indicative of contamination usually by an aerobic thermophilic spore former which has survived the sterilization. Broth culture should contain a minimum of 1×10^9 viable rhizobia per milliliter and should be completely free of contamination to be satisfactory for use in preparing inoculants.

C. Carrier Media

The qualities of a good carrier medium for rhizobia are: (1) nontoxic to *Rhizobium* species, (2) good absorption qualities, (3) easily pulverized and sterilized, (4) good adhesion to seeds, and (5) readily available at moderate cost. Peats from different areas of the same bog as well as peats from diverse areas differ greatly in their suitability as carriers for nodule bacteria. Chemical analysis is helpful, but other hidden qualities can be more important. In reality, the only way the suitability of a particular peat can be determined is by putting rhizobia into the carrier and monitoring their growth. Rhizobia should be able to increase in numbers by tenfold or more in 3–5 weeks.

Good-quality peat is not available in many countries. Roughley (1968, 1970, 1975) cites problems with variation in quality of peat from the same bog, high salt concentration, and in using excessive temperatures for drying. Strijdöm and Deschodt (1975) had good success with a mixture of coal dust, bentonite, and lucerne powder as a carrier for rhizobia. Bagasse was satisfactory when the sugar was removed before grinding. Corby (1975) had good success with a compost of corn cobs, 1000 kg; limestone, 27 kg; single superphosphate, 9 kg; and ammonium nitrate, 12 kg; but 6 months were needed for drying and grinding. Dube and Mandeo (1973) used finely pulverized lignite.

The composition of Demilco peat, the one used most commonly in the United States, is given in Table IV. After the wet peat is mined and screened to remove stones and roots, it is shredded and fed into a revolving drum through which very hot air is continuously passing. The peat is flash-dried. Moisture content drops from 60% to 8 or 9% in a matter of seconds. While still hot, the peat is pulverized in a hammer mill to the desired fineness and packaged in multiwalled paper bags until used.

TABLE IV. Analysis of Sedge Peat, a Carrier for *Rhizobium* spp.[a]

Sedge peat contents	Amounts
Total nitrogen	1.62 (%)
Organic matter	86.80 (%)
Ash (500°)	13.20 (%)
Exchangeable potassium	62 ppm
Nitrogen as NH_4 and NO_3	94 ppm
Available phosphorus	12 ppm
pH	4.5–5%
Moisture	7–8%
Analysis of ash	*Amount (%)*
Potassium	1.12
Phosphorus	0.33
Calcium	5.21
Magnesium	1.14
Iron	2.10
Silicon	28.00
Aluminum	6.32
Sodium	0.52
Sieve analysis	*Amount (%)*
Powder inoculant	
Through 200 mesh (ASTM)	80–90
On 200 mesh	18–8
On 50 mesh	Less than 1
Granular inoculant	
On 16 mesh (ASTM)	0
On 40 mesh	70–85
On 50 mesh	15–30
Through 50 mesh	Less than 4

[a] Peat mined and processed by Demilco, Waukesha, Wisconsin.

IV. INOCULANT MANUFACTURE

A. Types of Products

Based on use, there are two kinds of rhizobial inoculants: (1) those designed for application to leguminous seeds, and (2) those designed for direct application to the soil. Seed inoculants are more common.

1. Seed Inoculants

The practice of applying rhizobia to leguminous seeds, inoculation, was started soon after these bacteria were first cultured in the laboratory. Rhizobia were applied to seeds because this was an easy convenient way of implanting the bacteria in the root zone of the developing seedling. The laborious task of spreading tons of soil to provide a few rhizobia could thus be bypassed.

Currently, six forms of culture or inoculum are marketed: (1) bottle or agar slant, (2) peat-base, (3) liquid or broth, (4) lyophilized or freeze-dried, (5) frozen concentrates, and (6) oil-dried culture. Generally, the peat-base inocula are considered more dependable because of the protection the peat affords the bacteria in the package and on the seed (Burton, 1964, 1967; Vincent, 1974). The liquid culture from the fermentor is used in preparation of all forms except the agar slant. Concentration of bacterial cells by centrifugation is a second step in preparation of the frozen concentrates, the lyophilized, and the oil-dried forms.

Two different approaches are used in the large-scale manufacture of peat-base inoculants. The method commonly used in Australia and in Europe is to grow the rhizobia in broth and then inject this broth culture either into a pliable film package containing sterilized peat or into small bottles containing sterile peat enriched with alfalfa meal. The proper amount of broth culture to add is predetermined to provide optimum moisture in the package to ensure growth. The rhizobia proliferate on the substrate in the individual packages. This method has the advantage of eliminating or greatly reducing contaminants, but it is not readily adaptable to large-scale operations because of its excessive demands on labor and time. Also, it imposes difficulties in carrying out an effective quality-control program.

The method favored in the United States is to increase the rhizobia in broth to a minimum of 1×10^9 ml. A predetermined amount of extremely fine $CaCO_3$ is then added to the broth culture to neutralize the acid and assure attaining a pH of 6.8 in the peat-base inoculum. The broth–$CaCO_3$ suspension is then sprayed onto the pulverized heat-treated sedge peat while it is being agitated in a horizontal paddle-or ribbon-type mixer. The finished inoculant should have a moisture content of about 38% (wet basis).

After mixing, the inoculant is placed in this layers in roller tubs and allowed to incubate for a period of 48–72 hours at 26°–28°C. During this time, the heat of wetting is dissipated and the moisture is absorbed uniformly. The inoculant is then milled to break up lumps and is packaged. Viable rhizobia reach their peak in numbers 2 or 3 weeks with the fast-growing rhizobia and in 4–5 weeks for the slow-growing rhizobia (Fig. 5). Viable rhizobia decline in numbers after 5 or 6 weeks unless the inoculant is refrigerated. The decline is directly related to storage temperature. Inoculants kept at 4°C or lower will retain a high count of viable rhizobia for 12 months or longer.

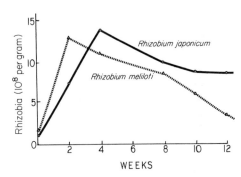

FIGURE 5. Growth of *Rhizobium meliloti* and *Rhizobium japonicum* in a peat-base carrier at 25°C for 12 weeks following preparation.

2. Soil Inoculants

Under some conditions inoculation of seeds is not practical. Seeds coated with toxic fungicides or insecticides cannot be successfully inoculated with rhizobia. Wetting the coats of fragile seeds, such as peanuts, with an inoculant slurry may reduce germination. Also, seeds with any particulate matter, such as peat-base inoculant or powdery chemicals, cannot be sown with the new type of air planters which depend upon a smooth seed coat for proper operation. Under these circumstances, the inoculant must be distributed separately from the seed.

Soil inoculants are of three types. In the granular peat-base type called "Soil Implant," the rhizobia actually grow in the peat particles. It is designed for application through an attachment on the drill which will meter out as little as 4 or 5 kg/ha of a granular material and place it in the furrow with the seed at planting time. The inoculant must consist of fine granules and must flow freely (Fig. 6). A second type, comprised of inert cores coated with peat-base inoculants and absorbant clays, was introduced by Taylor (1972) for mixing and sowing directly with seeds.

A third type of soil inoculant is prepared by centrifuging the broth culture to produce a creamy liquid of *Rhizobium* cells which is kept frozen until used by the farmer. At planting time the frozen concentrate is thawed and diluted with sufficient water to permit distribution in the seed furrow using the liquid fertilizer attachment on the seed drill (Scudder, 1975; Leffler, 1976). From 15 to 30 liters of bacterial suspension is usually applied per hectare.

Soil inoculants, both the granular and liquid types, have the following advantages: (1) They enable application of greater numbers of rhizobia than can be applied to most seed. This can help to overcome a flora of

FIGURE 6. *Rhizobium* inoculants, Left: conventional seed type; right: granular type applied directly to soil by drill attachment. Space = 1 cm.

dominantly ineffective rhizobia in the soil. (2) Soil inoculants do not interfere with use of chemical protectants on the seed. Inoculants applied separately from the seed are not adversely affected by toxic chemicals on the seed coat. (3) Soil inoculants are easy to apply. (4) Soil inoculants can be used to effect nodulation on growing plants which for some reason are not effectively nodulated.

B. Longevity

The prime objective in inoculating leguminous seeds is to provide a sufficient number of highly effective rhizobia to nodulate the host and keep it well supplied with nitrogen. Longevity of *Rhizobium* species is essential for success. The effective life of an inoculant is therefore a primary concern of the inoculant manufacturer, the seedsman, and the farmer.

The peat-base inoculant is the predominant form because it provides better protection for the rhizobia, both in the package and on the seed (Burton, 1964, 1967; Vincent, 1974; Roughley, 1975; Date and Roughley,

1977). Survival of *R. japonicum* in a peat-base inoculant packaged in polyethylene incubated at three different temperatures is shown in Fig. 7. Survival is inversely related to temperature. At 43°C only 0.01% of the rhizobia lived for as long as 2 weeks. Survival was good at 21 and 32°C.

According to Vincent (1958, 1965) lyophilized inoculants have a slightly longer shelf-life than do peat cultures at 28°–30°C, but death rate after application to seed is much higher. In the United States, commercial mixtures of lyophilized rhizobia and talc have not proved acceptable (Schall *et al.*, 1975).

Broth cultures are used successfully in Europe but the seed inoculated with the culture is planted soon afterward (Van Schreven, 1958). In the United States, liquid or broth inoculants for soybeans have become moderately popular because of the ease of application. However, under field conditions, good results require very large inocula and the seed must be planted soon after inoculating (Burton and Curley, 1965). Broth, lyophilized, and agar slant cultures share a common disadvantage; they lack the protection afforded by the peat on the seed and in the soil.

The number of rhizobia required to bring about effective nodulation depends upon age and form of inoculant, strain of organism, method of application, kind of seed, weather conditions, temperature, moisture, type of soil, seed bed preparation, and possibly other factors. However, as field data accumulate, the evidence is growing that far greater numbers of rhizobia are required for successful inoculation than was formerly suspected. Australian and Canadian workers now consider 1000 rhizobia per seed as a minimum (Date, 1976; Stevenson, 1976). Soybeans planted under field conditions require a minimum of 10^5 rhizobia per seed for effective nodulation (Burton and Curley, 1965).

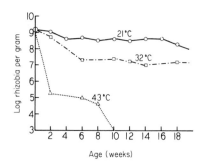

FIGURE 7. Effect of temperature on survival of *Rhizobium japonicum* in a peat-base inoculant packaged in polyethylene to prevent loss of water.

C. Quality Control

The quality of *Rhizobium* products (inoculants and preinoculated seeds) is monitored in only a few countries. The standards and methods of testing vary.

In Australia, a central organization (AIRCS, Australian Inoculant Research and Control Service) assures the quality of Australian inoculants. University or Department of Agriculture personnel (a) test and select rhizobial strains, (b) maintain and issue stock cultures of the chosen strains, (c) check quality of broth culture before it is mixed with peat, and (d) collect samples of the finished product from various distribution points and test them for rhizobia content. A minimum viable rhizobia count of 10^8 per gram up to the expiration date is considered satisfactory. The previous standard was 10^6 per gram. Samples of preinoculated seeds are also collected. A minimum of 10^3 viable rhizobia per seed is considered satisfactory, but few samples meet this standard (Brockwell et al., 1975).

Quality control in the United States is in sharp contrast to the Australian system. Manufacturers are responsible for selecting their own strains of rhizobia and for setting their own standards in manufacture. The generally poor quality of inoculants during the 1920s and early 1930s led the United States Department of Agriculture and several of the states to set up testing departments for legume inoculants, but the tests were not difficult. The minimal greenhouse test used served only to caution the farmer regarding use of questionable brands the following year. It did not function as a real measure of inoculant quality.

The introduction and widespread promotion of preinoculated seed in the United States has created a new interest in regulatory testing. In general, seeds inoculated longer than 2 or 3 weeks in advance of planting have not proved to be adequately inoculated when planted in soils devoid of rhizobia.

Without refrigeration, neither legume inoculants nor preinoculated seed will maintain high levels of viable rhizobia for long periods of time. It is necessary to specify the effective shelf-life. In Australia, inoculants are dated to expire 6 months after preparation as compared to 1 year in the United States, but the effective use period does not generally overlap two planting seasons. The net effect is almost the same, but the Australian system is more realistic.

The rule regarding preinoculated seed has not yet been specified. Certain states require that the date of inoculation be stamped on the seed tag and also the date beyond which the manufacturer does not consider the seed effectively inoculated.

V. USE OF *Rhizobium* INOCULANTS

Seed Application

1. Conventional Methods

Inoculants are generally designed for application to seeds, because this is a convenient practical way of placing the bacteria in the soil zone where the young seedling will develop. When the inoculant is mixed with water to form a slurry and then applied, the seed are more uniformly coated and the inoculant adheres to the seed firmly (Fig. 8). Seeds should be inoculated just before planting.

Inoculants have been introduced for dry application but the results have not been good. This can best be observed in a field devoid of effective rhizobia rather than in a field which has previously grown the particular legume.

Leguminous seeds differ greatly in size, shape, color, nature of seed coat, and many other characteristics. The amounts of inoculant and particularly water which can be applied safely depend upon the kind of seed. Caution must be exercised in inoculating not to get seeds so moist

FIGURE 8. Chickpea (*Cicer arietinum*) seeds. Left: noninoculated; right: inoculated with peat-base inoculant.

or sticky that they will not flow freely through a mechanical seeder. Small seeds can absorb more water than large seeds (Table V).

In the United States the normal rate of application is 4.4 gm of inoculant per kilogram of seed regardless of seed size, but the amount of water is greatly reduced for the larger seeds. The big seeds naturally receive a larger number of rhizobia per seed, but according to Burton and Curley (1965) they need more rhizobia for effective nodulation. Regulatory standards often specify a minimum of 10^3 viable rhizobia per seed regardless of size. An inoculant with 10^6 rhizobia per gram would meet this standard with peas, but with white clover, one would have to apply 400 times as much inoculant or an inoculant with 400 times as many rhizobia per gram. The minimum standard is much more difficult to attain with the small seeds.

2. "Custom Inoculation"

The modern farmer is extremely busy at planting time and prefers to have his seed inoculated in advance. Seedsmen who are equipped with treaters can apply the inoculant mechanically when the farmer is ready to plant. This practice is called "custom inoculation" (Carter, 1963). This

TABLE V. Amounts of Inoculant and Water Commonly Used in Inoculating Leguminous Seeds[a]

Seed type	Inoculant (gm/kg seed)	Water (ml/kg seed)	Rhizobia (per seed)[b]	Seeding rate (kg/ha)	Inoculant[c] (gm/ha)
White clover					
(*T. repens*)	4.4	22	500	2.2	10
Birdsfoot trefoil					
(*L. corniculatus*)	7.4	25	1,800	3.4	25
Red clover					
(*T. pratense*)	4.4	22	1,500	10.0	44
Alfalfa					
(*M. sativa*)	4.4	22	2,000	17.0	75
Vetch					
(*V. atropurpureal*)	4.4	13	40,000	43.0	190
Soybean					
(*Glycine max*)	4.4	13	150,000	65.0	285
Lupine blue					
(*L. angustifolius*)	4.4	13	160,000	82.0	360
Peas (edible)					
(*P. sativum*)	4.4	6	200,000	260.0	1,150

[a] Data from The Nitragin Co. The amounts of inoculant per gram of seed vary slightly with manufacturers.

[b] Based on rhizobia content of 200×10^6/gm inoculant.

[c] At specified seeding rate.

is in contrast to "preinoculation," the term used to describe inoculation of leguminous seeds months in advance of planting and before the seed are offered for sale.

Both peat-base and liquid cultures are used in "custom inoculation." Additives consisting of gums and sugars are often included in the inoculant slurries to improve adhesion to the seed and improve rhizobia survival on the seed.

3. Survival on Seeds

As pointed out earlier, inoculants are applied to seed because of convenience. Seeds generally provide a poor substrate for sustaining rhizobia (Burton, 1952, 1964, 1967, 1975; Vincent, 1958, 1965; Thompson et al., 1975). Like other nonsporulating gram-negative rods, rhizobia die rapidly under drying conditions, whether on the seed or in the soil. The death rate will depend upon form, age, size of inoculum, temperature, relative humidity, and to some extent *Rhizobium* species. Subterranean clover, white clover, and some alfalfa seeds have water-soluble antibiotics in their seed coats which adversely affect rhizobia (Bowen, 1961; Thompson, 1960, 1961).

Rhizobia in a peat carrier are able to survive in greater numbers on seed than the rhizobia in liquid or lyophilized inoculants. Strains of *R. japonicum, R. lupini,* and the cowpea rhizobia appear to die more rapidly than *R. meliloti, R. trifolii, R. leguminosarum,* or *R. phaseoli* (Burton and Curley, 1965; Carter, 1963). Scanty gum production has been offered as an explanation, but the soybean and cowpea rhizobia survive in soil for long periods (Reid et al., 1935).

4. Special Inoculation Techniques

Ordinary methods of inoculating seeds are adequate under many conditions, particularly when a good, quality peat inoculum slurry is applied just before planting. However, there are many other circumstances where special techniques are mandatory.

a. Lime Pellet. When leguminous seeds are to be mixed with acid fertilizers or planted in highly acidic soils, a lime coating is essential to protect the rhizobia (Roughley et al., 1966). An adhesive mixture containing peat inoculant is first applied to the seed which are then rolled in a finely pulverized limestone until all the seed are thoroughly coated. The large amount of gum slurry required to wet the seed, 5 liters/50 kg seed, permits application of a very large inoculum which gives added insurance of success. Two types of stickers, the synthetic substituted cellulose gums and plant resins, acacia or gum arabic, are used. The coating is usually done in a rotating drum equipped with paddles (Brockwell, 1972; Date, 1970).

Lime pelleting is not recommended for lupins, trefoils, and many of the tropical legumes, because the *Rhizobium* spp. for these plants will not tolerate the high pH of lime. Powdered rock phosphate or some other neutral powder is more suitable.

b. Inoculant Pellet. When leguminous seeds are to be sown on neutral or slightly acid soils and these soils are heavily infested with native ineffective rhizobia, it is very desirable to add as many of the effective inoculant rhizobia as possible to the seed. The seeds are first mixed with a gum arabic inoculant slurry (1 liter to 15 kg seed). The seed are then dried by quickly mixing with another 800 gm of the powdered inoculant (Burton, 1975). Plants from seeds inoculated this way are effectively inoculated even with a highly competitive native flora of rhizobia in the soil. Also seed inoculated by this method can better withstand the hazards of dry soils and high temperature for as long as 2 or 3 weeks following sowing.

c. Seed Coatings. Seed coatings under proprietary brand labels such as Prillcote, CelPril, Rhizo-Kote, Nu-Kote, and Gold-Coat, have recently been introduced. Special adhesives are used to stick various combinations of particulate substances such as clay, charcoal, gypsum, $CaCO_3$ micronutrients, and pesticides to the seed coat and give the seed a uniform smooth covering. Peat-base inoculants are often included. Claims are made for better stands with fewer seed, more available nutrients, beneficial growth factors, healthier seedlings, more effective nodulation, and other benefits, but whether these treatments are really superior to the conventional lime and phosphate coatings remains to be proved. The limited evidence is that rhizobia do not survive any longer than they do on regular lime-pelleted seed (Brockwell *et al.*, 1975). Coated seed is considerably more expensive.

VI. OUTLOOK

A world population of 7 billion people is predicted by the year 2000. With the diminishing supply of fossil fuels and the resulting increase in cost of nitrogen fertilizers, biological N_2 fixation is destined to supply a larger share of food protein. Of all the biological systems, the *Rhizobium*–leguminous plant association appears the most promising. Nonetheless, formidable barriers cloud the picture.

Grain legumes are our main source of plant protein for human consumption, but world production of grain legumes is only about 10% that of cereal grains, although they are planted on 15% of the total area (Hardy, 1976). Grain legumes, even with their nitrogen-fixing nodules, are not as productive as the cereal grains.

According to Hardy, about one-third of the total grain legume production comes from the soybean. Despite the fact that soybeans work symbiotically with rhizobia and fix almost one-half of the nitrogen they need, yields have increased only moderately during the past 25 years. Soybean production has increased fivefold in the past 25 years, but this has resulted largely from a fourfold increase in area used to grow soybeans.

Agronomists are puzzled and rhizobiologists are embarrassed that strains of *R. japonicum* cannot supply 75% or more of the soybean plants' need for nitrogen. Hardy and Havelka (1973, 1975) suggest that the plant is the limiting symbiont and that plants with greater photosynthetic capability would result in more nitrogen fixation and higher yields. Soybean physiologists are working on this problem.

Apparently the other grain legumes are not as productive as the soybean. Although the soybean occupies only about 30% of the total area planted to grain legumes (Table VI), it produces about one-half the food protein in addition to substantial yield of oils.

The greatest challenge to the rhizobiologist is to develop more effective *Rhizobium* strains for dry beans, chickpeas, peanuts, and dry peas. The limited data available indicate that only a small percentage of the

TABLE VI. World Areas Planted to Grain Legumes[a]

Leguminous crop	Hectares planted × 10³
Soybeans (*Glycine max*)	37,500
Peanuts (*Arachis hypogaea*)	18,000
Dry beans (*Phaseolus vulgaris*)	22,279
Chickpeas (*Cicer arietinum*)	10,543
Dry peas (*Pisum sativum*)	9,264
Cowpeas (*Vigna unguiculata*)	4,953
Broadbeans (*Vicia faba*)	4,683
Pigeon peas (*Cajanus cajan*)	2,587
Vetches (*Vicia* sp.)	1,879
Lentils (*Lens esculenta*)	1,717
Lupins (*Lupinus* sp.)	1,042
Other	6,324
Total	120,771

[a] From Food and Agricultural Organization (FAO), 1966.

seed is inoculated and that the benefits when obtained are of very low magnitude. The benefits or lack of them which have been noted could be attributed to poor nitrogen-fixing strains or a method of inoculating which could not possibly allow colonization of the roots and nodulation by the applied rhizobia. Working with soybeans, Johnson *et al.* (1965) found that an inoculum 800 to 1000 times the usual one had to be applied to attain even a moderate percentage of nodules from the applied rhizobia. Ecological problems in inoculating legumes have not received adequate attention. Certainly this is a fruitful area for research.

Data on world usage of *Rhizobium* inoculants are scanty. In the United States, *R. japonicum* inoculants constitute about 80% of the total. It is estimated that about 40% of the legume seeds planted are inoculated. Brazil, Australia, India, and most of the European nations manufacture *Rhizobium* inoculants.

There is good reason to suspect that *Rhizobium* inoculants will continue to be used and to increase in importance as the world population grows. Plant and *Rhizobium* geneticists will have to unite in purpose and work symbiotically to develop more efficient host plants and more efficient strains of rhizobia. Together they can accomplish what neither can achieve alone. Also, it will still be necessary to develop methods of inoculation which can assure nodulation by the improved strains.

REFERENCES

Allen, E. K., and Allen, O. N. (1950). *Bacteriol. Rev.* **14,** 273–330.
Badawy, F. H. (1965). Ph. D. Thesis, University of Wisconsin, Madison.
Badawy, F. H., and Allen, O. N. (1963). *Agron. Abst.* p. 31.
Beijerinck, M. W. (1888). *Bot. Ztg.* **46,** 726–804.
Bond, V. S. (1940). U.S. Patent No. 2,200,532.
Bowen, G. D. (1959). *Queens. J. Agric. Sci.* **16,** 267–282.
Bowen, G. D. (1961). *Plant. Soil*, **15n,** 155–165.
Brockwell, J. (1971). *CSIRO Div. Field St. Rec.* No. 10, pp. 51–58.
Brockwell, J. (1972). *Aust. Seed Rev.* **2,** 10–13.
Brockwell, J., and Hely, F. W. (1961). *Aust. J. Agric. Res.* **12,** 630–643.
Brockwell, J., Herridge, D. F., Roughley, R. J., Thompson, J. A., and Gault, R. R. (1975). *Aust. J. Exp. Agric. Anim. Husb.* **15,** 780–787.
Buchanan, R. E., and Gibbons, N. E. (eds.) (1974). "Bergey's Manual of Determinative Bacteriology," 8th ed. Williams & Wilkins, Baltimore, Maryland.
Burton, J. C. (1952). *Soil Sci. Soc. Am. Proc.* **16,** 356–358.
Burton, J. C. (1964). *In* "Microbiology and Soil Fertility" (C. M. Gilmour and O. N. Allen, eds.), pp. 107–134. Oregon State Univ. Press, Corrallis.
Burton, J. C. (1967). *In* "Microbial Technology" (H. J. Peppler, ed), pp. 1–33. Van Nostrand-Reinhold, Princeton, New Jersey.

Burton, J. C. (1975). *In* "Symbiotic Nitrogen Fixation in Plants" (P. S. Nutman, ed.), IBP Synth. Vol. pp. 175–189. Cambridge Univ. Press, London and New York.

Burton, J. C., and Briggeman, D. S. (1948). *Soil Sci. Soc. Am. Proc.*, **13**, 275–278.

Burton, J. C., and Curley, R. L. (1965). *Agron. J.* **57**, 379–381.

Burton, J. C., and Erdman, L. W. (1940). *J. Am. Soc. Agron.* **32**, 439–450.

Burton, J. C., and Wilson, P. W. (1939). *Soil Sci.* **47**, 293–303.

Burton, J. C., Martinex, C. J., and Curley, R. L. (1977). "Rhizobia Inoculants for Various Leguminous Species," Inf. Bull. 101. Nitragin Co., Milwaukee, Wisconsin.

Carter, A. S. (1963). *Soybean News* **14**, 4.

Christianson, D. D., Cavins, J. F. C., and Wall, J. S. (1965). *J. Agric. Food Chem.* **13**, 277–280.

Corby, H. D. L. (1975). *In* "Symbiotic Nitrogen Fixation in Plants" (P. S. Nutman, ed.), pp. 169–173. Cambridge Univ. Press, London and New York.

Date, R. A. (1970). *Plant Soil* **32**, 703–725.

Date, R. A. (1976). *In* "Symbiotic Nitrogen Fixation in Plants" (P. S. Nutman, ed.), Cambridge Univ. Press, London and New York.

Date, R. A., and Roughley, R. J. (1977). *In* "A Treatise on Dinitrogen Fixation" (R. W. F. Hardy and A. H. Gibson, eds.), Sect. IV. pp. 243–276. Wiley, New York.

Dube, J. N., and Mandeo, S. L. (1973). *Res. Industry,* **18**, 94.

Erdman, L. W., and Means, U. M. (1946). *Soil Sci. Soc. Am. Proc.* **11**, 255–259.

Erdman, L. W., and Means, U. M. (1949). *Soil Sci. Soc. Am. Proc.* **14**, 170–175.

Fred, E. B., Baldwin, I. L., and McCoy, E. (1932). "Root Nodule Bacteria and Leguminous Plants." Univ. of Wisconsin Press, Madison.

Graham, P. H. (1963). *J. Gen. Microbiol.* **30**, 245–248.

Graham, P. H. (1976). *In* "Symbiotic Nitrogen Fixation in Plants" (P. S. Nutman, ed.), pp. 99–112. Cambridge Univ. Press, London and New York.

Graham, P. H., and Parker, C. A. (1964). *Plant Soil* **20**, 383–395.

Hamdi, Y. A. (1965). Ph.D. Thesis, University of Wisconsin, Madison.

Hamdi, Y. A. (1968a). *Acta Microbiol. Pol.* **17**, 277–278.

Hamdi, Y. A. (1968b). *Arch. Microbiol.* **63**, 227–231.

Hamdi, Y. A. (1969). *Plant Soil* **31**, 111–121.

Hardy, R. W. F. (1976). *Proc. Int. Symp. Nitrogen Fixation*, Vol. 2, pp. 693–717.

Hardy, R. W. F., and Havelka, U. D. (1973). *Plant Physiol.* Suppl. **48**, 35.

Hardy, R. W. F., and Havelka, U. D. (1975). *Science* **188**, 633–643.

Hardy, R. W. F., and Holsten, R. D. (1972). "The Aquatic Environment: Microbial Transformations and Water Quality Management Implications." Environmental Protection Agency, Washington D.C.

Hardy, R. W. F., and Holsten, R. D. (1977). *In* "A Treatise on Dinitrogen Fixation" (R. W. F. Hardy and A. H. Gibson, eds.), Sect. IV, pp. 451–486. Wiley, New York.

Hellriegel, H., (1886). *Ztschr. Ver. R*übenzucker Industrie Deutschen Reichs, **36**, 863–877.

Holding, A. J., Tilo, S. N., and Allen, O. N. (1961). *Trans. Int. Congr. Soil Sci., 7th, 1960* Vol. II, pp. 608–616.

Jensen, H. L. (1967). *Arch. Microbiol.* **59**, 174–179.

Johnson, H. W., Means, U. M. and Weber, C. R. (1965). *Agron. J.* **57**, 179–185.

Kassim, M. (1976). M.Sc. Thesis, University of Missouri, Columbia.

Laird, D. G., and West, P. M. (1938). *Can. J. Res. Sect. C* **16**, 347–353.

Leffler, W. A. (1976) U.S. Patent 3,976,017.

Munns, D. N. (1976). "Proceeding of the NifTAL, Workshop on Nitrogen Fixation by Tropical Legumes," pp. 211–236. Kahului, Maui, Hawaii.

Munns, D. N. (1977). *In* "A Treatise on Dinitrogen Fixation" (R. W. F. Hardy and A. H. Gibson, eds.), Sect. IV, pp. 353–392, Wiley, New York.

Norris, D. O. (1958). *Aust. J. Agric. Res.* **9**, 629–632.

Norris, D. O. (1959). *Emp. J. Exp. Agric.* **27**, 87–97.

Norris, D. O. (1965). *Plant Soil* **22**, (2), 143–166.

Norris, D. O. (1965). *Proc. Int. Grassl. Congr., 9th, 1965* pp. 1087–1092.

Norris, D. O., and Date, R. A. (1976). *In* "Tropical Pasture Research, Principles and Methods" (N. H. Shaw and W. W. Bryan, eds.) pp. 134–174. Commonw. Agric. Bur., Harley Berkshire, England.

Norris, D. O., and 't Mannetje, L. (1964). *East Afr. Agric. For. J.* **29**, No. 3.

Parker, D. T., Allen, O. N., and Algren, H. L. (1949). *Crops Soils.* **1**, 10–11.

Purchase, H. F., and Vincent, J. M. (1949). *Proc. Linn. Soc. N. S. W.* **74**, 227–236.

Reid, J. J., Fred, E. B., and Baldwin, I. L. (1935). *J. Bacteriol.* **29**, 75–76.

Rice, W. A. (1976). "Alfalfa Production in the Peace River Region," pp. D1-D12. Can. Res. Stn. Beaverlodge, Alberta.

Roughley, R. J. (1968). *J. Appl. Bacteriol.* **31**, 259–265.

Roughley, R. J. (1970). *Plant Soil* **32**, 675–701.

Roughley, R. J. (1975). *In* "Symbiotic Nitrogen Fixation in Plants" (P. S. Nutman, ed.), IBP Vol. 7, pp. 125–136. Cambridge Univ. Press, London and New York.

Roughley, R. J., and Vincent, J. M. (1967). *J. Appl. Bacteriol* **30**, 362–376.

Roughley, R. J., Date, R. A., and Walker, M. H. (1966). *N.S.W. Dep. Agric., Div. Sci. Serv. Plant Ind. Bull.* **S-67**, 5.

Schall, E. D., Schenberger, L. C., and Swope, A. (1975). *Indiana* Agric. Exp. St., **106.**

Scudder, W. T. (1975). *Soybean Dig.* **35**, 16.

Steinberg, R. A. (1938). *J. Agric. Res.* **57**, 461–476.

Stevenson, C. L. (1976). *Legume Inoculant Insp. Program, Agric. Can.* **T-4-1.**

Strijdöm, B. W. (1963). Ph.D. Thesis, University of Wisconsin, Madison.

Strijdöm, B. W., and Allen, O. N. (1965). *Can. J. Microbiol.* **12**, 275–283.

Strijdöm, B. W., and Allen, O. N. (1967). *S. Afr. J. Agric. Sci.* **10**, 623–630.

Strijdöm, B. W., and Allen, O. N. (1969). *Phytophylatica* **1**, 147–152.

Strijdöm, B. W., and Deschodt, C. C. (1975). *In* "Symbiotic Nitrogen Fixation in Plant" (P. S. Nutman, ed.), pp. 151–168. Cambridge Univ. Press, London and New York.

Strong, T. H. (1937). *J. Counc. Sci. Ind. Res. Aust.* **10**, 12–16.

Taylor, G. G. (1972). U.S. Patent 3,672945.

Thompson, J. A. (1960). *Nature (London)* **187**, 619–620.

Thompson, J. A. (1961). *Aust. J. Agric. Res.* **12**, 578–592.

Thompson, J. A., Brockwell, J., and Roughley, R. J. (1975). *J. Aust. Inst. Agric. Sci.* **41**, 253–254.

Van Schreven, D. A. (1958). *In* "Nutrition of the Legumes" (E. G. Hallsworth, ed.), pp. 328–333. Butterworth, London.

Van Schreven, D. A., Otzen, D., and Lindenbergh, D. J. (1954). *Antonie van Leeuwenhoek* **20**, 33–57.

Vincent, J. M. (1945). *J. Aust. Inst. Agric. Sci.* **11**, 121–127.

Vincent, J. M. (1958). *In* "Nutrition of the Legumes" (E. G. Hallsworth, ed.), pp. 328–333, Butterworth, London.

Vincent, J. M. (1965). *Amr. Soc. Agron. Monogr.,* **10**, 384–435.

Vincent, J. M. (1970). *IBP Handb.* 15.

Vincent, J. M. (1974). *In* "Biology of Nitrogen Fixation" (A. Quispel, ed.), pp. 265–341. Am. Elsevier, New York.

Vincent, J. M. (1977). *In* "A Treatise on Dinitrogen Fixation" (R. W. F. Hardy and W. S. Silver, eds.), Sect. III, pp. 277–366. Wiley, New York.
Wacek, T. J., and Brill, W. J. (1976). *Crop Sci.* **16,** 519–523.
Wilson, J. B., and Umbreit, W. W. (1940). *Soil Sci. Soc. Am. Proc.* **5,** 262–263.
Wilson, P. W. (1940) "Biochemistry of Symbiotic Nitrogen Fixation," Univ. of Wisconsin Press, Madison.
Wright, W. H. (1925). *Soil Sci.* **20,** 95–129.

Chapter 3

Lactic Starter Culture Concentrates

RANDOLPH S. PORUBCAN
ROBERT L. SELLARS

I. INTRODUCTION

Commercial lactic starter culture concentrates have gained immense popularity in the United States food industry in the last 10 years. In particular, liquid nitrogen frozen concentrates have become the dominant form used by the dairy and cheese industry. Sold as either "bulk sets" or "direct sets" in approximately 70–500 ml quantities, such cultures are capable of inoculating from 300 to 1500 gal (1136–5678 liters) of milk directly. Subsequent incubation thereby produces either intermediate bulk starter when "bulk sets" are used, or the final product when "direct sets" are used. Cheddar cheese, cottage cheese, sour cream, buttermilk, and yogurt are some examples of these final products.

Other food industries also employ liquid nitrogen frozen lactic starter

59

MICROBIAL TECHNOLOGY, 2nd ed., Vol. I
Copyright © 1979 by Academic Press, Inc.
All rights of reproduction in any form reserved ISBN 0-12-551501-4

culture concentrates; namely, the meat, bakery, and fermented vegetable industries. Respective examples of cultured products produced in these industries are semidry sausage, sourdough French bread, and pickled cucumbers.

Freeze-dried and spray-dried concentrates of lactic starter cultures are also commercially available in the food industry. These have the advantageous feature of not requiring cryogenic* shipment and storage. However, performance limitations are usually experienced when such dry cultures are used directly without first making one or two intermediate propagations.

To clarify the terminology, lactic starter cultures or lactic acid starter cultures encompass several important genera of bacteria: Streptococci of group N, such as *Streptococcus lactis, Streptococcus cremoris, Streptococcus lactis* subsp. *diacetylactis,* and *Streptococcus thermophilus*†; Lactobacilli, such as *Lactobacillus bulgaricus, Lactobacillus acidophilus, Lactobacillus plantarum,* and *Lactobacillus helveticus;* and Pediococci, such as *Pediococcus cerevisiae.*

This chapter describes specifically the large-scale bioprocessing of concentrated lactic acid bacteria with associated emphasis on the commercial use of the resulting cultures. In the first edition of "Microbial Technology" Sellars (1967) discussed much of the general background information complementary to this chapter. Information on propagating lactic seed cultures, classification of cultures used in cheese manufacture, cultural and physiological characteristics of typical lactic cultures, starter metabolism, and growth curve data are given in this earlier chapter, all of which is still valid for most intents and purposes and, therefore, will not be reiterated here.

II. INDUSTRIAL BIOMASS PRODUCTION

The successful large-scale bioprocessing of lactic acid starter cultures depends on three interrelated and equally important elements: (1) processing equipment; (2) support facilities; and (3) operating procedures. We will limit our discussion in this section to processing where the desired end result is a highly concentrated starter culture destined to be used in preparing fermented food products.

Conventional fermentation equipment, such as that used by the phar-

---

* This term is used loosely here to indicate temperatures at or colder than −40°C.

† This organism may be more closely related to the group D streptococci (Buchanan and Gibbons, 1974).

maceutical industry and standard dairy processing equipment, do not satisfy the unique requirements for starter culture production. A compromise that hybridizes the pharmaceutical industry's ability to steam sterilize and maintain aseptic vessels with modern dairy ultrahigh-temperature (UHT) and clean-in-place (CIP) technology is required. Such a balance can be achieved by making appropriate modifications on conventional but modern dairy processing equipment. These modifications vary in their degree of sophistication and will be discussed in more detail later.

Figure 1 presents a general flow scheme in semiblock diagram form for the major components involved in the large-scale production of concentrated lactic acid starter cultures. The scheme begins at points A and B with media reconstitution followed by ultrahigh-temperature pasteurization (step C) and subsequent aseptic transfer to the fermentor (H). Here the sterile medium is inoculated and incubated; concomitantly, various growth parameters are controlled if required. The goal is to reach maximum biological concentration and activity in the fermentor prior to any further processing.

Once fermentation is complete the culture or fermentate is cooled and pumped into a "cold wall" storage tank (Fig. 1, item J). From here a variety of physical methods of concentration may be selected varying from lyophilization to centrifugal separation. Ultimately, the concen-

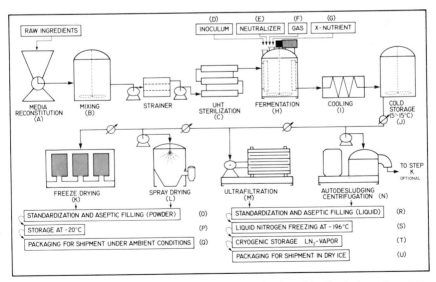

FIGURE 1. Flow scheme for the large-scale production of lactic starter culture concentrates.

trated starter culture* is marketed as either a powder or a frozen liquid concentrate. With this brief overview of the bioprocess we will proceed to a more detailed evaluation.

A. Growth Media

1. Formulation

Basal media ingredients for starter culture biomass production should be either soluble or able to remain in homogeneous suspension. With only a few exceptions this is generally the rule, particularly when centrifugal or membrane procedures are used to physically concentrate the starter bacteria after fermentation. Insoluble, nondispersible ingredients will be recovered with the cell concentrate in these procedures, consequently limiting the maximum level of cells attainable per unit volume (Stadhouders et al., 1969). This problem may be circumvented in certain cases by employing enzymatic predigestion of the otherwise insoluble ingredient(s). More specifically, cheese whey upon proteolytic digestion no longer precipitates during heat sterilization and is thus rendered more useful in preparing basal media for lactic Streptococcus starters (Hansen, 1969; Jost and Monti, 1977).

The use of totally soluble and/or dispersible ingredients does not in itself eliminate the possibility of recovering residual media solids with the cell concentrate (Lloyd, 1971). Data collected in our research laboratories demonstrate that the turbidity developed by Lactobacillus acidophilus (ATCC strain #4356), grown in initially clear MRS broth, is not entirely due to bacterial cells (Porubcan and Sellars, 1977). Certain unidentified initially soluble media substituents are apparently precipitated by the metabolism of the Lactobacillus. Reconstituted skim milk, which is a stable suspension, will coagulate during lactic culture growth; the acid-precipitated casein must be deflocculated with a suitable reagent, such as sodium citrate, before a reasonably clean centrifugal separation of cells can be made (Stadhouders et al., 1969). Reconstituted skim milk can be predigested with either papain or a suitable microbial protease prior to sterilization and inoculation with lactic Streptococcus starter(s) (Stanley, 1977). This procedure eliminates the need for citrate treatment after growth; however, certain Streptococcus cremoris strains, such as P_2 and AM_2, are retarded by such media (Stanley, 1977).

---

* The terms "starter cultures" and "cultures" are used as synonyms to "lactic acid starter cultures" in this chapter.

Insoluble, nondispersible ingredients, although often desirable from a price standpoint, have the additional drawback of being less available to the cell. Often the degree of "insolubility" is related to the heat treatment parameters employed. For example, when reconstituted sweet whey is ultrapasteurized or UHT heated at 290°F (143.3°C) for 4 seconds, it remains in suspension; however, when it is batch heat treated at 90°C (194°F) for 40 minutes, it precipitates dramatically and subsequently fails to support the prolific growth of lactic streptococci.

A variety of ingredients are in general use for the bulk cultivation of lactic acid bacteria. They can be classified into four broad groups, as listed and characterized in Table I.

To achieve maximum growth in batch fermentations with any particular strain or mixture of strains of lactic acid bacteria, it is necessary to formulate an advantageous basal medium. It may be as "simple" as reconstituted skim milk or a very complex mixture of several ingredients from each group in Table I. The importance of well-controlled fermentation research in establishing a correct formulation cannot be overemphasized. Most of the useful strains of lactobacilli, group N streptococci, and pediococci have rather fastidious growth requirements (Lawrence *et al.*, 1976; Etchells and Costilow, 1968; Thompson, 1976). Indeed, our technology has not advanced to the point where we can chemically define the growth media for these lactic bacteria and at the same time achieve maximum cell density and activity. Thus, we must continue to use such complex nutrients as skim milk, whey, yeast extract, peptones, and the like.

Unfortunately, complex nutrients such as those listed in group I of Table I are often subject to a great deal of variance from one manufacturer to another and even between different production lots from the same manufacturer. It is sensible to pretest samples of such ingredients prior to purchasing large quantities. In some cases where lot-to-lot uniformity is irregular it may be helpful to use two or more commercial brands simultaneously. For example, if a given basal medium formulation calls for 0.5% yeast extract, it may prove beneficial to use 0.25% from each of two different commercial brands, say brands A and B. Thus, the formulation would become more "poised" in that it would be less sensitive to a loss in quality in brand A than if brand A were used alone. The "poising" of a medium formulation so that it becomes less sensitive to lot-to-lot variations in the major ingredients is vital to obtaining uniform fermentations. Sometimes such "poising" can be effected by using an adjuvant ingredient(s) not directly related to the ingredient in question. For example, it has been shown that the unsatisfactory clotting of springtime milk by various strains of lactic streptococci can be over-

TABLE I. Ingredients Used in the Large-Scale Fermentation of Lactic Starter Cultures

Group I. Complex nutrients
A. Soluble (forming clear 1% solutions)[a]
   1. Yeast extracts or autolysates
   2. Casein hydrolysates
   3. Soy protein hydrolysates
   4. Peptones of animal origin
   5. Corn steep liquor
   6. Liver extracts
   7. Whey protein hydrolysates
B. Insoluble (forming stable aqueous suspensions)[b]
   1. Skim milk powder
   2. Sweet whey powder (from Swiss or mozzarella cheese)
   3. Acid whey powder (from Cheddar or cottage cheese)
   4. Whey protein concentrates
   5. Hydrolyzed cereal solids, dextrins, malto-dextrins
   6. Demineralized whey powder

Group II. Simple nutrients
A. Organic
   1. Glucose, fructose, and other monosaccharides
   2. Lactose, maltose, sucrose, and other disaccharides
   3. Tweens and related compounds
   4. Vitamins
B. Inorganic
   1. Mineral salts of Ca, Mn, Mg, Mo, Fe, Co, etc.

Group III. Buffers
A. Organic
   1. K, Na, or ammonium salts of citrate,[c] glycerol phosphate, and acetate
B. Inorganic
   1. Mainly the orthophosphates of K, Na, or ammonium;
      examples, monoammonium phosphate, disodium phosphate

Group IV. Defoamers
A. Silicone types
   1. Dow Corning FG-10
   2. Hansen's defoamer
B. Organic types
   1. Tweens (i.e., Tween 80)
   2. Acetylated monoglycerides

---

[a] Crude or unrefined varieties of the examples listed may be insoluble or may contain insoluble components.

[b] Stable suspensions may be dependent upon the method of preparation—see text.

[c] Citrate can also be a fermentable substrate and/or a deflocculating agent.

come by the addition of seven different amino acids at 50 ppm each (Erzinkyan *et al.*, 1966).

2. Reconstitution

Thorough and complete reconstitution of the various media substituents prior to sterilization or heat treatment is essential for two reasons: (1) incompletely hydrated proteins and peptones may heat denature differently compared to their hydrated counterparts (Porubcan, 1975; Babel 1976), and (2) nondispersed semidry aggregates of protein and/or carbohydrate may not sterilize during heat treatment due to inadequate heat penetration. This is particularly true when UHT processing is used where the thermal exposure time is so brief (Carlsen, 1977).

Proper reconstitution can be achieved by using modern dairy-type powder-to-liquid blenders, such as the Ladish Triblender or the Lanco Liquifier. Generally, the blender is connected to an agitated holding tank so that thorough recirculation back to the blender can be achieved during powder addition. (Refer to Fig. 1, parts A and B, for a diagrammatic representation of this process.)

3. Sterilization

By definition, sterilization is the complete destruction of all microorganisms, including mycotic and bacterial spores, by either wet steam at 120°C and 15 psig for 15 minutes or by dry heat at 160°–180°C for 3 hrs (Lawrence and Block, 1968). Sterilization per se cannot be achieved in complex media, such as milk or whey, by these conventional techniques without the associative destruction of certain growth-promoting properties, (Thomas and Turner, 1977; Webb, 1965). It is common knowledge, for example, that autoclaved (120°C, 15 minutes) skim milk is degraded to where it will not support active lactic culture growth. In the past, batch heat treatment at 85°–95°C for 40–120 minutes was used for such heat-sensitive media (Christensen, 1967). It was recognized, however, that although vegetative cells were destroyed under these conditions, it was not possible to guarantee the destruction of bacterial and mycotic spores. Thus, a condition less than sterile had to be accepted. The process of tyndallization (or sporulation), which employs the alternating application of batch heat treatment with appropriate incubation periods, can be effective but is not practical in large-scale bioprocessing.

The best way to effect the sterilization of complex media without inducing a concomitant loss in growth-supporting properties is through UHT* pasteurization. In our laboratory we have confirmed that UHT

---

* Also referred to as higher heat, shorter time (HHST) processing.

processing at 140°–145°C for 4–8 seconds will render 10% reconstituted skim milk (or sweet whey) sterile without any accompanying deleterious effects, such as caramelization. Such media support the prolific growth of lactic streptococci and popular lactobacilli such as L. *bulgaricus* and L. *acidophilus*. Understandably, chemical changes take place in milk and whey when they are UHT pastuerized (Hansen and Melo, 1977). Specifically, casein appears to be more resistant to structural alteration than β-lactoglobulin when raw skim milk is heated at 143° ± 1°C for 8–10 seconds (Hansen and Melo, 1977). These changes in the whey protein are potentially stimulatory to lactic acid bacteria (Webb, 1965; Woods, 1975), and current data from our laboratory support this contention.

Various types of industrial UHT pasteurizers are available. They can be classified as either indirect or direct referring to the method by which high-pressure steam heats the product. Of the indirect UHT pasteurizers, the tubular models are better engineered to withstand the high pressures and they do not encounter the gasketing problems that always plague the plate-type pasteurizers. The direct steam-injected UHT pasteurizers are not particularly suited for processing culture media. Direct injection of steam creates turbulence and physical stress that may induce media instability (Woods, 1975), not to mention the requirement for absolutely pure steam. In addition, moderate whey protein denaturation is enhanced by indirect UHT pasteurization, a potentially beneficial condition as previously discussed.

When skim milk is used as a substrate for lactic acid bacteria, the type of heat treatment to which it is subjected determines whether growth stimulation or inhibition will occur. For example, heating milk at 143°–160°F (62°–71°C) for 30–40 minutes stimulates the growth of *Streptococcus* starter cultures, whereas heating at 160°F (71°C) for 45 minutes or at 178°F (81°C) for 10–45 minutes inhibits growth (Webb, 1965; Woods, 1975). This "temperature conditioning" of milk and whey and other complex media should be treated as a separate entity apart from the time/temperature profile required to effect sterilization. A heat treatment that can optimize the growth response and also effect sterilization is the goal to achieve with any new medium formulation. Indirect tubular UHT processing systems, such as the Cherry-Burrell Spiratherm, can accomplish this difficult task. Large-scale Spiratherm systems with capacities over 2000 gal/hour regenerate input calories to between 55 and 65% depending on the unit.

As shown in Fig. 1 (C) reconstituted medium is pumped through a strainer to a ballast tank (not shown) associated with the UHT unit. This is usually done with automatic level control. The medium in the ballast tank is then fed to the UHT via a high-pressure pump, such as a homogenizer. After media sterilization, careful attention must be focused on the method of transfer to the presterilized fermentor(s) (Fig. 1, H). Here it is necessary that steam-locked sanitary pumps and valves be used and that the transfer piping be steam sterilized after conventional CIP.

B. Fermentation

1. Fermentor Design

The large-scale fermentation of lactic acid starter cultures requires a round dairy processor with some rather sophisticated modifications if both security against bacteriophage infection and maximum biomass production are to be achieved. An ideal processor/fermentor is diagrammatically indicated as H in Fig. 1; special design features include those listed below:

1. An ability to hermetically seal and pressurize the vessel at 15 psig (working) to allow in-place sterilization at 121°C when required.

2. A full pressure-rated jacket with zone controls for in-place fermentor sterilization control and temperature control during incubation. It should be possible to hold the fermentor contents at ± 0.5°C during incubation cycles.

3. The "round processor" design where the vessel's functional depth is about twice the inside diameter.

4. A variable-speed agitator with a full bottom sweep impeller. Mechanical speed reducing transmissions, such as a Reeves drive, are effective up to about 10 HP. Above 10 HP variable frequency AC control is recommended. Teflon agitator shaft seals are desirable.

5. Type 316 stainless steel on all product contact surfaces; type 304 stainless steel is acceptable for exterior surfaces.

6. Two, three, or four baffles installed at angles of 180°, 120°, or 90°, respectively, and extending above the liquid level perpendicular to the side wall with a separate CIP spray ball assembly for each baffle. Proper baffling ensures top-to-bottom turnover during agitation.

7. A high-temperature CIP system able to deploy a chlorinated alkaline detergent at 185°F (85°C) with subsequent water rinse, acid, and chlorination cycles. Single-use CIP systems, where fresh cleaning solutions are used for each independent operation, are preferred as cross-contamination possibilities are virtually eliminated by such systems. Economics Laboratory at Beloit, Wisconsin manufactures and installs an excellent system of this type.

8. An absolute 0.2–0.35 μm filter(s) on the headplate to assure that all air entering the headspace is bacteria free.* Filter units are attached to large sanitary butterfly valves that can be closed during hermetic (and CIP) operations. The filter(s) should be sized such that it will provide sufficient air displacement capacity to keep up with the unloading pump.

9. A vacuum/pressure safety relief valve on the headplate along with a separate rupture disk if double protection against vacuum collapse is desired.

10. Standard sanitary inlet ports for all transfer operations; one is usually a "no-foam" inlet for media fill (on headplate) and another can be steam jacketed for aseptic inoculation. Additional ports may be required on the headplate for various additives, such as neutralizer, gas, extra nutrient, and defoamer (see Fig. 1, E, F, and G). Ports are also required for electrodes such as pH electrodes. Thermistor or thermocouple probes for temperature sensing and control are generally installed in sidewall-mounted hermetically sealed wells that extend into the product by 6–10 inches.

11. A manhole with sight glass and a light fixture at 180°; both should be mounted on the headplate. Visual assessment of agitator rpm for a given starter culture is necessary to ensure optimization of this parameter. In addition, the manhole should be "safety interlocked" with the CIP system so that it must remain open with a louvered door in place during CIP.

12. Remote manometer readout of vessel volume.

The foregoing special design features truly convert an otherwise standard round processor into a functional fermentor for lactic acid bacteria. In contrast to typical pharmaceutical-type fermentors the parts and fittings here are of sanitary design; the agitator is low speed, usually 5–60 rpm on 1000–8000-gal vessels; electrodes are mounted through the headplate, not through the sidewall, and all connection fittings are sanitary welded and are, therefore, without threads.

2. Fermentation Process.

Illustration of the general fermentation process for lactic acid bacteria is best developed by referring once again to the example shown in Fig. 1. Ultrahigh-temperature sterilized growth media are aseptically transferred into a presterilized fermentor (Fig. 1, H) through a no-foam inlet in the headplate. Alternatively, the growth media can be batch heat treated

---

* Generally, this will be a safe filter for viruses such as bacteriophage since most airborne phage is associated with macroscopic dust (Jespersen, 1977). If phage contamination develops, incinerated air must be used instead of the filter.

at 85°–95°C for 40–120 minutes in the fermentor; however, UHT processing is usually the desired course. Subsequently, gentle agitation is started and the temperature is adjusted via the jacket controls until the exact incubation temperature is reached. The fermentor is then inoculated (usually at 1–2%) with fresh mother culture (Fig. 1, D) via an aseptic centrifugal pump* or by pressurized air unloading using sterile air. Fermentation is then commenced and various parameters are automatically controlled as required to achieve maximum biomass populations (Blaine, 1972; Gilliland, 1977; Efstathiou et al., 1975). These controls include agitation, temperature, pH, and in some instances redox potential (ORP) and nutrient concentration. After the fermentation is complete, usually by late log or early stationary phase, the culture or fermentate is cooled rapidly to 5°–15°C by pumping it through a tubular heat exchanger where 1°–2°C water is the cooling agent (Fig. 1, I). Slow cooling can be effected in the fermentor by circulating 1°–2°C water through the jacket. This may be a more advantageous approach with some cultures. The cooled culture can then be held in a cold wall dairy storage tank (Fig. 1, J) or in the actual fermentor if desired. From here it is further processed by one or more of four physical methods of concentration (Fig. 1, K, L, M, and N). These will be detailed in a following section.

The pH control function is of particular importance. Generally, "tight control" to within ± 0.1 pH unit of the set-point pH is most advantageous (Porubcan and Sellars, 1977; Richardson et al., 1977). This can be effected by using an automatic pH controller (or autotitrator) coupled with precise agitation control. Neutralizer, such as concentrated ammonium hydroxide, should be admitted into the fermentor headspace at 180° to the pH electrode(s) (Porubcan and Sellars, 1977). Providing that fermentor agitation is adequate in terms of top-to-bottom turnover, the pH electrode(s) can be submerged in the top 10–15% of the fermentor's working volume. Combination electrodes are preferred over separate glass and calomel electrodes, and pH signal preamplification up to approximately 1 V is highly desirable (Richardson et al., 1977). Preamplified pH signals are more stable and can be conveyed over longer distances. "Saw-toothed type" recordings of the pH control function (pH versus time) are important and should be obtained routinely.

The foregoing hypothetical descriptions depict a purely batch-type process. Continuous and/or dialysis culture fermentations with lactic streptococci have been studied (Osborne, 1977; Keen, 1972). From an industrial-scale standpoint, however, these are not practical. For example, Osborne's work with dialysis of lactic Streptococcus cultures

---

* A sanitary centrifugal pump with steam-jacketed shaft seals should be used.

against sterile media requires a medium-to-culture ratio of 30:1 to be effective. In our laboratory we have tried ultrafiltration as a means of removing metabolic waste products during the active log phase pH-controlled growth of lactic streptococci. When sterile medium was added to the fermentor at fixed intervals during ultrafiltration to compensate for the discarded permeate, the end result, with the cultures studied, was not an increase in biomass but an increase in the rate of lactic acid production in the fermentor (Porubcan and Sellars, 1977). This apparent increased rate of acid production during fermentation was not carried over by subsequent subcultures. No practical utility was demonstrated by these experiments.

An equation for batch bacterial growth discussed by Konak (1975) defines eight distinct growth phases: (1) lag phase, (2) accelerated growth, (3) logarithmic growth, (4) decelerated growth, (5) stationary phase, (6) accelerated death, (7) logarithmic death, and (8) survival phase. The overall result of the first five phases is a sigmoidal growth curve. The general equation for logarithmic growth is:

$$dN/dt = kN \tag{1}$$

where k is the specific growth rate, N is cell numbers per unit volume, and t is unit time. A straight line relationship is maintained during true log phase growth, but at the onset of decelerated growth dN/dt starts decreasing due to three growth-limiting factors: (1) limitations in nutrient supply, (2) accumulation of toxic metabolic waste products; and, (3) overcrowding—the "bump coefficient."

In phases 2 and 3 the rate of growth is directly proportional to N or the concentration of bacteria. In phases 4 and 5 collective retardation forces set in due to all or some of the above factors, and the growth rate becomes directly proportional to (N_s-N) at any given time t (Konak, 1975). N_s is the concentration of bacteria per unit volume in the stationary phase. Assuming that the relationships are governed by exponential law we can write:

$$\frac{1}{N_s^{a+b}} \frac{dN}{dt} = k \left(\frac{N}{N_s}\right)^a \left(1 - \frac{N}{N_s}\right)^b \tag{2}$$

where a, b, and k are constants and the sum $a + b$ may be considered the "overall order" of the bacterial growth process. In the special case where $a = 1$ and $b = 0$, Eq. (2) indicates logarithmic growth.

If we now define $F_P = N/N_s$ as the fractional increase in population density, we can write:

$$\frac{dF_P}{dt} = k N_s^{a+b-1} F_P^a (1 - F_P)^b \tag{3}$$

Maximum growth rate occurs when $F_P = a/(a + b)$; therefore, from the inflection point on the curve of F_P versus t (or N versus t) we can obtain one simple relationship between the constants a and b. This can be determined along with N_s and k in laboratory experiments where initial nutrient concentration, initial bacterial concentration, bacterial species, temperature, and hydrodynamics are all controlled. Equation (3), which in general form has proven validity in crystallization and precipitation applications, can be used here to guide scale-up operations for batch-type fermentations (Konak, 1974; Nielsen, 1964).

With lactic acid bacteria, particularly with group N streptococci and many lactobacilli, the most significant of the three growth-limiting factors seems to be the accumulation of toxic metabolic waste products. dN/dt for most *Streptococcus cremoris* strains grown at 22°C in 12% reconstituted skim milk under pH control at pH 5.40 (NH_4OH neutralization) will start to decrease once the ammonium lactate concentration reaches about 2–3% (w/w) (Porubcan and Sellars, 1977). There is still a surplus of nutrient (carbohydrate and protein) at this time and overcrowding is not a particularly significant issue under these specific growth conditions (viable plate counts run 5×10^9/ml or less after 16 hours of growth).

C. Physical Concentration

1. Centrifugal Separation

The concentration of bacteria by centrifugal separation depends on two general criteria: (1) the properties of the bacterial culture to be separated, and (2) the efficiency of the centrifuge. There are two bacterial properties that are directly proportional to separation efficiency: the cell diameter squared and the density difference between the cell and the liquid medium. The viscosity of the supernatant is inversely proportional to separation efficiency (Elsworth, 1962; Webb, 1964). Obviously, then, large, dense cells that are suspended in a low-viscosity medium are most easily separated; the composite effect of these properties determines the overall "centrifugability" of the particular strain(s). It is the cellular size and shape that most significantly influence separation because of the square function. Spherical cells separate more readily than rod-shaped cells (Elsworth, 1962). Thus, by this theory, a yeast cell having a diameter of 5 μm will separate 25 times more readily than a bacterial coccus cell having a diameter of 1 μm (everything else equal). Another way of considering this difference is that the feed rate to an industrial centrifuge can be 25 times faster for the yeast culture.

Temperature also affects the efficiency of separation; higher temperatures benefit separation but exert a negative effect on cell viability and activity. Thus, a compromise is required; usually most lactic cultures are held somewhere between 5° and 15°C during separation. The highest temperature that can be tolerated by a given strain should be determined; this temperature will depend on cell sensitivity and centrifuge design.

The efficiency of a centrifuge is often mistakenly confused with the relative centrifugal force (RCF)* or gravitational effect that can be generated. The sigma number for a particular machine is the correct function that defines efficiency. Sigma is proportional to $w^2 r^2 / S$, where w is the bowl rpm, r is the bowl radius, and S is the maximum distance that a particle must travel to reach the bowl wall or a collecting surface. In effect, sigma is the surface area of the theoretical gravity tank that is required to effect the same degree of separation as the machine (Elsworth, 1962).

For a number of years the only industrial centrifuges used for separating lactic acid bacteria were the Sharples and Carl-Padberg (CEPA) type tubular rotor machines. These machines spin at high rpm, exerting correspondingly high RCFs. The largest CEPA machine, the model Z-101, spins at 14,000 rpm, exerts 17,760 RCFs, and has a rotor sludge capacity of 10 liters. The major drawback of these machines is their manual nature. The rotor (or bowl) must be stopped, dismantled from the machine, and opened before the bacterial concentrate can be scraped out. Even when waxed parchment or polypropylene inserts are used as removable rotor liners, the process is not practical and is subject to contamination.

The development of modern automatic desludging CIP separators made the large-scale bioprocessing of concentrated lactic acid bacteria realistic. These disk–bowl type separators are very efficient machines because of their large effective surface areas (small S values in the sigma expression) and large bowl diameters. They spin at relatively low rpm (4500–6000) but nevertheless have sigma values comparable to the tubular rotor machines. There are two major types of automatic desludging CIP separators: first, there are those where a closed bowl is intermittently opened during operation to eject the concentrate; second, there are the nozzle machines that have continuously open ports in the bowl periphery for continuous concentrate ejection. Examples of the first type are De Laval's MRPX-213 and MRPX-217 and Westfalia's SA-60, SA-80, and SB-80. Of the second type, or the nozzle machines, the best example is Alfa-Laval's Hermetic Bactofuge. According to the manufacturer's

---

* $RCF = g \cdot w^2 \cdot r$ where g is the gravitational constant, w is the bowl rpm, and r is the bowl radius in centimeters.

specifications the Bactofuge Type D-3187M will remove 90% of the bacteria in raw milk when operating at 65°C.

Insofar as the concentration of lactic acid bacteria is concerned, the intermittent desludgers afford better control over concentration. With these machines the bowl can be precisely opened and closed whenever desired to effect a particular degree of concentration. With the nozzle-type machines the feed rate to the machine can be varied within certain limits (the nozzles must be kept satisfied) but beyond that it is a "take what you get" situation. In both cases, the machines require certain modifications for lactic acid bacteria processing. The conical disk plates should be highly polished and somewhat closer together than standard and they should have grooved channeling. The centripetal pump should be constructed with a smooth bore for low-capacity feed rates, and provisions for aseptic sludge removal should be considered. It is common to run these separators at 1/10 to 1/20 of their specified feed capacity when processing lactic acid bacteria. Variable-speed sanitary positive pumps are preferred over any other product-to-separator delivery system. With the intermittent automatic desludgers it is possible to concentrate certain lactic streptococci to between 20 and 100 times their initial concentration (Porubcan and Sellars, 1977).

A recent innovation pioneered by Chr. Hansen's Laboratory, Inc. in Milwaukee, Wisconsin, involves cooling the spinning separator bowl (Westfalia SA-80-06 autodesludger) with liquid nitrogen. Actually, liquid nitrogen per se does not contact the bowl, but the resulting supercold vapor makes the contact. The liquid nitrogen is sprayed onto the top of the spinning bowl; it is delivered through a cryogenic solenoid valve that is automatically opened and closed as a function of the supernatant discharge temperature. By cooling with liquid nitrogen in this manner it is possible to exactly counterbalance the frictional heat of the machine such that the supernatant and concentrate discharge at a temperature exactly equal to the fermentate input temperature. This degree of cooling efficiency cannot be achieved by using a water-cooled jacket on the centrifuge cover bonnet.

2. Ultrafiltration

The industrial-scale concentration of lactic *Streptococcus* starters by ultrafiltration was reported by Porubcan and Sellars (1976). In this work a skim milk culture of *S. lactis* mixed with *S. cremoris* (Hansen's #70) was controlled between pH 5.40 and 6.00 with concentrated ammonium hydroxide during a 16-hour incubation period at 24°C. Following this the culture was cooled to 12°C and concentrated on an Aqua Chem Ultra-Sep ultrafilter (150 ft² of membrane) for 2 hours with inlet and outlet pressures of 60 psig* and 30 psig, respectively. The result was a threefold increase in bacterial concentration with a proportional increase in lactic acid-producing activity.

Ultrafiltration (UF) is different from reverse osmosis in that the membrane is permeable to both water and soluble low molecular weight compounds rather than to water alone. Consequently, lower operating pressures of almost an order of magnitude are characteristic of UF systems (Payne *et al.*, 1973). In the concentration of bacteria it is advantageous to operate under conditions of lower pressure and lower osmolarity; thus, UF is the membrane process of choice. Depending on the nature of the bacterial culture and on the type of UF machine, the extent to which a culture can be concentrated varies from two to about twelvefold. Beyond this level of concentration the viscosity of the concentrate limits the attainable velocity across the membrane and thus the permeate flux through the membrane. A point of diminishing returns is ultimately reached. In addition, frictional heat buildup damages the cells at these higher concentrations. In any event, an efficient method of cooling the bacterial concentrate during ultrafiltration is a vital requirement.

To be of value in the concentration of starter bacteria an ultrafiltration system must be able to tolerate pH 13 caustic cleaning (CIP) and subsequent chemical sterilization. Aqua Chem's Ultra-Sep machines can be supplied with either polyvinyl or Teflon membranes. Both types of membranes tolerate pH 13 caustic; the former can be sterilized with either H_2O_2 or formaldehyde and the latter with hypochlorite as well. These Ultra-Sep machines are of a rectangular plate "sandwich" design. The degree of compression on this sandwich of membranes and plastic spacers is of critical importance. Hydraulic compression to between 5000 and 6000 psig is required for optimum performance on units having 150–600 ft² of membrane surface (Schantz, 1976). Another competitive UF machine of a plate-and-frame design is manufactured by DDS in Denmark; Chr. Hansen's Laboratory in Copenhagen has concentrated lactic streptococci with a pilot-scale (Als de Danske Sukkerfabrikker DDS) unit having 4 m² of membrane surface (Jespersen, 1974a).

3. Spray-Drying

Spray-drying is the transformation of a solution or suspension into a dry powder in a single operation. The feed liquid is atomized into a fine

---

* psig = pounds per square inch gauge—not absolute pressure, but gauge pressure.

spray which immediately contacts a flowing stream of hot air. The spray-drying of yogurt and related cultures was reported by Porubcan and Sellars (1975b). A mixture of *L. bulgaricus* and *S. thermophilus* (Hansen's CH-3) was grown in 12% reconstituted skim milk for 16 hours at 37°C. A Niro Mobile Minor laboratory-scale spray-drier was then used to dry the culture at an outlet temperature of 145°F (63°C), a culture feed rate of 12–15 ml/minute, and a nozzle velocity of 3.5 kg/cm^2. The resulting powder had a plate count on Hansen's yogurt agar (Porubcan and Sellars, 1973) of 12 × 10^8 viable colony-forming units (CFU) per gram and a moisture content of 4.5%.

The successful spray-drying of yogurt cultures is dependent upon a chemical stabilization process developed by Porubcan and Sellars (1975a). The cooled fermented culture is mixed with ascorbate and glutamate followed by pH adjustment to 6.10–6.60. This combination induces a chemical synergism that imparts partial protection to the cells during the drying process and significant stabilization to subsequent storage at ambient temperatures. The chemical mechanism involved is linked to the antioxidant properties of the ascorbate with glutamate (although aspartate works equally well), moderating the reaction by controlling the rate of ascorbate oxidation. After the ascorbate expires, the stabilization effect is no longer realized and the dry culture deteriorates. Air, moisture, and light accelerate the oxidation of ascorbate and, thus, should be minimized whenever possible.

Spray-drying per se is a concentration process by virtue of water removal; however, significant cellular destruction occurs even when the ascorbate/glutamate treatment is used. In fact, losses in viable plate count (in CFU) of about 70% are normal. The major advantage of the ascorbate/glutamate treatment process is the enhanced stability of the spray-dried powder to storage at 21°–23°C. Specifically, untreated spray-dried *L. acidophilus* will decay by more than 1000-fold (from 1.6 × 10^9 to 1.0 × 10^6 per gram) after 1 month at 21°C, whereas treated culture decays by only twofold (10.7 × 10^9 to 5.7 × 10^9 per gram) under the same conditions (Porubcan and Sellars, 1975a).

Two parameters are critical to the success of large-scale spray-drying of yogurt and related cultures: (1) The outlet temperature of the main chamber should be controlled between 63° and 68°C (145°–155°F) and (2) the relative humidity of the intake air should not exceed 60%. Different strains of *L. bulgaricus, L. acidophilus,* and *L. helveticus* exhibit different degrees of sensitivity to the spray-drying process. Acceptable results are achieved with most of these thermoduric lactics; however, mesophilic lactics are very sensitive to spray-drying.

4. Lyophilization

Batch freeze-drying of lactic starter cultures has been practiced commercially for over two decades; both tray- and cylinder shell-type driers have been used. Generally, the freeze-drying process is less destructive to starter culture cells than the spray-drying process. Mesophilic lactic streptococci can be freeze-dried with rather satisfactory results; losses in viable cell populations can run as high as 80%, but 50% is about normal. The ascorbate/glutamate treatment process (Porubcan and Sellars, 1975a) is also of significant value when used in conjunction with freeze-drying. As with spray-drying the stability of the resulting dry culture to storage at ambient temperatures is the most pronounced attribute gained by using this chemical treatment process.

Procedurally, the stabilized fermented culture is frozen in a thin film of 1-cm thickness or less, typically by slow rotation in a horizontally positioned stainless steel (10-liter) cylinder that is submersed in either liquid nitrogen ($-196°C$) or dry ice/alcohol ($-65°C$). Two liters of liquid culture containing 12–18% solids by weight require about 5–12 minutes for complete freezing in the dry ice/alcohol mixture. Subsequently, each cylinder is clamped to a manifold assembly where it is subjected to a strong vacuum that should reach 25 μm of mercury in less than 3 hours. The cylinders are maintained at 20°–24°C during vacuum drying. The vacuum is held for 12–16 hours and the condensate is collected at $-65°C$. The resulting crusted material is gently ground to a uniform powder while appropriate aseptic procedures are maintained. The moisture content in this culture powder will be less than 1% if proper procedures have been followed; this level may be difficult to achieve with thermoduric yogurt-type cultures.

For liquid capacities of 50 liters or more the tray-type driers, such as those manufactured by Stokes or Hull, are more appropriate than the cylinder shell driers. Here liquid nitrogen and/or its vapor is the preferred freezing agent. Compared to spray-drying, freeze-drying is considerably more expensive due to greater energy requirements.

D. Packaging

1. Containers

Seamless extruded aluminum cans in 70 to 360-ml capacities are ideal containers for liquid starter culture concentrates, particularly when liquid nitrogen freezing is employed (Rinfret and LaSalle, 1975). Extruded plastic containers, such as those used for frozen sherbet, and cylindrical cardboard containers are also used by some commercial starter culture suppliers. These containers cannot be subjected to submersion in liquid nitrogen, however.

Spray-dried or freeze-dried cultures are preferably packaged in moisture-proof pouches of a laminated foil–plastic construction. Larger quantities of 1–50 kg are commonly packaged in heavy guage plastic bags contained in fiber drums. Since oxygen, moisture, and light are deleterious to dry cultures, it is desirable to pack them under dry inert gas or under vacuum in opaque enclosures.

2. *Liquid Nitrogen Freezing*

The optimum method of freezing liquid starter culture concentrates is by direct immersion in liquid nitrogen or its vapor. A 360-ml aluminum can full of liquid culture concentrate at 12°C will freeze solid after about 3–5 minutes in liquid nitrogen at −196°C; complete freezing to −196°C will take about 12 minutes and will require from 1 to 1.3 liters of liquid nitrogen (Porubcan and Sellars, 1977; Reddy *et al.*, 1974).

The initial freezing rate for 350 ml of skim milk contained in a 360-ml aluminum can is 0.4°F/second in liquid nitrogen (complete immersion) and 0.04°F/second in static liquid nitrogen vapor (Shuster, 1976). Because of viscosity effects and altered colligative properties these rates are lowered anywhere from three- to fivefold when concentrated starter cultures containing over 15% solids (dry wt. basis) are frozen.

A newly developed process for preparing frozen pellets of concentrated starter culture is currently under investigation at Chr. Hansen's Laboratory in Milwaukee. By a proprietary process, droplets of liquid starter concentrate are "rained" into an agitated bath of LN_2. The resulting pellets range from 1–2 mm in diameter and, as such, permit a free-flowing bulk-packaged product to be prepared.

Generally, most lactic acid bacteria freeze well in liquid nitrogen (LN_2) and do not require special cryoprotective compounds such as glycerol. A satisfactory diluent for starter culture concentrates destined for LN_2 freezing is reconstituted skim milk (Porubcan and Sellars, 1977; Peebles *et al.*, 1969). Most lactic acid bacteria have an indefinite shelf-life when stored in liquid nitrogen (−196°C) or its vapor (−150°C). We have had Redi-Set brand bulk set cultures (*S. cremoris* and *S. lactis*) in test storage at −150° to −196°C for over 8 years without any measurable loss in viable plate count or lactic acid-producing activity. Storage at less optimum temperatures leads to completely different results. For example, these same bulk set cultures that last 8 years at −150° to −196°C will only keep 4–6 weeks at −40°C before showing a measurable loss in viability. In addition, Leach and Sandine (1976) have shown that strain dominance in mixed strain lactic streptococcus starters can change in prolonged storage (8 months) at −20°C. The data of Reddy *et al.* (1974) clearly indicate that LN_2 freezing and storage is the best procedure to maintain component balance in mixed strain lactic cultures.

Frozen starter cultures should be thawed just prior to use by direct submersion in water at 24°C for 3–20 minutes depending on the amount of agitation employed. Once thawed, the general rule is to use the culture immediately whenever possible; however, most thawed cultures will keep for 1–2 hours at 5°–10°C without a measurable loss in viability and/or activity. The particular manufacturer's instructions should be followed in all cases as these recommendations may vary from one supplier to another.

Liquid nitrogen frozen cultures can be shipped from manufacturer to user in special portable liquid nitrogen refrigerators, such as the Linde Model 30-10. It is possible to transport 88 × 70 ml size culture units in such a container which, in addition, contains enough liquid nitrogen to hold the cultures for about 28–30 days. More commonly, however, cultures are shipped in styrofoam boxes packed with dry ice (about −60°C effective temperature). Such containers retain the frozen cultures for 72–96 hours, depending on ambient conditions. Upon arrival the user immediately transfers the dry ice refrigerated cultures to a mechanical deep freeze at −40°C or below or to LN_2 vapor storage.

E. Building Facility

1. Design Concepts

A detailed discussion on the construction and design of the basic building facility for starter culture production is beyond the scope of this chapter. There are some important design concepts, however, that should be discussed here. Figure 2 illustrates a circular flow pattern for the production of frozen concentrated starter cultures. Raw materials enter the sequence at pie section 1 (media reconstitution) and process flow proceeds counterclockwise ending up at pie section 8 (cryogenic storage of frozen cultures). Warehouse facilities for the storage of raw materials are not shown. At the central core is the CIP system and the pipe chase for all major utilities. Although this circular design has some very desirable attributes, in actuality modifications are required to facilitate construction and enhance floor space economy.

Sanitary construction typical of modern dairy plants is an essential element here. In addition, various design sophistications are required, about ten of which encompass the most critical design aspects. For purposes of brevity these ten are listed below without attempts to cite and contrast all of the various alternatives.

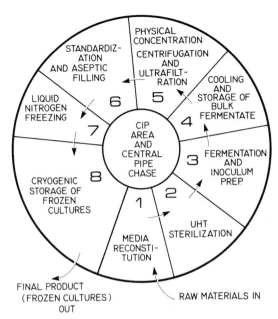

FIGURE 2. Circular design concept for the layout of processing equipment for the production of frozen lactic starter culture concentrates.

a. Air Handling System.   The air supply to process rooms (Fig. 2, pie sections 2 through 6) should be HEPA* filtered at 0.35 μm. Both Flanders and Cambridge manufacture such filters. Laminar flow HEPA filtered air is desirable in the aseptic filling areas and over fermentors (Austin, 1970). In these specific areas a Class 100 environment should be achieved; by definition this means that there are to be fewer than 100 particles of less than 0.5 μm in length per cubic foot of air (Anonymous, 1973). Filtered air is not required in the media reconstitution room since raw ingredients routinely contaminate this area; personnel working in this room should be prohibited from routinely entering process areas 2 through 6. Air lock vestibules are recommended where aseptic sections (Fig. 2, pie sections 2 through 6) access into warehouses and other nonfiltered areas.

b. Floors in Process Areas.   The floors in the process areas (all pie sections, Fig. 2) should be constructed of quarry tile of at least 0.75-inch thickness. An antiskid grit finish is desirable and flush epoxy grout and epoxy bedding is recommended. A waterproof membrane between the

* HEPA = high efficiency particulate air filters.

structural slab and the fill is a prerequisite. A minimum 2% pitch on the finished floor is usually required to direct water to floor drains.

c. Waterproof Insulation.   In general, everything should be waterproof in the process areas. Black iron, copper, and galvanized utility piping should be insulated and then hermetically sealed in polyvinyl chloride (PVC) casing. Where such piping is not insulated it can be painted with epoxy paint (except for steam pipes). Electrical conduit can be ordered encased in PVC as can all junction boxes, couplings, etc.

d. Walls in Process Areas.   Although ceramic tile is very acceptable, cement block sealed and painted with two coats of epoxy paint is completely impervious to water and is also very acceptable. Where the quarry tile floor meets the wall, a seal can be made with "10-year" silicone caulk such as that manufactured by Dow-Corning or General Electric. Bracket plates, electrical junction boxes, etc., should also be silicone caulked to wall surfaces.

e. Ceilings in Process Areas.   Conventional acoustical ceiling tile, if used, should be treated with a nonleaching bacteriostatic/fungistatic compound, such as Steridex manufactured by Pentagon Plastics of Springfield, Virginia.

f. Process Water.   In localities where it is not desirable to use the municipal water supply directly in processing, the best compromise treatment procedure is reverse osmosis (RO). Deionized water is expensive and very difficult to maintain free of bacterial contamination. Reverse osmosis water should be collected in a large reservoir (generally a reinforced fiberglass tank is used). In our operation we continuously circulate the collected RO water through ultraviolet sterilizing equipment. Milwaukee water has a hardness of 8 grains and commercial RO systems, such as those from Continental Water, Inc., are capable of reducing this to less than 1 grain in a single pass. Such water is very acceptable for culture media preparation (Porubcan and Sellars, 1977).

g. Process Cooling System.   All process cooling should be performed with refrigerated (1°C) water wherever possible. A central ice bank reservoir maintained by tandem screw compressors is generally used. Since the jackets of the various process vessels have multipurpose functions where steam, hot water, municipal cold water, and refrigerated water all share common channels at times, it is clear that glycol-type coolants would not be compatible.

h. Process Steam. Pure culinary steam generated by a "drinking water boiler" is most desirable, particularly if steam is to be directly injected into product, sanitary piping, or even cleaning solutions. If this type of boiler is not practical, then shell-and-tube-type heat exchangers and other indirect methods of heating can be used. Steam containing various boiler treatment compounds should not be allowed to chemically contaminate product contact surfaces.

i. Compressed Air. Instrument-grade compressed air should be used throughout the plant. At Chr. Hansen's Laboratories we define such air as having a dew point maximum of 35°F (1°–2°C) with absolute filtration of all particles larger than 0.5 μm.

j. Liquid Nitrogen (LN$_2$). A central liquid nitrogen dispensing system should be laid out such that distances from the source are absolutely minimized. Vacuum-insulated stainless steel transfer lines are best. If gaseous nitrogen is required for process applications, it can be obtained from the central LN$_2$ system. Both Minnesota Valley Engineering (MVE) Inc., Minneapolis, Minnesota and Linde Division of Union Carbide, Inc. can provide total cryogenic system engineering.

III. SPECIFIC STARTER CULTURES

A. Lactic Streptococci

1. Commercial Applications

Buttermilk, sour cream, hard and soft cheeses, cottage cheese, and cultured butter are now routinely produced using commercially available frozen concentrates of lactic streptococci. Available as either "bulk sets" or "direct sets," these concentrates are used to prepare intermediate bulk starter or the final product, respectively. In the former case the culture units range in size from 70 to 130 ml and contain from 5×10^9 to 50×10^9 CFU/ml. In the latter case the cultures are packaged in 130 to 500-ml units and contain from 50×10^9 to 400×10^9 CFU/ml. More specifically, a typical DVS (direct-vat-set; such as, Hansen's DVS #950 or DVS #960) culture for Cheddar cheese will contain 360 ml of concentrate containing 200×10^9 CFU/ml of S. cremoris (two to eight different strains) and will inoculate up to 5000 lb (2273 kg) of cheese milk. This represents an addition of 14.4×10^{13} CFU to a 10,000-lb vat which is approximately equivalent to using 1% of bulk starter that has a viable plate count of 4.0×10^9 CFU/ml.

Preparation of bulk starter from "bulk-set" frozen concentrates is rather straightforward. Reconstituted skim milk or phosphate-buffered starter medium is batch heat treated at 85°–95°C (alternatively, 230°F for 12–16 seconds can be used) for 40–60 minutes and subsequently cooled to 21°–28°C (for mesophilic starters) and inoculated with a defrosted culture. Typically, 70 ml of "bulk set" culture (such as Hansen's Redi-Set #253 or #70) will inoculate up to 300 gal. Incubation then proceeds at 21°–28°C for 12–18 hours or until the starter has developed sufficiently, i.e., when the pH reaches 4.90–5.00 for maximum activity. At this time the starter is cooled to 5°–10°C and held cold until used; it is not a good idea to hold "ripe" starter for more than 24 hours.

Richardson et al. (1977), Chen and Richardson (1977), and Ausavanodon et al. (1977) have proposed an improved method for preparing bulk starter. The procedure calls upon the culture user to prepare bulk starter utilizing a whey-based bacteriophage inhibitory medium and pH control. Fresh Cheddar cheese whey is first fortified with 1.0% monoammonium phosphate, 1.9% disodium phosphate, and 0.32% yeast extract and is then heated at 89°–95°C for 45 minutes. Inoculation follows with a suitable commercial lactic *Streptococcus* culture and incubation is carried out at 21°–27°C with pH control via anhydrous ammonia injection at pH 6.0 ± 0.1 pH unit. After about 16 hours of pH-controlled growth, a significant decrease in acid production is observed (by a decrease in neutralizer consumption) and the starter is essentially ready to use. Advantages claimed for this type of starter preparation are increased activity at reduced costs.

There are many disadvantages inherent to bulk starter systems in general, including the improved version just described, that are not encountered when liquid nitrogen frozen direct-vat-set (DVS) concentrates are used. First, frozen direct-vat-set cultures eliminate any need to cultivate starters in the cheese factory. By doing so there is a real savings in energy, labor, and raw material costs to the culture user. Second, frozen direct-vat-set cultures allow the cheese maker to rotate or use different starters, for purposes of bacteriophage control, on a vat-to-vat basis; or different cheese varieties can be made on a vat-to-vat basis. Third, DVS concentrates can be prepared with specific control over strain balance. For example, the culture manufacturer can grow select single strains of lactic streptococci in separate fermentors and combine them after subsequent concentration (Farr, 1969). Alternatively, the culture manufacturer can inoculate his large-scale fermentors with discreet proportions of compatible strains which would "meet each other" for only

10–12 generations before they must meet again in the cheese vat. Such procedures require the technological coordination and control of a sophisticated bacteriological laboratory. Fourth, DVS concentrates can be pretested for activity (ability to produce lactic acid), strain balance, bacteriophage, purity, and starter bacteria concentration. This is not at all possible with bulk starter systems. Pretesting helps to ensure positive daily performance and uniformity.

In addition to these four inherent advantages of DVS over bulk starter systems there are other potential advantages presently under investigation. For example, improved cheese yields in Cheddar cheese manufacturing and improved cheese quality have been reported (Rasmussen, 1977). More data is required in these areas, however, in order to ascertain statistical significance.

2. Strain Selection

One of the key criteria for selecting single strains of either S. cremoris or S. lactis destined for direct-vat-set use is "industrial centrifugability." It is important that whey-based cultures of lactic streptococci separate efficiently on industrial auto-desludging separators. Data from our laboratory indicate that "centrifugability" is largely an intrinsic characteristic of the particular strain, although growth media composition and various chemical treatments can influence this parameter.

A second important selection factor is activity or the ability to produce lactic acid. A potentially fast strain is one that will reduce litmus milk with acid coagulation within 16 hours at 21°C at a 1% inoculation or less (Babel, 1976).

Finally, a third selection factor is resistance to bacteriophage. Here, it now seems important to find strains that are not "readily lysogenic" or that do not spontaneously produce temperate phage and are induced to do so only with difficulty, if at all (Lawrence, 1977). Such a strain is not necessarily phage resistant but stands a better chance of being compatible with other lactic streptococci. Potentially good sources for lactic streptococci include commercial starter cultures and natural Cheddar cheese (Babel, 1976).

The above selection criteria do not necessarily hold true for "bulk set" culture strains. Some suppliers have allowed "natural selection" to govern the absolute composition of these starters. Here pseudolysogenic* and/or lysogenic strains perpetuate in such a way that variants to parent strains develop spontaneously. Although it is true that such strain com-

---

* Defined as a phage-carrying strain where the bacteria and phage can be separated.

binations are not "scientifically defined," it remains a fact of many years experience in the United States that such cultures have a definite resistance to bacteriophage infection. Since "bulk set" concentration levels can often be achieved by fermentation alone, such as fermentation in skim milk with pH control, "centrifugability" is not necessarily a concern (Jespersen, 1974). In addition, since "bulk set" starters are further propagated (often under less than ideal conditions) in the cheese factory for another 10–14 generations prior to use in the cheese vat, inherent stability to bacteriophage is of primary concern and may supersede the importance of scientifically defining the starter's composition in terms of individual strain percentage.

3. Concentrate Preparation

The preparation of concentrated S. cremoris and S. lactis cultures has been studied by Stadhouders et al. (1969), Peebles et al. (1969), Blaine (1972), Stanley (1977), Cogan et al. (1971), and Efstathiou et al. (1975). A recent general review on cheese starters has been prepared by Lawrence et al. (1976).

There is some disagreement regarding the optimum growth parameters for the lactic streptococci. In part, it stems from whether one is discussing the simultaneous propagation (in one fermentor) of different species and/or different genera (e.g., leuconostocs combined with streptococci) or whether the discussion relates to the propagation of single strains or simple "defined" mixtures of compatible strains (same species). Basically, the following information is valid, in a general sense, for most strains of lactic streptococci when they are grown separately or in simple compatible mixtures: An orthophosphate-buffered, yeast extract-fortified, sweet whey (preferably from Swiss cheese) based medium containing from 6 to 12% total solids by weight is a good general medium. A small amount of added skim milk powder (1–2%) would be of benefit if not for the problems encountered during separation of the cells as discussed previously. Ultrahigh-temperature and batch heat treatment options have already been discussed. Optimum incubation parameters are 25°–30°C for 8–18 hours depending on the strain(s). Generally, it is a good idea to automatically pH control the fermentation; optimum control is achieved between pH 5.40 and pH 6.30, with pH 6.00 most often preferred. Control should be maintained at ±0.1 pH unit. Neutralization is best effected with ammonium hydroxide or anhydrous ammonia, which are preferred over sodium, potassium, and calcium hydroxides (Blaine, 1972). Slow, constant agitation is generally the rule; nitrogen or CO_2, when sparged into the fermentor headspace, do not benefit the fermentation (Gilliland, 1977). At a temperature of 25°C with pH control at pH 6.00 (with ammonium hydroxide) the balance of component strains in a mixed strain culture is not altered appreciably (Efstathiou et al., 1975; Lamprech and Foster, 1963).

Keen (1972) has reported that inhibitory hydrogen peroxide develops during fermentation with some lactic streptococci and that catalase treatment can reverse the inhibition. Yeast extract has a reasonable amount of pseudocatalase-type activity and its presence obviates the need to add catalase per se (Porubcan and Sellars, 1977). Gilliland and Speck (1968) have demonstrated that D-leucine accumulates during fermentation and acts as an autoinhibitor. Data from our laboratory unequivocally point to ammonium lactate accumulation (due to the pH-control process) as the number one factor responsible for the slowing down of the growth rate. Again, the virtuous yeast extract seems to exert a therapeutic effect on the cells in that it somewhat increases resistance to ammonium lactate. This effect often varies considerably from one commercial yeast extract brand to another.

Subsequent to fermentation the cell crop is cooled to between 5° and 15°C and is treated with a defloculant, such as sodium citrate; from 2 to 4% by weight is usually required. The pH is then increased to about 6.6 and the cells are harvested by centrifugation or ultrafiltration as previously discussed. The cell concentrate can be suspended in a skim milk diluent or in a 7.5% lactose solution prior to freezing. The lactose is of particular benefit if the freezing is carried out at −37°C versus −196°C (Gilliland, 1977).

B. Lactobacilli

1. Commercial Applications

The genus Lactobacillus encompasses many species, only a few of which have significant commercial value at this time. In focusing our attention on these few it is appropriate to begin with Lactobacillus bulgaricus. In most commercial applications in the United States this organism is grown in symbiotic association with Streptococcus thermophilus. Concentrated liquid nitrogen frozen cultures containing these two organisms are used in the direct preparation of yogurt and in the preparation of bulk starter for Italian cheese varieties such as mozzarella, parmesan, provolone and romano. Direct-vat-set concentrates for these Italian cheeses are presently under development at Chr. Hansen's Laboratories.

Bulk starter for Italian cheese is prepared directly from frozen concentrates. Typically, a 70-ml unit of starter concentrate (such as Hansen's

Redi-Set R1 or CH-3) will inoculate up to 300 gal of either reconstituted skim milk (10–12% solids by weight) or thermophilic-type bulk starter media. Both types of growth media are batch heat-treated at 85°–95°C for 40–60 minutes and are subsequently cooled to between 37° and 45°C before inoculation and incubation. Incubation time is regulated accordingly; for example, at 37°C usually 12–14 hours are required, whereas at 45°C, 6 hours may be sufficient time to effect proper acid development (pH 4.90 or less). In general, the higher incubation temperatures somewhat favor *L. bulgaricus* over *S. thermophilus*. Since bacteriophages exist for both of these thermoduric lactics (Sozzi *et al.*, 1976) proper precautions with regard to sanitation and culture rotation should be followed just as for the mesophilic lactics.

Yogurt can be prepared directly from the same "bulk set" cultures that are used for Italian cheese bulk starter. A 70-ml unit, once again, is sufficient inoculum for up to 300 gal of milk. Milk for the manufacture of yogurt contains from 1.0 to 3.5% fat. The serum solids content of the milk is increased to 10.5–11.5% by addition of nonfat dry milk or condensed skim milk. The milk is then homogenized and heated to 185°–195°F (85°–90.5°C); holding time is 30–60 minutes. A high degree of heat treatment is essential because it improves the body of the product by limiting whey expulsion (Sellars and Babel, 1970). Following heat treatment, the milk is cooled to 104°–113°F (40°–45°C) and inoculated as stated above. Usually 8 hours are required at 104°F to effect proper acid development and coagulation. Yogurt which has a pH of 4.0 is very acceptable to the consumer. To attain this pH, cooling should be started when the pH is 4.3–4.4 (Sellars and Babel, 1970).

Lactobacillus acidophilus, because of its ability to implant the gastrointestinal tract, has recently become very popular in the form of Sweet Acidophilus. Here, cold, pasteurized low-fat milk is inoculated with a cell concentrate of *L. acidophilus* such that a final viable population of about 3×10^6 CFU/ml results in the milk. No incubation cycle is permitted and the inoculated milk is held refrigerated at 45°F (7°C) or below. The product tastes exactly like low-fat milk. Since incubated milk cultures of *L. acidophilus* have an "off flavor" due to the metabolism of the organism, this Sweet Acidophilus product has met with more general acceptance. Liquid nitrogen frozen concentrates of *L. acidophilus* in 70 to 130-ml units containing from 40×10^9 to 50×10^9 CFU/ml are now commercially available. One 130-ml unit, when added to 500 gal of cold milk, will effect a viable population of about 3×10^6 CFU/ml in the milk.

Various thermoduric lactobacilli, such as *L. acidophilus, L. bulgaricus,* and *L. lactis,* have been fed to livestock for the purpose of controlling disease, such as scours (Metchnikoff, 1907; Sandine *et al.*, 1972; Mitchell and Kenworthy, 1976; Muralidhara *et al.*, 1977). Renewed interest in this old concept has developed due to the general concern over the widespread prophylactic use of broad-spectrum antibiotics in animal feed. In addition, the present availability of commercial frozen concentrates of these lactobacilli has improved the functionality of the entire concept. Spray-dried thermoduric lactobacilli are also being evaluated commercially in this regard (Porubcan *et al.*, 1977). In terms of human health and nutrition there is a growing general consensus that these lactobacilli may perform a beneficial function (Metchnikoff, 1908; Meyers, 1931; Mitsuoka, 1971; Sandine *et al.*, 1972; Speck, 1976).

The use of frozen concentrates of *Lactobacillus plantarum* (such as Hansen's LP-1) in the fermented vegetable industry is gaining in popularity. A process for the pure culture fermentation of cucumbers was patented by Etchells and Bell (1968). This process produces a firm textured, fermented cucumber, whereas previous state-of-the art processing often resulted in bloated and/or softened cucumbers. Extraneous fermentations involving nonlactic acid-producing bacteria and/or yeasts and molds would often cause these aforementioned defects. The Etchells and Bell process involves several steps; (1) heat blanching of the cucumbers to destroy surface asporogenous microorganisms and inactivate any deleterious heat-labile enzyme systems that may be present; (2) packing of the cucumbers aseptically into sanitized containers and covering with a preacidified (pH 4.6–4.2) brine that gives an equilibrated brine strength (% NaCl by wt.) of about 1.6–10%; (3) incubation of the containers with a suitable pure culture, such as *L. plantarum* or *Pediococcus cerevisiae,* at a favorable temperature until the fermentation has progressed to completion. If 1 billion viable cells are added per quart, about 72 hours at 90°F will be required.

For practical reasons this process is more or less limited to small containers such as No. 10 tin cans and quart or gallon glass jars. Etchells (1973) reported an adapted process suitable for the controlled fermentation of cucumbers and other vegetables brined in bulk. Here chemical (chlorine) sanitation of the washed green cucumbers replaced the heat shocking step; suitable buffering (to ensure complete fermentation of all sugar) and nitrogen purging (to reduce the CO_2 content in the brine) steps were added. When the 25° salometer (6.6% NaCl) brining procedure is followed, the fermentation reaches completion in 7–12 days when brine temperatures are held between 78° and 85°F (26–29°C). A single 130-ml liquid nitrogen frozen concentrated culture unit (Hansen's LP-1) of *L. plantarum* contains over 6.5×10^{12} viable cells. When added to 2000 gal of brined cucumbers it will provide 3–5 billion viable organisms per gallon, a very adequate inoculum level.

2. Concentrate Preparation

Published articles on the preparation of concentrates of lactobacilli suitable for commercial use are not as numerous as for the lactic streptococci. It has been our experience that pH-controlled milk cultures of thermoduric lactobacilli such as *L. bulgaricus, L. lactis, L. acidophilus,* and *L. helveticus* are difficult to concentrate effectively by industrial centrifugation even when copious amounts (2–6% by wt.) of sodium citrate and/or other casein deflocculating reagents are added after growth. Clarified sweet whey fortified with yeast extract is a satisfactory medium for some of these lactobacilli but not for others. Centrifugal separation of the denatured whey proteins along with the cells is usually a severe problem if unclarified whey is used, and, once again, treatment with sodium citrate does not usually remedy the problem (Porubcan and Sellars, 1977).

Various broth-type growth formulas are commercially available that can, with some modification, be rather effective. BBL's (Baltimore Biological Laboratory Division, Bioquest, Inc., Baltimore, Maryland) LBS broth and MRS broth are good examples of these. Etchells and Costilow (1968) have patented an improved version of the LBS broth formula originally developed by Rogosa (1951). Their improvement involves an upward adjustment of the final pH of the medium. *Lactobacillus plantarum* grows very well in LBS broth modified as such. In spite of the strong acetate buffer capacity of LBS it may be desirable to control the pH during growth for some strains of *L. plantarum. Lactobacillus plantarum* strain 442 (Fleming, N.C. State), for example, does very well at 30°C in LBS broth pH controlled at 5.60 with 50% NaOH. A total incubation time of 18 hours produces a viable plate count of between 50 and 100×10^8 cells per ml in the fermentor. Subsequent concentration by either centrifugation or ultrafiltration produces a very acceptable concentrate (Porubcan and Sellars, 1977).

Lactobacillus acidophilus (ATCC 4356) grows well in commercial MRS broth at 37°C. The acetate level in LBS broth is too high for prolific *L. acidophilus* growth. Control at pH 6.0 (±0.1 pH unit) with either ammonium hydroxide or sodium hydroxide improves the cell yield. On an industrial scale it is necessary to adapt the MRS formulation by substituting industrial peptones and yeast extracts for the laboratory types given. For example, Humko Sheffield's Pepticase (tryptic digest of casein) at approximately one-fourth the cost per pound is a very acceptable substitute for BBL's Trypticase.

Some thermoduric lactobacilli are particularly sensitive to changes in the oxidation–reduction potential (ORP) of the media. This effect may or may not be related to dissolved oxygen or fermentor headspace gas composition. An anaerobic headspace, such as a N_2 purged headspace, will stimulate certain strains of *L. bulgaricus* (Porubcan and Sellars, 1977).

C. Pediococci

1. Commercial Applications

The use of *Pediococcus cerevisiae* as a starter culture for fermented sausage was reviewed by Niven *et al.* (1955). Their process was patented (Niven *et al.*, 1959).

Previous state-of-the-art processing relied on certain microorganisms, usually indigenous to the sausage ingredients, to produce satisfactory levels of lactic acid during fermentation. Because of the unpredictability of this "natural process," spoilage often resulted.

Frozen concentrates of *P. cerevisiae,* (such as Hansen's PC-1 and PC-2) in quantities of 70–130ml provide sufficient inoculum for up to 300 lb of the meat–sodium chloride–nitrite–seasonings–dextrose mixture, also known as the emulsion. Viable plate counts on either BBL's APT or BBL's LBS agar run as high as 100×10^9 CFU/ml in these commercial concentrates.

A general processing schedule for all sausage products cannot be given; however, the starter culture technique reduces traditional production times by about 80% for thuringer, cervelat, or summer sausage (Everson *et al.*, 1970). At controlled smokehouse temperatures of 100°–110°F, freshly stuffed casings of inoculated sausage emulsion will usually reach pH 4.80 in 24 hours or less depending upon the kind of sausage and the activity of the particular starter strain. Nitrite levels should not exceed ⅛ oz/cwt of emulsion; and dextrose additions should be at least 0.75 lb/cwt (Niven *et al.*, 1958).

2. Concentrate Preparation

Pediococcus cerevisiae is a particularly desirable microorganism for sausage fermentation because of its salt tolerance and ability to autoinactivate around pH 4.80, a very desirable end-point pH for many types of fermented sausage. The commercial production of this organism can be effected in fermentors charged with industrially modified versions of APT, MRS, or LBS broth. pH control with either concentrated ammonium hydroxide or 50% sodium hydroxide solution (between pH 5.40 and 6.00) significantly benefits biomass yield in the fermentor. Incremental additions of fermentable carbohydrate(s) may also encourage higher biomass yields. Optimum incubation conditions are 16–18 hours at 25°–30°C, depending on the particular strain of *P. cerevisiae* being processed. Bacterial concentrations in the fermentor should average between 5×10^9 and 15×10^9 CFU/ml after about 16–18 hours of incubation (Porubcan and Sellars, 1977).

The appropriate method of physical concentration subsequent to fermentation is ultrafiltration. With a fermentate concentration of 10×10^9 CFU/ml it is only necessary to reduce the fermentate volume by four-fifths to achieve the desired level of 50×10^9 CFU/ml in the commercial concentrate. The "broth-based" fermentation substrate facilitates the ultrafiltration process. Culture temperatures should be held between 10° and 15°C during concentration. Once concentrated, the culture can be dispensed into 70- or 130-ml aluminum cans and subsequently flash frozen in liquid nitrogen at $-196°C$. Essentially 100% of the cells survive this method of preservation and, when cryogenic freezing as such is employed, no glycerol or other cryoprotectants need be added (Porubcan and Sellars, 1977).

REFERENCES

Anonymous (1973). *Dairy & Ice Cream Field,* July.

Ausavanodom, N., White, R. S., Young, G., and Richardson, G. H. (1977). *J. Dairy Sci.* **60,** 1245–1251.

Austin, P. R. (1970). "Design and Operation of Clean Rooms." Business News Publ. Co., Detroit, Michigan.

Babel, F. J. (1976). Purdue University (personal communication).

Blaine, J. W. (1972). Ph.D. Thesis, Oregon State University, Corvallis.

Buchanan, R. E., and Gibbons, N. W., eds. (1974). "Bergey's Manual of Determinative Bacteriology," 8th ed. Williams & Wilkins, Baltimore, Maryland.

Carlsen, V. R. (1977). Cherry-Burrell, Inc. (personal communication).

Chen, Y. L., and Richardson, G. H. (1977). *J. Dairy Sci.* **60,** 1252–1255.

Christensen, V. W. (1967). U.S. Patent 3,354,049.

Cogan, T. M., Buckley, D. J., and Condon, S. (1971). *J. Appl. Bacteriol.* **34,** 403.

Efstathiou, J. D., McKay, L. L., Morris, H. A., and Zottola, E. A. (1975). *J. Milk Food Technol.* **38**(8), 444–448.

Elsworth, R. (1962). "Phase Separation of Bacterial Suspensions by Centrifuge." New Brunswick Sci. Co., New Jersey.

Erzinkyan, L. A., Pakhlevanyan, M. Sh., and Akopyan, L. G. (1966). *Dairy Sci. Abstr.* **21,** 386.

Etchells, J. L. (1973). *Pickle Pak. Sci.* **3,** 4–14.

Etchells, J. L., and Bell, T. A. (1968). U.S. Patent 3,403,032.

Etchells, J. L., and Costilow, R. N. (1968). U.S. Patent 3,410,755.

Everson, C. W., Danner, W. E., and Hammes, P. A. (1970). *J. Agric. Food. Chem.* **18,** (4), 570.
Farr, S. M. (1969). U.S. Patent 3,420,742.
Gilliland, S. E. (1977). *J. Dairy Sci.* **60** (5), 805–809.
Gilliland, S. E., and Speck, M. L. (1968). *J. Dairy Sci.* **51,** 1573.
Hansen, A. P., and Melo, T. S. (1977). *J. Dairy Sci.* **60,** 1368–1373.
Hansen, R. (1969). *Nord. Mejeri-tidsskr.* **9,** 192–194.
Jespersen, N. J. T. (1974). Chr. Hansen's Laboratory, Inc. (personal communication).
Jespersen, N. J. T. (1974). *S. Afr. J. Dairy Technol.* **6,** 63.
Jespersen, N. J. T. (1977). Chr. Hansen's Laboratory, Inc. (personal communication).
Jost, R., and Monti, J. C. (1977). *J. Dairy Sci.* **60,** 1387–1393.
Keen, A. R. (1972). *J. Dairy Res.* **39,** 151–159.
Konak, A. R. (1974). *Krist. Tech.* **3,** 243.
Konak, A. R. (1975). *Biotechnol. Bioeng.* **17,** 271–272.
Lamprech, E. D., and Foster, E. M. (1963). *J. Appl. Bacteriol.* **26,** 359.
Lawrence, C. A., and Block, S. S. (1968). "Disinfection, Sterilization, and Preservation," Lea & Febiger, Philadelphia, Pennsylvania.
Lawrence, R. C. (1977). "Cheddar Cheese Starters." A paper presented at the annual meeting of the American Dairy Science Association.
Lawrence, R. C., Thomas, T. D., and Terraghi, B. E. (1976). *J. Dairy Res.* **43,** 141–193.
Leach, R. D., and Sandine, W. E. (1976). *J. Dairy Sci.* **59** (8), 1293–1397.
Lloyd, G. T. (1971). *Dairy Sci. Abstr.* **33,** 411.
Lyster, R. L. (1971). *J. Dairy Res.* **38,** 403.
Metchnikoff, E. (1907). "Quelques remarques sur le lait aigni." Putnam, New York.
Metchnikoff, E. (1908). "Prolongation of Life." Putnam, New York.
Meyers, R. P. (1931). *Am. J. Public Health,* **21,** 867–872.
Mitchell, I., and Kenworthy, R. (1976). *J. Appl. Bacteriol.* **41,** 163–174.
Mitsuoka, T. (1971). *Proc. Int. Symp. Conversion & Manuf. Food Stuffs Microorg., Kyoto, Japan.*
Muralidhara, K. S., Sheggeby, G. G., Elliker, P. R., England, D. C., and Sandine, W. E. (1977). *J. Food Prot.* **40** (5), 288–295.
Nielsen, A. E. (1964). "Kinetics of Precipitation." Pergamon, Oxford.
Niven, C. F., Jr., Deibel, R. H., and Wilson, G. D. (1955). "The Use of Pure Culture Starters in the Manufacture of Summer Sausage," Annu. Meet. Am. Meat Inst., Chicago, Illinois.
Niven, C. F., Jr., Deibel, R. H., and Wilson, G. D. (1958). *Am. Meat Inst. Found. Cir.* **41.**
Niven, C. F., Jr., Deibel, R. H., and Wilson, G. D. (1959). U.S. Patent 2,907,661.
Osborne, R. J. W. (1977). *J. Soc. Dairy Technol.* **30** (1), 40–44.
Payne, R. E., Hill, C. G., Jr., and Amundson, C. H. (1973). *J. Milk Food Technol.* **36** (7), 359–363.
Peebles, M. M., Gilliland, S. E., and Speck, M. L. (1969). *Appl. Microbiol.* **17,** 805.
Porubcan, R. S. (1975). M.S. Thesis, Dept. of Chemistry, University of Wisconsin, Milwaukee.
Porubcan, R. S., and Sellars, R. L. (1973). *J. Dairy Sci.* **56,** 634.
Porubcan, R. S., and Sellars, R. L. (1975a). U.S. Patent 3,897,307.
Porubcan, R. S., and Sellars, R. L. (1975b). *J. Dairy Sci.* **58,** 787.
Porubcan, R. S., and Sellars, R. L. (1976). *Pap. ADSA Meet., 1976.*
Porubcan, R. S., and Sellars, R. L. (1977). Chr. Hansen's Laboratory, Inc. (unpublished research).
Porubcan, R. S., Sellars, R. L., Daines, T., and Rovics, J. (1977). Unpublished studies on the feeding of *L. acidophilus* to calves. (Data collected at FAR, Inc., Juneau, Wisconsin, supported by Chr. Hansen's Laboratory, Inc.)

Rasmussen, H. (1977). *Dairy Field* Sept., p. 70H.

Reddy, M. S., Vedamuthu, E. R., Washam, C. J., and Reinbold, G. W. (1974). *J. Dairy Sci.* **57** (1), 124–127.

Richardson, G. H., Cheng, C. T., and Young, R. (1977). *J. Dairy Sci.* **60** (3), 378–386.

Rinfret, A. P., and LaSalle, B. (1975). "Round Table Conference on the Cryogenic Preservation of Cell Cultures." Natl. Acad. Sci., Washington, D.C.

Rogosa (1951). *J. Bacteriol.* **62,** 132–133.

Sandine, W. E., Muralidhara, K. S., Elliker, P. R., and England, D. C. (1972). *J. Milk Food Technol.* **35,** 691–702.

Schantz, R. N. (1976). Aqua Chem, Milwaukee, Wisconsin (personal communication).

Sellars, R. L. (1967). *In* "Microbial Technology" (H. J. Peppler, ed.), pp. 34–75. Van Nostrand-Reinhold, Princeton, New Jersey.

Sellars, R. L., and Babel, F. J. (1970). "Cultures for the Manufacture of Dairy Products." Chr. Hansen's Laboratory, Inc., Milwaukee, Wisconsin.

Shuster, J. M. (1976). Minnesota Valley Engineering, Inc., Minneapolis (personal communication).

Sozzi, T., Maret, R., and Poulin, J. M. (1976). *Appl. Environ. Microbiol.* **32,** (1), 131–137.

Speck, M. L. (1976). *J. Dairy Sci.* **59,** 338–343.

Stadhouders, J., Jansen, L. A., and Hup, G. (1969). *Neth. Milk Dairy J.* **23,** 182–199.

Stanley, G. (1977). *J. Soc. Dairy Technol.* **30** (1), 36–39.

Thomas, T. D., and Turner, K. W. (1977). *N. Z. J. Dairy Sci. Technol.* **12,** 15–21.

Thompson, J. (1976). *J. Bacteriol.* **127** (2), 719–730.

Webb, B. H. (1965). "Fundamentals of Dairy Chemistry," Avi Publ. Co., Westport, Connecticut.

Webb, F. C. (1964). "Biochemical Engineering," Van Nostrand-Reinhold, Princeton, New Jersey.

Woods, W. C. (1975). "Trends in High Temperature Sterilization of Dairy Products—A Symposium," Annu. Conf. Am. Cultured Dairy Products Inst.

Chapter 4

Production of Single-Cell Protein for Use in Food or Feed

JOHN H. LITCHFIELD

I. INTRODUCTION

What are single-cell proteins and what are their present and potential uses for human food or animal feed? In this chapter, we discuss the various groups of microorganisms that have been considered for food or feed use, including algae, bacteria, yeasts, molds, and higher fungi. The dried cells of these organisms are collectively referred to as "single-cell proteins" (SCP). This term was coined by C. L. Wilson at the Massachu-

MICROBIAL TECHNOLOGY, 2nd ed., Vol. I
Copyright © 1979 by Academic Press, Inc.
All rights of reproduction in any form reserved ISBN 0–12–551501–4

setts Institute of Technology in 1966 and has been used widely since that time in referring to microbial cells grown for food or feed applications (Scrimshaw, 1968).

People have eaten certain microorganisms as a portion of their diet since ancient times. Top-fermenting yeast (*Saccharomyces* species) was recovered as a leavening agent for bread as early as 2500 B.C. (Frey, 1930). Bread and yeast-leavened baked products contain 1–4% of bakers' yeast based on the weight of flour used (Reed and Peppler, 1973).

Fermented milks and cheeses produced by lactic acid bacteria of the genera *Streptococcus* and *Lactobacillus* were consumed by the early Egyptians and Greeks and reached a higher state of development during the Roman era (100–50 B.C.) (Kosikowski, 1977). These cultured dairy products contain from 10^7 to 10^{10} lactic acid bacteria per gram (Lampert, 1965; Pederson, 1971).

Wild mushrooms were prized as a delicacy by the pharaohs of Egypt. During the First Century B.C., the epicures of the Roman Empire held wild mushrooms in high esteem, as indicated in the verse of the poet Horace. The Romans also regulated the grading and selling of mushrooms. In addition, they recognized the danger resulting from consumption of poisonous wild mushrooms (Seelig, 1959). Mushrooms were also eaten in ancient times in the Orient. The current practices utilized in cultivating the various species used in China, Japan, Malaya, and the Philippines undoubtedly originated from ancient lore.

Bluegreen algae of the genus *Spirulina* grow in alkaline lakes in certain regions of the world. People in the Lake Chad region of Africa have eaten *Spirulina* species for many generations. Also the Aztecs in Mexico were consuming *Spirulina* species as a major source of protein at the time of arrival of the Spanish explorers in the sixteenth century (Clement, 1968; Ad Hoc Panel, 1975).

The purposeful cultivation of microorganisms for direct use in human food or animal feed is a fairly recent development. Bakers' yeast, (*Saccharomyces cerevisiae*) was grown in an aerated molasses–ammonium salts medium for food use in Germany during World War I (1914–1918). Aerobic yeasts, particularly, *Candida utilis* (torula yeast) were produced in Germany for food and feed use during World War II (1939–1945) (Prescott and Dunn, 1959; White, 1954).

Considerable effort has been expended since World War II to develop processes for mass cultivation of microbial cells (SCP). Numerous reviews and symposia describe the development of processes for utilizing raw materials, including simple sugars, starches, cellulose, agricultural and food-processing wastes, and hydrocarbons by bacteria, yeasts, molds, and higher fungi, and photosynthetic conversions of wastes by algae (Davis, 1974; Gutcho, 1973; Humphrey, 1974; Kihlberg, 1972; Lipinsky and Litchfield, 1970, 1974; Litchfield, 1974, 1977a,b,c, 1978; Mateles and Tannenbaum, 1968; Rockwell, 1976; Tannenbaum and Wang, 1975).

II. PRODUCTION

A. Algae

Algae can be grown either photosynthetically and autotrophically using either artificial lighting or sunlight, or heterotrophically in the dark with organic carbon and energy sources. Most of the research on mass cultivation of algae has been conducted under photosynthetic conditions particularly for waste treatment in sewage oxidation ponds using sunlight (Oswald, 1969; Oswald and Golueke, 1968) or using artificial illumination in conjunction with the development of bioregenerative systems for use in life-support systems for extended space exploration (Casey et al., 1963; Lachance, 1968; Litchfield, 1967d).

On the other hand, Japanese investigators have conducted research on growth of algae under heterotrophic conditions in the dark using organic carbon and energy sources. The goal of this research has been to identify algal strains that have higher growth rates under heterotrophic conditions than under photosynthetic–autotrophic conditions (Endo and Shirota, 1972).

1. Growth Conditions

Table I summarizes the results of studies on the growth of selected species of algae of interest in SCP production. More important strains include *Chlorella sorokiniana, Scenedesmus acutus, Scenedesmus quadricauda, Scenedesmus obliquus, Spirulina maxima,* and *Spirulina platensis.* Limited studies have been conducted with *Oocystis polymorpha. Chlorella regularis* has been grown under both autotrophic and heterotrophic conditions.

In open-circulation systems, particularly sewage oxidation ponds, mixed cultures of algae tend to predominate rather than a single strain. For example, a mixed culture of *Chlorella ellipsoidea* and *S. obliquus* developed in open pond systems in Japan. Mixed cultures of algae dominated by *S. quadricauda* was used in sewage oxidation pond studies at the University of California experiment station at Richmond (Oswald, 1969).

TABLE I. Growth of Selected Algae

Organism	Carbon and energy source	Scale	temperature (°C)	pH	Specific growth (hr⁻¹)	Culture density	Yield (dry wt. basis)	References
Chlorella pyrenoidosa 71105	CO_2(7% in air), urea (0.4 gm/liter) fluorescent lamp, 15,000 lm	600-gal tank	39	4.5–5.8	0.0625	0.28 v/v%	1 g/day	Casey et al. (1963)
C. pyrenoidosa	CO_2(10%), fluorescent lamp, 52,000 lm	2.7 liters	38	6.4–6.5	—	8.98 gm/liter	36.5 gm/day	Matthern and Koch (1964)
Chlorella regularis	(1) Autotrophic: CO_2 (5% in air), fluorescent lamp, 40 klux	1 liter	36	6.5	0.26	6 ml/liter	—	Endo and Shirota (1972)
	(2) Heterotrophic (dark): acetic acid (100 ml/liter)	20-liter jar fermentor (13 liters medium)	36	6.8	0.28	15 ml/liter	0.48 gm/gm acetate utilized	
Chlorella sorokiniana	CO_2 (5% in air), 70 ml/min, 20mmol M KNO_3/liter, fluorescent lamps	405-ml annular chemostat	39	—	—	—	1.59 gm/day	Richardson, et al. (1969)
Oocystis polymorpha	CO_2 (5% in air), 70 ml/min, fluorescent lamps	405-ml annular chemostat	39	—	—	—	1.25 gm/day	Richardson, et al. (1969)
Scenedesmus acutus 276-3A	CO_2, sunlight,	shallow tanks 55m², 30cm deep	Ambient (30)	7–8	—	—	20gm/m²/day	Anonymous (1970); Becker and Venkataraman (1978)
Scenedesmus quadricauda	SO_2, sunlight	10⁶ liter pond	25–35	8.5–9.5	—	1 gm/liter	20 tons/acre/year (12.6 gm/m²/day)	Oswald and Golueke (1968)
S. quadricauda and Scenedesmus obliquus	CO_2 (0.5%), sunlight	54,000 liters (900 m²)	34	6.5–7.0	—	1.5–2.0 gm/liter	10 gm/m²/day	Enebo (1969)
Spirulina maxima	CO_2 (0.5%) combustion gases, bicarbonate, sunlight	700-m² pond	Ambient	—	—	—	15 gm/m²/day	Durand-Chastel and Clement (1972); Clement (1975)

Organism	Conditions	Dimensions	Temperature (°C)	pH		Concentration	Productivity	Reference
S. maxima	CO$_2$ (combustion gas), sunlight	20 × 5 × 0.1 m basin	Ambient	—	—	—	12 gm/m^2/day	Clement (1968); Clement *et al.* (1967)
S. maxima	CO$_2$-enriched air, 250-ml Erlenmeyer flasks 100 ml medium) 4000 lux, 2 v/v/min; (a) synthetic medium,		30	9.5		2 gm/liter (12 days)	—	Kosaric *et al.* (1974)
	(b) sewage effluent		30	9.5		0.77 gm/liter (9 days)		
Spirulina platensis	CO$_2$, 10-liter jar fermentor (20 cm diameter), 10 klux		35–37	8–10.5	0.0189	4.2 gm/liter		Ogawa *et al* (1972)

97

The limiting factor in the growth of algae on a large scale is illumination. In outdoor pond cultivation length of day and night, cloud cover, and the fixed nature of the spectrum and intensity of sunlight limit algal cultivation to latitudes between 35° north and south.

Estimated energy requirements for algal growth at a 20% efficiency indicate a requirement of 35kWh for producing 1 kg of algae (Enebo, 1969). In studies with C. sorokiniana (pyrenoidosa 71105), Shuler and Affens (1970) obtained maximum efficiency of conversion of light energy to chemical energy of 12% in an annular cylindrical culture unit illuminated with artificial light at 15,700 lm (36,000 fc). In other studies with the same organism, Matthern et al. (1969) obtained cultured densities in the range of 25.5 gm/liter (dry weight basis) at an illumination of 300,000 lm which is 30 times the intensity of sunlight as measured at the Earth's surface. Taking into account projected future costs of electricity, it is clear that artificial illumination for the production of algae as a source of SCP is not feasible and the outdoor culture in open ponds is the only possible alternative.

Carbon dioxide is the source of carbon for photosynthetic–autotrophic growth of algae. Air contains only 0.03% CO_2. Accordingly, additional CO_2 must be supplied to the culture. Combustion gas was used as a source of CO_2 for growing Spirulina maxima by the Institute Français du Petrole. In regions where naturally alkaline lakes are present, such as Lake Texcoco in Mexico, S. maxima can be grown without enrichment with CO_2 because of the naturally high concentrations of sodium carbonate present in these waters (Clement, 1975).

In sewage ponds, algal growth is limited by the extent of liberation of CO_2 and ammonia by bacterial action. Slow and uniform liberation of CO_2 is necessary to provide a uniform supply of the carbon source to the growing culture (Oswald and Golueke, 1968).

A heterotrophic culture of algae has been investigated in Japan to determine means of producing algae as a source of SCP from organic carbon sources without light (Endo and Shirota, 1972; Nakayama et al., 1974). Specific growth rates in the range of 0.3 per hr are obtained with Chlorella regularis strain S-50 using glucose as a substrate in the dark (Endo and Shirota, 1972).

Nitrogen sources suitable for algal growth include ammonium salts or nitrates. The nitrogen sources together with phosphorus and mineral nutrients may be readily available in sewage ponds. However, in other water supplies where synthetic media are used in culturing algae these nutrients may have to be supplied.

A number of pond and tank systems for large-scale algal cultivation have been developed in various countries including Japan, Czechoslovakia, West Germany, India, Algeria, and Mexico (Anonymous, 1970; Becker and Venkataraman, 1978; Clement, 1975; Enebo, 1969). In the United States extensive studies have been conducted on a pilot scale by Oswald and his co-workers at the University of California, Richmond, for a number of years.

Aseptic conditions are not maintained during the large-scale growth of algae in ponds. However, the potential contamination of algal cultures grown in sewage oxidation ponds by enteric pathogenic bacteria and viruses must be given serious consideration (Cooper, 1962). *Chlorella* cultures also excrete water-soluble substances into the growth medium that provide nutrients for bacterial growth (Ward *et al.*, 1965; Vela and Guerra, 1966).

The key factors in large-scale algal cultivation are agitation to maintain cells in suspension and prevent sedimentation and adjusting flow rates so that the detention period in the pond will exceed the generation time of the particular algal culture to allow maximum population development.

2. Yields

Yields of algal cultures given in Table I that were obtained under artificial light conditions are of only academic interest since sunlight is the only economically feasible energy source for algal cultivation. The yields in the range 12–15 gm/m²/day (dry weight basis) obtained with *Spirulina maxima* and *Scenedesmus quadricauda* are typical of those that can be obtained in outdoor pond conditions (Clement, 1968, 1975; Durand-Chastel and Clement, 1972; Oswald and Golueke, 1968).

3. Cell Recovery

A major problem in algal culture is harvesting the culture. Experience with *Chlorella*–Scenesdesmus cultures indicates that large volumes of water must be handled because of the low cell density in the range of 1 to 2 gm (dry weight) per liter obtained with these organisms. Also these cultures settle out slowly. Cells must be recovered by concentration, dewatering, and drying. Inorganic compounds, such as aluminum sulfate and calcium hydroxide, and cationic polymers have been investigated as flocculants for algae (Oswald and Golueke, 1968). However, these flocculants cannot be separated from the harvested cells, which makes this method unsatisfactory for recovering algae for food or feed applications. Algae may autoflocculate in shallow ponds at pH 9.5 or above without flocculants (Oswald and Golueke, 1968).

Ion-exchange resins effect a good separation of algae in the pH range 2.8–3.5 (Golueke and Oswald, 1970). However, the cost of the hydrochloric or sulfuric acid regenerants are sufficiently great that this process would not be economically feasible. The cost of separating and concentrating algae by centrifugation, flocculation, or a combination of these steps appears to be economically prohibitive at the present time.

Spirulina maxima floats on the surface in clumps when maximum growth is attained. It can be harvested by skimming at much lower cost than would be incurred by centrifugation (Durand-Chastel and Clement, 1972).

B. Bacteria and Actinomycetes

Bacteria are of interest for use in SCP production because of their high growth rate (20–30 minutes generation time) as compared to 2–3 hours for yeast and 16 hours or more for algae, molds, and higher fungi. Various species of bacteria utilize a wide range of carbon and energy sources including those from renewable resources, such as carbohydrates (sugars, starch, cellulose) in either pure form or in agricultural and forest product wastes, or those from nonrenewable resources, such as hydrocarbons and petrochemicals. Certain species of actinomycetes have growth rates and substrate utilization patterns similar to those of bacteria and consequently are of interest for use in SCP production.

1. Growth Conditions

A wide range of bacterial species has been considered for SCP production. Important considerations in selecting bacterial cultures suitable for use in SCP production include specific growth rate, yield on a given substrate, pH and temperature tolerance, aeration requirement, genetic stability, and freedom from associated bacteriophages. The strain also should not be pathogenic to plants, animals, or humans either by direct infection or by excretion of toxins. Also certain bacterial members of the Enterobacteriaceae may contain endotoxins which may effect man or animals adversely even after the cells are killed by heat. Another important consideration is that the strain must not have the potential for mating with known pathogenic bacteria to yield potentially pathogenic hybrids. This type of mating can occur quite readily within the Enterobacteriaceae.

Substrate and nitrogen concentrations are important factors affecting growth of bacteria. Substrate concentrations for both cultures are in the 1–5% range but may be considerably lower for continuous cultures. Carbon-to-nitrogen ratios in the growth medium should be maintained in the range of 10 : 1 or less to favor high protein contents in the cells and prevent accumulation of lipids or storage substances, such as poly-β-

hydroxybutyrate. Anhydrous ammonia or ammonium salts are suitable nitrogen sources for SCP production by bacteria and actinomycetes. Phosphorus is required by bacteria and should be supplied as a feed-grade phosphoric acid to avoid contamination of the product with arsenic or fluoride that is found in crude industrial phosphoric acid. Generally water supplies provide adequate quantities of mineral salts for bacterial growth; however, some natural waters may require supplemental minerals, such as iron, magnesium, and manganese, to make up deficiencies. In such cases minerals should be added as sulfates or hydroxides rather than as chlorides to avoid corrosion problems in stainless steel fermentation equipment.

During growth, pH may be controlled in the 5–7 range preferred by most bacteria by the rate of ammonia and phosphoric acid addition to the medium. In continuous processes, instrumentation can be designed to effect feedback control of pH and nitrogen content of the medium.

Temperature tolerance is an important characteristic of bacterial strains for SCP production from hydrocarbons or alcohols. For example, in the production of bacterial SCP from methane, heat production is in the order of 10 kcal/gm of cells at a yield coefficient of 1.0 gm of cells per gram of methane (Hammer et al., 1975). Therefore, cooling cost for bacterial SCP production can be significant unless high-temperature-tolerant strains in the range of 35°–45°C can be used. Holve (1976) and Wang et al. (1976) discuss methods for calculating the heat production associated with bacterial growth on hydrocarbon and carbohydrate substrates based on the elemental composition of the cells.

Oxygen transfer to growing cells is also an important factor in bacterial SCP production under aerobic conditions, especially from hydrocarbons, methanol, or ethanol. Mateles (1971) presented an empirical equation for estimating oxygen requirements for cell production based on the elemental compositions of the carbon source and cells, the molecular weight of the carbon source, and the yield of cells based on the carbon source consumed. Also, cooling requirements can be gauged from the estimated oxygen requirement by use of the factor 0.11 kcal/mmole of oxygen consumed, developed by Cooney et al. (1969).

Maintaining sterility is important in the mass cultivation of bacteria. Undesirable contaminants such as enteric pathogens may grow equally as well as the desired strain since most bacterial SCP processes operate in a pH range of 5–7. Metabolic excretion products, such as amino acids, may provide sufficient nutrients for the growth of contaminants. Both growth media and equipment should be sterilized in bacterial SCP production processes and aseptic conditions must be maintained throughout the growth.

Table II presents growth characteristics of selected bacteria and actinomycetes having potential utility for SCP production. Some of the most important processes will be discussed here.

During the 1960s there was considerable interest in the potential for producing SCP from bacteria using gaseous and liquid hydrocarbons and chemicals derived from them, such as methanol and ethanol. In particular, there was interest in utilizing methane in regions remote from markets where this gas would otherwise be flared at the wells. The option also existed for manufacturing methanol from the methane and shipping the methanol to the site of SCP production (Rosenzweig and Ushio, 1974).

Research on a pilot plant-scale process for producing bacterial SCP from methane using *Methylococcus capsulatus* (Harrison *et al.*, 1972; Hamer *et al.*, 1975), or a mixed culture of *Pseudomonas* sp., *Hyphomicrobium* sp., *Acinetobacter* sp., and *Flavobacterium* sp. (Wilkinson *et al.*, 1974; Harrison, 1976) has been conducted at Shell Research Limited in the United Kingdom. High productivities and yield coefficients can be obtained in these processes. However, the productivity of the processes for continuous production of bacterial SCP for methane is limited by transfer of oxygen and methane from the gas phase to the bacterial cells, occurrence of single or double gaseous substrate limitation, high heat production necessitating cooling, and possible formation of inhibitory products (Hamer *et al.*, 1975; Harwood and Pirt, 1972).

In addition, potential explosive hazards require operation below 12.1% by volume of oxygen (Hamer *et al.*, 1967). Therefore capital costs for production of bacterial SCP from methane may be significantly higher than from other substrates.

Kyowa Hakko Kogyo Company, Limited in Japan has developed several processes for the production of bacterial SCP from gaseous hydrocarbons. Methane, ethane, propane, *n*-butane, isobutane, propylene, butylene, or mixtures of these hydrocarbons were utilized by *Brevibacterium ketoglutamicum* ATCC No. 15, 587 (Tanaka *et al.*, 1972).

Other bacteria that have been investigated for growth on gaseous hydrocarbons substrates include various mixed cultures on methane (Wolnak *et al.*, 1972; Sheehan and Johnson, 1971), *Arthrobacter simplex* on propane and butane (Orgel *et al.*, 1971); and *Corynebacterium hydrocarbonoclastus* on propane (Akiba *et al.*, 1973). The actinomycete *Nocardia paraffinica* utilizes *n*-butane readily (Sugimoto *et al.*, 1972).

Liquid *n*-alkanes are suitable carbon and energy sources for a wide range of bacteria and actinomycetes. Only a few examples are cited here.

Acinetobacter cerificans has been investigated on a small pilot plant-scale for the production of SCP from purified normal alkanes by Exxon

Corporation in a joint venture with Nestle Alimentana, S.A. (Perkins and Furlong, 1967; Guenther and Perkins, 1968). The Chinese Petroleum Corporation, Taiwan has developed a pilot-scale process for producing *Achromobacter delvacvate* from diesel oil (Ko *et al.*, 1964) and *Pseudomonas* No. 5401 from fuel oil (Ko and Yu, 1968). However, none of these processes has been practiced on a full commercial scale.

Actinomycetes, including *Mycobacterium phlei* and *Nocardia* sp., have been grown on liquid C_{10}–C_{20} hydrocarbons on a laboratory scale (Wagner *et al.*, 1969). However, no further development of these processes has occurred.

Bacterial SCP production from methanol is an interesting recent development. Methanol has the advantages of high solubility in water, lack of explosive hazards, freedom from traces of hydrocarbons, and ease of removal from the cell product. Pathways of C_1 compound metabolism, microorganisms, and the utilization of methanol by bacteria are discussed in recent reviews (Cooney and Levine, 1972; Kosaric and Zajic, 1974) and in a symposium (Organizing Committee, 1975). Bacteria of interest in producing SCP from methanol include *Methylomonas methanolica* (Dostalek *et al.*, 1972; Dostalek and Molin, 1975), *Methylophilus (Pseudomonas) methylotrophus* (MacLennan *et al.*, 1973; Young, 1973; Gow *et al.*, 1975; Anonymous, 1976b, 1977a), *Pseudomonas* sp. (Goldberg *et al.*, 1976; Mateles *et al.*, 1976), *Pseudomonas utilus* and *P. inaudita* (Yoshikawa *et al.*, 1975), and a mixed culture consisting of *Methylomonas methylovora* and species of *Flavobacterium, Pseudomonas,* and *Xanthomonas* (Cremieux *et al.*, 1977; Ballerini *et al.*, 1977).

The most advanced process for the production of SCP from methanol has been developed by Imperial Chemical Industries Limited. The organism used is *Methylophilus (Pseudomonas) methylotrophus* (MacLennan *et al.*, 1976; Anonymous, 1976b, 1977a). Gow *et al.* (1975) have described the novel "pressure cycle" airlift fermentor used in this process. The advantages claimed for this system include high oxygen transfer rates without oxygen limitation, ready removal of heat liberated during growth at high productivity, a homogeneous liquid phase, and lack of contamination through drive shafts and mechanical seals encountered with conventional fermentors.

Initial studies were accomplished on a 1000 metric tons/year pilot plant over a 3-year period. Construction has started on a 50,000–75,000 metric ton annual capacity plant, which is scheduled for startup in late 1979 (Anonymous, 1976b, 1977a).

In Japan, Mitsubishi Gas Chemical Company has investigated the growth of bacteria of the genus *Pseudomonas* that will grow on methanol at 38°C (Nagai, 1973). Also, Mitsubishi Petrochemical Company, Ltd.,

TABLE II. Growth of Selected Bacteria and Actinomycetes on Various Substrates

Organism	Carbon and energy source	Scale aeration, agitation	Temperature (°C)	pH	Specific growth rate or dilution rate (D) (hr⁻¹)	Culture density (gm/liter) (dry wt. basis)	Yield (dry wt. basis) (gm/gm substrate used)	References
Achromobacter delvacvate	Diesel oil	6000-liter fermentor, 3000 liters of medium, 1 vol/vol/min	35–36	7.0–7.2	—	10–15 (48 hours)	—	Ko et al. (1964)
Acinetobacter (Micrococcus) cerificans	(a) Gas oil	7.5-liter fermentor,	30	7.0	(a) 0.4–1.0	8–10	0.10–0.12	Ertola et al. (1969)
	(b) n-Hexadecane	4.5 liters of medium, 1 vol/vol/min 500-rpm,			(b) 1.1–2.0	8–10	0.80–0.90	
	(c) n-Hexadecane	7.5-liter fermentor 3.5–7.0 mMO$_2$/liter/min	30	6.8	(c) 1.33	—	1.20	Guenther and Perkins (1968)
Bacillus megaterium	Collagen meat packing waste	2.3 liter working volume fermentor, 3 v/v/min (continuous), 400 rpm	34	7.0	0.25 (D)	2.3	—	Bough et al. (1972)
Brevibacterium sp.	Mesquite wood	14-liter fermentor, 1.0–1.5 v/v/min, 1350–1500 rpm	30–37	6.45–7.2	—	—	0.444	Fu and Thayer (1975); Thayer et al. (1975)
Cellulomonas sp. Alcaligenes	Bagasse	7-,14-,530-liter fermentors, batch and continuous	34	6.6–6.8	0.2–0.29 0.08–0.1 (D)	16 (batch) 10 (continuous)	0.44–0.50	Han et al. (1971)
Methylococcus capsulatus	Methane	2.8 liters of medium, 6.7% CH$_4$, 17.1–19.4% O$_2$ v/v, 38.6–49.5 ml/min, 1450-rpm (continuous)	37	6.9	0.14	0.4	1.00–1.03	Harwood and Pirt (1972)

Organism	Substrate	Fermentor conditions						Reference
Methylomonas clara	methanol	1000-m³ reactor volume (continuous)	39	6.8	0.5 0.3–0.5(D)	—	0.50	Faust et al. (1977)
Methylomonas methanolica	Methanol	4-liter working volume fermentor, 300–400 mM O₂/liter/min 1.5 v/v/min, 1200–1500 rpm, continuous	30	6.0	0.24 (D)	9.6	0.48	Dostalek et al. (1972); Dostalek and Molin (1975)
Methylophilus (Pseudomonas) methylotrophus	Methanol	Pressure cycle air-lift fermentor, continuous	35–40	—	0.38–0.5	30	0.5	MacLennan et al. (1973); Young (1973); Gow et al. (1975); Anonymous (1976, 1977)
Nocardia sp. NBZ-23	n-Alkanes	7.5-liter fermentor, 6–8 liters of medium, 1500 rpm	30	6.8	1.25	14.7	0.98	Wagner et al. (1969)
Pseudomonas sp. No. 5401	Fuel oil	6000-liter fermentor: (a) batch (b) continuous	36–38	7.0	0.16 0.12 (D) 0.25 (D)	16 (24–26 hr) 10 8	1.00 — —	Ko and Yu (1968)
Pseudomonas sp. Hyphomicrobium sp. Acinetobacter sp. Flavobacterium sp. (mixed culture)	Methane	1.0-liter fermentor, 0.9 liters of medium 1 atm; 470 rpm	32	5.7	0.06 (D)	0.8	0.99	Wilkinson et al. (1974)
Rhodopseudomonas gelatinosa	Bicarbonate, wheat bran	4-liter working volume fermentor, 75-W incandescent lamp, 1100 rpm	40	0.2	(a) batch, 0.31 (b) continuous, 0.028 (D)	4.33 3.15	— —	Shipman et al. (1975)
Thermomonospora fusca	Cellulose pulping fines	10-liter fermentor, 3 liter/min, 60 rpm	55	7.4	—	—	0.35–0.40	Crawford et al. (1973)

has developed a bacterial SCP process based on the continuous culture system utilizing methanol in a 500-metric tons/year pilot plant (Masuda and Yoshikawa, 1973; Tung, 1975).

Hoechst/Uhde has developed a pilot-plant-scale process for growing *Methylomonas clara* on methanol. A 1000 metric tons/year unit has been constructed near Frankfurt, West Germany (Faust *et al.*, 1977).

Actinomycetes, particularly *Streptomyces* sp., will also grow on methanol as a substrate (Kato *et al.*, 1974). However, only laboratory-scale studies have been reported.

Ethanol has characteristics similar to methanol insofar as SCP production is concerned. Exxon Corporation has conducted small-scale pilot-plant studies on the growth of *Acinetobacter calcoaceticus* on ethanol (Laskin, 1975).

Recent price increases of petroleum and refined petroleum products have made hydrocarbons and chemicals derived from them, such as methanol and ethanol, less attractive as raw materials for SCP production than renewable resources, such as agricultural waste or by-products. In particular, sugars and starches are readily available from tropical crops, such as sugar cane or cassava, respectively. Also, cellulose is present in agricultural or forest products residues and food-processing waste. Certain chemical industry wastes, such as those from maleic anhydride and phthalic anhydride production, may provide utilizable substrates for bacterial growth (Edwards *et al.*, 1972). Some of the bacteria and corresponding substrates that have been investigated for utilizing waste material for SCP production include *Bacillus megaterium* on collagen meat packing waste (Bough *et al.*, 1972), *Brevibacterium* sp. and *Cytophaga* sp. on mesquite wood (Chang and Thayer, 1974; Thayer *et al.*, 1975; Fu and Thayer, 1975), *Cellulomonas* sp. on bagasse (Han *et al.*, 1971; Dunlap, 1975a, b; Srinivasen *et al.*, 1977), *Pseudomonas dentrificans* on sulfite waste liquor (Camhi and Rogers, 1976), sequential cultures of *Bacillus* sp. and *Lactobacillus* sp. on collagen- and starch-containing wastes (Busta *et al.*, 1977), and mixed cultures of lactic acid bacteria and yeasts on swine waste–corn mixtures (Weiner, 1977a, b).

Shipman *et al.* (1975) have grown the photosynthetic bacterium *Rhodospeudomonas gelatinosa* in continuous culture using wheat bran infusion as the substrate. Although the cells contained significant quantities of protein and vitamins, productivities did not appear sufficiently attractive to justify further development of this process.

Celluloytic, thermophilic actinomycetes offer an opportunity to produce SCP from cellulosic wastes without requiring extensive cooling systems for fermentors. Laboratory studies on *Thermomonospora fusca* indicated that 60–65% degradation of paper mill fines could be achieved in 96 hours, yielding a 30% protein product (Crawford et al., 1973). General Electric Company constructed a small pilot plant at Casa Grande, Arizona for producing SCP from feed lot wastes using thermophilic actinomycetes (Bellamy, 1974, 1975). However, this pilot plant is not now in operation because of adverse economic conditions.

2. Yield

A number of factors affect growth rate, yield, and productivity (dry weight of cells per unit volume-unit time) of bacteria used for SCP production. Oxygen transfer, mass transfer of substrate to the cell, and heat production are important limiting factors for bacterial growth on either carbohydrates or hydrocarbons as carbon and energy sources. Also, inhibitory substances may be produced during growth particularly in continuous cultures.

Table II shows that typical yields of bacteria and actinomycetes grown on hydrocarbons range from 0.80 to 1.20 gm dry matter/gm substrate utilized. The maximum theoretical yields of bacteria from methane assuming the involvement of a mixed function oxidase is 0.91 gm dry cells/gm substrate according to Van Dijken and Harder (1975).

Harwood and Pirt (1972) attained experimental yields for *Methylococcus capsulatus* of 1.00–1.03 gm (dry wt.)/gm methane. These differences could be accounted for by the difficulties in determining carbon balances with gaseous substrates in continuous culture systems.

Experimental values of yields of bacteria from methanol given in Table II range from 0.48 to 0.50 gm. In studies with various *Pseudomonas* sp. Goldberg et al. (1976) obtained yields from methanol of 0.38–0.54 gm dry cells/gm substrate. The theoretical values calculated by Van Dijken and Harder (1975) for the yield of bacterial cells from methanol was 0.73 gm (dry wt.)/gm substrate. They assumed a yield of adenosine triphosphate (Y_{ATP}) from 3-phosphoglycerate as a key intermediate of 10.5 and a cell composition of C_4H_8ON for their calculations.

Table II shows yields of bacteria and actinomycetes from complex carbohydrates of 0.35–0.50 gm(dry wt.)/gm substrate. A typical yield value for *Pseudomonas fluorescens* grown aerobically on glucose is 0.51 gm (dry wt.)/gm substrate (Mennett and Nakayama, 1971).

During the 1960s there was interest in bioregenerative systems based on hydrogen-fixing bacteria for recycling expired carbon dioxide and urinary excretion products during extended space missions. Breathable oxygen would be produced by the electrolysis of water. The hydrogen also produced by electrolysis, along with a portion of the oxygen, the expired carbon dioxide, and urea, could be used by *Alcaligenes (Hy-*

drogenomonas) species to produce a microbial cell product to be utilized as food. Foster and Litchfield (1964, 1967, 1968a,b, 1969) designed a continuous culture system for *Alcaligenes eutrophus* which utilizes hydrogen, oxygen, and carbon dioxide in a 6 : 2 : 1 ratio. Productivities were in the range of 2.0 gm (dry wt.)/liter/hour (Litchfield, 1972). Yields of 5.2 gm (dry wt.) of *A. eutrophus* cells per gram-mole of hydrogen at a specific growth rate of 0.42 have been obtained (Bongers, 1969, 1970).

3. Cell Recovery

A number of problems arise in recovering bacterial cells from the growth medium. In most SCP processes, cell densities are in the order of 10–20 gm (dry wt.)/liter. Consequently, large volumes of water must be handled. Also, bacterial cells have a smaller size (1–2 μm) than other microorganisms and bacterial cell densities are very close to that of water (1.003 gm/cm^3). Taking into account these factors, the cost of separating bacterial cells by centrifugation would be prohibitively high.

Wang (1968a,b, 1969) estimated that total costs for centrifugal separation of bacteria from the growth medium would be approximately four times as great as for yeasts.

Labuza (1975) pointed out that plate-and-frame filter presses are not amenable to continuous processing as used in SCP production. Also, in vacuum filters, the cell size and packing density of bacteria results in compression on screens and filtration ceases because the void volume decreases to nearly zero. The use of filter aids and flocculants would contaminate the cell product and make it undesirable for food or feed applications.

As a result of the high cost of centrifugation consideration has been given to alternative methods of bacterial cell recovery in SCP production.

Acinetobacter cerificans can be concentrated by a two-zone froth-flotation process according to Perkins and Furlong (1967). Concentration ratios of bacterial cells in the froth were significantly greater than those in the slurry.

Imperial Chemical Industries, Ltd. have developed a proprietary process for separating *Methylophilus methylotrophus* grown on methanol from the growth medium without using flocculating agents. A dewatered cell product 25 gm (dry wt.)/liter or greater is then concentrated by conventional decanter-type centrifuges (Gow *et al.*, 1975; Mac-Lennan *et al.*, 1976). The resulting concentrated product is then dried.

Hoechst/Uhde have operated a pilot-plant-scale process for separating *Methylomonas clara* grown on methanol from the culture medium based on electrochemical coagulation and centrifugation. The cell product is then spray-dried (Faust *et al.*, 1977).

There is a lack of information in the literature on initial concentration of bacterial cells by evaporation prior to spray-drying or drum-drying. Also, information is not readily available on flow characteristics of bacterial cell suspensions or of drying characteristics of initially dewatered bacterial cells of interest in SCP production. A further discussion of dewatering and drying of cells will be given subsequently in the section on yeasts.

Cell recovery processes from SCP production result in wastes from the spent growth medium and cell wash water. To the extent possible, spent media should be recycled since their biochemical oxygen demands (BOD) are substantial. In the case of methane, the excess substrate can be recycled readily and the spent medium and cell washings contain primarily inorganic salts and small amounts of cells. These residues must be treated by conventional waste treatment processes.

C. Yeasts

The technology of yeast production has developed since the nineteenth century. The reader should consult Reed and Peppler (1973) and Chapter 5, on yeast production, in this volume for information on the historical background and the current status of bakers' yeast production processes. This section covers the current status of food and fodder yeast production.

1. Growth Conditions

Table III lists some of the yeasts and substrates that are of potential interest for use in SCP production other than those in the genus *Saccharomyces,* such as *S. cerevisiae* and *S. carlsbergensis (S. uvarum),* which can be grown as primary yeast from cane or beet molasses as substrates.

Bakers' yeast, *S. cerevisiae,* was used for food and fodder yeast in Germany during World War I (Prescott and Dunn, 1959). However, this yeast does not utilize pentoses and requires supplementation with amino acids and B vitamins, mainly thiamin, niacin, pyridoxine, pantothenic acid, and inositol, which are usually furnished by blends of cane and beet molasses.

During World War II, *Candida utilis* was produced as a source of feed or food yeast in Germany from wood hydrolyzates which contain pentoses. This yeast does not require amino acids or B vitamins for growth and utilizes ammonium salts as a source of nitrogen. This development

TABLE III. Selected Yeasts and Substrates of Interest for SCP Production

Yeast	Substrate	References
Candida spp.	n-Alkanes	Kanazawa (1975); Sonoda *et al.* (1973)
Candida boidinii	Methanol	Reuss *et al.* (1975)
Candida ethanothermophilum	Ethanol	Masuda *et al.* (1975)
Candida guilliermondii	Diesel oil	
Candida kofuensis	n-Alkanes	Ueno *et al.* (1974, a, b, c)
Candida lipolytica	n-Alkanes	Evans and Shennan, (1974)
	Gas oil	Champagnat and Filosa (1971); Laine and du Chaffaut (1975)
	Oxanone wastewater	Wiken (1972)
Candida methanolica	Methanol	Goto *et al.* (1976)
Candida pseudotropicalis	Waste polyethylene	Brown *et al.* (1975)
Candida rigida	Liquefied petroleum gas	Imada *et al.* (1972)
Candida steatolytica	Brewery wastes	Shannon and Stevenson (1975 a, b)
Candida tropicalis	n-Alkanes	Cooper *et al.* (1975)
	Ethanol	Rychtera *et al.* (1977)
	Sulfite waste liquor	Rychtera *et al.* (1977)
	Starch	Spencer-Martins and Van Uden (1977)
Candida utilis	Ethanol	Ridgeway *et al.* (1975)
	Sulfite waste liquor	Holderby and Moggio (1960); Anderson *et al.* (1974); Camhi and Rogers (1976)
	Anaerobic digester supernatant	Irgens and Clarke 1976)
	Molasses	Mian *et al.* (1976)
	Tapioca starch	Thanh and Wu (1975)
Candida utilis and *Endomycopsis fibuligera*	Potato starch waste	Wiken (1972)
Debaryomyces kloeckeri	Soybean spent solubles	Sugimoto (1974)
Hansenula polymorpha	Methanol	Cooney and Levine (1975); Cooney *et al.* (1975)
Kluyveromyces fragilis	Cheese whey	Wasserman *et al.* (1961); Bernstein and Everson (1973); Moulin *et al.* (1976)
	Crude lactose	Vananuvat and Kinsella (1975 a, b)
	Lactose permate, ultrafiltrate	Delaney *et al.* (1975); Lane (1977)

(*continued*)

TABLE III (Continued)

Yeast	Substrate	References
Rhodotorula glutinis	Domestic sewage	Fustier and Simard (1976)
Rhodotorula rubra	Potato hydrolyzate	Tong et al. (1973a, b)
Torulopsis methanosorba	Methanol	Yokote et al. (1974)
Torulopsis methanolovescens	Methanol	Goto et al. (1976)
Trichosporon cutaneum	Cheese whey	Akin et al.(1967)
	Oxanone wastewater	Wiken (1972)

led to the postwar advances and present-day practices in SCP production processes utilizing a variety of yeasts and substrates.

Many of the same characteristics cited for bacteria are important criteria in selecting yeasts for use in SCP production, particularly specific growth rate, yield on a given substrate, pH and temperature tolerance, aeration requirements, and genetic stability. Also, the yeast strain must not be pathogenic to plants, animals, or humans.

Substrate and nitrogen concentrations for yeast growth should be adjusted to provide a C : N ratio in the range of 7 : 1 to 10 : 1 to favor high protein contents. Concentrations of carbohydrates in batch cultures range from 1 to 5%. In continuous cultures, with hydrocarbons or alcohols, lower concentrations are used. At higher C : N ratios many yeasts, particularly those of the genus Rhodotorula, accumulate a substantial portion of the cell weight in the form of lipids. Again, anhydrous ammonia or ammonium salts are suitable nitrogen sources for yeasts of interest in SCP production.

Anhydrous ammonia can be used in combination with phosphoric acid to effect pH control. It is desirable to select a yeast having a high specific growth rate and yield in the pH range of 3.5–4.5 to minimize the possibility of growth of any bacterial contaminants.

Carbohydrate substrates are soluble in aqueous media. Mass transfer is not limited by solubility. The same situation applies to methanol which is completely miscible with water. However, C_{10}–C_{18} hydrocarbons are soluble in water only to the extent of 6–16 mg/ml. The uptake of hydrocarbons by yeasts such as C. lipolytica may take place from either the liquid or the gas phase. These substrates are believed to penetrate into the cells by passive diffusion (Prokop and Sobotka, 1975).

As with bacteria, heat is liberated by yeasts growing on carbohydrates, hydrocarbons, or alcohols. For example, C. utilis, growing on sulfite waste liquor, which is representative of carbohydrate substrates, generates approximately 3740–8000 kcal/kg dry weight (Holderby and

Moggio, 1960). On the other hand, for *Candida* species growing on hydrocarbons, calculated values of heat liberated range from 4400–8000 kcal/kg for yields of 1.2–0.9 gm cells (dry wt.)/gm substrate utilized, respectively (Kanazawa, 1975; Laine and du Chaffaut, 1975). Actual observed values of heat liberated during growth are in this range. A value of 7600 kcal/kg dry weight of yeast is cited by Bennett *et al.* (1969) for *Candida* sp. growing on *n*-alkanes. Cooney *et al.* (1975) estimated a heat load of 37 kcal/liter/hour for production of yeast on methanol assuming a productivity of 5 gm/liter/hour, a yield of 0.5 gm cells per gram oxygen and using a factor of 0.12 kcal/mmole of oxygen.

It is clear that heat tolerance is a desirable attribute for a yeast to be used in SCP production. Most yeasts have the highest specific growth rates in the range 30°–34°C. A yeast capable of growing well in the range 40°–45°C would be of considerable interest. *Hansenula polymorpha* ATCC 26012 has an optimum growth temperature range of 37°–42°C (Cooney and Levine, 1975). This organism gives good cell yields from methanol at 42°C.

In most regions where SCP production facilities are located, cooling water is rarely available at a temperature below 20°C. Mechanical refrigeration would be required for cooling fermentors to maximum desired temperatures for growth.

The growth rate of yeasts under aerobic conditions depends upon the rate of mass transfer of oxygen and substrates to and across the cell surface. For growth on carbohydrates, the oxygen requirement is 1 gm or less per gram (dry weight) of cells (Reed and Peppler, 1973). On *n*-alkanes the oxygen requirement is approximately 2 gm of oxygen per gram (dry weight) of cells (Johnson, 1967).

According to Laine and du Chaffaut (1975) yeast grown on hydrocarbons requires an oxygen transfer coefficient of 10–15 kg/m³/hour. A Waldhof-type agitated and aerated fermentor has been used extensively for achieving desired levels of aeration in food and fodder yeast production from both carbohydrate and hydrocarbon substrates (Prokop and Sobotka, 1975). However, in recent years, air-lift-type fermentors of the type developed by LeFrançois (1955) have been used to an increasing extent to achieve increased oxygen transfer rates and/or decreased power requirements for aeration and agitation, particularly in the production of yeast SCP from hydrocarbons (Laine, 1972; Hatch, 1975; Kanazawa, 1975).

Yeast SCP production may or may not take place under sterile conditions. In either batch or long-term continuous yeast production, one must balance the need for contamination control, by maintaining sterile conditions, with the capital and operating costs of the equipment required. In producing *C. utilis* (torula yeast) from sulfite waste liquor, the growth medium is passed through heat exchangers to effect sterilization and then charged into clean Waldhof-type fermentors. However, sterile conditions are not maintained during growth and contamination control depends upon maintaining a pH below 6.0, aerating with a sterile air supply, and maintaining a large population of yeast in the fermentor (Holderby and Moggio, 1960; Anderson *et al.*, 1974).

In the case of yeasts grown on hydrocarbons, clean but nonsterile conditions were used by British Petroleum Company, Limited in an air-lift fermentor pilot plant for producing *Candida* sp. from gas oil at Lavera, France. However, in producing *Candida* sp. from purified *n*-alkanes in a conventional agitated–aerated fermentor at Grangemouth, Scotland, British Petroleum used a completely sterile system and maintained sterile conditions throughout operations (Evans, 1968; Bennett *et al.*, 1969).

Table IV summarizes the growth characteristics of selected yeasts of importance in SCP production. Only the more important current processes will be discussed here.

Production of *C. utilis* dates from the 1940s. As mentioned previously, sulfite waste liquor from the paper industry was used as the substrate to convert to a salable product what would otherwise be a difficult to treat waste (Harris *et al.*, 1951; Holderby and Moggio, 1960).

A 10 million lb/year capacity torula yeast plant has been constructed and operated by Boise-Cascade Corporation at Salem, Oregon, to treat 85,000 tons of sulfite liquor from an ammonia-base pulp process that would otherwise be burned as fuel. Both food-grade and feed-grade yeasts are produced. Two continuous agitated and aerated 60,000-gal (227,124-liter) Waldhof fermentors are used to process 48,000–50,000 gal (181,700–189,270 liters) of medium (Anderson *et al.*, 1974).

The temperature is maintained at 36°–37°C by passing the medium through a shell and tube heat exchanger. The yeast is harvested by centrifugation, washed three times, pasteurized in a plate heat exchanger at 95°C, and spray-dried.

Candida utilis has been produced from a wide range of other substrates, as indicated in Table III. Ethanol, produced from ethylene, is a suitable substrate for this yeast. Amoco Foods Company has constructed a 10,000,000-lb (454,000-kg) capacity plant for continuous production of *C. utilis* from ethanol at Hutchinson, Minnesota. All liquid streams are sterilized at 149°C. Ammonia is sterilized by filtration and is used for pH control. The cells are harvested continuously, washed, and spray-dried to give a food-grade product (Anonymous, 1977b; Ridgeway *et al.*, 1975).

TABLE IV. Growth of Selected Yeasts on Various Substrates

Organism	Carbon and energy source	Scale, aeration, agitation	Temperature (°C)	pH	Specific growth rate or dilution rate (D) (hr⁻¹)	Culture density (gm/liter) (dry wt. basis)	Yield (dry wt. basis) (gm/gm substrate used)	References
Candida ethanothermophilum ATCC 20380	Ethanol	30-liter fermentor, 17 liters of medium, 17 liters/min, 500 rpm (continuous)	40	3.5	0.20 (D)	8.0	0.95	Masuda et al. (1975)
Candida lipolytica	(a) n-Alkanes	1800-liter working volume fermentor, 1.5 v/v/min, continuous	32	5.5	0.16 (D)	23.6	0.88	Evans and Shennan (1974)
	(b) Gas oil	12-m³ fermentor, 2400 liters of medium, 1 v/v/min, continuous	30	4.0	—	25	0.18	Champagnat and Filosa (1971) Laine and du Chaffauf (1975)
Candida tropicalis	n-Alkanes	Air-lift 50,000-liter fermentor	30	3.0	0.15–0.24	10–30	1.0–1.1	Silver and Cooper (1972a, b), Cooper et al. (1975)
Candida utilis	(a) Sulfite waste liquor	Walhof fermentor 0.50 cfm	32	5.0	0.5	—	0.39	Harris et al. (1951)
	(b) Ethanol	100–140 mM O₂/liter/hour (continuous)	30	4.6	0.3	6–7	—	Ridgeway et al. (1975)
Hansenula polymorpha ATCC 26012	Methanol	1-liter fermentor, 375 ml of medium	37–42	4.5–5.5	0.22 (0.13,D)	1.2	0.36	Cooney and Levine (1975); Cooney et al. (1975)
Kluyveromyces (Saccharomyces) fragilis	Cheese whey	1600-gal (6057-liter) fermentor, 1.5 mM O₂/gal/min	32	5.5	0.66	14.6	0.55	Wasserman et al. (1961)

			28	7.4	0.158–0.211	0.08	—	
Rhodotorula glutinis	Domestic sewage	14-liter fermentor, 0.7 v/v/min, 300 rpm, continuous						Fustier and Simard (1976)
Saccharomyces cerevisiae	Cane molasses	(a) 75–225 m^3 fermentors, 2 mM O_2/liter/min (l v/v/min)	30	4.5–5.0	0.20–0.25	40–45	0.50	Reed and Peppler (1973)
		(b) 5,900-gal (34,068-liter fermentors, continuous	30	4.5–5.0	0.14	70	—	Olson (1961)

115

Candida utilis has also been grown on ethanol in a 1000-metric ton/ year pilot plant at Kojetin, Czechoslovakia. Scaleup is planned to a 60,000 ton/year scale (Wells, 1975).

In Japan, Mitsubishi Petrochemical Company, Limited has conducted research on a pilot-plant-scale process for growing *Candida acidother-mophilum* and *Candida ethanothermophilum* on ethanol. These organisms are claimed to tolerate temperatures up to 40°C (Masuda *et al.*, 1975).

Over 25 × 10^9 lb (11.4 × 10^9 kg) of sweet and acid cheese whey are produced in the United States each year. Early research on the utilization of this whey, which contains 69–77% lactose (dry basis), by *Kluyveromyces (Saccharomyces) fragilis* was conducted at the Eastern Regional Research Center of the U.S. Department of Agriculture (Wasserman *et al.*, 1961).

Dry food-grade *Kluyveromyces fragilis* yeast has been produced in the United States for more than a decade. A 5000 short ton/year plant for producing either yeast aerobically or ethanol anaerobically from cheese whey is operated by Amber Laboratories at Juneau, Wisconsin (Bernstein and Everson, 1973; Bernstein *et al.*, 1975). Also, research has been conducted on the growth of *K. fragilis* on lactose permeates and ultrafiltrates from whey protein recovery (Delaney *et al.*, 1975; Lane, 1977).

The Symba process developed by the Swedish Sugar Corporation utilizes two yeasts symbiotically for the treatment of wastes containing starch. *Endomycopsis fibuligera* produces α- and β-amylases which hydrolyze starch to lower saccharides, predominantly glucose and maltose, and *C. utilis* utilizes these sugars for growth. The process was originally designed to treat potato-processing wastes but can be applied to other wastes that contain starch, such as rice-processing waste.

Data obtained on a pilot plant providing for the production of 40–100 kg dry yeast per day were used to design a plant to handle 126 m^3 of waste per hour having a biochemical oxygen demand (BOD) of 15,000 mg/liter. BOD reductions are estimated to be 85% or greater (Wiken, 1972).

Research at the Dutch State Mines demonstrated that *C. lipolytica* or *Trichosporon cutaneum* can be used to treat oxanone water, a waste mixture of organic acids resulting from the manufacture of caprolactam used as the raw material for producing Nylon 6. Special strains of these yeasts have high resistance to the toxic effects of organic acids at pH 6.5. A yeast production of 4–4.5 kg/m^3/hour and 80% removal of organic acids are obtained (Wiken, 1972).

During the early 1960s, British Petroleum Company, Limited introduced the concept of using *C. lipolytica* and related yeasts for the production of SCP from hydrocarbon fractions including gas oil and purified n-alkanes prepared by molecular sieve purification (Champagnat et al., 1963; Champagnat and Filosa, 1965, 1971; Evans, 1968; Bennett et al., 1969; Evans and Shennan, 1974; Laine and du Chaffaut, 1975). In the case of gas oil, the objective was to utilize the n-alkane fraction, effect dewaxing, and concurrently produce SCP. About 10% of the gas oil is utilized by the yeast and the remaining 90% is returned to an adjacent refinery.

British Petroleum constructed two pilot plants for producing *Candida* sp. from hydrocarbons. A 16,000 metric ton/year plant at Lavera, France was designed to utilize gas oil containing 25% $C_{15}-C_{30}$ n-alkanes having a boiling range of 300°–380°C. A continuous air-lift-type fermentor was used under clean but nonsterile operating conditions. This plant ceased operation in 1976. A 4000 metric ton/year pilot plant has been constructed and operated at Grangemouth, Scotland. It utilized purified n-alkanes prepared by a molecular sieve process (97.5–99% $C_{10}-C_{23}$, boiling range 175°–300°C). A continuous aerated and agitated fermentor was operated under aseptic conditions. The mineral nutrients were sterilized by heating, while air and ammonia are sterilized by filtration (Evans, 1968; BP Proteins Ltd., 1974).

The British Petroleum processes led to the development of similar yeast-based SCP processes utilizing hydrocarbon substrates by petroleum and chemical industries throughout the world. These processes use the following yeasts: *Candida* sp. (Kanegafuchi Chemical Industry Company, Limited; Takata, 1969; Kanazawa, 1975; Kanazawa et al., 1975), *Candida kofuensis* (Mitsui-Toatsu Chemicals, Inc.; Ueno et al., 1974a,b,c), *Candida novellus* (Liquichimica Biosintesi, S.p.a.; Giacobbe et al., 1975), and *Candida tropicalis* (Gulf Research and Development Company; Cooper et al., 1975).

A 100,000 metric ton/year yeast SCP plant constructed in Sardinia by British Petroleum and Ente Nazionale Idrocarburi has not yet been placed into operation because of a dispute with Italian governmental authorities over hydrocarbon residues and product quality (Anonymous, 1976a, 1978a,b). Another 100,000 metric ton yeast SCP plant constructed by Liquichimica Biosintesi S.p.a. at Saline, di Montebello, Italy using the Kanegafuchi Chemical Industry process has not been operated for the same reason (Anonymous, 1978a,b).

Other pilot-plant yeast SCP processes have been operated by Institute Français du Petrole, Soleige, France (Decerle et al., 1969), Sovnaft, Kojotin, Czechoslovakia (Anonymous, 1973b), and All-Union Research Institute of Biosynthesis of Protein Substances, Bashkira and Gorki, USSR (Anonymous, 1974b). Moo-Young (1976) surveyed existing or planned SCP production facilities throughout the world. His compilation should be consulted for further information.

2. Yield

Factors affecting the growth rate, yield, and productivity of yeasts used for SCP production from various substrates are similar to those for bacteria.

Table IV presents yields of selected yeasts grown on various substrates. Experimental yields on carbohydrates range from 0.39 to 0.55 gm cells (dry wt.)/gm substrate utilized. These values are similar to the 0.51 gm (dry wt.)/gm substrate reported by Hernandez and Johnson (1967) for *C. utilis* grown on glucose. Data on yields reported in the literature should be evaluated carefully to determine the extent to which nonsugar carbon sources in complex natural products and wastes contribute to yield.

In the case of *C. utilis* grown on ethanol, Rychtera *et al.* (1977) obtained a yield of 0.75 gm (dry wt.)/gm substrate. They cited a previous published yield value in the range of 0.68–0.72 gm (dry wt.)/gm ethanol. The 0.95 gm (dry wt.)/gm ethanol yield reported for *Candida ethanothermophilum* by Masuda *et al.* (1975) appears to be unusually high and the reasons for these differences are not readily discernible from the information given in their patent.

The yield of *Hansenula polymorpha* grown on methanol of 0.36 gm (dry wt.)/gm substrate obtained by Cooney and Levine (1975) is lower than a calculated theoretical yield of 0.6 gm (dry wt.)/gm substrate based on the number of electrons available for transfer to oxygen. They point out that this discrepancy could probably be explained when the pathway of methanol oxidation and incorporation into cell mass is better understood. They cite other published yield data for yeasts grown on methanol ranging from 0.29 to 0.45 gm (dry wt.)/gm substrate.

The yields of various *Candida* species grown on purified *n*-alkanes range from 0.88 to 1.1gm (dry wt.)/gm substrate. These yields were obtained in pilot-plant-scale operations and are representative of the values attainable in large-scale production facilities. These values agree with the calculations of Guenther (1965), which indicate that yields greater than 1.0 gm (dry wt.)/gm substrate are theoretically possible from hydrocarbons on an energy basis.

Various equations have been developed for showing the stoichiometry of yeast growth on carbohydrates or hydrocarbons (Darlington, 1964; Guenther, 1965; Bennett *et al.*, 1969; Kanazawa, 1975). Typical equations are given by Bennett *et al.* (1969) for carbohydrates [Eq. (1)] and hydrocarbons [Eq. (2)]:

$$8n \ CH_2O + 0.8n \ O_2 + 0.19n \ NH_4 + \text{essential elements} \rightarrow$$
$$n(CH_{1.7}O_{0.5}N_{0.19}Ash) + 0.8n \ CO_2 + 1.3n \ H_2O + 80,000n \ kcal \quad (1)$$

$$2n \ CH_2 + 2n \ O_2 + 0.19n \ NH_4 + \text{essential elements} \rightarrow$$
$$n(CH_{1.7}O_{0.5}N_{0.19}Ash) + n \ CO_2 + 1.5 \ H_2O + 200,000 \ n \ kcal \quad (2)$$

These equations will vary with the composition of the yeast cell product and the same considerations apply to other microorganisms used for SCP production, including algae, bacteria, molds, or higher fungi.

Holve (1976) presented a series of equations for describing the stoichiometry of conversion of carbohydrates, hydrocarbons, alcohols, or organic acids to SCP by microorganisms. His model depends upon assumed chemical compositions of microorganisms based on literature values. In practice, actual analyses of the composition of the cells being grown should be used in such calculations.

3. Cell Recovery

Yeast cells range in size from 5 to 8 μm and have a density of 1.04–1.09 gm/cm^3. They can be recovered readily from the growth medium by continuous centrifugation. Usually, the cells are centrifuged in a first stage to initially dewater, yielding a yeast cream. This is followed by two subsequent washing centrifugations. The final washed yeast cream usually contains 15–20% solids. This procedure is typical of that used in the recovery of C. utilis after growth on sulfite waste liquor (Holderby and Moggio, 1960; Anderson et al., 1974).

In the case of Candida sp. grown on n-alkanes, a wash with a surfactant may be required to remove traces of hydrocarbon following initial centrifugation (Wang, 1968; Labuza, 1975). With gas oil, a much more complex separation and cleanup procedure must be used to remove residual hydrocarbons and lipids. Decantation, phase separation with solvents, washing with sufactants, and solvent extraction have been used alone or in combination. Also hydrocarbons and lipids can be depleted by allowing the yeast to remain in contact with the growth medium in absence of added substrate for varying periods. Solvents can be removed by steam (Litchfield, 1977b).

For recovering Kluyveromyces fragilis after growth on cheese whey, the entire growth medium can be passed through a three-stage evaporation to concentrate the solids from 8 to 27% to give a feed-grade product. If a high-protein–low-ash product is desired, multistage centrifugation is used (Anonymous, 1975; Bernstein and Everson, 1973).

Wang (1968b, 1969) has analyzed methods and costs for recovering microbial cells including yeasts. He points out that the cost of cell separation could be reduced if the size of the cells could be increased or if flocculants compatible with a food- or feed-grade product could be developed.

The separated cells can either be drum- or spray-dried. According to Labuza (1975) drum-drying brings about two to five logarithmic cycles of kill of yeast cells. Associated bacteria having similar thermal resistances would be killed to the same extent. However, drum-drying, if not carefully controlled, may result in excessive darkening of the product.

Under the temperature and time conditions used in spray-drying, 5–8 logarithmic cycles of kill of yeast are obtained and cells of pathogenic bacteria, such as *Salmonella* and *Staphylococcus,* should be killed to the same extent. The color and finished characteristics of the final yeast product prepared by spray-drying are superior to those obtained after drum-drying (Labuza, 1975).

D. Molds and Higher Fungi

As mentioned earlier in this chapter, certain higher fungi, the mushrooms, have been used as human food since ancient times. Also, many Asian fermented food products are produced by mold fermentations.

Molds and higher fungi grow at much slower rates than bacteria and yeasts. Doubling times may range from 6 to 16 hours, or longer.

Mushrooms are usually grown as the fruiting body in manure or synthetic compost beds (see Lambert, 1938; Stoller, 1954; Litchfield, 1967a; Volume II, Chapter 8, for reviews of mushroom cultivation practices).

In the United States, the basidomycete *Agaricus campestris* (*A. bisporus*) is the sole species produced. In Japan, the Shiitake mushroom, *Cortinellus berkelyanus* Ito and Inci, is produced by a 2-year incubation of logs of certain trees of the Fagaceae and Betulaceae inoculated with spore suspensions. The Chinese padistraw mushroom, *Volvaria volvacea* (Bull.) Fr., is cultured by inoculating moistened rice straw with spawn (Lambert, 1938).

The Perigord truffle, *Tuber melanosporum* Vitt, or black truffle, which belongs to the family Ascomycetes, is produced in southern France by taking soil from the vicinity of truffle-bearing oaks and raking this soil beneath newly planted oak trees (truffieres or truffle orchards). About 5–10 years after the oaks have been planted, truffles appear and yields range up to 100 lb/acre (Seelig, 1959).

This section covers production of mold or fungal mycelium for food or feed applications, since the mycelium of these organisms can be grown in submerged culture in a manner similar to growing yeasts.

1. Growth Conditions

Fungal mycelium was grown in submerged culture in Germany during World War II for use as food. Numerous species have been investigated for their potential utility as food or feed since that time and the reader is referred to previous reviews for background information (Litchfield, 1967a,b, 1968; Gray, 1970).

The same general considerations for selecting yeast cultures for SCP production apply to molds and higher fungi, including specific growth rate, yield, pH and temperature tolerance, aeration requirements, and genetic stability. Lack of pathogenicity, or toxigenicity, is very important. Strains of some species of molds such as *Aspergillus fumigatus* and *Fusarium graminearum* are very hazardous to humans (Wilson, 1971; Sherwood and Peberdy, 1974). Yet these organisms have been proposed for SCP production (Anderson *et al.*, 1975; Gregory *et al.*, 1976; Khor *et al.*, 1976). The use of organisms of this type should be avoided even though the particular strain selected may be judged "safe" after preliminary toxicological evaluation.

Numerous species of molds and higher fungi have been evaluated for production as SCP or as food flavor additives on carbohydrates or on agricultural, food-processing, or forest products wastes that contain carbohydrates. Table V presents some typical examples of both organisms and substrates.

The concentrations of carbohydrate substrates used are in the 1–10% range. Carbon-to-nitrogen (C : N) ratio in some species, when as high as 20 : 1, have given cells with satisfactory protein contents, but in most cases C : N values of 5 : 1 to 15 : 1 are preferred.

Anhydrous ammonia or ammonium salts can be used as nitrogen sources in continuous cultures and at the same time effect pH control. In many waste treatment applications, fungal cultures are grown in either a batch or semicontinuous mode and ammonium phosphate, ammonium sulfate or ammonium nitrate are added to the waste as supplemental nitrogen sources.

Phosphoric acid is suitable as a source of phosphorus. Mineral nutrients, such as potassium, sulfur, magnesium, calcium, iron, manganese, zinc, copper, and cobalt, are required for growth of most fungi. However, the concentrations of specific inorganic constituents in the growth medium may vary considerably when complex organic materials of variable composition are used as substrates. In the case of mushroom mycelia produced for food flavoring purposes, the concentrations of inorganic salts in the medium must be limited to levels actually required for growth to prevent the development of a bitter flavor in the mycelium (Litchfield, 1967c).

The mycelium of certain mushrooms will not grow in synthetic media unless supplemented with vitamins. Some strains of *Coprinus comatus* have an absolute requirement for thiamin, and the addition of thiamin also enhances growth of other mushrooms. *Agaricus* spp. and *Morchella* spp. will grow in a simple synthetic medium. However, the rate of growth of the latter group of organisms is enhanced by supplements, such as

TABLE V. Growth of Selected Molds and Higher Fungi on Various Carbon and Energy Sources

Organism	Carbon and energy source	Scale, aeration, agitation	Temperature (°C)	pH	Specific growth rate or dilution rate (D)(hr⁻¹)	Culture density (gm/liter)	Yield of Mycelium gm/100 gm substrate (dry wt. basis) Supplied	Used	References
Agaricus blazei	Glucose, citrus press water, orange juice	250-ml Erlenmeyer shake flask, 2-inch stroke, 80 cpm, 1-liter, 40-liter bottle	Ambient	3.5–7.5	—	15.2 18.8 20.6	30 31 34	42 46 41	Block et al. (1953)
Agaricus campestris	Glucose	20-liter fermentor, air flow, 2–3 v/v/min, 400 rpm	25	4.5		20	—	—	Humfeld and Sugihara (1949)
	Malt syrup– cane molasses	2/liter fermentor, 13 mM O_2/liter/hour, 400 rpm	27	5.0–5.5	—	7.2	15.8	44.6	Moustafa (1960)
Aspergillus niger	Carob bean extract	3000-liter fermentor (total capacity)	30–36	3.4	0.25	31.5	45	—	Imrie and Vitos (1975)
Aspergillus oryzae	Coffee waste water	5000-gal (18,927-liter tank, air flow. 1 v/v/min	28	4.0–4.5	0.037 (D)	—	—	45	deCabrera et al. (1974)
Boletus edulis	Glucose	300-ml Erlenmeyer flask,100-ml medium, 150 rpm	25	4.5–5.5	0.00166	2.5	—	25	Van Eybergen and Scheffers (1972)
Calvatia gigantea	Brewery waste: (a) Grain press (b) Trub press liquor	250-ml Erlenmeyer flasks, 100-ml medium, 200 rpm	25	6.0	— —	(a) 6.25 (b)27.72	— —	24.8 74.9	Shannon and Stevenson (1975b)
Fusarium moniliforme	Carob bean extract	14-liter fermentor (8.5 liter working volume), (semicontinuous). 400–700 rpm. 0.25 v/v/min	30	5.5–6.5	0.18	8.8	—	0.384	Drouliscos et al. (1976)
Geotrichum sp.	Corn and pea waste	Aeration pool 10,000 gal (37,854 liter), continuous,	Ambient	3.7	—	0.75–1.0	—	—	Church et al. (1973)

Organism	Substrate	Fermentation system	Temperature	pH					Reference
deliquescens	milling waste	(189,270 liter continuous)				3.2-3.5	—	—	Church et al. (1973)
	Soy whey	18-liter fermentor	Ambient	4.6	—	7.5	—	26	Church et al. (1972)
Lentinus edodes	Glucose	30-liter fermentor, 2 v/v/min, 400-600 rpm	25	5.5	0.12				Teramoto et al. (1966)
Morchella crassipes	Glucose,	10-liter carboy, 7 liters of medium, 0.08 mM O_2/liter/min	25	6.5	—	8.02	33.6	48.6	Litchfield et al. (1963a)
	maltose,					3.55	31.4	47.8	
	lactose,					3.30	29.2	46.3	
	cheese whey,					2.38	5.96	32.7	Litchfield and Overbeck (1966)
	corn canning waste,					0.75	13.9	27.8	
	pumpkin canning waste,					5.51	20.0	42.6	
	sulfite liquor (NH₃)	5-gal carboy, 16 liters of medium, 0.25 v/v/min	Ambient	5.0-7.0		2.0		78	Kosaric et al. (1973)
Morchella deliciosa	Sulfite waste liquor	5-gal carboy, 16 liters of medium, 0.25 v/v/min	Ambient	6.0		1.0		32	Kosaric et al. (1973)
Morchella esculenta	Glucose,	10-liter carboy, 7 liters of medium, 0.08 mM O_2/liter/min	25	6.5		7.85	32.8	48.1	Litchfield et al. (1963a)
	maltose,					3.40	30.1	47.5	
	lactose,					1.65	14.6	43.4	
	cheese whey,					1.28	3.21	32.1	Litchfield and Overbeck (1966)
	corn canning waste,					0.85	15.3	33.3	
	pumpkin canning waste,					7.92	28.8	50.4	
	sulfite waste liquor (NH₃)	5-gal carboy, 16 liters of medium, 0.25 v/v/min	Ambient	5.0-6.0		1.9		65	Kosaric et al. (1973)
Morchella hortensis	Glucose,	10-liter carboy, 7 liters of medium, 0.08 mM O_2/liter/min	25	6.5		8.20	34.3	48.8	Litchfield et al. (1963a)
	maltose,					3.75	33.2	49.0	
	lactose,					3.60	31.8	47.9	
	cheese whey,					8.65	23.5	43.6	Litchfield and Overbeck (1966)
	corn canning waste,					1.23	22.3	33.5	
	pumpkin canning waste					8.25	27.5	48.1	

(continued)

TABLE V (Continued)

Organism	Carbon and energy source	Scale, aeration, agitation	Temperature (°C)	pH	Specific growth rate of dilution rate (D)(hr^{-1})	Culture density (gm/liter)	Yield of Mycelium gm/100 gm substrate (dry wt. basis) Supplied	Yield of Mycelium gm/100 gm substrate (dry wt. basis) Used	References
Paecilomyces varioti (Pekilo)	Spent sulfite liquor	360-m³ fermentor	—	—	0.2(D)	17	—	55	Romantschuk (1975); Romantschuk and Lehtomaki (1978)
Trichoderma harzianum	Coffee waste	5000 (18,927-liter) 7500-gal (28,391-liters) tanks (continuous). 2 v/v/min	Ambient	3.5	0.19–0.10	—	—	31–47	Aquirre et al. (1976); Espinosa et al. (1975)
Trichoderma viride	Corn and pea waste	Aeration pool 10,000 gal (37,854 liters), continuous 11,000-gal ditch (41,640 liter)	Ambient	4.6	—	1.2	—	—	Church et al. (1973)
T. viride	Ball-milled cellulose	14-liter fermentor, 6.1 liters of medium, 0.3 v/v/min, 350 rpm (continuous)	30	5.0	0.033–0.08(D)	4.0	—	70	Peiterson (1977)
Tricholoma nudum	(a) Glucose	3-liter fermentor, 1 liter/min, 400 rpm	25	5.0	—	22.6	—	37.6	Reusser et al. (1958b)
	(b) Sulfite waste liquor	30-liter fermentor, 8.2 liter/min, 300 rpm	25	5.5	—	10.6	—	64.0	Reusser et al. (1958b)

corn steep liquor, which are rich in trace nutrients including vitamins (Litchfield et al., 1963a).

A pH range of 3.0–7.0 is suitable for most fungi. It is generally desirable to operate at a pH of 5 or below to minimize bacterial contamination problems, particularly in growing fungi on wastes under nonaseptic conditions. In growing mycelia of morel mushrooms (Morchella sp.) a pH of 6.0–6.5 is used to maximize growth rate, yield, and flavor. This requires that the culture be grown under aseptic conditions (Litchfield, 1964).

Most fungi grow best in the range 25–36°C. Growth rates decrease markedly above this range. Mushroom mycelia appear to be less tolerant of temperatures above 30°C than other fungi. Other fungi grow well at higher temperatures. For example, Aspergillus niger grown on carob bean extract grows well up to 36°C (Imrie and Vlitos, 1975).

Cooling must be provided to control the heat generated by aerobic growth of fungal cultures, and the same considerations noted previously for bacteria and yeasts apply here.

The oxygen requirements for growth of fungal cultures are quite complex. Fungal mycelium generally grows either in a filamentous or a pellet form in submerged culture. A recent review discussed the factors affecting the growth of molds in pellet form (Metz and Kossen, 1977).

The formation of pellets is largely determined by the extent of agitation. In general, pellet formation is favored by low agitation and aeration rates. For example, with Morchella spp. sparging to give aeration rates of 0.08–0.15 mmole oxygen/liter/minute gave highest yields in the pellet form. Above 0.15 mM oxygen/liter/minute growth was filamentous and yields decreased (Litchfield et al., 1963a). This decrease in yield with increased aeration is not always observed with other fungi and in many cases higher yields are obtained when growth is in the filamentous form.

In many cases, the mycelium will develop around the impeller in agitated vessels. In such instances, it is preferable to use fermentors in which the aeration system also provides agitation. In most reports of mass cultivation of fungal mycelium, aeration rates are given as air flow per volume of medium per minute, rather than as oxygen absorption rates, which makes it difficult to compare the conditions used in different studies.

Only a few reports of growth of fungi on noncarbohydrate substrates have appeared in the literature. A strain of Graphium grows on natural gas (methane and ethane), with ethane being the preferred substrate (Volesky and Zajic, 1971). Lentinus edodes, Pleurotus ostreatus, and Schizophyllum sp. grow on a variety of carbon sources, particularly C_{10}–C_{16} hydrocarbons, methanol, and ethanol (Sugimori et al., 1971).

Only selected fungal SCP processes listed in Table V will be discussed here. Of particular interest are processes for utilizing wastes, particularly those containing cellulose and those for producing edible fungi.

Edible mushroom mycelium is primarily of interest as a food flavoring rather than a primary source of protein. Laboratory and small pilot plant-scale studies have been conducted on species of *Agaricus, Boletus, Calvatia, Cantharellus, Collybia, Lentinus, Morchella, Tricholoma,* and *Xylaria* (Block, 1960; Litchfield, 1967a, 1968; Reusser *et al.*, 1958a,b; Sugihara and Humfeld, 1954).

Morchella mycelium has been produced on a commercial scale for sale as a food flavoring (Heinemann, 1963; Klis, 1963; Litchfield, 1964, 1967a,b,c). Typical growth conditions are shown in Table V for *Morchella crassipes, Morchella esculenta* and *Morchella hortensis.*

In a large pilot-plant run, *Morchella* sp. were grown in a 2000-gal (7570-liter) tank containing 1500 gal (5678 liters) of medium. Approximately 1.5–2 short tons of fresh mycelium were obtained containing 90% moisture or 24–30 gm (dry wt.)/liter (Robinson and Davidson, 1959).

Carob beans are widely grown in the Mediterranean region for edible gum production. The waste pods contain sugars which can be extracted and used as a substrate for growth of fungi. A process for utilizing the carbohydrates in carob bean extract using *Aspergillus niger* has been developed by Tate and Lyle, Limited in the United Kingdom. Cell densities of 31.5 gm/liter were obtained in 3000-liter capacity fermentors (Imrie and Vlitos, 1975). In other work in Greece, *Fusarium moniliforme* has been grown on carob bean extract in a semicontinuous laboratory fermentor system. A harvest of 8.8 gm/liter/day was obtained (Droulisos *et al.*, 1976).

Corn and pea canning wastes have been treated in aeration pools of 10,000 gal (37,854 liters), or in a 11,000-gal (41,639-liter) oxidation ditch using *Trichoderma viride* as the inoculum. However, a *Geotrichium* sp. that was naturally present tended to predominate over the season in corn wastes apparently because of an oversupply of nitrogen and phosphorus, while the pea waste was dominated by a *Fusarium* sp. Biochemical oxygen demand reductions of 96% and 95% were obtained for corn and pea wastes, respectively (Church *et al.*, 1973).

In other studies, *Gliocladium deliquescens* was used to treat corn wet milling wastes in a 50,000-gal aerated pilot unit and soy-processing waste in a laboratory fermentor scale study (Church *et al.*, 1972, 1973). A 87% reduction in chemical oxygen demand (COD) of the corn wet milling waste and greater than a 97% reduction in the BOD of soy whey waste were obtained.

Ranks, Hovis, McDougall, Limited in the United Kingdom have performed extensive studies on a wide range of fungi as a potential source of edible protein (Anderson et al., 1975). *Fusarium graminearum* was selected as the organism for process development (Solomons and Scammell, 1976). The strain used is claimed to be nontoxic (Towersey *et al.*, 1976, 1977). Pilot plant-scale studies are in progress but detailed operating data have not been released.

ICAITI in Guatemala has investigated growth of various fungi on coffee-processing wastes (deCabrera *et al.*, 1974; Rolz, 1975; Aguirre *et al.*, 1976). *Trichoderma harzianum* was used to treat wastes from coffee processing in El Salvador in a system consisting of a 2000-gal (7571-liter) equalization tank, two 5000-gal (18,927-liter) fermentors, and two 450-gal (1734-liter) seed fermentors. Aeration was supplied by sparging and the fermentation was run under nonaseptic conditions at pH 3.5. A 73% reduction of COD was observed in 24 hours with net gain of 3.2 gm/liter of solids. Approximately 56% of the dry solids was protein.

Large quantities of cellulose are present in agricultural wastes and in pulp and paper industry wastes. The conversion of cellulose in bagasse from sugar cane operations by bacteria and in animal feed lot wastes by thermoactinomycetes to SCP was covered previously.

Research at the U.S. Army Natick Laboratories has shown that *Trichoderma viride* has a complete cellulase system that catalyzes the conversion of cellulose to cellobiose, and cellobiose in turn to glucose (Mandels and Sternberg, 1976). Cellulose must be pretreated to achieve satisfactory yields of glucose by the *T. viride* cellulase system because its insolubility, crystallinity, and intimate association with lignin limit access of this enzyme to glycosidic bonds. Alkali treatment and/or ball milling to a fine particle size (70–100 μm) enhances enzyme action.

Alkali or photochemical treatment, electron irradiation, and hydrolysis were investigated for increasing the biodegradability of cellulose by various fungi, including *A. fumigatus, Penicillium* sp., *T. viride, Chaetomium* sp., and *Geotrichium candidum* (Rogers *et al.*, 1972). Best results were obtained by preheating the cellulose with dilute solutions of sodium nitrite followed by irradiation of the cellulose nitrite slurries with ultraviolet light at 3650 Å for 1–5 days.

Either *T. viride* alone or a mixed culture of *T. viride* and *C. utilis* converted barley straw to a protein product containing 21–26% protein (Peitersen, 1975a,b). The time required for maximum cellulase and cell production was less when the combined cultures were used. In subsequent studies *T. viride* was grown on ball-milled cellulose in a 14-liter continuous culture system; cellulose was consumed to the extent of 50–75% (Peitersen, 1977).

According to Eriksson and Larsson (1975) the rot fungus *Sporotrichum pulverulentum* utilized waste mechanical fibers from a newsprint mill only to a limited extent. The protein content of these fibers after treatment was only 14%.

Moo-Young et al. (1977) found that *Chaetomium cellulyticum* gave 50–100% greater growth rates and 80% more cellular protein than *T. viride* when it was grown on a medium containing 1% of a purified form of amorphous cellulose (Solka-floc) or partially delignified sawdust. However, *T. viride* produced greater quantities of free cellulases at higher rates and degraded greater amounts of both substrates.

Fungal degradation of cellulose and concomitant SCP production have not yet been practiced successfully on a large pilot-plant or commercial scale. Pretreatment of cellulose by ball milling or other physical or chemical means to make it amenable to utilization by fungi appears to be excessively costly at the present time.

The Pekilo process developed by the Finnish Pulp and Paper Research Institute utilizes *Paecilomyces varioti* for converting sulfite waste liquor to SCP. A 10,000-metric ton capacity plant has been constructed (Romantschuk, 1975; Romantschuk and Lehtomäki, 1978). A 360-m^3 continuous fermentor system is used, having a retention time of 4.5–5 hours. The mycelial concentration in fermentors reaches 17 gm (dry wt.)/liter. Productivity is in the range 2.7–2.8 kg/m^3/hr, and daily production ranges from 15 to 16.5 metric tons/24hr. The process is designed to be installed between the cooking operation and the evaporation plant of the pulp mill to reduce pollutional loads.

2. Yield

Table V presents yield data on selected molds and higher fungi grown on various substrates. Yields of mycelium vary widely, depending upon organism and substrate. As would be expected, the highest yields were obtained in laboratory-scale studies. Lower conversions of substrate to cell mass are often obtained with many of the fungi listed than with bacteria or yeasts.

A yield of 0.45 gm (dry wt.)/gm substrate for *Aspergillus niger* grown on carob bean extract is typical for molds and yields of 0.32–0.50 gm (dry wt.)/gm substrate is typical for mushroom mycelium such as the *Morchella* sp.

Factors affecting growth rate and yield productivity of molds and higher fungi are similar to those of bacteria and yeasts. The problem of oxygen transfer to filamentous or pellet mycelium, discussed previously, is a major limiting factor in improving yields of fungi in commercial-scale production facilities.

3. Cell Recovery

Molds and higher fungi are more easily recovered from the growth medium because of the filamentous or pellet form of the mycelium which permits the use of screens or filter presses. Basket centrifuges, rotary vacuum filters, and inclined screens have been evaluated for their utility in recovery of mycelium of *Morchella* sp.

In one study a Tolhurst basket centrifuge operated at 1179 rpm for 10 minutes yielded a 2-inch-thick *Morchella* mycelial cake containing 74% moisture. The pellets are ruptured, but this is not a problem when the mycelium is to be dried for food use into powdered form. Dewatering with a Byrd-Young vacuum filter was unsuccessful because the pellets did not cling to the filter and the spaces between the pellets were large enough to make vacuum operation ineffective (Robinson and Davidson, 1959). Inclined screens effectively separate mycelial pellets from the spent medium and aid in washing the mycelium.

In studies with corn and pea canning wastes, the mycelium was collected on vibratory 120-mesh screen. A rotary vacuum filter was used successfully in collecting *G. deliquescens* mycelium after growth on corn wet milling wastes with solid contents ranging up to 22% (Church *et al.*, 1973).

In the Pekilo process for producing *Paecilomyces* sp. on sulfite waste liquor, the mycelium is collected and washed on a drum filter and the spent liquor passes to the evaporator at the pulp mill. The mycelium is mechanically dewatered to 35–45% dry weight content (Romantschuk, 1975).

Aspergillus niger grown on carob bean extract is in a filamentous rather than a pellet form. Rotary vacuum filters can be used to separate the mycelium from the medium without a precoat. In contrast to *Morchella* mycelium, the porosity of the mycelium permits washing in the filter without resuspending it, and the cake can be scraped from the rotary drum and disrupted mechanically (Imrie and Vlitos, 1975).

Fungal mycelium can be dried in steam-tube rotary, tray, or belt driers. Mushroom mycelium for food use must be washed thoroughly to remove excess growth medium which may contribute undesirable flavor to the dried product. Drying temperatures must be controlled carefully to avoid damaging heat-sensitive flavor components and darkening the product excessively.

III. PRODUCT QUALITY AND SAFETY

Three aspects of SCP product quality will be considered here: (1) nutritional value of SCP products for use in human food or animal feeds,

(2) safety of SCP products for human or animal feeding, and (3) production of functional protein concentrates and isolates free from nucleic acids and toxic factors. Although the primary use of SCP products is for their protein content, cells of microorganisms also contain carbohydrates, lipids, vitamins, and minerals that contribute to nutritional value. On the other hand, microbial cells also contain nonprotein nitrogen materials, such as nucleic acids, that have harmful effects when consumed by humans or nonruminant animals in quantities above certain limits. This problem is discussed subsequently.

A. Nutritional Value

There is considerable information available on the composition of microbial cells, including protein, amino acid, vitamin, and mineral contents (Waslien, 1975). Table VI summarizes proximate analysis of selected microorganisms of interest in SCP production. Crude protein values are based on nitrogen values multiplied by the factor 6.25. Protein content computed in this manner is used for formulating protein products into feeds. However, it should be emphasized that these values do not reflect the true protein content of microorganisms since nitrogen or nonprotein nitrogenous substances, including nucleic acids, is erroneously included as protein.

Protein and lipid contents of microorganisms reflect the composition of the medium and growth conditions. Yeasts, molds, and higher fungi have higher cellular lipid contents and lower nitrogen (and protein contents) when grown in media having high amounts of available carbon as energy source and a deficiency of nitrogen. The importance of C : N ratio in the medium for producing various microorganisms has been discussed previously.

Ash contents of SCP products also vary depending upon the mineral content of the growth medium. Also, procedures used in product separation, including washing, can result in large variations in ash contents of a given organism. Table VII presents amino acid contents of selected microorganisms of interest in SCP production as compared with the Food and Agricultural Organization (FAO) reference values. In general, only a few microorganisms, particularly the bacteria, have amino acid profiles that compare favorably with the FAO values as regards methionine content. With yeasts, in addition to methionine deficiencies, lysine/arginine ratios are wider than desirable and must be adjusted in poultry rations (Shacklady and Gatumel, 1973). Also, in algae, glycine and methionine deficiencies have been observed in chick feeding studies (Leveille et al., 1962).

However, actual performance of SCP products as determined in feeding studies is the most important measure of nutritional value. For human food applications, protein digestibility and protein efficiency ratio (PER), biological value (BV), or net protein utilization (NPU) determined in rats are used as measures of the utility of an SCP product. For animal feed applications metabolizable energy, protein digestibility, and feed conversion ratio (weight of ration consumer/weight gain) are usual measures of performance in broiler chickens, swine, and calves (Shacklady and Gatumel, 1973; Smith and Palmer, 1976), while egg production is used for laying hens (Shannon et al., 1976; Waldroup and Hazen, 1975).

Table VIII shows typical values obtained from animal feeding studies of SCP products. Supplementation with methionine is necessary to obtain a PER or BV equivalent to animal proteins, such as casein, for most microorganisms. (Shacklady and Gatumel, 1973; Waldroup and Flynn, 1975). This amino acid, or its hydroxy analog, is available commercially at low enough cost to permit supplementation of feedstuffs.

The results of feeding studies with broiler chickens, using *Methylophilus methylotrophus* (Gow et al., 1975), *Pseudomonas* sp. grown on methanol (Waldroup and Payne, 1974), or *Aspergillus niger* grown on carob bean extract (Imrie and Vlitos, 1975), do not show significant differences between experimental and control diets at the 10% level of SCP in the diet. Similar results were also observed with broiler chickens in feeding studies with the yeasts *C. lipolytica* (Shacklady and Gatumel, 1973) and *C. tropicalis* (Waldroup et al., 1971) up to the 15% level in the diet. In contrast, an SCP product grown on methanol gave greater and more efficient gains in broiler chickens at the 15% level in pelleted diets as compared with soybean meal (White and Balloun, 1977). In swine feeding studies with yeast grown on *n*-alkanes feed conversion ratios tend to decrease as percentages of SCP in the diet increase above about 10% (Litchfield, 1977b). For bacteria grown on methanol, efficiencies of feed conversion were equivalent to those of whitefish meal at levels up to 20% of the diet of pigs (Whittemore et al., 1976). When yeast SCP grown on hydrocarbons is used as a milk replacer for calves and steers, the limit appears to be approximately 7.5% in the ration (Shacklady and Gatumel, 1973).

The older literature on animal feeding studies with molds and higher fungi has been summarized previously (Litchfield, 1968). More recent studies, summarized in Table VIII, confirm earlier observations that supplementation with DL-methionine is necessary to obtain satisfactory performance with molds and higher fungi. In the United States, the Food and Drug Administration regulations allow the use of the dried yeasts *S. cerevisiae, K. (Saccharomyces) fragilis,* and *C. utilis* in foods provided

TABLE VI. Proximate Composition of Selected Microorganisms of Interest in SCP Production

Organism	Substrate	Composition (gm/100 gm dry wt)							References
		Nitrogen	Protein	Fat	Total CHO	Crude fiber	Ash	Energy (kcal/gm)	
Algae									
Chlorella sorokiniana	CO_2	9.6	60	8	22	3	9	5.2	Lubitz (1963)
Chorella regularis S-50	CO_2	9.3	58	16	—	4.4	6.7	—	Endo and Shirota (1972) Clement et al. (1967)
Spirulina maxima, synthetic medium	CO_2	10	62	3	—	—	—	—	
Spirulina maxima, sewage	CO_2	8.5	53	4.8	28	—	15	—	Kosaric et al. (1974); Nguyen et al. (1974)
Bacteria and actinomycetes									
Acinetobacter (Micrococcus) cerificans	Hexadecane	11	72	—	—	—	—	—	Guenther and Perkins (1968)
Cellulomonas sp.	Bagasse	14	87	8	—	—	7	—	Han et al. (1971)
Methalomonas clara	Methanol	12–13	80–85	8–10	—	—	8–12	—	Faust et al. (1977)
Methylophilus (Pseudomonas) methylotrophus	Methanol	13	83	7	—	<0.05	8.6	3.0	Gow et al. (1975)
Thermomonospora fusca	Pulping fines	4.8–5.6	30–35	—	—	—	—	—	Crawford et al. (1973)
Yeasts									
Candida lipolytica (Toprina)	n-Alkanes	10	65	8.1	—	—	6	—	Shacklady (1972)
Candida lipolytica (Toprina)	Gas oil	11	69	1.5	—	—	8	—	Shacklady (1972)
Candida utilis	Ethanol	8.3	52	7	—	5	8	—	Amoco Foods Co. (1974)
Candida utilis	Sulfite	9	55	5	—	—	8	—	Peppler (1968)
Hansenula polymorpha	Methanol	—	50	—	—	—	—	—	Cooney and Levine (1975)
Kluyveromyces (Saccharomyces) fragilis	Cheese whey	9	54	1	—	—	9	—	Peppler (1968)
Saccharomyces cerevisiae	Molasses	8.4	53	6.3	—	—	7.3	—	Peppler (1968)
Trichosporon cutaneum	Oxanone wastes	8.6	54	8	31	—	7	11	Wiken (1972)

Molds and higher fungi									
Agaricus campestris, white variety	Glucose	—	36	3	49	6.9	4.5	—	Humfeld and Sugihara (1949)
Agaricus campestris, brown variety	Glucose	—	45	—	—	—	5.2	—	Humfeld and Sugihara (1949)
Aspergillus niger	Molasses	7.7	50	—	—	—	—	—	Anderson et al. (1975)
Fusarium graminearum	Starch	8.7	54	—	—	—	—	—	Anderson et al. (1975)
Morchella crassipes	Glucose	5.0	31	3.1	—	—	—	—	Litchfield et al. (1963b)
Morchella crassipes	Sulfite waste liquor	4.1	26	4.4	39	—	5.9	—	Kosaric et al. (1973)
Morchella esculenta	Glucose	5.0	31	1.9	—	—	—	—	Litchfield et al. (1963b)
Morchella hortensis	Glucose	5.4	34	1.4	—	—	—	—	Litchfield et al. (1963b)
Paecilomyces varioti (Pekilo)	Sulfite waste liquor	8.8	55	1.3	25	7	6	—	Romantschuk and Lehtomaki (1978)
Trichoderma viride	Starch	10.2	64	—	—	—	—	—	Anderson et al. (1975)

TABLE VII. Amino Acid Content of Selected Organisms of Interest for SCP Production[a, b]

Organism	Substrate	Ala	Arg	Asp	Cys	Glu	Gly	His	Ile	Leu	Lys	Met	Phe	Pro	Ser	Thr	Try	Tyr	Val	References
Algae																				
Chlorella sorokiniana	CO_2	5.9	5.6	5.9	—	9.3	4.8	1.4	3.4	4.0	7.8	1.8	2.7	4.0	2.2	3.2	1.4	2.7	5.1	Lubitz (1963)
Chlorella regularis S-50 Autotrophic	CO_2	7.3	5.8	8.8	0.7	11.8	5.4	1.8	4.2	8.1	7.7	1.3	5.1	4.3	3.0	3.6	1.5	2.6	5.9	Endo and Shirota (1972)
S-50 heterotrophic	Glucose	7.4	10.2	7.7	0.9	9.9	4.9	3.0	3.4	7.0	9.4	1.8	3.2	2.2	3.4	4.0	1.4	3.0	5.1	Endo and Shirota (1972)
Spirulina maxima	CO_2	6.5	6.6	8.6	0.4	13.6	4.6	1.7	5.8	7.8	4.8	1.5	4.6	3.8	4.3	4.6	1.3	3.9	6.3	Clement et al. (1967)
Bacteria and Actinomycetes																				
Acinetobacter (Micrococcus) cerificans	*n*-Alkanes	6.7	4.7	8.0	—	10.7	5.0	1.7	4.3	6.5	5.2	1.8	3.6	3.1	2.8	4.1	2.8	1.3	5.4	Ertola et al. (1971)
Cellulomonas–Alcaligenes	Bagasse	—	6.5	—	—	—	—	7.8	5.4	7.4	7.6	2.0	4.7	—	—	5.5	—	—	7.1	Han et al. (1971)
Methylococcus capsulatus	Methane	—	6.2	—	0.6	—	5.1	2.2	4.3	8.1	5.7	2.7	4.6	—	—	4.6	—	3.8	6.5	Hamer et al. (1975)
Methylophilus (Pseudomonas) methylotrophus	Methanol	6.8	4.5	8.5	0.6	9.6	4.9	1.8	4.3	6.8	5.9	2.4	3.4	3.0	3.4	4.6	0.9	3.1	5.2	Gow et al. (1975)
Thermomonospora fusca	Cellulose pulping fines	13.9	5.6	6.7	0.4	18.0	4.4	2.0	3.2	6.1	3.6	2.0	2.6	6.1	2.6	4.0	—	1.9	13.0	Crawford et al. (1973)
Yeasts																				
Candida lipolytica (Toprina)	*n*-Alkanes	7.4	4.8	10.2	1.1	11.3	4.8	2.0	4.5	7.0	7.0	1.8	4.4	4.4	4.8	4.9	1.4	3.5	5.4	Evans (1968)
Candida lipolytica	Gas oil	5.8	5.0	10.0	0.9	12.1	4.5	2.1	5.3	7.8	7.8	1.6	4.8	3.7	5.1	5.4	1.3	4.0	5.8	Evans (1968)
Candida utilis	Sulfite waste	5.8	5.4	9.2	—	15.6	3.6	1.2	3.8	7.6	4.8	1.1	8.6	6.0	5.0	5.4	2.4	6.2	3.8	Peppler (1965)
Candida utilis	Ethanol	5.5	5.4	8.8	0.4	14.6	4.5	2.1	4.5	7.1	6.6	1.4	4.1	3.4	4.7	5.5	1.2	3.3	5.7	Amoco Foods Co. (1974)
Hansenula polymorpha ATCC 26012	Methanol	6.1	5.6	10.5	0.7	14.4	5.2	2.4	5.1	8.3	8.1	1.5	5.0	—	—	5.2	—	4.8	6.2	Levin and Cooney (1973)
Kluyveromyces (Saccharomyces) fragilis	Cheese whey	—	—	—	—	—	—	2.1	4.0	6.1	6.9	1.9	2.8	—	—	5.8	1.4	2.4	5.4	Bernstein and Everson (1973)

Organism	Substrate	Ala	Arg	Asp	Cys	Glu	Gly	His	Ile	Leu	Lys	Met	Phe	Pro	Ser	Thr	Try	Tyr	Val	Reference
Rhodotorula glutinis	Domestic sewage	5.1	4.5	6.0	—	9.3	3.5	2.2	3.3	5.4	7.1	—	3.2	3.0	3.3	3.1	—	3.1	3.6	Fustier and Simard (1976)
Saccharomyces cerevisiae	Molasses	—	5.0	—	1.6	—	—	4.0	5.5	7.9	8.2	2.5	4.5	—	—	4.8	1.2	5.0	5.5	Reed and Peppler (1973)
Fungi																				
Aspergillus niger	Carob bean extract	—	—	—	1.1	—	—	—	4.2	5.7	5.9	2.6	3.8	—	—	5.0	2.1	3.2	5.2	Imrie and Vlitos (1975)
Fusarium graminearum	Starch	6.6	6.8	10.4	1.2	11.0	6.5	3.3	4.0	6.0	6.9	1.8	4.0	4.8	5.3	4.6	—	3.8	6.6	Anderson et al. (1975)
Gliocladium deliquescens	Waste starch	6.1	4.8	7.9	1.0	8.6	4.7	2.3	3.8	6.2	5.3	1.3	3.0	4.2	3.9	4.3	1.4	3.6	4.7	Smith et al. (1975)
Morchella crassipes	Glucose	6.2	3.0	4.8	0.4	14.1	3.1	2.0	2.9	5.6	3.5	1.0	1.9	5.1	3.1	3.0	1.5	1.7	3.0	Litchfield et al. (1963b)
	Sulfite waste liquor (NH₃)	5.3	3.9	5.9	0.6	8.3	3.8	2.0	2.9	6.1	5.6	1.2	5.8	4.0	4.0	4.1	—	—	4.1	LeDuy et al. (1974)
Morchella esculenta	Glucose	4.8	8.0	5.0	0.3	14.8	2.9	2.1	2.7	5.1	3.8	0.9	2.5	4.2	3.1	3.0	0.9	1.7	3.4	Litchfield et al. (1963b)
Morchella hortensis	Glucose	4.5	4.0	4.6	0.4	15.4	3.0	1.9	2.4	5.0	3.0	0.7	2.3	4.5	2.8	2.7	1.0	1.9	2.9	Litchfield et al. (1963b)
Paecilomyces varioti	Spent sulfite liquor	—	—	—	1.1	—	—	—	4.3	6.9	6.4	1.5	3.7	—	—	4.6	1.2	3.4	5.1	Romantschuk (1975)
Trichoderma harzianum	Coffee-processing wastes	—	—	—	—	—	—	—	4.0	5.8	4.7	1.1	3.5	—	—	3.8	—	—	4.9	Aguirre et al. (1976)
Trichoderma viride	Barley straw	4.8	4.0	7.8	1.5	8.0	4.5	1.9	3.5	5.8	4.4	1.4	8.7	3.8	4.2	4.9	—	3.3	4.4	Peitersen (1975a)
Trichosporon cutaneum	Oxanone wastewater	—	—	—	1.0	—	—	—	5.2	7.1	7.3	1.6	4.5	—	—	5.5	1.4	4.2	6.1	Wiken (1972)
FAO reference		—	—	—	—	—	—	—	4.2	4.8	4.2	2.2	2.8	—	—	—	1.4	—	4.2	Food and Agricultural Organization (1957)

a Amino acid content, gm/16 gm N

b Key to amino acid abbreviations:

Ala, alanine; Arg, arginine; Asp, aspartic acid; Cys, cystine; Glu, glutamic acid; Gly, glycine; His, histidine; Ile, isoleucine; Leu, leucine; Lys, lysine; Met, methionine; Phe, phenylalanine; Pro, proline; Ser, serine; Thr, threonine; Try, tryptophan; Tyr, tyrosine; Val, valine.

TABLE VIII. Performance of Selected Single-Cell Protein Products in Animal Feeding Studies

Single-cell protein and substrate	Treatment and percent in diet	Animal	Performance			Feed conversion ratio (kg/kg weight gain)	References
			Protein digesti-bility (%)	Protein efficiency ratio (PER)	Biological value (BV)		
Algae							
Chlorella sorokiniana	(1) Dried, 10	Rat	86	2.19	—	—	Lubitz (1963)
	(2) Dried + 0.2 L-methionine; 10	Rat	86	2.90	—	—	—
Spirulina maxima	Dried, 10	Rat	84	2.3–2.6	72	—	Bourges *et al.* (1972); Durand-Chastel and Clement (1972)
Bacteria							
Acinetobacter (Micrococcus) cerificans (n-hexadecane)	Dried	Rat	83.4	—	67	—	Ertola *et al.* (1969)
Bacillus megaterium (collagen waste)	Dried	Rat	—	1.88[a]	—	—	Bough *et al.* (1972)
Methylophilus methylotrophus (methanol)	(1) Dried; 9.8	Broiler chicken	—	—	—	2.30 (2.33)[b]	Gow *et al.* (1975)
	(2) Dried; 6.7	Pig	—	—	—	3.13 (3.34)[b]	
Pseudomonas sp. (methanol)	Dried; 10	Broiler chicken	—	—	—	1.74 (1.64)[b]	Waldroup and Payne (1974)

Yeasts							
Candida lipolytica (n-alkanes)	(1) Dried	Rat	96	—	61	—	Shacklady and Gatumel (1973)
	(2) Dried + 0.3 DL-methionine	Rat	96	—	91	—	
	(3) Dried; 10	Broiler chicken	88	—	—	2.58 (2.68)[a]	
	(4) Dried; 7.5	Pig	92	—	—	3.04 (3.11)	
C. lipolytica (gas and oil)	(1) Dried	Rat	94	—	54	—	Shacklady and Gatumel (1973)
	(2) Dried + 0.3 DL-methionine	Rat	95	—	96	—	
Candida tropicalis (n-alkanes)	Dried; 15	Broiler chicken	—	—	—	2.03 (2.06)[b]	Waldroup et al. (1971)
Candida utilis (sulfite waste liquor)	(1) Dried	Rat	85–88	0.9–1.4	32–48	—	Bressani (1968)
	(2) Dried + 0.5 DL-methionine	Rat	90	2.0–2.3	88	—	
Kluyveromyers fragilis (cheese whey)	Dried	Rat	—	—	—	1.5	Booth et al. (1962)
Saccharomyces cerevisiae (molasses)	Dried	Rat	80–90	1.7	58–69	—	Bressani (1968)

(continued)

TABLE VII (*Continued*)

Single-cell protein and substrate	Treatment and percent in diet	Animal	Protein digesti-bility (%)	Performance: Protein efficiency ratio (PER)	Performance: Biological value (BV)	Feed conversion ratio (kg/kg weight gain)	References
Fungi							
Aspergillus niger (carob bean extract)	Dried; 20	Broiler chicken	— —	2.50 —	— —	2.02 (2.12)[a]	Imrie and Vlitos (1975)
Fusarium graminearum	(1) Dried; 10.5 (2) Dried; 40	Rat Pig	— —	1.89 —	— —	— 1.76 (2.04)[a]	Duthie (1975)
Fusarium moniliforme (carob bean extract)	(1) Dried (2) Dried + 6.3 DL-methionine	Rat Rat	— — —	1.15 2.38 —	— — —	— — —	Drouliscos et al. (1976)
Gliocladium deliquescens (waste starch)	Dried —	Rat —	— —	— —	49 —	— —	Smith et al. (1975)
Trichoderma viride (waste starch)	Dried —	Rat —	— —	— —	48 —	— —	Smith et al. (1975)

[a] Adjusted to casein PER of 2.5.
[b] Control ration value.

that the folic acid content of the yeast does not exceed 0.04 mg/gm (0.008 mg pteroyl glutamic acid/gm) (Code of Federal Regulations, 1978). Use of dried yeast in foods has been limited to flavor enhancement and B vitamin supplementation. High levels of *C. utilis* in foods have led to poor acceptance (Klapka *et al.*, 1958).

Bressani (1968) discussed earlier studies on supplementation of foods with yeasts and on their value in human nutrition. More recently Young and Scrimshaw (1975) point out that yeast and algae have low protein digestibility values that can be improved considerably if these SCP products are processed suitably. However, nucleic contents are a limiting factor in the extent to which SCP products can be used in human foods. The Protein Advisory Group (PAG) of the United Nations (1970a,b,c, 1971) has developed guidelines for preclinical testing of novel protein sources and supplementary foods which should be followed in any development program on SCP products for human food use.

B. Safety

There are several key problem areas in using SCP products for human foods. These include the following: (1) high nucleic acid contents of many microorganisms that would result in kidney stone formation or gout if consumed as a significant portion of the protein intake; (2) poor digestibility and possible adverse gastrointestinal and skin reactions; and (3) possible presence of toxic or carcinogenic compounds from residues of substrates, from biosynthesis by the organism, or from chemical reactions during processing and drying.

Calloway (1974) pointed out that adverse responses may be observed in human subjects fed algae, bacteria, yeast, or fungal SCP products even though favorable results were obtained previously in animal studies. Published values of nucleic acid contents on a dry weight basis of algae are in the range of 4–6% (Waslien *et al.*, 1970; Durand-Chastel and Clement, 1972), bacteria, up to 16% (Senez, 1973; Gow *et al.*, 1975); yeasts, 6–11% (Trevelyan, 1975; Waslien *et al.*, 1970); and molds, 2.5–6% (Smith *et al.*, 1975; Trevelyan, 1975). Scrimshaw (1975) states that a human intake of 2 gm of yeast nucleic acid per day would be within safe limits. However, there would be a significant risk of kidney stone formation or gout at intakes greater than 3 gm/day.

Gastrointestinal disturbances, including nausea and vomiting, have been encountered in human feeding studies with algae (Powell *et al.*, 1961; Dam *et al.*, 1965; Lee *et al.*, 1967); with bacteria, including *A. eutrophus (H. eutropha)* and *Klebsiella (Aerobacter aerogenes)* (Waslien *et al.*, 1970); and with yeast grown on ethanol (Scrimshaw, 1973). Also, skin lesions were observed in human volunteers who consumed *C. utilis* grown on sulfite waste but not with glucose-grown cells (Scrimshaw, 1975). In bacteria the toxic materials were either intracellular or bound to the cell, since washed cells were used in feeding studies (Calloway, 1974).

It has been mentioned previously that strains of some molds, particularly *S. fumigatus* and *F. graminearum*, may produce substances that are toxic to humans and animals.

From the standpoint of animal feeding, high nucleic acid contents in SCP are not a problem with ruminants and are better tolerated by nonruminant livestock than by humans. De Groot (1974) discusses minimal evaluations for establishing the safety of SCP. Taylor *et al.* (1974) present criteria for the evaluation of SCP for feeding nonruminants.

Questions have arisen on possible hazards from the presence of traces of residual *n*-alkanes, odd-carbon fatty acids, or polycyclic aromatic hydrocarbons in yeast SCP grown on hydrocarbon substrates. The results of extensive studies have shown that the polycyclic hydrocarbon contents of *Candida* sp. grown on hydrocarbons were no greater than those in other commercial yeast products or other representative foods. Also no carcinogenic, mutagenic, or teratogenic effects were noted in appropriately designed studies (Engel, 1973).

The PAG Ad Hoc Working Group on Single-Cell Proteins concluded that low levels of residual *n*-alkanes and presence of odd-carbon fatty acids do not constitute a hazard (Protein Advisory Group, 1976a,b). The PAG has developed guidelines for production of SCP for human consumption and on the nutritional and safety aspects of novel proteins for animal feeding (Protein Advisory Group, 1972, 1974).

C. Functional SCP Products

As a result of high nucleic acid contents and possible presence of toxic factors in microbial cells, there is an alternative of processing the cells to remove cell walls, toxic factors, and nucleic acids. To be useful as functional food additives, the resulting SCP concentrates or isolates should have desirable physical–chemical functional properties, including water and fat binding, emulsion stability, dispersibility, gel formation, whippability, and thickening.

Both chemical and enzymatic processes can be used to reduce nucleic acid contents of SCP to acceptable levels of 2–3% or less. Chemical processes include acid precipitation at pH 5 (Hedenskog and Mogren, 1973) and acid or alkaline hydrolysis (Daly and Ruiz, 1975; Newell *et al.*, 1975). Enzyme processes are based on heat shock and incubation or on incubation alone to degrade nucleic acids by endonucleases within the cells (Sinskey and Tannenbaum, 1975; Robbins, 1976) or by autolysis (Trevelyan, 1976).

Methods for disrupting microbial cells include mechanical means, such as ball mills or homogenization at elevated pressures, autolysis, or acid, alkaline, or enzyme hydrolysis (Litchfield, 1977a). Concentrates and isolates can be prepared by precipitation from acid or alkaline solutions (Kalina and Nicholas, 1974; Mogren et al., 1974; Newell et al., 1975; Robbins, 1976). A preliminary attempt to improve the functional characteristics of C. utilis protein concentrates by succinylation was not successful (McElwain et al., 1975).

Both bacterial protein and C. utilis protein can be spun into fibers (Heden et al., 1971; Huang and Rha, 1971). Fibers can be formed by extrusion of the disrupted cell extract into an acid coagulating medium such as acetate buffer, pH 4.5 (Akin, 1973, 1974; Ridgeway, 1974; Litchfield, 1977a). These fibers would then be used for forming textured protein-type products.

Bakers' yeast protein concentrate is now approved by the Food and Drug Administration for use as a functional food additive (Code of Federal Regulations, 1978). In addition, it is reasonable to expect that other yeast concentrates or isolates will be approved in the future when their safety and utility are demonstrated to the satisfaction of regulatory authorities. Extensive evaluations will be required to establish the safety of concentrates or isolates from algal, bacterial, or fungal cells where there is no past history of safe use for human consumption.

IV. ECONOMIC AND ENERGY CONSIDERATIONS

Economic and energy considerations in SCP production are intimately linked together. Single-cell protein production processes are capital and energy intensive. Marketing considerations are also important aspects of the overall economics of SCP production.

A. Manufacturing Costs

Manufacturing costs include costs of the raw materials (carbon and energy source, nitrogen source, mineral nutrients, and other chemicals) materials and supplies, utilities, labor including fringe benefits, maintenance, taxes and insurance, working capital and interest, depreciation, and general administration. Several cost comparisons of various SCP processes have been published (Litchfield, 1977b,c; Moo-Young, 1977).

Manufacturing costs of SCP are very dependent upon the cost of the carbon and energy source to the producer of SCP products. Various estimates for the costs of hydrocarbon substrates, as a percentage of total manufacturing cost, range from 13 to 57.5% depending upon the process and the year of the estimate (Litchfield, 1977b). SCP production costs from methanol show a much greater dependency upon increases in substrate costs than in the case of n-paraffins (Dimmling and Seipenbusch, 1978).

Rising costs of petroleum and products derived from this raw material have made the economics of SCP processes based on hydrocarbons tenuous. This has already been reflected in a lack of announcements of new production facilities, suspension of operations on existing facilities, and abandonment of announced plans for new facilities (Anonymous, 1978a,b).

For waste-type materials, raw materials costs are a lower percentage of total costs than for hydrocarbons, methanol, or ethanol. Moo-Young (1977) cites a range of 17–26% of total costs for sulfite waste liquor and bagasse.

Waste materials must be generated in large quantities in close proximity to the SCP production plant over most of the year to be economically attractive. Otherwise, the costs associated with collecting and transporting wastes may be excessive (Burch et al., 1963).

Next to materials costs, utilities are the second most important element of production costs for most SCP processes. Utilities required include water, electricity, and steam. Ratledge (1975) estimated the following water requirements in millions of liters for 100,000 metric tons/year for plants of different processes: yeasts, n-alkanes, 18.2; bacteria, methanol, 45.5; and bacteria, methane, 18.2. Energy in the form of electricity and steam is used for sterilization, aeration, cooling, product separation, and drying. For example, electric power requirements for producing yeasts from n-alkanes on a 100,000-metric ton/year scale range from 500 to 3620 kWh/metric ton of SCP; and steam, from 4600 to 10,600 kg/metric ton of SCP (Litchfield, 1977b). Energy requirements for cooling are considerable for hydrocarbon- and methanol-based processes. British Petroleum estimates a cooling requirement of 100×10^6 kcal/hour for producing yeast from purified n-alkanes (BP Proteins, Ltd., 1974). The requirements for carbohydrate-based processes would be significantly lower as discussed previously in the sections on various microorganisms used in SCP production.

It is difficult to compare manufacturing cost estimates given in the literature for various processes because differing cost bases are used for raw materials, utilities, labor, maintenance, and depreciation. In some instances, cost estimates do not include working capital requirements or realistic interest charges.

Location is an all-important factor in determining manufacturing costs for SCP products. Raw materials availability and costs, labor, and utilities costs vary widely in different geographical regions. Also proximity to markets is an important factor in plant location.

Lewis (1976) considered both indirect and direct energy inputs for SCP processes, including energy used for carbohydrate substrate cultivation and transportation, for production and construction of capital equipment, and for production and processing of SCP and treatment of wastes. Typical gross energy requirements in megajoules (MJ) per kilogram of protein for selected organisms and substrates were: (1) *Candida* sp.; methanol, ethanol, or gas oil, 190; *n*-paraffins, 185; distilling industry wastes, 155; molasses, 75; (2) Pseudomonads; methane, 165; methanol or ethanol, 145; and (3) *Aspergillus niger*: solid agricultural waste, 80; agricultural process effluent, 30.

He concluded that processes based on hydrocarbons compared very unfavorably to those based on carbohydrates. The energy requirements for SCP production compared with those for equivalent protein production by conventional agriculture indicate limited applicability of SCP processes to less developed countries.

B. Capital Costs

Costs for constructing SCP production facilities that require aseptic operation, complex product separation, and cleaning, as is the case of hydrocarbons and methanol, have increased markedly. Even SCP plants based on carbohydrates now require clean operations to meet the requirements of regulatory agencies for food- or feed-grade materials; and costs for stainless steel equipment, suitable centrifuges, etc., to meet these requirements are also increasing.

Estimates of fixed capital costs in 1975 for 100,000-metric tons/year SCP plants (U.S. Gulf Coast) were $90.5 million and $66.0 million for processes based on *n*-alkanes and methanol, respectively (Brownstein and Constantinides, 1975). For a 50,000–75,000 metric tons/year SCP plant based on methanol, ICI estimated a 1976 cost of $70 million (Anonymous, 1976b). Future plants in advanced countries will be even more costly because of rapid increases in construction costs.

C. Market Considerations

At the present time, food-grade yeasts *(S. cerevisiae, C. utilis,* and *K. fragilis)* have a position in food ingredient markets for their flavor enhancement characteristics in foods. Soybean concentrates (70–72% protein) and isolates (90–95% protein) are the major competitive products to food-grade SCP products for either functional or nutritional uses at the present. Most SCP products and soybean products have lower methionine contents than are desirable in human nutrition and would require supplementation to make a balanced source of protein for human nutrition.

The SCP isolates and concentrates developed to date do not have the wide range of functional properties exhibited by soy protein concentrates and isolates. However, some yeast SCP products, particularly *C. utilis,* do have interesting functional effects in baked and pasta products. Any new SCP products will have to compete on cost and functional equivalent bases with soy protein products and perhaps additional protein concentrates or isolates from plant protein sources, such as cottonseed and peanuts, or animal proteins such as fish, milk, or meat.

Food industry and consumer acceptance of SCP products in foods will be an important factor in their future success. Stone and Sidel (1974) describe a consumer survey on SCP acceptance.

Algae have a bitter lingering flavor and impart a dark undesirable color to foods (Cook *et al.*, 1963). Further processing, such as solvent extraction, or formulating into other foods may minimize flavor problems, but the color is difficult to mask. Also, further processing is necessary to improve digestibility of algae.

There is a lack of published information on human sensory studies with bacterial SCP, although bland colorless powders can be made. The primary focus of bacterial SCP has been toward feedstuff applications.

The reports of adverse gastrointestinal reactions in human feeding studies with bacteria cited previously will limit further development of food-grade bacterial SCP products. A process for removing endotoxins from bacterial SCP has been described (Dasinger and Nasland, 1972) but its effectiveness has not been verified in published toxicological studies.

The flavor characteristics of food-grade yeasts are well known. Dried food yeasts and yeast autolysates have been used for many years (Reed and Peppler, 1973).

Dried mycelia of certain higher fungi, the mushrooms, have pleasant flavors. These products can be used in soups, sauces, or gravy formulations as replacements for dried cultured mushrooms. The high costs of producing dried morel mushroom mycelium made this product uncompetitive with imported dried mushrooms in the United States.

Similar considerations apply to SCP products in feedstuff markets. Soybean meal and fish meal are the major competitive feedstuffs. It has been pointed out previously that wide fluctuations in the prices of these feedstuffs may make feed-grade SCP products noncompetitive for substantial periods in Western Europe and Japan because of the high fixed capital and operating costs of SCP plants (Lipinsky and Litchfield, 1974; Litchfield, 1977c).

The producer of formulated feedstuffs uses price, acceptability, and nutritional performance as factors in deciding which protein ingredients to use. Least cost linear programming is now used in the feedstuff industry to determine the lowest cost protein that will give the desired performance. Feed-grade SCP products will have to compete with other feed proteins on this basis.

V. CONCLUSIONS

Processes are available for producing a wide range of microorganisms as protein sources for food or feed applications. In general, most algae have serious shortcomings for either food or feed use, including poor digestibility and sensory characteristics, high costs of harvesting, and ease of contamination with pathogenic organisms. *Spirulina* species are claimed to have more desirable characteristics than many algae species, but their requirements for alkaline conditions for growth limit production to a few geographical regions, such as Mexico and Central Africa.

Bacteria and actinomycetes have many desirable characteristics for SCP production, including high growth rates and protein contents and favorable amino acid profiles, as compared with other microorganisms. However, the possible presence of toxic substances in bacterial cells limits human food applications. Capital and energy costs for bacterial SCP production are also high. Although substantial experimental animal feeding studies have been conducted on bacterial SCP produced from methanol, the potential place of this product in feedstuff markets will not be established until after the one plant now under construction comes into production.

Certain species of yeasts have had established positions in food and feed markets. Increasingly higher costs of hydrocarbons make processes based on these raw materials appear unattractive in the future. Carbohydrates and wastes that contain carbohydrates are the lowest cost substrates for yeast SCP production. Future growth of markets for food and feed yeast and for yeast protein concentrates and isolates will depend upon their economic competitiveness with alternative proteins from plant or animal sources.

Molds and higher fungi have relatively slow growth rates and are easily contaminated by other microorganisms unless aseptic conditions are maintained during growth. However, they are easily harvested by filtration or screening, which is an advantage over other microorganisms. Use of molds or higher fungi as food or feed will be governed by the same economic, market, nutritional, and safety factors that effect the acceptability of other SCP products.

REFERENCES

Ad Hoc Panel (1975). "Underexploited Tropical Plants with Promising Economic Value." Nat. Acad. Sci., Washington, D.C.
Aguirre, F., Maldonado, O., Rolz, P., Menchu, J. F., Espinosa, R., and deCabrera, S. (1976). *ChemTech* **6**(10) 636–642.
Akiba, T., Kajiyama, S., and Fukimbara, T. (1973). *J. Ferment. Technol.* **51**, 343–347.
Akin, C. (1973). U.S. Patent 3,781,264.
Akin, C. (1974). U.S. Patent 3,833,552.
Akin, C., Witter, L. D., and Ordal, Z. J. (1967). *Appl. Microbiol.* **15**, 1339–1344.
Amoco Foods Co. (1974). "Torutein Product Bulletin." Chicago, Illinois.
Anderson, C., Longton, J., Maddix, C., and Scammel, G. W. (1975). *In* "Single Cell Protein II" (S. R. Tannenbaum and D. I. C. Wang, eds.), pp. 314–329. MIT Press, Cambridge, Massachusetts.
Anderson, R., Weisbaum, R. B., and Robe, K. (1974). *Food Process.* **35**(7), 58–59.
Anonymous (1970). *Chem. Eng. (N.Y.)* **77**(14), 30.
Anonymous (1973a). *Eur. Chem. News* **23** (573), 14.
Anonymous (1973b). *Eur. Chem. News* **24** (591), 12.
Anonymous (1974a). *Eur. Chem. News* **25** (622), 18.
Anonymous (1974b). *Chem. & Eng. News* **52** (33), 30.
Anonymous (1975). *Chem. Eng. (N.Y.)* **82**(6), 36–37.
Anonymous (1976a). *Eur. Chem. News* **28** (720), 6.
Anonymous (1976b). *Chem. Ind. (London)* **20**, 859.
Anonymous (1977a). *Process Biochem.* **12**(1), 30.
Anonymous (1977b). *Food Eng.* **49**(6), 95–97.
Anonymous (1978a). *Chem. Eng. News* **56**(12), 10.
Anonymous (1978b). *Chem. Eng. News* **56**(38), 12–13.
Ballerini, D., Parlouar, D., Lapeyronnie, M., and Sri, K. (1977). *Eur. J. Appl. Microbiol.* **4**, 11–19.
Becker, E. W., and Venkataraman, L. V. (1978). "A Manual on the Cultivation and Processing of Algae as a Source of Single Cell Protein." Central Food Technological Research Institute, Mysure, India.
Bellamy, W. D. (1974). *Biotechnol. Bioeng.* **16**, 869–880.
Bellamy, W. D. (1975). *In* "Single Cell Protein II" (S. K. Tannenbaum and D. I. C. Wang, eds.), pp. 263–372. MIT Press, Cambridge, Massachusetts.
Bennett, I. C., Hondermarck, J. D., and Todd, J. R. (1969). *Hydrocarbon Process.* **48**(3), 104–108.
Bernstein, S., and Everson, T. C. (1973). *Proc. Natl. Symp. Food Process. Wastes, 4th, 1973* EPA 660/2-73-031.
Bernstein, S., Tzeng, C. H., and Sisson, D. (1975). Abstract BMPC 68, *Chem. Congr. North Am. Continent, 1st, 1975*, pp. 151–166.

Block, S. S. (1960), *J. Biochem. Microbiol. Technol. Eng.*, **2**, 243–252.
Block, S. S., Stearns, T. W., Stephens, R. L., and McCandless, R. F. J. (1953). *J. Agric. Food Chem.* **1**, 890–893.
Bongers, L. (1969). *Dev. Ind. Microbiol.* **11**, 241–255.
Bongers, L. (1970). *J. Bacteriol.* **104**, 145–151.
Booth, A. N., Robinson, D. J., and Wasserman, A. E. (1962). *J. Dairy Sci.* **45**, 1106–1107.
Bough, W. A., Brown, W. L., Porsche, J. D., and Doty, D. M. (1972). *Appl. Microbiol.* **24**, 226–235.
Bourges, H., Sotomayor, A., and Mendoza, E. (1972). *Abstr. Int. Congr. Nutr., 9th, 1972* p. 85.
BP Proteins, Ltd. (1974). "The Toprina Cycle." London, England.
Bressani, R. (1968). *In* "Single Cell Protein" (R. I. Mateles and S. R. Tannenbaum, eds.), pp. 90–121. MIT Press, Cambridge, Massachusetts.
Brown, B. S., Jones, J. C., and Hulse, J. M. (1975). *Process Biochem.* **10**, 3–7.
Brownstein, A. M., and Constantinides, A. (1975). *169th Natl. Meet., Am. Chem. Soc.,* Paper No. 86.
Burch, J. E., Lipinsky, E. S., and Litchfield, J. H. (1963). *Food Technol.* **17**(10), 54–60.
Busta, F. F., Schmidt, B. E., and McKay, L. L. (1977). U.S. Patent 4,018,650.
Calloway, D. H. (1974). *In* "Single Cell Protein" (P. Davis, ed.), pp. 129–146. Academic Press, New York.
Camhi, J. D., and Rogers, P. L. (1976). *J. Ferment. Technol.* **54**, 437–449 and 450–458.
Casey, R. P., Lubitz, J. A., Benoit, R. J., Weissman, B. J., and Chau, H. (1963). *Food Technol.* **17**, 1039–1043.
Champagnat, A., and Filosa, J. A. (1965). U.S. Patent 3,193,390.
Champagnat, A., and Filosa, J. A. (1971). U.S. Patent 3,560,341.
Champagnat, A., Vernet C., Laine, B., and Filosa, J. (1963). *Nature (London)* **197**, 13–14.
Chang, W. T. M., and Thayer, D. W. (1974). *Dev. Ind. Microbiol.* **16**, 456–464.
Church, B. D., Nash, H., and Brosz, W. (1972). *Dev. Ind. Microbiol.* **13**, 30–46.
Church, B. D., Erickson, E. E., and Widmer, C. M. (1973). *Food Technol.* **27**(2), 36–42.
Clement, G. (1968). *In* "Single-Cell Protein" (R. I. Mateles and S. R. Tannenbaum, eds.), pp. 306–308. MIT Press, Cambridge, Massachusetts.
Clement, G. (1975). *In* "Single-Cell Protein II" (S. R. Tannenbaum and D. I. C. Wang, eds.), pp. 467–474. MIT Press, Cambridge, Massachusetts.
Clement, G., Giddey, C., and Menzi, R. (1967). *J. Sci. Food Agric.* **18**, 497–501.
Code of Federal Regulations (1978). "Title 21, Food and Drugs; 172.325, 172.896, Bakers Yeast Protein; Dried Yeasts." U.S. Govt. Printing Office, Washington, D.C.
Cook, B. B., Lau, E. W., and Bailey, B. M. (1963). *J. Nutr.* **81**, 23–29.
Cooney, C. L., and Levine, D. W. (1972). *Adv. Appl. Microbiol.* **15**, 337.
Cooney, C. L., and Levine, D. W. (1975). *In* "Single-Cell Protein II" (S. R. Tannenbaum and D. I. C. Wang, eds.), pp. 402–423. MIT Press, Cambridge, Massachusetts.
Cooney, C. L., Wang, D. I. C., and Mateles, R. I. (1969). *Biotechnol. Bioeng.* **11**, 269–281.
Cooney, C. L., Levine, D. W., and Snedecor, B. (1975). *Food Technol.* **29**(2), 33–42.
Cooper, P. G., Silver, R. S., and Boyle, J. P. (1975). *In* "Single-Cell Protein II" (S. R. Tannenbaum and D. I. C. Wang, eds.), pp. 454–466. MIT Press, Cambridge, Massachusetts.
Cooper, R. C. (1962). *Am. J. Public Health* **52**, 252–257.
Crawford, D. L., McCoy, E., Harkin, J. M., and Jones, P. (1973). *Biotechnol. Bioeng.* **13**, 833–843.
Cremieux, A., Chevalier, J., Combet, M., Dumenil, G., Parlouar, D., and Ballerini, D. (1977). *Eur. J. Appl. Microbiol.* **4**, 1–9.

Daly, W. H., and Ruiz, L. P., Jr. (1975). "Fabrication of Single Cell Protein from Cellulosic Wastes, EPA" Rep. No. EPA-6702–75–032.

Dam, R., Lee, S., Fry, P. C., and Fox, H. (1965). *J. Nutr.* **86**, 376–382.

Darlington, W. A. (1964). *Biotechnol. Bioeng.* **6**, 241–242.

Dasinger, B. L., and Nasland, L. A. (1972). U.S. Patent 3,644,175.

Davis, P., ed. (1974). "Single-Cell Protein." Academic Press, New York.

deCabrera, S., Mayorga, H., Espinosa, R., and Rolz, C. (1976). *Proc. Int. Congr. Food Sci. Technol., 4th, 1974* Vol. 4, pp. 296–301.

Decerle, C., Franckowiak, S., and Gatellier, C. (1969). *Hydrocarbon Process.* **48**(3), 109–112.

DeGroot, A. P. (1974). *In* "Single-Cell Protein" (P. Davis, ed.), pp. 75–92. Academic Press, New York.

Delaney, R. A. M., Kennedy, R., and Walley, B. D. (1975). *J. Sci. Food Agric.* **26**, 1177–1186.

Dimmling, W., and Seipenbusch, R., *Process Biochem.* **13**(3), 9–12, 14, 15, 34.

Dostalek, M., and Molin, N. (1975). *In* "Single-Cell Protein II" (S. R. Tannenbaum and D. I. C. Wang, eds.), pp. 385–401. MIT Press, Cambridge, Massachusetts.

Dostalek, M., Haggström, L., and Molin, N. (1972). *Ferment. Technol. Today, Proc. Int. Ferment. Symp., 4th, 1972,* pp. 497–501.

Drouliscos, N. J., Macris, B. J., and Kokke, R. (1976). *Appl. Environ. Microbiol.* **31**, 691–694.

Dunlap, C. E. (1975a). *In* "Single-Cell Protein II" (S. R. Tannenbaum and D. I. C. Wang, eds.), pp. 244–262. MIT Press, Cambridge, Massachusetts.

Dunlap, C. E. (1975b). *Food Technol.* **29**(12), 62–67.

Durand-Chastel, H., and Clement, G. (1972). *Int. Congr. Nutr., 9th, 1972* Paper C-3-6.

Duthie, I. F. (1975). *In* "Single-Cell Protein II" (S. R. Tannenbaum and D. I. C. Wang, eds.), pp. 505–544. MIT Press, Cambridge, Massachusetts.

Edwards, V. H., Kinsella, J. E., and Sholiton, D. B. (1972). *Biotechnol. Bioeng.* **14**, 123–147.

Endo, H., and Shirota, M. (1972). *Ferment. Technol. Today, Proc. Int. Ferment. Symp., 4th, 1972* pp. 533–541.

Enebo, L. (1969). *Chem. Eng. Prog., Symp. Ser.* **65**(93), 60–65.

Engel, C. (1973). *In* "Proteins from Hydrocarbons" (H. Gounelle de Pontanel, ed.), pp. 53–81. Academic Press, New York.

Eriksson, K. E., and Larsson, K. (1975). *Biotechnol. Bioeng.* **17**, 327–348.

Ertola, R. J., Mazza, L. A., Balatti, A. P., and Sanahuja, J. (1969), *Biotechnol. Bioeng.* **11**, 409–416.

Ertola, R. J., Segovia, R. F., Gambaruto, M., Monosiglio, J. C., and Artuso, C. (1971). *Dev. Ind. Microbiol.* **12**, 72–76.

Espinosa, R., Maldonado, O., Menchu, J. F., and Rolz, C. (1975). *Chem. Congr. North Am. Continent, 1st, 1975* Abstract BMPC 65.

Evans, G. H. (1968). *In* "Single-Cell Protein" (R. I. Mateles and S. R. Tannenbaum, eds.), pp. 243–254. MIT Press, Cambridge, Massachusetts.

Evans, G. H., and Shennan, J. L. (1974). U.S. Patent 3,846,238.

Faust, U., Präve, P., and Sukatsch, D. A. (1977). *J. Ferment. Technol* **55**, 609–614.

Food and Agricultural Organization (1957). *FAO Nutr. Stud.* No. 16.

Foster, J. F., and Litchfield, J. H. (1964). *Biotechnol. Bioeng.* **6**, 441–456.

Foster, J. F., and Litchfield, J. H. (1967). *NASA Spec. Publ.* SP–134, 201–212.

Foster, J. F., and Litchfield, J. H. (1968a). *NASA Spec. Publ.* **165**, 67–74.

Foster, J. F., and Litchfield, J. H. (1968b). *SAE Trans.* 2639–2646.

Foster, J. F., and Litchfield, J. H. (1969). *NASA Contract. Rep.* **NASA CR-1296.**

Frey, C. N. (1930). *Ind. Eng. Chem.* **22,** 1154–1162.

Fu, T. T., and Thayer, D. W. (1975). *Biotechnol. Bioeng.* **17,** 1749–1960.

Fustier, P., and Simard, R. E. (1976). *Can. Inst. Food Sci. Technol.* **9,** 182–185.

Giacobbe, F., Puglisi, P., and Longbardi, G. (1975). *169th Natl. Meet., Am. Chem. Soc.* Paper No. 74.

Goldberg, I., Rock, J. S., Ben-Bassat, A., and Mateles, R. I. (1976). *Biotechnol. Bioeng.* **18,** 1657–1668.

Golueke, C. G., Jr., and Oswald, W. J. (1970). *J. Water Pollut. Control Fed.* **42,** R304.

Goto, S., Okamoto, R., Kuwajima, J., and Takemetsu, A. (1976). *J. Ferment. Technol.* **54,** 213–224.

Gow, J. S., Littlehailes, J. D., Smith, S. R. L., and Walter, R. B. (1975). *In* "Single Cell Protein II" (S. R. Tannenbaum and D. I. C. Wang, eds.), pp. 370–384. MIT Press, Cambridge, Massachusetts.

Gray, W. D. (1970). *Crit. Rev. Food Technol.* **1,** 225–229.

Gregory, K. F., Reade, A. E., Khor, G. L., Alexander, J. C., Lumsden, F. H., and Losos, G. (1976). *Food Technol.* **30**(3), 30–35.

Guenther, K. R. (1965). *Biotechnol. Bioeng.* **7,** 445–446.

Guenther, K. R., and Perkins, M. B. (1968). U.S. Patent 3,384,491.

Gutcho, S. (1973). "Proteins from Hydrocarbons." Noyes Data Corp., Park Ridge, New Jersey.

Hamer, G., Heden, C. G., and Carenberg, C. D. (1967). *Biotechnol. Bioeng.* **9,** 499–514.

Hamer, G., Harrison, D. E. F., Harwood, J. H., and Topiwala, H. H. (1975). *In* "Single-Cell Protein II" (S. R. Tannenbaum and D. I. C. Wang, eds.), pp. 357–369. MIT Press, Cambridge, Massachusetts.

Han, Y. W., Dunlap, C. E., and Callihan, C. D. (1971). *Food Technol.* **25,** 130–133 and 154.

Harris, E. E., Hajny, G. J., and Johnson, M. C. (1951). *Ind. Eng. Chem.* **43,** 1593–1596.

Harrison, D. E. F. (1976). *ChemTech,* **6,** 570–574.

Harrison, D. E. F., Topiwala, H. H., and Hamer, G. (1972). *Ferment. Technol. Today, Proc. Int. Ferment. Symp., 4th, 1972* pp. 491–495.

Harwood, J. H., and Pirt, S. J. (1972). *J. Appl. Bacteriol.* **35,** 597–607.

Hatch, R. T. (1975). *In* "Single-Cell Protein II" (S. R. Tannenbaum and D. I. C. Wang, eds.), pp. 46–68. MIT Press, Cambridge, Massachusetts.

Heden, C. G., Molin, N., Olsson, U., and Rupprecht, A. (1971). *Biotechnol. Bioeng.* **13,** 147–150.

Hedenskog, G., and Mogren, H. (1973). *Biotechnol. Bioeng.* **15,** 129–142.

Heinemann, B. (1963). U.S. Patent 3,086,320.

Hernandez, E., and Johnson, M. J. (1967). *J. Bacteriol.* **94,** 991–1001.

Holderby, J. M., and Moggio, W. A. (1960). *J. Water Pollut. Control Fed.* **32,** 171–181.

Holve, W. A. (1976). *Process Biochem.* **11**(10), 2–7.

Huang, F., and Rha, C. K. (1971). *J. Food Sci.* **36,** 1131–1134.

Humfeld, H., and Sugihara, T. F. (1949). *Food Technol.* **3,** 355–356.

Humphrey, A. E. (1974). *Chem. Eng. (N.Y.)* **81**(26), 98–112.

Imada, O., Hoshiai, K., and Tanaka, M. (1972). U.S. Patent 3,635,796.

Imrie, F. K. E., and Vlitos, A. J. (1975). *In* "Single-Cell Protein II" (S. R. Tannenbaum and D. I. C. Wang, eds.), pp. 223–243. MIT Press, Cambridge, Massachusetts.

Irgens, R. L., and Clarke, J. D. (1976). *Eur. J. Appl. Microbiol.* **2,** 231–241.

Johnson, M. J. (1967). *Science* **155,** 1515–1519.

Kalina, V., and Nicholas, P. (1974). U.S. Patent 3,821,080.

Kanazawa, M. (1975). *In* "Single-Cell Protein II" (S. R. Tannenbaum and D. I. C. Wang, eds.), p. 438–453. MIT Press, Cambridge, Massachusetts.

Kanazawa, M., Maruyama, T., and Katoh, S. (1975). *169th Natl. Meet., Am. Chem. Soc.* Paper No. 85.

Kato, N., Tsuji, K., Tani, Y., and Ogata, K. (1974). *J. Ferment. Technol.* **52**, 917–920.

Khor, G. L., Alexander, J. C., Santos-Nuñez, J., Reade, A. E., and Gregory, K. F. (1976). *Can. Inst. Food Sci. Technol. J.* **9**(3), 139–143.

Kihlberg, R. (1972). *Annu. Rev. Microbiol.* **26**, 427–466.

Klapka, M. R., Duby, G. A., and Pavcek, P. L. (1958). *J. Am. Diet. Assoc.* **34**, 1317–1320.

Klis, J. (1963). *Food Process.* **24**(9), 99.

Knecht, R., Prave, P., Seipenbusch, R., and Sukatsch, D. A. (1977). *Process Biochem.* **12**(4), 11–14.

Ko, P. C., and Yu, Y. (1968). *In* "Single-Cell Protein" (R. I. Mateles and S. R. Tannenbaum, eds.), pp. 255–262. MIT Press, Cambridge, Massachusetts.

Ko, P. C., Yu, Y., and Li, C. S. (1964). "Protein from Petroleum by Fermentation Process." Chinese Solvent Works, Chinese Petroleum Corp., Taiwan.

Kosaric, N., and Zajic, J. E. (1974). *Adv. Biochem. Eng.* **3**, 89–125.

Kosaric, N., LeDu, A., and Zajic, J. E. (1973). *Can. J. Chem. Eng.* **51**, 186–190.

Kosaric, N., Nguyen, H. T., and Bergougnou, M. A. (1974). *Biotechnol. Bioeng.* **16**, 881–896.

Kosikowski, F. (1977). "Cheese and Fermented Milk Foods," 2nd ed. Edwards, Ann Arbor, Michigan.

Labuza, T. P. (1975). *In* "Single-Cell Protein II" (S. R. Tannenbaum and D. I. C. Wang, eds.), pp. 69–104. MIT Press, Cambridge, Massachusetts.

Lachance, P. A. (1968). *In* "Single-Cell Protein" (R. I. Mateles and S. R. Tannenbaum, eds.), pp. 122–152. MIT Press, Cambridge, Massachusetts.

Laine, B. M. (1972). *Can. J. Chem. Eng.* **50**, 154–156.

Laine, B. M., and du Chaffaut, J. (1975). *In* "Single-Cell Protein II" (S. R. Tannenbaum and D. I. C. Wang, eds.), pp. 424–437. MIT Press, Cambridge, Massachusetts.

Lambert, E. B. (1938). *Bot. Rev.* **4**, 397–426.

Lampert, L. M. (1965). "Modern Dairy Products," Chapter 15, p. 20. Chem. Publ. Co., New York.

Lane, A. G. (1977). *J. Appl. Chem. Biotechnol.* **27**, 165–169.

Laskin, A. I. (1975). Abstract BMPC39. *Chem. Congr. North Am. Continent, 1st, 1975*

LeDuy, A., Kosaric, N., and Zajic, J. E. (1974). *Can. Inst. Food Sci. Technol. J.* **7**, 44–49.

Lee, S. K., Fox, H. M., Kies, C., and Dam, R. (1967). *J. Nutr.* **92**, 281–285.

LeFrançois, L., Marillen, C. G., and Mejane, T. V. (1955). French Patent 1,102,200.

Leveille, G. A., Sauberlich, H. E., and Shockley, J. W. (1962). *J. Nutr.* **76**, 423–428.

Levine, D. W., and Cooney, C. L. (1973), *Appl. Microbiol.* **26**, 982–990.

Lewis, C. W. (1976). *J. Appl. Chem. Biotechnol.* **26**, 568–575.

Lipinsky, E. S., and Litchfield, J. H. (1970). *Crit. Rev. Food Technol.* **1**, 580–618.

Lipinsky, E. S., and Litchfield, J. H. (1974). *Food Technol.* **28**(5), 16–22 and 40.

Litchfield, J. H. (1964). *In* "Global Impacts of Applied Microbiology" (M. P. Starr, ed.), pp. 327–337. Wiley, New York.

Litchfield, J. H. (1967a). *In* "Microbial Technology" (H. J. Peppler, ed.), pp. 107–144. Van Nostrand-Reinhold, New Jersey.

Litchfield, J. H. (1967b). *Food Technol.* **21**(2), 55–57.

Litchfield, J. H. (1967c). *Biotechnol. Bioeng.* **9**, 289–304.

Litchfield, J. H. (1967d). *Proc. Int. Congr. Nutr. 7th, 1966* Vol. 4, pp. 1068–1074.

Litchfield, J. H. (1968). *In* "Single-Cell Protein" (R. I. Mateles and S. R. Tannenbaum, eds.), pp. 309–329. MIT Press, Cambridge, Massachusetts.

Litchfield, J. H. (1972). *Dev. Ind. Microbiol.* **13**, 317–331.

Litchfield, J. H. (1974). *Chem. Process. (London)* **20**(9), 11–18.

Litchfield, J. H. (1977a). *Food Technol.* **31**(5), 175–179.

Litchfield, J. H. (1977b). *Biotechnol. Bioeng. Symp.* **7**, 77–90.

Litchfield, J. H. (1977c). *Adv. Appl. Microbiol.* **22**, 267–305.

Litchfield, J. H. (1978). *Chem Tech* **8**, 218–223.

Litchfield, J. H., and Overbeck, R. C. (1966). *Food Sci. Technol. Proc. Int. Congr. 1st, 1962*, Vol. 2, pp. 511–520.

Litchfield, J. H., Overbeck, R. C., and Davidson, R. S. (1963a). *J. Agric. Food Chem.* **11**, 158–162.

Litchfield, J. H., Vely, V. G., and Overbeck, R. C. (1963b). *J. Food Sci.* **28**, 741–743.

Lubitz, J. A. (1963). *J. Food Sci.* **28**, 229–232.

McElwain, M. D., Richardson, T., and Amundson, C. H. (1975), *J. Milk Food Technol.* **28**, 521–526.

MacLennan, D. G., Gow, J. S., and Stringer, D. A. (1973). *Process Biochem.* **8**(6), 22–24.

MacLennan, D. G., Ousby, J. C., Owen, T. R., and Steer, D. C. (1976). U.S. Patent 3,989,594.

Mandels, M., and Sternberg, D. (1976). *J. Ferment. Technol.* **54**, 267–286.

Masuda, Y., and Yoshikawa, K. (1973). *Expert Work. Group Meet. Manuf. Proteins Hydrocarbons, U. N. Ind. Dev. Organ.* Paper ID/WG 164/20.

Masuda, Y., Kato, K., Takayama, Y., Kida, K., and Nakanishi, M. (1975). U.S. Patent 3,868,305.

Mateles, R. I. (1971). *Biotechnol. Bioeng.* **13**, 581–582.

Mateles, R. I., and Tannenbaum, S. R., eds. (1968). "Single-Cell Protein." MIT Press, Cambridge, Massachusetts.

Mateles, R. I., Goldberg, I., and Battat, E. (1976). U.S. Patent 3,989,595.

Matthern, R. O., and Koch, R. B. (1964). *Food Technol.* **18**(5), 58–62 and 64–65.

Matthern, R. O., Kostick, J. A., and Okada, I. (1969). *Biotechnol. Bioeng.* **11**, 863–874.

Mennett, R. H., and Nakayama, T. O. M. (1971). *Appl. Microbiol.* **22**, 772–776.

Metz, B., and Kossen, N. W. F. (1977). *Biotechnol. Bioeng.* **19**, 781–799.

Mian, F. A., Ajdery, A., and Fazeli, A. (1976). *J. Ferment. Technol.* **54**, 76–81.

Mogren, H. L., Hedenskog, G. O., and Enebo, L. E. (1974). U.S. Patent 3,848,812.

Moo-Young, M. (1976). *Process Biochem.* **11**(10), 32–34.

Moo-Young, M. (1977). *Process Biochem.* **12**(4), 6–10.

Moo-Young, M., Chahal, D. S., Swan, J. E., and Robinson, C. W. (1977). *Biotechnol. Bioeng.* **19**, 527–538.

Moulin, G., Galzy, P. and Joux, J. L. (1976). *Food Sci. Technol., Proc. Int. Congr., 4th, 1974* Vol. 3, pp. 47–53.

Moustafa, A. M. (1960). *Appl. Microbiol.* **8**, 667.

Nagai, I. (1973). *Expert Work. Group Meet. Manuf. Proteins Hydrocarbons, U. N. Ind. Dev. Organ* Paper ID/WG 164/28.

Nakayama, O., Ueno, T., and Tsuchiya, F. (1974). *J. Ferment. Technol.* **52**, 225–232.

Newell, J. A., Robbins, E. A., and Seeley, R. D. (1975). U.S. Patent 3,867,555.

Nguyen, H. T., Kosaric, N., and Bergougnou, M. A. (1974). *Can. Inst. Food Sci. Technol. J.* **7**, 114–116.

Ogawa, T., Kozasa, H., and Terui, G. (1972). *J. Ferment. Technol.* **50**, 143–149.

Olson, A. J. C. (1961). *SCI Monogr.* **12**, 81–93.

Organizing Committee (1975). "Microbial Growth on C-1 Compounds." Microbiol., Washington, D.C.

Orgel, G., Pietrusza, E. W., and Joris, G. G. (1971). U.S. Patent 3,622,465.

Oswald, W. J. (1969). *Chem. Eng. Prog., Symp. Ser.* **65**(93), 87–92.

Oswald, W. J., and Golueke, C. G. (1968). *In* "Single-Cell Proteins" (R. I. Mateles and S. R. Tannenbaum, eds.), pp. 271–305. MIT Press, Cambridge, Massachusetts.

Pederson, C. S. (1971). Microbiology of Food Fermentations." Avi Publ. Co., Westport, Connecticut.

Peitersen, N. (1975a). *Biotechnol. Bioeng.* **17**, 361–374.

Peitersen, N. (1975b). *Biotechnol. Bioeng.* **17**, 1291–1299.

Peitersen, N. (1977). *Biotechnol. Bioeng.* **19**, 337–348.

Peppler, H. J. (1965). *J. Agric. Food Chem.* **13**, 34–36.

Peppler, H. J. (1968). *In* "Single-Cell Protein" (R. I. Mateles and S. R. Tannenbaum, eds.), pp. 229–242. MIT Press, Cambridge, Massachusetts.

Perkins, M. B., and Furlong, L. E. (1967). U.S. Patent 3,355,296.

Powell, R. C., Nevels, E. M., and McDowell, M. E. (1961). *J. Nutr.* **75**, 7–12.

Prescott, S. C., and Dunn, C. G. (1959). "Industrial Microbiology," 3rd ed., Chapter 3. McGraw-Hill, New York.

Prokop, A., and Sobotka, M. (1975). *In* "Single-Cell Protein II" (S. R. Tannenbaum and D. I. C. Wang, eds.), pp. 127–157. MIT Press, Cambridge, Massachusetts.

Protein Advisory Group (1970a). "PAG Statement No. 4 on Single Cell Protein." FAO/WHO/UNICEF, United Nations, New York.

Protein Advisory Group (1970b). "PAG Guideline No. 6 for Pre-Clinical Testing of Novel Sources of Protein." FAO/WHO/UNICEF, United Nations, New York.

Protein Advisory Group (1970c). "PAG Guideline No. 7 for Human Testing of Supplementary Food Mixtures." FAO/WHO/UNICEF, United Nations, New York.

Protein Advisory Group (1971). "PAG Guideline No. 8 on Protein-Rich Mixtures for Use as Weaning Foods." FAO/WHO/UNICEF, United Nations, New York.

Protein Advisory Group (1972). "PAG Guideline No. 12 on the Production of Single Cell Protein for Human Consumption." FAO/WHO/UNICEF, United Nations, New York.

Protein Advisory Group (1974). "PAG Guideline No. 15 on Nutritional and Safety Aspects of Novel Protein Sources for Animal Feeding." FAO/WHO/UNICEF, United Nations, New York.

Protein Advisory Group (1976a). *PAG Bull.* **6**(2), 1–6.

Protein Advisory Group (1976b). *PAG Bull.* **6**(2), 6–9.

Ratledge, C. (1975). *Chem. Ind. (London)* **21**, 918–920.

Reed, G., and Peppler, H. J. (1973). "Yeast Technology." Avi Publ. Co., Westport, Connecticut.

Reuss, M., Gniesser, J., Reng, H. G., and Wagner, F. (1975). *Eur. J. Appl. Microbiol.* **1**, 295–305.

Reusser, F., Spencer, J. F. T., and Sallans, H. R. (1958a). *Appl. Microbiol.* **6**, 1–4.

Reusser, F., Spencer, J. F. T., and Sallans, H. R. (1958b). *Appl. Microbiol.* **6**, 5–8.

Richardson, B., Orcutt, D. M., Schwertner, H. A., Martinez, C. L., and Wickline, H. E. (1969). *Appl. Microbiol.* **18**, 245–250.

Ridgeway, J. A., Jr. (1974). U.S. Patent 3,843,807.

Ridgeway, J. A., Jr., Lappin, T. A., Benjamin, B. M., Corns, J. B., and Akin, C. (1975). U.S. Patent 3,865,691.

Robbins, E. A. (1976). U.S. Patent 3,991,215.

Robinson, R. F., and Davidson, R. S. (1959). *Adv. Appl. Microbiol.* **1**, 261–278.

Rockwell, P. J. (1976). "Single Cell Proteins from Cellulose and Hydrocarbons, 1976." Noyes Data Corp., Park Ridge, New Jersey.

Rogers, C. J., Coleman, E., Spino, D. F., and Purcell, T. C. (1972). *Environ. Sci. Technol.* **6**, 715–718.

Rolz, C. (1975). *In* "Single-Cell Protein II" (S. R. Tannenbaum and D. I. C. Wang, eds.), pp. 273–313. MIT Press, Cambridge, Massachusetts.

Romantschuk, H. (1975). *In* "Single Cell Protein II' (S. R. Tannenbaum and D. I. C. Wang, eds.), pp. 344–355. MIT Press, Cambridge, Massachusetts.

Romantschuk, H., and Lehtomäki, M. (1978). *Process Biochem.* **13**(3), 16, 17, 23.

Rosenzweig, M., and Ushio, S. (1974). *Chem. Eng. (N.Y.)* **81**(1), 62–63.

Rychtera, M., Barta, J., Feichter, A., and Einselle, A. A. (1977). *Process Biochem.* **12**(2), 26–30.

Scrimshaw, N. S. (1968). *In* "Single-Cell Protein" (R. I. Mateles and S. R. Tannenbaum, eds.), pp. 3–7. MIT Press, Cambridge, Massachusetts.

Scrimshaw, N. S. (1973). *In* "Proteins from Hydrocarbons" (H. Gounelle de Pontanel, ed.), pp. 189–213. Academic Press, New York.

Scrimshaw, N. S. (1975). *In* "Single-Cell Protein II" (S. R. Tannenbaum and D. I. C. Wang, eds.), pp. 24–45. MIT Press, Cambridge, Massachusetts.

Seelig, R. A. (1959). "Cultivated Mushrooms," 3rd ed. United Fresh Fruit and Vegetable Association, Washington, D. C.

Senez, J. (1973). *In* "Proteins from Hydrocarbons" (H. Gounelle de Pontanel, ed.), pp. 3–26. Academic Press, New York.

Shacklady, C. A. (1972). *World Rev. Nutr. Diet.* **14**, 154–179.

Shacklady, C. A., and Gatumel, E. (1973). *In* "Proteins from Hydrocarbons." (H. Gounelle de Pontanel, ed.), pp. 27–52. Academic Press, New York.

Shannon, D. W. F., McNab, J. M., and Anderson, G. B. (1976). *J. Sci. Food Agric.* **27**, 471–476.

Shannon, L. J., and Stevenson, K. E. (1975a). *J. Food Sci.* **40**, 826–826.

Shannon, L. J., and Stevenson, K. E. (1975b). *J. Food Sci.* **40**, 830–832.

Sheehan, B. T., and Johnson, M. J. (1971). *Appl. Microbiol.* **21**, 511–515.

Sherwood, R. F., and Peberdy, J. F. (1974). *J. Sci. Food Agric.* **25**, 1081–1087.

Shipman, R. H., Kao, I. C., and Fan, L. T. (1975). *Biotechnol. Bioeng.* **17**, 1561–1570.

Shuler, R. L., and Affens, W. A. (1970). *Appl. Microbiol.* **19**, 76–86.

Silver, R. S., and Cooper, P. G. (1972a). *72nd Natl. Meet., AIChE, 1972* Paper 16d.

Silver, R. S., and Cooper, P. G. (1972b). *Pap., Div. Pet. Chem. 164th Natl. Meet., Am. Chem. Soc.* Paper No. 31.

Sinskey, A. J., and Tannenbaum, S. R. (1975). *In* "Single-Cell Protein II" (S. R. Tannenbaum and D. I. C. Wang, eds.), pp. 158–178. MIT Press, Cambridge, Massachusetts.

Smith, R. H., and Palmer, R. (1976). *J. Sci. Food Agric.* **27**, 763–770.

Smith, R. H., Palmer, R., and Reade, A. E. (1975). *J. Sci. Food Agric.* **26**, 785–795.

Solomons, G. L., and Scammell, G. W. (1976). U.S. Patent 3,937,654.

Sonoda, Y., Someya, J., Futai, N., Tagaya, N., and Murakami, T. (1973). *J. Ferment. Technol.* **51**, 479–483.

Spencer-Martins, I., and Van Uden, N. (1977). *Eur. J. Appl. Microbiol.* **4**, 29–35.

Srinivasan, V. R., Fleenor, M. B., and Summers, R. J. (1977). *Biotechnol. Bioeng.* **19**, 153–154.

Stoller, B. B. (1954). *Econ. Bot.* **8**, 48–95.

Stone, H., and Sidel, J. (1974). *In* "Single-Cell Protein" (P. Davis, ed.), pp. 147–160. Academic Press, New York.

Sugihara, T. F., and Humfeld, H. (1954). *Appl. Microbiol.* **2**, 170–172.

Sugimori, T., Oyama, Y., and Omichi, T. (1971). *J. Ferment. Technol.* **49**, 435–446.

Sugimoto, H. (1974). *J. Food Sci.* **39**, 934–938.

Sugimoto, M., Yokoo, S., and Imada, O. (1972). *Ferment. Technol. Today, Proc. Int. Ferment. Symp., 4th, 1972* pp. 503–507.

Takata, T. (1969). *Hydrocarbon Process.* **48**(3), 99–103.

Tanaka, K., Ohshima, K., and Yamamoto, M. (1972). U.S. Patent 3,639,210.

Tannenbaum, S. R., and Wang, D. I. C., eds. (1975). "Single-Cell Protein II." MIT Press, Cambridge, Massachusetts.

Taylor, J. C., Lucas, E. W., Gable, D. A., and Graber, G. (1974). In "Single Cell Protein" (P. Davis, ed.), pp. 179–186. Academic Press, New York.

Teramoto, S., Taguchi, H., Veda, R., and Yoshida, T. V. (1966). Proc. Int. Congr. Food Sci. Technol. 2nd, 1966 Paper A 3.9.

Thanh, N. C., and Wu, J. S. (1975). Can. Inst. Food Sci. Technol. J. **8**, 202–205.

Thayer, D. W., Yang, S. P., Key, A. B., Hang, H. H., and Barker, J. W. (1975). Dev. Ind. Microbiol. **16**, 465–474.

Tong, P. Q., Riel, R. R., and Simard, R. E. (1973a). Can. Inst. Food Sci. Technol. J. **6**, 212–215.

Tong, P. Q., Simard, R. E., and Riel, R. R. (1973b). Can. Inst. Food Sci. Technol. J. **6**, 239–243.

Towersey, P. J., Longton, J., and Cockram, G. N. (1976). U.S. Patent 3,937,693.

Towersey, P. J., Longton, J., and Cockram, G. N. (1977). U.S. Patent 4,041,089.

Trevelyan, W. E. (1975). J. Sci. Food Agric. **26**, 1673–1680.

Trevelyan, W. E. (1976). J. Sci. Food Agric. **27**, 753–762.

Tung, T. (1975). Food Eng. **47**(4), 24.

Ueno, K., Asai, Y., Shimada, M., and Goto, S. (1974a). J. Ferment. Technol. **52**, 861–866.

Ueno, K., Asai, Y., Shimada, M., and Kametani, T. (1974b). J. Ferment. Technol. **52**, 867–872.

Ueno, K., Asai, Y., Yonemura, H., and Kametani, T. (1974c). J. Ferment. Technol. **52**, 873–877.

Vananuvat, P., and Kinsella, J. E. (1975a). J. Food Sci. **40**, 336–342.

Vananuvat, P., and Kinsella, J. E. (1975b). J. Food Sci. **40**, 823–825.

Van Dijken, J. P., and Harder, W. (1975). Biotechnol. Bioeng. **17**, 15–30.

Van Eybergen, G. C., and Scheffers, W. A. (1972). Antonie van Leeuwenhoek **38**, 448–450.

Vela, G. R., and Guerra, C. N. (1966). J. Gen. Microbiol. **42**, 123.

Volesky, B., and Zajic, J. E. (1971). Appl. Microbiol. **21**, 614–622.

Wagner, F., Kleeman, T., and Zahn, W. (1969). Biotechnol. Bioeng. **11**, 393–408.

Waldroup, P. W., and Flynn, N. W. (1975). Poult. Sci. **54**, 1129–1133.

Waldroup, P. W., and Hazen, K. K. (1975). Poult. Sci. **54**, 635, 637.

Waldroup, P. W., and Payne, J. R. (1974). Poult. Sci. **53**, 1039–1042.

Waldroup, P. W., Hillard, C. M., and Mitchell, R. J. (1971). Poult. Sci. **50**, 1022–1029.

Wang, D. I. C. (1968a). Chem. Eng. (N.Y.) **75**(18), 99.

Wang, D. I. C. (1968b). In "Single-Cell Protein" (R. I. Mateles and S. R. Tannenbaum, eds.), pp. 217–228. MIT Press, Cambridge, Massachusetts.

Wang, D. I. C. (1969). Chem. Eng. Prog., Symp. Ser. **65**(93), 66.

Wang, H. Y., Mou, D.-G., and Swartz, J. R. (1976). Biotechnol. Bioeng. **18**, 1811–1814.

Ward, C. H., Moyer, J. E., and Vela, G. R. (1965). Dev. Ind. Microbiol. **6**, 213–222.

Waslien, C. I. (1975). Crit. Rev. Food Sci. Nutr. **6**, 77–151.

Waslien, C. I., Calloway, D. H., Margen, S., and Costa, F. (1970). J. Food Sci. **35**, 294–298.

Wasserman, A. E., Hampson, J. W., and Alvare, N. F. (1961). J. Water Pollut. Control Fed. **33**, 1090–1094.

Weiner, B. A. (1977a). Eur. J. Appl. Microbiol. **4**, 51–57.

Weiner, B. A. (1977b). Eur. J. Appl. Microbiol. **4**, 59–65.

Wells, J. (1975). 169th Natl. Meet., Am. Chem. Soc. Paper No. 72.

White, J. (1954). "Yeast Technology," Chapter 2, p. 23. Wiley, New York.

White, W. B., and Balloun, S. L. (1977). Poultry Sci. **56**, 266–273.

Whittemore, D. J., Moffat, I. W., and Taylor, A. B. (1976). J. Sci. Food Agric. **27**, 1163–1170.

Wiken, T. O. (1972). *Ferment. Technol. Today, Proc. Int. Ferment. Symp., 4th, 1972* pp. 569–596.

Wilkinson, T. G., Topiwala, H. H., and Hamer, G. (1974). *Biotechnol. Bioeng.* **16,** 41–59.

Wilson, B. J. (1971). *Microb. Toxins* **6,** 207–295.

Wolnak, B., Andreen, B., and Wen, C. V. (1972). U.S. Patent 3,649,459.

Yokote, Y., Suginoto, M., and Abes, S. (1974). *J. Ferment. Technol.* **52,** 201–209.

Yoshikawa, J. Katsur, T., Fukita, Y., Wada, H., and Tanigawa, Y. (1975). U.S. Patent 3,901,762.

Young, R. J. (1973). *Expert Work. Group Meet. Manuf. Proteins Hydrocarbons, U. N. Ind. Dev. Organ.* Paper ID/WG 164/9.

Young, V. R., and Scrimshaw, N. S. (1975). *In* "Single-Cell Protein II" (S. R. Tannenbaum and D. I. C. Wang, eds.), pp. 564–586. MIT Press, Cambridge, Massachusetts.

Chapter 5

Production of Yeasts and Yeast Products

H. J. PEPPLER

I. INTRODUCTION

Yeast technology comprises those propagations involving several types of yeast used in the alcoholic fermentation of foods, beverages, and feeds as well as those processes which enrich, alter, or extract concentrates of living yeast cells. Among the derived products are nutritional supplements, flavor carriers, enzymes, nucleosides, nucleotides, proteins, and polysaccharides. Yeasts and their derivatives are manufactured on every continent. The most common product, bakers' yeast, is grown by at least 73 companies in 32 countries, excluding (for lack of data) the eastern European nations (Perlman, 1977). The market for yeast products sold to the baking industry in the United States is close to $120 million annually; in 1976 about 15.7 billion lb of yeast-

MICROBIAL TECHNOLOGY, 2nd ed., VOL.I
Copyright © 1979 by Academic Press, Inc.
All rights of reproduction in any form reserved. ISBN 0-12-551501-4

leavened bread-type products were shipped (*Bakery Trends*, 1977). By 1980 shipments are expected to increase 4.7% in this category of products which includes white pan bread, hearth bread and rolls, variety breads, rolls (hamburger, wiener, brown and serve, Kaiser, Parkerhouse, and other types), English muffins, bread stuffing, and crumbs.

Of the 349 species of yeast described by taxonomists only a few are of economic importance (Lodder, 1970). The more commonly used species and the processes and products usually associated with them are listed in Table I. Outstanding in this group of yeasts are the versatile strains of *Saccharomyces cerevisiae* used universally in food-related fermentations and beverage production. For protein biosynthesis, however, *Candida utilis* production is growing relatively rapidly, mainly to provide dried biomass (nonfermentative) as grain-sparing supplements for formulated livestock feeds. How these yeasts are grown and how the cells are utilized are reviewed in this chapter.

II. YEAST PRODUCTION

Yeast biomass is grown on a variety of energy-yielding carbon compounds and phosphate and nitrogen sources in aerobic batch and continuous culture systems. During yeast growth in the fed-batch process, the addition of sugar-containing media, usually diluted molasses, is

TABLE I. Principal Industrial Yeasts and Their Usage

Yeast species	Products/applications
Saccharomyces cerevisiae	Bread-type products, beer brewing, wine making, distilled beverages, ethanol, cider, food yeast, feed yeast, yeast-derived products, (autolysates, hydrolysates, protein, biochemicals), invertase
Saccharomyces uvarum (syn. *S. carlsbergensis*)	Beer brewing
Saccharomyces saké	Sake brewing
Saccharomyces bayanus	Sparkling wines
Saccharomyces lactis	Lactase
Kluyveromyces fragilis (syn. *S. fragilis*)	Food yeast, feed yeast, ethanol
Candida utilis	Food yeast, nucleic acids, feed yeast
Candida tropicalis	Food yeast, feed yeast
Candida pseudotropicalis	Food yeast, feed yeast
Candida lipolytica	Feed yeast

metered incrementally to minimize ethanol accumulation. It is a flexible system, suitable for small and large fermentors [up to 60,000 gal (225 m³)], and readily permits switching to alternate substrates and yeast cultures. The fed-batch process is the principal method of producing bakers' yeast, some food yeasts, and biomass for the extraction of enzymes and the preparation of autolysates.

In continuous and semicontinuous yeast propagating systems at steady state, the input of nutrients and sugar source (usually molasses, wood sugars, or spent sulfite liquor) and the withdrawal of yeast emulsion are in balance. Modified Waldhof fermentors with capacities of around 100,000 gal (375 m³) provide the aeration needed by *Candida* species for efficient assimilation of hexoses, pentoses and other carbonyls. While lacking some of the flexibility of the fed-batch process, the continuous culture systems developed for *Candida utilis, C. tropicalis,* and other species have made possible the marked increases in annual tonnage of fodder yeast and food yeast. Estimates of yeast production throughout the world in 1977 are shown in Table II.

Changes in yeasting practices for leavening and beverage fermentations have influenced advances in industrial microbiology since the 1850s. Among the many contributions are (1) Pasteur's discovery in 1874 that yeast utilizes ammonia, and yeast growth is stimulated by aeration which represses ethanol formation; (2) the pure-culture concept and methods introduced by Hansen in 1896; (3) the fed-batch (Zulauf) process invented by Hayduck in 1919; and (4) the replacement of grain mashes with molasses media in the 1920s. The historical and modern eras of yeast technology have been reviewed often: Kiby (1912) published a production manual for bakers' compressed yeast; Walter (1953)

TABLE II. Estimated Annual Yeast Production, 1977 (Dry Tons)[a]

Location	Bakers' yeast	Dried yeast[b]
Europe	74,000	160,000
North America	73,000	53,000
The Orient	15,000	25,000
United Kingdom	15,500	[d]
South America	7,500	2,000
Africa	2,700	2,500

[a] Based in part on surveys by Moo-Young (1976); World Wide Survey of Fermentation Industries, 1967 (1971); and private sources.

[b] Dried yeast includes food and fodder yeasts; data for petroleum-grown yeasts are not available.

[c] Production figures for USSR not reported.

[d] None reported.

and Irvin (1954) outlined the traditional Vienna process and the productive bakers' yeast modifications developed from 1870 to 1915; and advances in modern yeast making were published by Frey (1930, 1957), White (1954), Pyke (1958), Butschek and Kautzmann (1962), Peppler (1967, 1970, 1978), Burrows (1970), Harrison (1971), Reed and Peppler (1973), Rosen (1977), and Hoogerheide (1977). Descriptive information covering over 160 yeast-related patents granted in the United States since 1970 was compiled by Johnson (1977).

A. Active Yeasts

Viable biomass, the primary product of all yeast propagations, is obtained for use as (1) active yeast for a variety of food fermentations, (2) active yeast for autolysis or enzyme extraction, or (3) inactive (nonfermentative) dried yeast (95–96% solids). Furthermore, the yeast crop may be used directly, without separation of cells and medium, as starter yeast in breweries, distilleries, and wineries. The cells can also be removed from the growing medium, washed, and concentrated. Such concentrates, known generally as yeast cream, may be (1) dewatered to form press or filter cake (as bakers' compressed yeast), (2) dehydrated further to produce active dry yeast (ADY) for bread making and other fermentations, (3) extracted to recover invertase and other cell constituents, (4) autolysed to prepare flavor and nutritional concentrates, or (5) pasteurized and desiccated by drum- or spray-drying processes to obtain inactive yeast for use as food and feed ingredients, and as raw material for extraction of ribonucleic acid and other cellular components.

The major process streams for separated yeast, shown in Fig. 1, outline the pathway to three major end products: compressed yeast and active dry yeast, forms of active yeast; and primary-grown food yeast, a common type of inactive yeast.

1. Bakers' Yeast

a. Cultures. The production of bakers' yeast is a multistage, aerobic, fed-batch process. It begins in the laboratory with pure cultures of *Saccharomyces cerevisiae* carried on agar media. The naturally occurring strains have been selected for their vigorous growth, high yields on molasses, storage stability, and functional properties in doughs and the breads made from them. Stock cultures of the proved strains are maintained in the diploid state. Yeast hybrids for leavening have been patented, but none has come into general use (Burrows and Fowell, 1961; Langejan and Khoudokormoff, 1976).

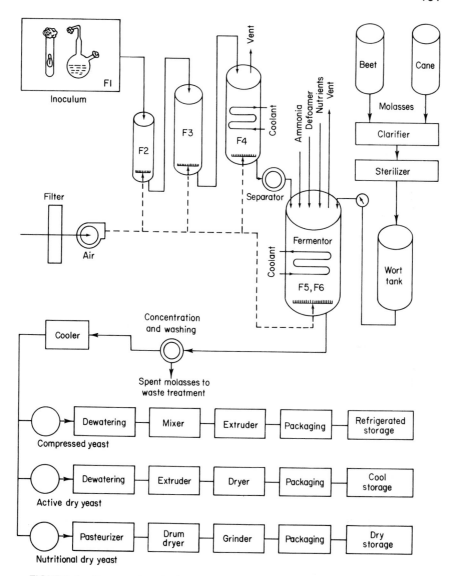

FIGURE 1. Process steps in the production of bakers' yeast and food yeast. (After a sketch by W. H. Chan, Universal Foods Corp., Milwaukee, Wisconsin.)

b. Propagation. Beet and cane molasses are the principal sources of sugars which bakers' yeast can assimilate. In a few countries, however, grain extracts, such as beer wort, continue to be used for bakers' yeast production.

Following inoculation of Pasteur flasks (Fig. 1, F1), the yeast is grown in progressively larger fermentors in which yeast solids increase from about 1 lb in the initial plant stage (F2) to more than 1500 lb in the preproduction stage (F5). The initial generations, F1–F3, are batch propagations; F2 and F3 are aerated lightly with sterile air. Generally the Pasteur flask culture is grown in a malt syrup medium up to 2 days at 30°C. It is the inoculant for culture tank F2 which in turn is transferred totally to the larger pure-culture tank, F3. Both of these fermentors are closed systems in which the substrate, usually a molasses medium enriched with corn steep liquor, is sterilized in place. The fermented contents of the final pure culture tank (F3) are pumped aseptically to a larger tank (F4), a fed-batch, aerated, intermediate stage frequently designated the stock fermentation. At the end of this propagation stage, the yeast is concentrated by centrifugation to a creamy suspension. This stock yeast is the inoculum for the following stage (F5), in which are used the largest fermentors available, molasses is apportioned at rates balancing sugar consumption, and aeration is intensified. All propagations are carried out at 30°C.

In the trade fermentation (F6), the final stage, the yeast is grown in high-capacity (up to 200 m^3) fermentors fabricated of stainless steel or other corrosion-resistant materials. The fermentor is equipped with an aeration system, cooling coils, nutrient metering system, sampling port, and foam- and pH-controlling devices. Generally the cleaned and sanitized vessels and their fittings are sterilized with steam at atmospheric pressure and charged with potable water, pitching yeast, nutrients, and sufficient sterilized molasses solution (wort) to provide about 1% fermentable sugar. Aeration, begun immediately, mixes the ingredients as they are pumped in. Heated and filtered air enters the fermentor through perforated pipes arranged in gridiron fashion near the bottom of the tank below the cooling coils. Metering of the molasses wort begins early in the propagation and continues at predetermined rates throughout most of the schedule. Efficient biosynthesis demands careful management of sugar and air inputs to avoid excess carbon losses. Overfeeding is avoided by controlling the available sugar concentration in the fermentor well below 0.1%, and by adjusting aeration to provide a minimum oxygen absorption rate of 0.2 m moles O_2/liter/min. Maximum yeast yield is obtained when the increase in yeast solids in the fermentor is limited to 20–25% per hour. Phosphate and ammonia are added in the early hours of the schedule in amounts according to the yeast crop expected, and ammonia or other alkali subsequently to hold the pH between 4 and 6.

(1) Nutrients. Beet molasses and cane molasses (blackstrap) supply fermentable sugars, essential minerals (potassium, phosphorus, magnesium, zinc, iron, and copper), amino nitrogen (asparagine, aspartic acid, alanine, glutamic acid, and glycine), and vitamins (biotin, thiamin, pyridoxine, pantothenic acid, and inositol). These nutrients vary according to the type of molasses, its geographical origin, management of the sugar crop, and efficiency of sugar mill operations. In general, beet molasses is fivefold higher in organic nitrogen than cane, but half of it is betaine which is not assimilated by *Saccharomyces.* Cane molasses is substantially richer in biotin, pantothenic acid, thiamin, magnesium, and calcium. Both types of molasses are nearly equal in their content of fermentable sugar, potassium, trace minerals, niacin, pyridoxine, and inositol. For bakers' yeast production, beet molasses is mixed with at least 20% cane molasses to assure an adequate supply of biotin.

To prepare the mixed molasses solutions, usually called wort, beet and cane are processed separately; i.e., diluted, pH adjusted, heated, clarified, sterilized, and blended.

Molasses contains only a small portion of the mineral nutrients required to grow bakers' yeast of desired composition. All propagations must be supplemented heavily with phosphorus and nitrogen to sustain optimum growth, maximum yield, and quality. Phosphoric acid, its ammonium salts, anhydrous ammonia, and ammonium sulfate are the usual additives. Other supplements may be added to assure highest yields, performance, and stability of bakers' yeasts. These additions include magnesium sulfate, potassium chloride, zinc sulfate, and thiamin. Where urea is used, biotin supplements may be required to obtain maximum yields of yeast.

(2) Aeration. Of the numerous aeration systems that have been developed for bakers' yeast propagations, the sparger type is most practical and in general use. It is a simple, horizontal network (grid) of parallel pipes, with numerous drilled openings (0.8–1.5 mm diameter), attached to a large, central, vertical air conduit. The assembly is secured near the bottom of the fermentor. With proper design, this aeration system easily supplies the high air inputs required for yeast biosynthesis (1 v air/1 v broth/min). In a new Danish yeast plant, which began operations in 1973, the sparger system in the trade fermentors (150 m^3) has 24 side pipes containing about 30,000 holes of 1.5-mm diameter (Rosen, 1977).

Alternate systems with increased air utilization efficiencies have come into commercial use. They combine mechanical agitation and novel air distributors and are described as agitator–aerator system, deep jet aerator (Vogelbusch), self-priming aerator (Frings), and turbine aerator system (Chemapec). The operating characteristics of such systems were discussed by Reed and Peppler (1973).

(3) Process Control. Most parameters of yeast propagation can be regulated automatically, generally by sensors which signal the activation of pumps and pneumatic devices. Temperature, pH, defoamer demand, and nutrient rate may be regulated automatically. Molasses feed rate may be controlled by monitoring the ethanol concentration in the effluent air (Rosenquist and Egnell, 1966; Rungaldier and Braun, 1961; Bach et al., 1978). Yeast growth may be regulated by combining an on-line computer with process sensors which determine indirectly cell concentration, yield, and growth rate (Wang et al., 1977). An alternate system for optimizing yeast production, described by Nyiri et al. (1977), regulates the ratio of carbon : nitrogen in the nutrient feed on the basis of the respiratory quotient (RQ) which is calculated from determinations of rates of CO_2 evolution and O_2 consumption of the culture. For maximal yeast productivity the RQ ($= dCO_2/dO_2$) should be around 1.0.

In the modern bakers' yeast plant in Denmark, an electronic control system automatically monitors startup, propagation, and harvesting (Rosen, 1977). The problems of process measurements in computer-coupled fermentation systems, based on experience with a yeast-producing process, have been discussed by Swartz and Cooney (1978); also see Armiger and Humphrey, Volume II, Chapter 15.

c. Harvesting Yeast. Depending upon the amount of pitching yeast used, the length of trade propagations vary from 10 to 20 hours, and the cell solids from 3.5 to 5% in the fermentor broth. In some European plants, bakers' yeast production generally passes through more stages than described above and requires more time to reach the trade stage (Suomalainen, 1963; Butschek and Kautzmann, 1962). In the new Danish bakers' yeast plant described by Rosen (1977), a trade propagation of about 10 hours is used. At the end of the trade propagation, the broth is cooled as yeast cells are recovered by concentration in centrifugal separators, diluted with water, and reseparated. The washed suspension of cells, known in the trade as yeast cream, usually contains 17–19% solids. Some nozzle-type centrifuges with special bowl design can concentrate yeast to 23% solids. Yeast cream is chilled and stored at 0°C until processed further, either as bakers' compressed yeast, active dry yeast, or dried inactive yeast. For bakers' compressed yeast and active dry yeasts, the cream is dewatered in custom-modified plate-and-frame filter presses. A friable cake of 27–30% solids, known as press cake, is generally obtained. Some plants use rotary vacuum filters to dewater yeast cream intended for the compressed yeast market.

d. Compressed Yeast. This perishable product may be packaged in a variety of market forms: foil-wrapped blocks ($^3/_5$ to $^5/_8$ oz) and waxed paper wrapped and sealed blocks (1–2 oz) for household use; larger blocks (0.5, 1, and 5 lb) wrapped in waxed paper and 25- to 50-lb bulk yeast (crumbled) in sealed, polyethylene-lined, multiwall bags (Schuldt and Seeley, 1966).

Except for the bulk (crumbled) product, compressed yeast is prepared for packaging by mixing press cake with a small amount of water containing emulsifiers to facilitate smooth extrusion, even cutting, a standard consistency, and uniform plasticity in storage. The preferred emulsifiers bring about a whiter yeast cake and reduce surface water spotting due to water migration. Certain processed vegetable oils, glyceryl monostearates, and lactylated monoglycerides are suitable for this purpose. Further shelf-life enhancement of the household yeast products is provided by the incorporation of cereal starch and mold inhibitors; both ingredients are declared on the label.

Crumbled yeast is prepared from press cake which is broken into small, irregular pieces by forcing it through a suitable screen or perforated plate.

Domestic brands of compressed yeast contain from 8 to 9% nitrogen, 1–1.4% phosphorus, and 29–31% solids. In other countries, compressed yeast of 27% solids are common but may reach 30% in some trade areas. All compressed yeast products are generally stored at 4°C or lower.

e. Active Dry Yeast. Dehydrated or active dry yeast is a viable product produced from yeast press cake extruded directly through a perforated plate, forming spaghetti-like strands which are broken into short pieces and broadcast to uniform depth on a continuous, perforated, stainless steel belt. Dehydration is by stages over a 6-hour period (Thorn and Reed, 1959). Warm air (25°–45°C) is forced at moderate velocities through the bed of cylindrical fragments as they move from chamber to chamber in a tunnel dryer. The air humidity is so regulated that yeast moisture can equilibrate at about 8%. The light-tan, porous granules obtained are friable, have a mild odor, and may vary in moisture from 7.5 to 8.3%, and contain about 6.5–7.2% nitrogen. The performance characteristics of active dry yeast and its uses in baking are discussed in reviews by Thorn and Reed (1959) and Reed and Peppler (1973).

Domestic active dry yeast (ADY) is marketed in a variety of forms and packages. As unground, irregular cylinders, it is air-packed in fiber drums, cartons, and multiwall bags containing 10–300 lb product. Coarsely ground, free-flowing ADY is vacuum-packed in glass jars (4 oz), metal cans (8 oz, 2 lb), 25-lb nitrogen-pack cans, and also in pouches of paper–foil–pliofilm construction (5–42 gm ADY) packed in nitrogen or in air-pack protected with butylated hydroxyanisole. The smallest units in each category are generally for household use and for insertion in packages of prepared mixes (bread, hot rolls, coffee cake, pizza). In Europe active dry yeast is also vacuum-packed with flexible film in 500-gm, 1-kg, and 10-kg blocks.

Active dry yeast produced as described above may be packaged in nitrogen or other oxygen-free atmosphere to prolong shelf-life. When antioxidants and nonionic surfactants are emulsified with the yeast cream before dehydration, an active dry yeast of enhanced storage stability is obtained (Chen and Cooper, 1962). This means of protecting active dry yeast permits its use in bakery product prepared mixes of yeast-leavened cereal flours in which the yeast is an integral part of the ingredients (Chen et al., 1966).

Numerous methods have been developed to dehydrate yeast press cake with high viability (95–99%). One alternate batch process, the rotolouver dryer, yields semihard, smooth-surfaced pellets. This old process, still in limited use, consists of tumbling the fragments of extruded press cake in a double-walled, rotating drum while warm air streams through the fluted inner shell. The light-tan, firm ovals formed vary in size, usually 0.5–3 mm in diameter and 1–4 mm in length.

Fluidized bed drying methods for producing active dry yeast have been described by Burrows (1976), Taylor (1975), and Trevelyan (1973). A spray-drying process for viable yeast was developed by Fantozzi and Trevelyan (1971).

2. Brewing Yeasts

For the production of lager beer and ale, selected and adapted strains of S. cerevisiae and S. carlsbergensis (S. uvarum), respectively, are used (Thorne, 1975). Pure cultures are commonly employed. After brewing, most of the yeast is recovered and prepared to pitch another fermentor or is processed and dried for use as food or feed supplements.

Yeast handling in a modern brewery involves the preparation and keeping in readiness of pure yeast cultures for plant use at all times, as well as the recovery of yeast from finished fermentations, storing it, and determining its suitability for pitching new fermentations. Following a weekly schedule, according to Strandskov (1965), pooled colonies of the selected yeast strain are grown in tubes of hopped wort and transferred through successive stages until a mass culture is obtained in 10 liters of wort. The yeast crop from this fermentation is transferred to a final pure-culture vessel of 6-barrel (715-liter) capacity and diluted to a count of 6 million cells per ml with sterile plant wort. This process is repeated until the working capacity of the fermentor is reached. Then the buildup begins, also stagewise, to supply pitching yeast to the plant. Fermentors are usually pitched at the rate of 1–2 lb yeast slurry (15–17% yeast solids) per barrel of wort, or 10–14 million yeast/ml wort (see Volume II, Chapter 1).

Thorne (1970) described a yeasting system for brewing yeast which begins with a single cell in a drop of wort and ends, after numerous cultivations, in progressively increased volumes of wort, with enough pure yeast to pitch a 100-hl fermentor twice every week.

Yeast recovered from beer may be recycled after washing, cooled (0°–2°C), and held overnight. After about 20 recyclings, replacement with pure culture, as described above, is recommended (Strandskov, 1965). Before recycling plant yeast, its bacterial contaminants may be inactivated by washing the yeast with phosphoric acid solution (pH 2.8), by suspension in 0.75% ammonium persulfate solution, or by a combination of both treatments (Bruch et al., 1964).

Methods of determining brewing yeast vigor and other aspects of its behavior during fermentation have been discussed by Gilliland (1971), Reed and Peppler (1973), Thorne (1975), and MacLeod (1977).

3. Distillers' Yeasts

The variety of yeasts selected for brewery, distillery, and winery operations function primarily as catalysts to effect a maximum ethanol yield efficiently and to impart distinctive and uniform characteristics to the final product. A wide range of practices are found. For example, distillers' yeast may be produced in modern closed systems with meticulous control to deliver the inoculum free of interfering microorganisms (Coulter, 1964; also see Volume II, Chapter 3; Lyons and Rose, 1977). A new mash may be inoculated with a previously finished yeast mash, thereby reducing the preparation of a laboratory culture to weekly intervals. For spirits production, the mash may be seeded directly with bakers' yeast (Stark, 1954). For the latter practice, special strains in the form of active dry yeast are produced by bakers' yeast manufacturers.

In the modern yeasting system described by Coulter (1964), plant cultures are started daily from purity-checked wort agar slants of proved strains of S. cerevisiae. After building up the yeast concentration in three laboratory stages of sterilized malt syrup (25-ml, 500-ml, and 2-gal volume), the yeasting continues in sterilized mash through three successive plant stages of 45-, 450-, and 9000-gal batches. A rye–barley malt medium is soured with *Lactobacillus delbrückii* to facilitate sterilization. When a yeast population of approximately 100 million viable cells per ml is reached, the culture is used to inoculate plant fermentors at the rate of 3% by volume. Following fermentation and distillation, the yeast is recovered in combination with the grain residues (see Volume II, Chapter 3; Harrison and Graham, 1970).

When barley malt and rye were replaced with less expensive corn or milo and the mash was treated with glucamylase (produced with *Aspergillus awamori*), Van Lanen *et al.* (1975) obtained higher than usual cell counts and longer retention of viability of the stored seed yeast.

4. Wine Yeasts

The wine maker selects the strain of yeast which will exhibit the alcoholic fermentation capacity and flavor properties judged to be proper for the kind of grapes used. Although many cultures with characteristic names (Burgundy, Tokay, Montrachet, Epernay, etc.) are traditional in wine making, they are predominantly strains of *S. cerevisiae* (Kunkee and Amerine, 1970; also see Volume II, Chapter 5). Suitable pure cultures may be obtained from commercial laboratories serving the wine industry and from culture collections maintained by government agencies in wine-producing areas. The production of yeast starter has been described by Thoukis *et al.* (1963) and DeSoto (1955). When the crushing season begins, a sample of pasteurized or sterilized must (about 1–5 liters grape juice) is inoculated from a laboratory pure culture slant and built up to bulk fermentation size by serial transfers to increased volumes of fresh must treated with bisulfite or pasteurized to destroy unwanted yeast and bacteria. During the cultivations, the yeast grows to a peak population near 25 million cells per ml, and adapts itself to sulfur dioxide and to ethanol. The vigorous master culture, or starter, is charged to fresh batches of must at the rate of 2–5% by volume. At the end of fermentation, the yeast settles with the lees and cannot be recovered economically for reuse.

Three species of wine yeasts are produced in compressed and active dry form and marketed in North America: *Saccharomyces cerevisiae* (Montrachet strain) for table wines, *S. bayanus* for secondary fermentations of sparkling wines, and *S. beticus* for sherry fermentations. These products, available mainly as bulk, active dry wine yeast, are produced by propagating the selected yeast strain aerobically in molasses wort and employing equipment and procedures used in the manufacture of bakers' yeast (Thoukis *et al.*, 1963; Reed and Peppler, 1973).

5. Sake Yeast

Yeast starters, called *moto*, are prepared for the sake brewing process from a mash of steamed rice, mold stock (*koji*) and water. Two procedures are in general use: a natural sequential bacteria and yeast fermentation at 10°C, which requires about 30 days to complete (Ohwaki and Lewis, 1970), and a pure-culture modification, which is ready for pitching the main fermentation (*moromi*) in about 9 days. In the shorter yeasting process, the mash is acidified with lactic acid and inoculated with a pure culture of sake yeast, a selected strain of *S. cerevisiae* (Kodama, 1970).

6. Yeast Culture for Livestock Feeds

Yeast culture is one of eight feed yeast products defined by the Association of American Feed Control Officials (AAFCO, 1977):

Section 96.8 Yeast Culture is the dried product composed of yeast and the media in which it is grown, dried in such a manner as to preserve the fermenting activity of the yeast. The media must be stated on the label.

In general, the production of yeast culture begins with a culture medium, such as diluted molasses, which is inoculated with a suitable yeast, usually a selected bakers' yeast. Following several hours of fermentation, part of the brew is mixed with a mixture of moistened cereal grains, incubated for varying periods, and dried. Mild drying conditions are used to avoid damage to the yeast, enzymes, and other heat-sensitive products formed during the fermentation stages. Yeast culture is fed to livestock and poultry primarily as a digestive aid (Peppler and Stone, 1976). Rations for ruminants are generally fortified with 1–1.5% yeast culture, 1.5–2% in hog feeds, and 2–3% in poultry rations. The addition of 0.5% yeast culture to corn silage at ensiling markedly stimulated the natural lactic acid fermentation (Lane et al., 1974). Increased lactic acid contents of silage media were also noted by Woolford (1976) when unidentified yeasts isolated from various silages were added with the inoculum used.

B. Inactive Yeasts

Inactive (heat-killed, nonfermentative) yeasts, generally designated "dried yeast," are obtained either by recovery of residual brewers' yeast or by harvesting yeast grown primarily for the purpose of supplementing human diets and animal rations as sources of protein, flavor, and vitamins of the B complex. Such food and feed (fodder) yeasts are grown aerobically to high nitrogen contents (above 8%) under conditions comparable to those used in bakers' yeast propagations. However, a greater variety of carbon substrates may be used, notably molasses, sulfite mill waste liquor, wood hydrolysate, whey, hydrocarbons, and ethanol (Moo-Young, 1976; Litchfield, 1977a,b).

Food and feed yeasts are known by many synonyms: dried yeast, inactive dried yeast, dry yeast, dry inactive yeast, levure-aliment sèche, dried torula yeast, sulfite yeast, wood sugar yeast, xylose yeast, levadura alimenticia, la levure alimentieri, Nahrungshefe, and Saccharomyces Siccum, single-cell protein (SCP), and microbial protein. Since 1968 the term single-cell protein (SCP), introduced by M.I.T. scientists (Scrimshaw, 1968), has been identified worldwide with the efforts of microbial technology to provide an abundance of low-cost protein. Such programs initially concentrated on growing yeast on hydrocarbons, but they soon expanded rapidly to include the propagation of many species of bacteria, fungi, algae, and yeasts on a wide variety of agricultural and industrial by-products and waste materials (Laskin, 1977). The developments in this field are discussed by Litchfield in this volume, Chapter 4, and in numerous reviews: GIAM V (1978, also abstracted by Behrman, 1978), O'Sullivan (1978), Litchfield (1977a,b), Moo-Young (1976, 1977), Rockwell (1976), and Lipinsky and Litchfield (1974).

1. Primary-Grown Yeasts

The principal dried yeast products in the United States are comprised of two species of Saccharomyces, Candida utilis, and Kluyveromyces (Sacch.) fragilis (Table I): S. cerevisiae, molasses-grown strains of bakers' yeast, and recovered brewing yeast strains of this species; K. fragilis cultured on cheese whey; C. utilis propagated on sulfite mill waste liquor and ethanol; and S. uvarum (S. carlsbergensis) recovered from beer and ale. In some countries Candida tropicalis is grown on wood sugars and sulfite waste liquor (Butschek, 1962), Candida pseudotropicalis is cultured on whey, and C. utilis is propagated on molasses (Chien, 1960).

Quality guidelines for dried yeast grown for human use have been established by governmental agencies, professional associations, and food processors. Food yeast must conform to specifications of yeast type, color, flavor, microbial content, chemical composition, and vitamin potency as published by the International Union of Pure and Applied Chemistry (IUPAC), The British Pharmaceutical Codex, the National Formulary (N.F. XIII) of the American Pharmaceutical Association, and the food additive regulations of the Food and Drug Administration (FDA), U.S. Department of Health, Education and Welfare. Some food manufacturers impose additional chemical and bacteriological specifications for dried yeast added to products for thermal processing and special formulations.

As defined by N.F. XIII, dried yeast is "the dried cells of any suitable strain of Saccharomyces cerevisiae . . . or Candida utilis . . . grown in media other than those required for beer production. . . . Such yeasts, properly designated as to species, are commonly known as Primary Dried Yeast . . . and Torula Dried Yeast." "Brewer's dried yeast" is defined as "a by-product from the brewing of beer . . . washed free of beer prior to drying." When the washing step includes one or more alkaline treatments to remove hop resins adsorbed on the yeast cells, the product is designated "debittered brewer's dried yeast." Each type of dried yeast, N.F. XIII, must be free of fillers (starch, corn meal, etc.) and contain a minimum of 45% protein (N × 6.25); not more than 7% moisture and 8% ash; and, in each gram, the equivalent of not less than 120 μg thiamin hydrochloride, 40 μg riboflavin, and 300 μg niacin. Live bacterial and mold count maxima are 7500 and 50/gm respectively. In addition, the U.S. Food and Drug Administration has established a zero tolerance for Salmonella.

Dietary usage of food yeast is further restricted by the FDA on the basis of its folic acid content. When added to foods as a flavor ingredient, dried yeast may be used to the extent that the folic acid content it contributes does not exceed 40 μg/gm yeast [about 8 μg of free folic acid (pteroylglutamic acid)/gm]. Additions to foods in special dietary usage are limited to 400 μg total folic acid per day (100 μg free folic acid/ adult). Analyses of primary dried yeast on the market show both total and free folic acid levels to be well within the limits prescribed by the FDA. For each type of food yeast, the average total folic acid content was found to be 12 μg/gm. The free folic acid content of both dried yeasts was consistently less than 1 μg/gm yeast (Peppler, 1965).

Standards and methods specified by IUPAC in 1966, describe dried yeast as "the whole organism of one individual yeast or a mixture of several yeasts belonging to the family Saccharomycetaceae . . . and to the family Cryptococcaceae . . . obtained either as a by-product of fermentation processes or by special culture and conforming to such standards as may be laid down." Specifically excluded are yeasts which have been extracted, those containing more than 20% fat, and yeasts which carry inert fillers or substances that are not incorporated components of normal yeasts cells. Nine standards are set: Upper limits are specified for moisture (10%), ash (10%), lead (5 μg/gm), arsenic (5 μg/gm), live bacteria (7500/gm), and molds (50/gm). Minimum levels are set for nitrogen (7.2%, equivalent to 45% protein; i.e., N × 6.25), thiamin (10 μg/gm), riboflavin (30 μg/gm), and niacin (300 μg/gm). Dried yeast must also be free of starch and bacteria of the genus Salmonella.

The British Pharmaceutical Codex in 1949 stated that food yeast be dried by "a process which avoids decomposition of the vitamins present" and in 1954 required, in each gram, not less than 100 μg thiamin, 40 μg riboflavin, and 300 μg niacin (Pyke, 1958). The Swedish Pharmacopoeia in 1948, according to Pyke (1958), specified only the minimum content for thiamin (150 μg/gm).

Color, flavor, granulation, and packaging of food yeast are not specified in official standards. In the trade, however, these properties assume economic and esthetic importance according to the requirements of dietetic preparations, pharmaceutical products, and food-processing applications. The National Formulary (N.F. XIII) merely states, "Dried Yeast occurs as yellowish white to weak yellowish orange flakes, granules or powder, with an odor and taste characteristic of the type." Dried torula yeast is described in identical terms but "with a characteristic odor and bland taste."

Food yeast marketed in the United States generally conforms to the specifications prescribed in N.F. XIII, the food additive orders of the FDA, and the requirements of food processors.

Dried yeasts are readily compounded with a variety of spices, extracts, and imitation food flavors. One such flavor carrier, hickory-smoked food yeast, is used to impart a distinctive baconlike flavor and aroma to foods, especially snack-type products. Special processing and roasting of brewers' yeast produces a novel cocoa powder replacement (Andres, 1978).

Official specifications for dried yeasts used to supplement animal feeds are published by the Association of American Feed Control Officials (AAFCO, 1977). Crude protein (N × 6.25) values of 40% (minimum) are specified for primary dried yeast, grain distillers' dried yeast, molasses distillers' dried yeast, and torula dried yeast. The minimum for brewers' dried yeast is 35%. Most of the feed yeast available comes from two sources: breweries and fermentation of pulp mill sulfite liquor (torula dried yeast). Average values, in percentage, for major components are as follows (torula dried values in parentheses): crude protein 44 (47), crude fat 0.9 (1.6), crude fiber 2.7 (2.5), ash 6.6 (8.1), calcium 0.1 (0.5), phosphorus 1.4 (1.6). The vitamin values (in μg/gm) are: thiamin 90 (6), riboflavin 30 (40), niacin 440 (550), pantothenic acid 110 (65), choline 3700 (2800), folic acid 9 (20), and biotin 0.9 (1.0) (Peppler and Stone, 1976).

a. Molasses-Grown Dried Yeast.   Dried yeast may be produced on molasses in bakers' yeast plants using procedures described above (Fig. 1), or in factories designed expressly for food and feed yeast production, as in Taiwan's facility for continuous propagation of *C. utilis* in a battery of six Waldhof fermentors (Chien, 1960). In both aerobic systems, the yeasts are grown under sugar-limiting conditions to a nitrogen content above 8% (dry basis), separated, chilled and stored, or processed directly. After enrichment with vitamins (if needed to meet customer specifications), the yeast cream is pasteurized, usually drum-dried to about 95% dry matter, pulverized, and packed in plastic-lined fiber drums. In addition to the chemical and bacteriological standards discussed above, food processors may specify color and particle size. A detailed review of dried yeast composition was published by Peppler (1970).

Propagations of food and feed yeasts may also be fed selected minerals to enrich dried yeast with organically bound essential trace elements, notably zinc, selenium, and chromium. Such custom-grown yeast products are being developed in the wake of recent dietary trace element research (Mertz, 1969, 1975, 1977; Toepfer et al., 1973; Schwarz, 1974, 1976; Levander, 1975; Frost and Lish, 1975; Ammerman and Miller, 1975; Sakurai and Tsuchiya, 1975; Tuman et al., 1978).

A fluidized bed system for drying inactive yeast has been developed by Bardot (1972). Press or filter cake is mixed with dried yeast to a solids content which can be pelleted and is dried by fluidization with air or nitrogen at 150°C. Large-scale production at costs below those for spray-drying and improved digestibility of the product are claimed. The costs of removing water from yeast cells were compared by Labuza (1975).

b. Wood Sugar-Grown Dried Yeast. *Candida utilis* and *C. tropicalis* assimilate a variety of organic acids, aldehydes, and sugars (2–3% pentoses and hexoses) occurring in sulfite liquor. In the production of *C. utilis* in the United States, as described by Kaiser and Jacobs (1957), a blend of sulfite liquor from several pulp digesters is stripped of sulfur dioxide, adjusted to pH 4.5 with ammonia, sampled for sugar and phosphorus analysis, and supplemented with the required phosphorus and potassium. No biotin is added since both species, unlike most yeasts, synthesize this vitamin. Continuous, automatic, on-stream monitoring of sugar and phosphorus contents assures accurate and economical proportioning of added phosphate based on the sugar input. Ammonia additions and pH adjustments are also made and recorded automatically.

At the rate of 100 gal/min, the fortified liquor is introduced at the top of the draft tube of a 60,000-gal (225-m^3) modified Waldhof fermentor. Rapid mixing of liquor and yeasty emulsion occurs at the bottom of the draft tube as air (about 2300 ft^3/minute) is distributed by the rotor (about 390 rpm). Fermentor temperature is maintained near 32°C by continuously circulating the emulsion through an external cooler coupled with a refrigeration system. Yeast is harvested by withdrawing emulsion at a rate to balance the substrate input, defoamed mechanically, separated, washed, concentrated to about 15% solids, pasteurized, and dried to less than 6% moisture on atmospheric, steam-heated, double-drum dryers.

For production of food yeast, the yeast cream is washed several times to remove most of the residual lignosulfonates. This results in a high-nitrogen (above 8%) dried yeast unusually low in sodium content (less than 0.002%).

c. Whey-Grown Dried Yeast. As whey is drawn from the cheese vat it contains about 4% lactose, 1% protein, 1% ash, and a small amount of lactic acid. It will support the growth of the lactose-fermenting yeasts *Saccharomyces lactis, K. fragilis* and *C. pseudotropicalis.* Only *K. fragilis* is used in the United States for the production of dried yeast.

In the process reported by Bernstein and Plantz (1977), diluted whey is supplemented with ammonia, phosphoric acid, minerals, and yeast extract. After the medium is adjusted to pH 4.5, it is pasteurized (80°C for 45 minutes), cooled to 30°C, and transferred to a 15,000-gal (55-m^3) stainless steel fermentor equipped with an aeration system and jacketed for cooling. Following inoculation with *K. fragilis*, the cell count is increased to 1×10^9/ml in a period of 8–12 hours before the continuous yeast propagating phase is started. The addition of diluted whey and nutrients (1250 gal/hour) is monitored automatically over the 5- to 7-day growing period. Rapid analysis of broth samples for lactose and glucose concentration aids in maintaining the desired fermentation efficiency and in balancing whey input with yeast withdrawal. The yeast is recovered either as feed yeast or food yeast. For feed yeast the broth and cells are concentrated in a three-stage evaporator and spray-dried; the product contains about 45% crude protein. For the production of food yeast (50% protein), the yeast is harvested by centrifugation and the cream is spray-dried. Ethanol is recovered from the evaporator condensate and the supernatant streams from the centrifuges. Annual production of dried yeast is estimated at 5000 tons.

d. Ethanol-Grown Dried Yeast. Since 1975 *C. utilis* has been produced commercially in a solution of inorganic nutrients containing 100–500 ppm ethanol (Ridgway *et al.*, 1975). With two fermentors in parallel, operation of the closed propagating system is continuous, aseptic, fully automated, and computer-monitored. Fermentor broth is withdrawn continuously and separated by centrifugation. After pasteuri-

zation, the yeast cream is spray-dried. Modifications in substrate composition and cream processing yield four yeast products which differ in
emulsion capacity, water binding, nutritional value, and flavor (Andres,
1977). The entire annual production, about 7000 tons, is processed for
use in the food industry.

Weatherholtz and Holsing (1976) evaluated the acceptance of test
meals prepared with these materials (20 gm/day) and found no major
differences, as compared with soybean and potato supplements in the
control meals consumed by the student test group.

e. Hydrocarbon-Grown Dried Yeast. Since the early 1960s, the demand and subsequent research for new sources of protein led to the
development of numerous processes for producing high-protein (above
50%) microbial biomass on a variety of carbonaceous raw materials.
Bioconversion of petroleum fractions for use in feeds and foods was most
attractive while supplies of crude oil were cheap and plentiful and the
cost of protein from agricultural sources was climbing. Since yeasts and
bacteria proved to be superior protein synthesizers, processes for both
were the first to reach pilot stage and advance to commercial-scale
development.

Petroprotein from current production, principally in the form of dried
yeast (*Candida* species), is used directly in animal feeds. It partially
replaces protein which is in short supply and is needed for human
consumption. Direct human use of hydrocarbon-grown yeasts and bacteria awaits the completion of clinical studies aimed at reconfirming the
safety of microbial protein products. Until the public health issue of
residual hydrocarbons is fully resolved, two new yeast factories in Italy
remain on standby until authorities grant operating permits (O'Sullivan,
1978; Behrman, 1978). Each plant expects to produce 100,000 metric
tons of yeast annually in aseptic systems. The facility at Sarroch (Sardinia) is designed to grow *C. lipolytica* on *n*-paraffins with a protein
content of more than 60% in the dried yeast. The other production
complex, located at Saline di Montebello (Calabria), will grow *Candida
maltosa*.

Although the petroprotein industry may be set back by the present
high cost of hydrocarbon feedstocks and the low price of soybeans, the
evaluation of SCP processes on a variety of substrates continues. These
studies include, for example, biomass production on cellulosic materials, starch, molasses, paper mill wastes, and the by-product and waste
streams of food plants, livestock feeder lots, municipal garbage, and
industrial plants (Gillies, 1978; Han, 1978). This worldwide interest in the
SCP field has been the subject of many excellent recent reviews: Chap-

ter 4 in this volume and Litchfield (1977a,b); global activity (GIAM V, 1978); United States patents issued since 1970 (Johnson, 1977); update of developments and problems (Laskin, 1977); survey of manufacturers and economics (Moo-Young, 1976, 1977); utilization of cellulose and hydrocarbons (Rockwell, 1976); symposium (Tannenbaum and Wang, 1975); and developments in France (Senez, 1975).

f. Dried Yeast Produced on Methanol. In addition to the raw materials discussed above, microbial protein production on methanol and methane has received the most attention. Processes with yeasts and bacteria have reached the pilot stage (Laskin, 1977). In the most advanced yeast-based process, *Candida boidinii* is grown on methanol in continuous culture (Reuss et al., 1975; Laskin, 1977).

2. By-Product Yeast

Yeast grown during beer brewing exceeds the amounts required for pitching fresh wort. This surplus yeast may be separated from beer by centrifugation. The concentrated yeast cream may be pasteurized and drum-dried as feed yeast, mixed with spent brewers' grains for fodder, debittered and dried as food yeast, or converted to products used for nutritional and flavoring purposes (Andres, 1978).

The bitter components (hop resins and tannin materials) adsorbed on brewers' yeast are removed by making the slurry slightly alkaline and recovering the yeast by centrifuging. After one or two more alkaline washes, if needed, the final suspension of debittered yeast is acidified (about pH 5.5) and dried (East et al., 1966).

III. YEAST-DERIVED PRODUCTS

Debittered brewers' yeast and primary-grown species of *Saccharomyces, Candida,* and *Kluyveromyces* are the raw materials for the manufacture of yeast autolysates, hydrolysates, invertase, lactase, ribonucleic acid, isolated protein, and yeast glycan. These products meet specific applications in the food, beverage, pharmaceutical, and fermentation industries.

A. Autolysates and Hydrolysates

Active and inactive dried yeasts can be partially solubilized and converted to nutritional products and condiments in the form of syrups (pastes) and powders. These products are generally known as yeast extract paste, dried yeast extract, dired yeast autolysate, yeast hydro-lysate paste, and yeast hydrolysate powder. The principal raw materials are debittered brewers' yeast and S. cerevisiae grown on molasses and whey.

1. Autolysed Yeast

Autolysis is a process of self-digestion in which the cellular enzymes of viable yeast cells are induced to ferment some of the glycogen and solubilize a portion of the yeast protein. Fermentation and hydrolysis begin rather quickly in an agitated slurry of viable yeast (15–18% yeast solids) as it is slowly warmed to 50°C. Initiation of autolysis is stimulated and accelerated by small amounts (3–5%) of sodium chloride, ethyl acetate, or chloroform (not permitted in the United States). The reaction continues at about 45°C, with moderate agitation, for 12–24 hours. When the desired concentration of soluble nitrogen is reached, the autolysate may be pasteurized and drum-dried without removing the yeast cell walls, glycogen, and other undigested constituents of the cell. In the usual practice, however, a clear extract is obtained by centrifugation and then concentrated by evaporation to a thick syrup or paste (80% solids). It is used in this form, or spray-dried to form hygroscopic, water-soluble powders (East et al., 1966).

Autolysed dried yeast extracts differ in composition. When the au-tolysis of primary-grown yeast is induced with NaCl, the solids (96%) contain 6.8% nitrogen, 5% monosodium glutamate, and 37% NaCl. Au-tolysates prepared with organic solvents as modifiers of cell wall per-meability generally contain 97% solids, 11% ash, 5% monosodium glutamate, 2.5% NaCl, and 8% nitrogen (about 30% as α-amino nitro-gen). Some low-salt autolysates can be drum-dried without separation of undigested yeast cell matter. Such nonhygroscopic powders are only 50% soluble and contain 92% solids, 7–9% nitrogen, 6–8% ash, and 2–3% free amino acids.

The water-soluble, spray-dried, autolysed yeast extracts are important food ingredients, contributing a meaty flavor and aroma to soups, sau-sages, meats, gravies, sauces, seafoods, and vegetables.

2. Hydrolysed Yeast

Hydrolysis by controlled cooking of yeast in a strong acid solution is the most efficient method of manufacturing yeast extracts. Dried yeast is reslurried with water to a solids content of 65–80% and concentrated hydrochloric acid is added. The mixture is heated to 100°C in a wiped-film evaporator with reflux condenser. Hydrolysis continues until the desired level of amino nitrogen is reached, generally 6–12 hours for converting 50–60% of the total nitrogen to amino nitrogen, depending upon the initial acidity (Ziemba, 1967). Neutralization of the hydrolysate to pH 5–6, usually with sodium hydroxide, is followed by filtration, decolorization, and concentration to a syrup (45% solids) or paste (85% solids), or spray-dried (95% solids). Yeast hydrolysate syrups contain about 42% solids, 18% NaCl, 2.5–3% nitrogen, and 3.5% free glutamic acid.

Liquid hydrolysates diluted with two parts of yeast cream can be dried successfully on drum-dryers. While hydrolysis results in the highest yields of yeast extract, there is some loss of vitamins, protein, and flavor. The high final salt content of hydrolysates limits their use in foods.

To meet the special needs of food processors and health food formulators, yeast extracts and yeast autolysates may be fortified with monosodium glutamate and 5′-nucleotides, combined with extracts of vegetable or animal proteins, or blended with dried yeast. One such dietary product combines, prior to drying, equal weights of yeast autolysate and enzyme-hydrolysed lactalbumin. This high-nitrogen (9.5–10.5%) powder is also high in α-amino nitrogen (35–40% of the total nitrogen).

B. Enzymes

Numerous enzymes and coenzymes can be extracted from yeasts commercially. Those obtained in substantial quantities include invertase, lactase, coenzyme A, and nicotinamide adenine dinucleotide (NAD). Two of these, invertase and lactase, are produced in the largest volume and primarily for food use.

1. Invertase

Invertase (sucrase, saccharase, E.C. 3.2.1.26) is a β-fructofuranosidase whose action converts sucrose to fructose and glucose, and raffinose is hydrolysed to fructose and melibiose. Although the enzyme is present in many yeasts, S. cerevisiae, as bakers' yeast, is the principal commerical source (Meister, 1965).

Two commercial forms are produced: an invertase-rich dry yeast used to prepare high-test molasses from sugar cane juice, and a clear liquid concentrate extracted from autolysed yeast for the candy industry. The liquid product is the concentrate obtained by evaporation, under vacuum, of filtrates or extracts obtained from bakers' yeast autolysates. An invertase-enriched dry product is obtained by separating the yeast cells, after a short period of autolysis, dewatering the cream, extruding the press cake, and drying the pellets as is done in dehydrating active dry yeast (Peppler and Thorn, 1960).

Suomalainen *et al.* (1967) found invertase activity of bakers' yeast was nearly tripled when grown with mannose as carbon source instead of glucose. Highly purified invertase for biochemical research use is obtained by processing *C. utilis* cells. This yeast is also the preferred source for uricase and 6-phosphogluconate dehydrogenase.

2. Lactase

Lactase (β-galactosidase, β-D-galactopyranosidase, β-D-galactosidegalactohydrolase E.C. 3.2.1.23) catalyzes the hydrolysis of lactose to glucose and galactose. The hydrolytic reaction is used to alter the chemical and physical properties of milk, milk products, and whey (Bouvy, 1975; Holsinger, 1978; Guy and Bingham, 1978). Lactase is present in a number of common yeasts, including *S. lactis, K. fragilis* and *C. pseudotropicalis.* The commercial product, an odorless, tasteless, high-activity [40,000 oNPG (ortho-nitrophenyl-β-D-galactoside) units/gm] white powder, is extracted from *S. lactis* (Bouvy, 1975).

The process, described by Mahoney *et al.* (1975), compared four strains of *K. fragilis* grown aerobically in a lactose–salts medium. The harvested cells were autolysed at 37°C in 1 M potassium phosphate buffer (pH 7.0) containing manganese chloride (0.1 mM) and magnesium sulfate (0.5 mM). Lactase was isolated from the extract, after concentration, by precipitation with acetone. Stable preparations were obtained, but the lactase activity was low (860–1220 oNPG units/gm, dry wt.). Commercial, whey-grown, spray-dried yeast (*K. fragilis*) could also be used for lactase extraction. The highest enzyme yield (about 500 oNPG units/gm dry yeast) was also obtained with the phosphate buffer extraction procedure used for wet cells.

C. Ribonucleic Acid

Ribonucleic acid (RNA) is present in all living organisms, and microorganisms are exceptionally rich in nucleic acids. Yeasts, mainly species of *Candida* grown aerobically on molasses or sulfite waste liquor, are the preferred industrial raw material for the extraction of RNA. Potassium-sensitive strains of *Candida,* especially *C. utilis,* contain 10–15% RNA and are low in deoxyribonucleic acid (DNA) (Miyata, 1976; Akiyama *et al.*, 1975). Modifications in culture media composition (additions of zinc, antibiotics, etc.) and growing conditions influence the RNA biosynthesis (consult this volume, Chapter 10).

RNA is recovered from dried yeast (*C. utilis*) by extracting slurries (10% solids) with hot sodium hydroxide solution containing potassium chloride (about pH 11). After centrifuging, the supernate is treated with acid or ethanol and the crude precipitate is dried. Bakers' yeast may be extracted by a similar procedure using hot KCl–NaOH (Kuninaka *et al.*, 1976).

The crude RNA powder is converted enzymatically to the flavor-potentiating nucleotides 5'-guanosine monophosphate (5'-GMP) and 5'-inosine monophosphate (5'-IMP). For use in food applications the salts of these flavor accentuators are generally mixed, e.g., a 1 : 1 blend of disodium inosinate and disodium guanylate, with 5% of this mixture diluted in monosodium glutamate.

Japan is the principal producer of yeast nucleic acids, RNA, and its derivatives (Yamada, 1977).

D. Other Products from Yeasts

Candida utilis, and bakers' and brewers' yeasts, are the principal sources of biocatalysts, biochemicals, and metabolic intermediates used primarily for analytical and biochemical research, and in medical applications. Some of these are alcohol dehydrogenase, hexokinase, L-lactate dehydrogenase, glucose-6-phosphate dehydrogenase, glyceraldehyde-3-phosphate dehydrogenase, inorganic pyrophosphatase, the phosphoric acid esters of adenosine (AMP, ADP, ATP), the cytochromes, adenine, cytidine, guanosine, uridine, amino acids, glutathione (Horie *et al.*, 1976), vitamins, and enzymes of the citric acid cycle. Most of these products are isolated and purified in small quantities; their preparation has been reviewed by Harrison (1968, 1970).

As the predominant source of microbial protein (SCP), the isolation and characterization of protein from yeasts has attracted universal attention (Chen and Peppler, 1978). However, the commercial production of yeast protein concentrates is uneconomical at present (see Litchfield in this volume, Chapter 4). A recently developed process, described below, indicates the technology and feasibility of protein isolation from bakers' yeast.

1. Isolated Yeast Protein

A bakers' yeast cream containing 7–10% solids is passed through an industrial homogenizer (8000 psig) three times (Newell *et al.*, 1975; Robbins *et al.*, 1975). The homogenization ruptures more than 90% of the yeast cells and releases about 90% of the protein. After adjustment to pH 9.5, the homogenate is centrifuged. Two streams are collected: One is an extract of soluble protein and the other is a slurry of cell wall fragments. The soluble extract, following acidification, is incubated to initiate yeast nuclease activity and to precipitate the protein (Robbins, 1976). The protein is concentrated by centrifugation and spray-dried (Seeley, 1977). The final product, a free-flowing white powder, contains 75% protein and less than 1% nucleic acid. Its functional and nutritional properties have been described by Seeley (1977) and Robbins and Seeley (1977).

The proteins of *C. utilis* grown on ethanol have been isolated and modified for extrusion by Akin (1973, 1974). Vananuvat and Kinsella (1975) extracted protein from *K. fragilis* and studied its functional properties.

TABLE III. Principal United States Manufacturers of Yeasts and Yeast Products

	Bakers' yeast	Dried yeast	Wine Distillers, etc.	Extracts, lysates, etc.
Fleischmann Division Standard Brands, Inc. New York, New York	1	1	1	1
Red Star Yeast Division Universal Foods Corp. Milwaukee, Wisconsin	1	1	1	1
Anheuser-Busch, Inc. St. Louis, Missouri	1	2		2
Federal Yeast Division Diamond-Shamrock Corp. Baltimore, Maryland	1			
Yeast Products, Inc. Clifton, New Jersey		1, 2		2
Amber Laboratories Juneau, Wisconsin		2, 4		2
Lake States Yeast Division St. Regis Paper Company Rhinelander, Wisconsin		3		
Boise-Cascade, Inc. Salem, Oregon		3		
Amoco Food Company Chicago, Illinois		5		
Stauffer Chemical Co. Westport, Connecticut		4		4
Kraft Inc. Chicago, Illinois		4		

[a] Numbers identify carbon source on which yeast was grown: (1), molasses; (2), beer; (3), sulfite liquor; (4), whey; (5), ethanol.

2. Isolated Yeast Glycan

Yeast glycan is the concentrated, insoluble material separated from slurries of mechanically ruptured yeast cells (Newell *et al.*, 1975). After it is spray dried, the free-flowing white powder (6% moisture) contains about 87% carbohydrate, 11% protein, less than 0.5% nucleic acid, and 1% ash. In aqueous food systems yeast glycan functions as a thickener (Seeley, 1977).

IV. OUTLOOK

The market for yeast and yeast products sold to the food industry is relatively mature. It is expected to maintain a unit growth of 2–4% annually in the next 3–5 years. Production of yeast as a protein source for animal feeds will increase as plant proteins become more expensive and the safety of yeasts grown on uncommon substrates is established. Table III lists the manufacturers of yeast and yeast products operating in the United States.

REFERENCES

AAFCO (1977). Official Publication. Association of American Feed Control Officials, Inc., Baton Rouge, Louisiana.

Akin, C. (1973). U.S. Patent 3,781,264.

Akin, C. (1974). U.S. Patent 3,833,552.

Akiyama, S., Doi, M., Arai, Y., Nakao, Y., and Fukuda, H. (1975). U.S. Patent 3,909,352.

Ammerman, C. B., and Miller, S. M. (1975). *J. Dairy Sci.* **58,** 1561–1577.

Andres, C. (1977). *Food Process.* **38**(11), 87–88.

Andres, C. (1978). *Food Process.* **39**(11), 22–23.

Bach, H. P., Woehrer, W., and Roehr, M. (1978). *Biotechnol. Biochem.* **20,** 799–807.

Bakery Trends (1977). *Bakery Production and Marketing,* June 1977, pp. 79–100. Gorman Publ. Co., Chicago, Illinois.

Bardot, P. (1972). U.S. Patent 3,660,908.

Behrman, D. (1978). *ASM News* **44,** 102–106.

Bernstein, S., and Plantz, P. E. (1977). *Food Eng.* **49**(11), 74–75.

Bouvy, F. A. M. (1975). *Food Prod. Dev.* **9**(2), 10–14.

Bruch, C. W., Hoffman, A., Gosine, R. M., and Brenner, M. W. (1964). *J. Inst. Brew.* **70,** 242.

Burrows, S. (1970). *In* "The Yeasts" (A. H. Rose and J. S. Harrison, eds.), Vol. 3, pp. 349–420. Academic Press, New York.

Burrows, S. (1976). U.S. Patent 3,962,467.

Burrows, S., and Fowell, R. R. (1961). British Patents 868,133 and 868,621.

Butschek, G. (1962). *In* "Die Hefen" (F. Reiff *et al.*, eds.), Vol. II, pp. 611–664. Verlag Hans Carl, Nürnburg.

Butschek, G., and Kautzmann, R. (1962). *In* "Die Hefen" (F. Reiff *et al.*, eds.), Vol. II, pp. 501–610. Verlag Hans Carl, Nürnburg.

Chen, S. L., and Cooper, E. J. (1962). U.S. Patent 3,041,249.

Chen, S. L., and Peppler, H. J. (1978). *Dev. Ind. Microbiol.* **19**, 79–94.

Chen, S. L., Cooper, E. J., and Gutmanis, F. (1966). *Food Technol.* **20**, 1585.

Chien, H. C. (1960). *Food Eng., Feb.,* pp. 92–95.

Coulter, C. J. (1964). *Proc. Am. Soc. Brew. Chem.* pp. 64–69.

DeSoto, R. T. (1955). *Am. J. Enol. Vitic.* **6**, 26–30.

East, E., Smith, B. J., and Borsden, D. G. (1966). *Food Manuf.* Sept.

Fantozzi, E., and Trevelyan, W. E. (1971). U.S. Patent 3,615,685.

Frey, C. N. (1930). *Ind. Eng. Chem.* **22**, 1154–1162.

Frey, C. N. (1957). *In* "Yeast—Its Characteristics, Growth and Function in Baked Products—a Symposium" (C. S. McWilliams and M. S. Peterson, eds.), pp. 7–33. Quartermaster Food and Container Institute for the Armed Forces, Chicago, Illinois.

Frost, D. V., and Lish, P. M. (1975). *Annu. Rev. Pharmacol.* **15**, 259–284.

GIAM V (1978). "Global Impacts of Applied Microbiology" (P. Matangkasombut, ed.), *Proc. Int. Conf., 5th, 1977,* Mahidol University, Bangkok, Thailand. John Wiley, New York.

Gillies, M. T. (1978). "Animal Feeds from Waste Materials." Noyes Data Corp., Park Ridge, New Jersey.

Gilliland, R. B. (1971). *In* "Modern Brewing Technology" (W. P. K. Findlay, ed.), pp. 108–128. Macmillan, New York.

Guy, E. J., and Bingham, E. W. (1978). *J. Dairy Sci.* **61**, 147–151.

Han, Y. W. (1978). *Adv. Appl. Microbiol.* **23**, 119–153.

Hansen, E. C. (1896). "Practical Studies in Fermentation." C. Griffin & Co., London.

Harrison, J. S. (1968). *Process Biochem.* **3**(8), 59–62.

Harrison, J. S. (1970). *In* "The Yeasts" (A. H. Rose and J. S. Harrison, eds.), Vol. 3, pp. 529–545. Academic Press, New York.

Harrison, J. S. (1971). *Prog. Ind. Microbiol.* **10**, 129–177.

Harrison, J. S., and Graham, J. C. J. (1970). *In* "The Yeasts" (A. H. Rose and J. S. Harrison, eds.), Vol. 3, pp. 283–348. Academic Press, New York.

Hayduck, F. (1919). German Patents 300,662 and 303,221.

Holsinger, V. H. (1978). *Food Technol.* **32**(3), 35–40.

Hoogerheide, J. C. (1977). *CHEMTECH* **7**(2), 94–97.

Horie, Y., Katayama, H., Uchida, T., and Kuroiwa, Y. (1976). Japan Kokai 76/76,483.

Irvin, R. (1954). *In* "Industrial Fermentations" (L. A. Underkofler and R. J. Hickey, eds.), pp. 273–306. Chem. Publ. Co., New York.

I.U.P.A.C. World Wide Survey of Fermentation Industries 1963 (1966). *Pure Appl. Chem.* **13**, 405–417.

Johnson, J. C. (1977). "Yeasts for Food and Other Purposes." Noyes Data Corp., Park Ridge, New Jersey.

Kaiser, W., and Jacobs, W. (1957). *Food Process.* **1**, 22 and 58.

Kiby, W. (1912). "Handbuch für Presshefenfabrikation." Friedrich, Viehweg & Son, Braunschweig, West Germany.

Kodama, K. (1970). *In* "The Yeasts" (A. H. Rose and J. S. Harrison, eds.), Vol. 3, pp. 225–282. Academic Press, New York.

Kuninaka, A., Fujimoto, M., and Yoshino, H. (1976). Japan Kokai 76/86,188.

Kunkee, R. E., and Amerine, M. A. (1970). *In* "The Yeasts" (A. H. Rose and J. S. Harrison, eds.), Vol. 3, pp. 5–71. Academic Press, New York.

Labuza, T. P. (1975). *In* "Single Cell Protein II" (S. R. Tannenbaum and D. I. C. Wang, eds.), pp. 69–104. MIT Press, Cambridge, Massachusetts.

Lane, G. T., Spiegel, R. E., and Botts, R. L. (1974). *J. Dairy Sci.* **57**, 135.

Langejan, A., and Khoudokormoff, B. (1976). U.S. Patent 3,993,783.

Laskin, A. I. (1977). *Annu. Rep. Ferment. Processes* **1**, 151–175.

Levander, O. A. (1975). *J. Am. Diet. Assoc.* **66,** 338–344.

Lipinsky, E. S., and Litchfield, J. H. (1974). *Food Technol.* **28**(5), 16.

Litchfield, J. H. (1977a). *Food Technol.* **31**(5), 175–179.

Litchfield, J. H. (1977b). *Adv. Appl. Microbiol.* **22,** 267–305.

Lodder, J. (1970). "The Yeasts, a Taxonomic Study," 2nd ed. North-Holland Publ., Amsterdam.

Lyons, T. P., and Rose, A. H. (1977). *In* "Alcoholic Beverages" (A. H. Rose, ed.), Vol. 1, p. 635–692. Academic Press, New York.

MacLeod, A. M. (1977). *In* "Alcoholic Beverages" (A. H. Rose, ed.), Vol. 1, pp. 43–137. Academic Press, New York.

Mahoney, R. R., Nickerson, T. A., and Whitaker, J. R. (1975). *J. Dairy Sci.* **58,** 1620–1629.

Meister, H. (1965). *Wallerstein Lab. Commun.* **28,** 7–15.

Mertz, W. (1969). *Physiol. Rev.* **49,** 163–239.

Mertz, W. (1975). *Nutr. Rev.* **33,** 129–135.

Mertz, W. (1977). *Food Prod. Dev.* **11**(6), 62 and 67–69.

Miyata, S. (1976). Japan Kokai 76/54,975.

Moo-Young, M. (1976). *Process Biochem.* **11**(10), 32–34.

Moo-Young, M. (1977). *Process Biochem.* **12**(4), 6–10.

Newell, J. A., Robbins, E. A., and Seeley, R. D. (1975). U.S. Patent 3,867,555.

Nyiri, L. K., Toth, G. M., Parmenter, D. V., and Krishnaswami, C. S. (1977). U.S. Patent 4,064,015.

Ohwaki, K., and Lewis, M. J. (1970). *Wallerstein Lab. Commun.* **33,** 105–116.

O'Sullivan, D. A. (1978). *Chem. & Eng. News* **56**(12), 12; 56(38), 12–13.

Pasteur, L. (1874). *Compt. Rend.* **78,** 213.

Peppler, H. J. (1965). *J. Agr. Food Chem.* **13,** 34–36.

Peppler, H. J. (1967). *In* "Microbial Technology" (H. J. Peppler, ed.), pp. 145–171. VanNostrand-Reinhold, Princeton, New Jersey.

Peppler, H. J. (1970). *In* "The Yeasts" (A. H. Rose and J. S. Harrison, eds.), Vol. 3, pp. 421–462. Academic Press, New York.

Peppler, H. J. (1978). *Annu. Rep. Ferment. Processes* **2** (in press).

Peppler, H. J., and Stone, C. W. (1976) *Feed Manag.* **27**(8), 17–18.

Peppler, H. J., and Thorn, J. A. (1960). U.S. Patent 2,992,748.

Perlman, D. (1977). *CHEMTECH* **7,** 434–443.

Pomper, S., and Akerman, E. (1971). U.S. Patent 3,617,306.

Pyke, M. (1958). *In* "The Chemistry and Biology of Yeasts" (A. H. Cook, ed.), pp. 535–586. Academic Press, New York.

Reed, G., and Peppler, H. J. (1973). "Yeast Technology," Avi Publ. Co., Westport, Connecticut.

Reuss, M., Gneiser, J., Reng, H. G., and Wagner, F. (1975). *Eur. J. Appl. Microbiol.* **1,** 295–305.

Ridgway, J. A., Jr., Lappin, T. A., Benjamin, B. M., Corns, J. B., and Akin, C. (1975). U.S. Patent 3,865,691.

Robbins, E. A. (1976). U.S. Patent 3,991,215.

Robbins, E. A., Sucher, R. W., Schuldt, E. H., Sidoti, D. R., Seeley, R. D., and Newell, J. A. (1975). U.S. Patent 3,887,431.

Rockwell, P. J. (1976). "Single Cell Proteins from Cellulose and Hydrocarbons." Noyes Data Corp., Park Ridge, New Jersey.

Rosen, K. (1977). *Process Biochem.* **12**(3), 10–12.

Rosenquist, S. O., and Egnell, E. R. (1966). Swedish Patent 209,600.

Rungaldier, K., and Braun, E. (1961). U.S. Patent 3,002,894.

Sakurai, H., and Tsuchiya, K. (1975). *Environ. Physiol. & Biochem.* **5,** 107–118.
Schuldt, E. H., and Seeley, R. D. (1966). *Baker's Dig.* **40**(2), 42.
Schwarz, K. (1974). *Fed. Proc., Fed. Am. Soc. Exp. Biol.* **33,** 1748–1757.
Schwarz, K. (1976). *Med. Clin. North Am.* **60**(4), 745–758.
Scrimshaw, N. S. (1968). *In* "Single-Cell Protein" (R. I. Mateles and S. R. Tannenbaum, eds.), pp. 3–7. MIT Press, Cambridge, Massachusetts.
Seeley, R. D. (1975). *Food Prod. Dev.* **9**(7), 45.
Seeley, R. D. (1977). *MBAA Tech. Q.* **14**(1), 35–39.
Senez, J. C. (1975). "Producing Food Protein from Hydrocarbons," pp. 29–33. CNRS, Paris.
Stark, W. H. (1954). *In* "Industrial Fermentations" (L. A. Underkofler and R. J. Hickey, eds.), Vol. I, pp. 17–72. Chem. Publ. Co., New York.
Strandskov, F. B. (1965). *Wallerstein Lab. Commun.* **28,** 29–33.
Suomalainen, H. (1963). *Pure Appl. Chem.* **7,** 639.
Suomalainen, H., Christiansen, V., and Oura, E. (1967). *Suom. Kemistil. B* **40,** 286–287.
Swartz, J. R., and Cooney, C. L. (1978). *Process Biochem.* **13**(2), 3–7 and 24.
Tannenbaum, S. R., and Wang, D. I. C., eds. (1975). "Single-Cell Protein II." MIT Press, Cambridge, Massachusetts.
Taylor, R. (1975). U.S. Patent 3,885,049.
Thorn, J. A., and Reed, G. (1959). *Cereal Sci. Today* **4,** 198.
Thorne, R. S. W. (1970). *Process Biochem.* **5**(4), 15–16 and 22.
Thorne, R. S. W. (1975). *Process Biochem.* **10**(7), 17–20 and 25–28.
Thoukis, G., Reed, G., and Bouthilet, J. R. (1963). *Am. J. Enol. Vitic.* **14,** 148–154.
Toepfer, E. W., Mertz, W., Roginski, E. E., and Polanski, M. M. (1973). *J. Agric. Food Chem.* **21,** 69–73.
Trevelyan, W. E. (1973). U.S. Patent 3,780,181.
Tuman, R. W., Bilbo, J. T., and Doisy, R. J. (1978). *Diabetes* **27,** 49–56.
Vananuvat, P., and Kinsella, J. E. (1975). *J. Agric. Food Chem.* **23,** 216–221 and 613–616.
Van Lanen, J. M., Smith, M. B., and Maisch, W. F. (1975). U.S. Patent 3,868,307.
Walter, F. G. (1953). "The Manufacture of Compressed Yeast," 2nd ed. Chapman & Hall, London.
Wang, H. Y., Cooney, C. L., and Wang, D. I. C. (1977). *Biotechnol. Bioeng.* **19,** 69–86.
Weatherholtz, W. M., and Holsing, G. C. (1976). *Ecol. Food Nutr.* **5**(3), 153–159.
White, J. (1954). "Yeast Technology." Chapman & Hall, London.
Woolford, M. K. (1976). *J. Appl. Bacteriol.* **41,** 29–36.
World Wide Survey of Fermentation Industries, 1967 (1971). *I.U.P.A.C. Inf. Bull.* No. 3.
Yamada, K. (1977). "Industrial Fermentation." Int. Tech. Inf. Inst., Tokyo.
Ziemba, J. V. (1967). *Food Eng.* **19**(1), 82–85.

Chapter 6

Production of
Butanol–Acetone
by Fermentation

M. T. WALTON
J. L. MARTIN

MICROBIAL TECHNOLOGY, 2nd ed., VOL. I
Copyright © 1979 by Academic Press, Inc.
All rights of reproduction in any form reserved. ISBN 0-12-551501-4

I. INTRODUCTION

From its start as a wartime venture in England in 1916 until the establishment of the synthetic butanol process in the late 1930s, the fermentation process for the manufacture of butanol was the major method for its production. The plants were large operations in comparison to other fermentations then or now. A plant at Peoria, Illinois, contained ninety-six 50,000-gal fermentors and was capable of grinding 25,000 bu of corn per day for use as raw material.

Because of the unfavorable economics in comparison to the synthetic process using petroleum-based feed stocks, the fermentation process ceased to operate in the early 1960s in the United States and in most other countries. However, as price and availability of ethylene and propylene as feed stocks for the synthetic processes have become subjects of concern, there is renewed interest in examining the fermentation processes as a means of producing all or a portion of our future needs of butanol and acetone.

This chapter outlines the procedures used in the fermentation processes and presents data and suggestions which should be considered in evaluating the economics of these processes, based on replenishable carbohydrate raw materials, and of the synthetic processes which use diminishing hydrocarbon feed stocks.

II. HISTORY

The interesting history of the butanol–acetone process has been described by several authors and will be only briefly summarized here (Gabriel, 1928; Gabriel and Crawford, 1930; McCutchan and Hickey, 1954).

The original observation that bacteria produce butanol was made by Pasteur in 1861. The interest in commercializing the process was accelerated in 1909 in England primarily as a means to obtain butadiene as raw material for synthetic rubber. In 1912 Dr. Chaim Weizmann broke from the group that was developing the process but continued to study the fermentation. After about 2 years he found an organism, which he called *Clostridium acetobutylicum,* which successfully fermented starchy grains to produce acetone, butanol, and ethanol. He applied for a patent on the process and British Patent 4845 was issued to him (Weizmann, 1915).

With the outbreak of World War I in 1914, the production of acetone was of great interest to England for the manufacture of cordite, the explosive used in naval warfare. A plant was built at Kings Lynn, England, in 1914 for the production of acetone by fermentation, but the operation was a failure until Dr. Weizmann was placed in charge of the plant and installed his process. Because of the shortage of corn in England, the process was transferred in August 1916 to Canada where it was operated until November 1918. To supplement the output of acetone from Canada, a butanol–acetone plant was built in Terre Haute, Indiana, which operated from May to November of 1918.

While the Terre Haute plant was operating during World War I, there was no use for butanol and it was stored. Shortly after the end of the war, E. I. DuPont de Nemours & Company developed nitrocellulose lacquers for the automobile industry, and it was found that butyl acetate was the solvent of choice for this coating system. In order to supply butanol for conversion to butyl acetate, the butanol–acetone (ABE) fermentation unit at Terre Haute was started in 1920 as a private venture by Commercial Solvents Corporation, which had an exclusive license under the United States Weizmann patent (Weizmann, 1919).

The validity of the patents covering the Weizmann fermentation process was challenged in England in 1926 and in the United States in 1931. In landmark decisions, British Patent 4845 and U.S. Patent 1,315,585 were held to be valid and infringed. One of the key issues in these suits was whether the patent specifications gave proper directions for the isolation of *Clostridium acetobutylicum* (BY). In his decision on the English suit Lord Romer wrote,

> . . . the description in the Specification is intended for the guidance of those who are desirous of obtaining BY and not for people who wish to avoid getting it. (Reports of Patent Design and Trademark Cases, 1926; *U.S. Patent Quarterly,* 1932.)

After the expiration of the Weizmann patent in 1936, new ABE fermentation plants were built in Philadelphia, Pennsylvania; Baltimore, Maryland; Puerto Rico; and Japan.

Many early attempts were made to produce acetone and butanol by the fermentation of molasses since it furnished carbohydrates at a cost lower than did corn. However, it was not until 1938, when new microorganisms were discovered, that molasses was used successfully as a substrate for this fermentation (Arzberger, 1938; Carnarius and McCutchan, 1938; Legg and Tarvin, 1938; Woodruff et al., 1937; Müller, 1938).

III. VERSATILITY OF THE PROCESSES

A. Solvent Ratios

Acetone and butanol are produced from starchy materials such as corn by use of the species *C. acetobutylicum* with a solvents yield of about 26.5% based on dry corn. The solvent ratio usually runs 30% acetone, 60% butanol, and 10% ethanol (Gabriel, 1928; Gabriel and Crawford, 1930; McCutchan and Hickey, 1954). An unusual variant of *Clostridium saccharo-butyl-acetonicum-liquefaciens* (Code C-12) ferments molasses rather poorly but gives excellent yields on corn with a high butanol ratio. The solvent yield is 25% based on dry corn, with a ratio of 19% acetone, 78% butanol, and 3% ethanol.

The species of microorganism most widely used to ferment molasses is named *C. saccharo-butyl-acetonicum-liquefaciens* (Arzberger, 1938; Carnarius and McCutchan, 1938). With the exception of the variant (Code C-12), discussed in the preceding paragraph, microorganisms of this species give poor results on starchy grains. Selected variants of the foregoing species give solvent yields of 30–33% based on sucrose, with solvent ratios which vary from 20 to 35% acetone, 76–61% butanol, and 4% ethanol. (Carnarius and McCutchan, 1938; Legg and Tarvin, 1928; Woodruff *et al.*, 1937). Baba (1954a,b) reported data obtained by fermenting with *Clostridium toanum* a medium containing 0.3% defatted rice bran, 0.3% $(NH_4)_2SO_4$, and 7.0–7.8% sugar derived from cane juice, or blackstrap molasses, or a mixture of the two. The yield of solvents was about 30% based on glucose with an average solvents ratio of 52.92% butanol, 42.47% 2-propanol, 3.20% ethanol, and 1.41% acetone. Baba (1953) stated that the volume of gas evolved from the fermentation of a 6% sucrose solution with *C. toanum* was about 26 times the volume of mash and that the composition of the gas was about 50% CO_2–50% H_2.

Variations in solvent ratios have been accomplished largely by isolation of different cultures. Little success has been achieved by genetic manipulation of the cultures, but no work with butanol cultures has been reported in the last few years using newer techniques that have been successfully employed with other species of bacteria.

The ratio of solvents can be modified to some extent by manipulating the temperature of the mash and the composition of the nutrients in it. Tarvin (1941) reported that the addition of nitrates instead of ammonia to a molasses fermentation increased the amount of acetone at the expense of the butanol, e.g., in a typical experiment the addition of $NaNO_3$ at a concentration of 4.5 gm/liter increased the acetone percentage from 30.7 to 40.9. Hongo (1958) stated that the addition of neutral red in a concentration of 6 mM to a *C. acetobutylicum* fermentation of either xylose or glucose increased the butanol ratio at the expense of acetone, giving about 10–15% more butanol. Carnarius (1940) found that the butanol ratio could be increased 3–5% at the expense of acetone by cooling a molasses mash from 30°C to 24.5°C 16 hours after inoculation.

B. Raw Materials

Most of the acetone and butanol made by fermentation has used one or more of three sources of carbohydrates, i.e., corn, blackstrap molasses,* or high-test molasses.* Other common sources of carbohydrates such as wheat, rice, horse chestnuts, Jerusalem artichokes, potatoes, and beet molasses can be readily used. However, when we consider these fermentation processes as possible replacements for those based on hydrocarbon feed stocks, we must look carefully to carbohydrates derived from sources now considered to be waste products. These include hydrolyzed wood, hydrolyzed corncobs, hydrolyzed garbage, whey, sulfite liquor, and hydrol, a by-product from the manufacture of glucose from corn.

Leonard and Peterson (1947) found that sugar solutions prepared by the hydrolysis of maple, oak, and fir could be fermented with *C. acetobutylicum* to give a solvents yield of 24.5–38.5% based on the sugar utilized if the sugar solutions were first steam stripped of furfural and other impurities and then adjusted with lime to pH 6.5. The sugar utilization was about 80–85% with a solvent ratio of approximately 72% butanol, 25% acetone, and 3% ethanol. Further work needs to be done on this raw material since yields fell off when the initial sugar concentration of the mash increased much beyond 3% (Tsuchiya *et al.*, 1949).

The production of butanol and acetone from the sugars present in waste sulfite liquor was studied by Wiley *et al.* (1941), who treated the liquor first with lime to remove sulfites, then with additional lime to precipitate lignin, followed by removal of excess calcium salts as the sulfate. After the pH was adjusted to 5.8, the following nutrients were added: 0.05% $(NH_4)_2HPO_4$, 0.1% molasses, and 0.1% $CaCO_3$. In the fermentation about 80% of the sugars were utilized with a yield of 25–30% based on the sugars used. The ratio of solvents was 75% butanol, 20% acetone, and 5% ethanol. This process suffers from the same deficiency as that described above in that the sugar concentration of the sulfite liquor runs from 1 to 3%.

Langlykke *et al.* (1948) studied the production of butanol and acetone by fermenting xylose and glucose derived from the acid hydrolysis of corncobs. Their sugar solution, which was made in a bronze hydrolysis unit, contained about 18 ppm of copper, and this toxic element was removed by treating the liquor with powdered iron. The purified hy-

---

* Blackstrap molasses is the concentrated mother liquor from the crystallization of sucrose from sugar cane juice and contains 50–55% sucrose. High-test molasses is the concentrate of sugar cane juice in which 67% of the sucrose has been inverted to glucose and fructose to prevent crystallization and to avoid sugar import duties.

drolysate with an initial sugar concentration of 5.03% gave a sugar utilization of 90% and a 30.6% yield of solvents based on the sugar used. The ratio of solvents was butanol 61.7%, acetone 31.8%, and ethanol 6.5%. The hydrolysis work showed that from 2000 lb of corncobs one could obtain 606 lb of glucose and 484 lb of xylose (Dunning and Lathrop, 1945). It may be noted that the toxic effect of copper could be eliminated by a proper choice of materials of construction for the hydrolyzer. The problem of furfural toxicity would yield to steam stripping of the hydrolysate since furfural forms an azeotrope with water (Horsley, 1973).

Some recent results on the butanol fermentation of pentoses are exhibited in Table I.

IV. FERMENTATION PROCEDURES AND EQUIPMENT

This section discusses procedures geared to a plant equipped with fermentors of 50,000-gal capacity.

A. Overview of Procedures and Results

1. Corn and Other Grains

The fermentation of corn with *C. acetobutylicum* does not require any additives. Since corn oil is a valuable product and is not beneficial to the fermentation, it is recovered by steeping the corn followed by grinding with roller mills which flatten the germ so that it can be removed by screens. The oil is extracted from the germ and the degerminated meal used for the fermentation. Depending upon the value of the oil-cake meal as livestock feed, it is either sold for this purpose or is added with the degerminated meal to the fermentation mash. The starch portion of the oil-cake meal is used by the bacteria and the protein portion enhances the feed value of the recovered solids.

The corn meal is added to 60% water and 40% stillage (fermented liquor from which solvent has been distilled) to a calculated concentration of 8.5% of original dried corn. The concentration of solvents in the finished fermentation liquor (beer), after 50–56 hours, is about 22.5 gm/liter, with a solvent ratio of 30% acetone, 60% butanol, and 10% ethanol. About 0.25–0.26 lb of solvents are produced per pound of dry corn equivalent.

In order to obtain complete utilization of the starch, its starting concentration in the mash must be chosen so the butanol concentration in the final beer is no higher than 13 gm/liter. Depending on the butanol ratio

TABLE I. Recent Reports on the A.B.E. Fermentation of Hydrolysates of Corncobs, Cornstalks, and Wood

Culture	Substrate	Time (hours)	Yield (%)	Solvent Composition (%)			References[a]
				B	A	E	
C. acetobutylicum S$_{25}$	Hydrolysate of corncobs, 7%		26.2				(1)
	Hydrolysate of sawdust, 7%		22.2				(1)
Butyl culture	One part hydrolysate of cornstalks Three parts molasses	50–55	31–37				(2)
Butyl culture	Wood and plant hydrolysates, 8%		35.5	62	32	6	(3)
C. acetobutylicum 314	Corncob hydrolysate, 40–60%, (3.7% total) Molasses, 60–40% (sugar)	48–72	40	67.5			(4)
Butyl culture	Pentoses, 13.5%		25.4	67	33		(5)

[a] Key to references: (1) Taha et al. (1974); (2) Nakhamonivich et al. (1965); (3) Tornescher Hefe, G.m.b.H. (1954); (4) Nakhmanovich et al. (1960); (5) Moldenhauer and Lechner (1951).

characteristic of the culture being used, the required initial concentration of starch will be in the range of 6–8% on dry corn equivalent.

Differing greatly from the Weizmann organism, the Code C-12 culture has poor proteolytic powers and the medium must be fortified during the fermentation with ammonia, adding about 1.0% on dry corn equivalent to maintain a pH of 5.5–5.8. The fermentation is slower, requiring 72–80 hours, giving yields of about 25–25.5% on dry corn equivalent with a ratio of 19% acetone, 78% butanol, and 3% ethanol.

2. Molasses

As in the corn fermentation, a butanol concentration of 13 gm/liter in the final fermentation will limit the fermentation. Thus, the sugar concentration used in the starting mash varies from 5.5 to 7.5% depending upon the solvent ratio produced by the culture selected. The mash is augmented by the addition of superphosphate and ammonia. The amount of phosphate varies from 0.05 to 0.2% of P_2O_5 based on sugar. Blackstrap molasses requires the lesser amount and high-test molasses requires amounts in the higher range. The exact amount is determined by laboratory fermentation tests. The ammonia is added to high-test molasses mashes in hourly increments as needed to maintain the pH at 5.6–6.0, about 1.2–1.3% NH_3 based on sugar concentration being required. Because of the higher buffering properties of blackstrap molasses, all or nearly all of the ammonia can be added to mashes of this molasses before cooking.

3. Procedures Common to Fermentation of Both Types of Raw Material

a. Slopback. *Slopback* is the term used for the procedure of adding hot still residue from the solvent stripping column to the mash makeup. This procedure is used in all of the butyl fermentation processes. The amount of slopback varies from 25 to 50% of the total mash, the percentage used being the highest tolerated by the culture without giving reduced yields. The advantages are an increase in yield of solvents, a decrease in the amount of nutrients required, a saving in heat required in cooking, a saving in cooling water, and a substantial saving in steam required to evaporate the residue.

b. Sterilization. The fermentation equipment (tanks and lines) is sterilized 4–6 hours at 110°–115°C and is then cooled before use while maintaining it under a pressure of 15 psig (pounds per square inch gauge) with sterile fermentor gas.

Sterile fermentor gas is preferred to air to replace steam during the cooling of the fermentors to ensure an anaerobic atmosphere during the start of the fermentation in the plant. The absence of air in the fermentor gas reduces the cleanup problem when the gas is to be used for manufacture of methanol, ammonia, or Dry Ice. If the gas is not to be used for further processes, sterile air may be used for pressure maintenance.

The gas is sterilized by sending it through a tower of raschig rings in which a 5% solution of cresol in water is circulating. Other satisfactory methods, such as the use of glass wool filters, can be employed.

B. Culture Maintenance

Since the butanol–acetone organisms produce true spores, the cultures are maintained as spores, preferably with soil as a carrier. Each fermentor is started from the spore stock. The method of preparing the spore stock is as follows.

Sand, soil, volcanic ash, or a mixture of these, is screened and about 250 gm is placed in a 500-ml Erlenmeyer flask. Sufficient water is added to moisten the material. The flasks are plugged with cotton and sterilized four times at 24-hour intervals for 3 hours at 15 psig steam pressure. Excess moisture is removed in an oven.

Sufficient second- or third-generation culture (a generation for this discussion is a 24-hour transfer) is allowed to incubate until the fermentation is complete (about 72 hours) and then is added aseptically to the dried sterile soil until it is moderately wet. The flasks are then placed in a desiccator containing $CaCl_2$, the pressure is reduced, and they are allowed to remain until dry (about 3–4 days).

To check for contamination, a liberal amount of the soil–spore mixture is added to a molasses or corn medium, incubated, and plated daily for 3 days. The incubating flasks are plated aerobically and if no contaminating organisms appear, the cultures are then tested for solvent yield. If yields and solvent ratios are satisfactory, plant tests of several fermentors are made before the culture is adopted for general use.

C. Laboratory Development of Cultures

The first, second, and third generations are carried in the laboratory. For molasses fermentations and aided corn fermentations using *C. saccharo-butyl-acetonicum-liquefaciens*, the first generation is carried in 15 ml of peeled potato medium. This medium is prepared by blending in a Waring Blendor 250 gm wet weight of potatoes, 5 gm glucose, and 2 gm $CaCO_3$ with water to make 1000 ml of medium. This is distributed 15 ml to a tube and the tubes are plugged with cotton and autoclaved 1 hour at 15 psig steam pressure. The tubes are inoculated with about 1 gm of stock culture. The inoculated tubes are immediately heated for 1 minute in boiling water or in a steamer to inactivate vegetative cells and to "shock" the spores and are immediately cooled to inoculating temperature. The tubes are then incubated at 31°C for 24 hours.

For unaided corn fermentations (*C. acetobutylicum*), the first-generation medium is made with 5% dry corn equivalent in water. No additives are necessary. The medium is sterilized for 2 hours at 20 psig steam pressure. The inoculation procedure is the same as above, but the incubation is carried out at 37°C.

One first-generation tube is used to inoculate 300 ml of medium in a 500-ml flask. The inoculum medium for aided molasses contains 5.6% sucrose, 0.3% each of $(NH_4)_2SO_4$ and $CaCO_3$, and 0.01% P_2O_5 as superphosphate. The medium for aided corn fermentations is similar except that corn is substituted for molasses. The inoculated medium is incubated at 31°C. Medium for the unaided corn process is made with 5.5% dry corn equivalent and is incubated at 37°C.

After 24 hours, the one second-generation flask is used to inoculate two 4-liter flasks containing 2900 ml of the same medium as that for the second generation. One of these flasks is used to inoculate the first vessel in the plant. Since the fermentors in the plant are set in groups of four or more, at least one more flask is developed than needed to allow a choice of the flasks to be used. The flasks used should be gassing actively, have a pH of about 5.4, and pass a microscopic examination showing actively growing bacteria. The carbohydrate should be about one-half fermented when the flask is used.

D. Seed Stage

The first plant stage is usually a 1000-gal tank, although with certain cultures an intermediate 80-gal tank is used before the 1000-gal tank. The tanks are filled to 75% capacity with the desired medium. For the molasses fermentation cultures, the medium is 5.5% sucrose as molasses, 0.3% each of $(NH_4)_2SO_4$ and $CaCO_3$, and 0.02% P_2O_5 as superphosphate. The aided corn fermentation uses the same medium, except that corn on a dried equivalent basis is substituted for the molasses. Unaided corn seeds are made to 6.5% dry corn equivalent concentration and require no nutrients.

The medium is generally cooked in a batch cooker. The molasses medium is cooked at 118°C for 60 minutes and the corn mash is cooked at 133°C for 90 minutes. Either one is pumped hot through steam-sterilized lines to previously sterilized tanks still under steam and held a few minutes before being cooled to fermenting temperature by water in the jackets of the tanks.

After the tanks are cooled to the desired temperature, each is inoculated with the contents of one 4-liter flask. In the molasses fermentation, pH and Brix measurements are made on seeds immediately after inoculation and again after 24 hours to determine activity. The pH should be about 6.0 at the start and fall to about 5.4 at 24 hours. The Brix will drop about 1.5 points in 24 hours. In seeds with unaided corn, the activity is measured by titratable acidity. Gas rates measured on samples taken at 24 hours of age, using a Smith gas-rate tube, should show a displacement of at least 3 inches in 60 minutes in all three types of media (Salle, 1939). Aerobic contamination tests are started at 12 hours of age and read before the seeds are used at 26–28 hours. If these tests are negative, the seeds are used. An extra seed or two is usually carried with the group and if all are good, any extras are added to the fermentors.

The seeds are fermented under 15 psig gas pressure. Before the seeds generate sufficient gas, the gas pressure is maintained with fermentor gas from the sterile system, or with sterile air if gas is not available.

E. Fermentors

The final liquid volume in a 50,000-gal fermentor is in the range 45,000–47,500 gal, this volume being largely dependent on the extent of the foaming encountered.

After the fermentor has been washed free of residue from the previous fermentation, it is sterilized 4–5 hours with steam at 15 psig. It is then cooled, using sterile fermentor gas to hold a pressure of 10–15 psig. The molasses medium ranges from 5.5 to 7.5% sugar as molasses, 0.0025–0.015% P_2O_5 as superphosphate, and 0.07–0.1% NH_3 depending upon the culture used and on whether the medium contains blackstrap or high-test molasses. Aided corn mash consists of 6.8% dry corn equivalent, 0.014% P_2O_5 as superphosphate, and about 0.07% NH_3. The ammonia is added during the fermentation in hourly increments to maintain the pH in the range 5.5–6.0.

The ingredients, including 25–50% "slopback," are added to horizontal batch cookers with horizontal shafts to which are attached vertical rakes. Steam is added directly to the mash through several check valves in the bottom of the cooker. The molasses medium is cooked 60 minutes at 107°C; the corn medium is cooked 90 minutes at 133°C. After cooking, the mash is pumped through sterilized lines and cooled with jacketed water coolers to the fermenting temperature. Four cooks are required to fill one fermentor. Two seeds are added to the fermentor early in the filling period. A slightly positive pressure of sterile fermentor gas is maintained on the fermentors during the filling period (2–3 psig) and this is increased to 10–15 psig after filling and maintained at this value during fermentation by the gas evolved by the fermentation.

Samples are withdrawn from the fermentor at frequent intervals to check the pH of the aided medium and the titratable acidity of unaided corn medium. About every 8 hours samples are streaked on agar plates and incubated aerobically to check for contaminants. The molasses fermentation is completed in 40–45 hours. The unaided corn fermentation is completed in 50–60 hours, and the aided corn requires 70–80 hours.

F. Recovery of Solvents

After fermentation, the fermented liquor (beer) is sent to a beer well about twice as large as a fermentor at intervals sufficient to allow for a constant flow to the beer still. This is a continuous still, containing about 30 perforated plates, which concentrate the solvents to a 50% solution in water. The mixed solvents can then be separated by batch fractionation in the following steps: (1) Charge a kettle equipped with steam scrolls and a 40-plate column; (2) first take off the acetone fraction and collect it into the acetone running tank; (3) after a 5°C rise in outflow temperature, turn the flow into the ethyl fraction tank; (4) when the distillate starts to separate into two layers, turn the top layer to the crude butyl tank and return the bottom layer to the kettle. When there is no longer a top layer in the distillate, the steam is shut off and the liquid in the kettle is discarded.

The various crude fractions are further fractionated to remove traces of the contaminating solvents. The butanol fraction is dried by taking off the distillate through a decanter, returning the top layer to the kettle and the lower layer to the crude solvent fraction tank. When there is no longer separation and the butanol tests water-free, it is collected in the finished butanol tank. The kettle residue containing higher alcohols and esters is generally esterified and sold separately.

Of course, suitable reflux ratios are used throughout the separation process. A continuous fractionation system can be used, but its lack of flexibility might limit the choice of fermentation raw materials and the choice of cultures producing different solvent ratios.

G. Production Controls

Sugar in molasses is determined by means of Fehling's solution by any one of several common procedures. The sugar analysis of the cargo is used to calculate the amount of molasses needed. A hydrometer calibrated in degrees Brix is used to check the sugar content of the mash at the start of the fermentation and at the end to determine completeness of fermentation.

The starch content of the meal is determined by converting to glucose and analyzing by Fehling's solution. The moisture of the meal is obtained by determining loss in weight of a sample under desiccation in a drying oven. The presence of starch in the fermentation residue is determined by the iodine reaction.

Since the fermentations produce gas, the rate of gas evolution is a good measure of the fermentation rate. This rate is checked at intervals of about 8 hours using a Smith gas-rate tube (Salle, 1939). Since the harmful bacterial contaminants are facultative and since the butanol–acetone producers are strict anaerobes, the presence of growth on a nutritive agar plate streaked with a fermenting sample indicates the presence of contaminants. This tool is used in all fermentation stages.

A pH determination is used to indicate the need for ammonia addition during fermentation, and a large drop in pH indicates the presence of contaminants. The titratable acidity in the unaided corn fermentation is used to follow the course of that fermentation.

The concentration of solvents in the beer is estimated by comparing the specific gravity of the distillate from a measured sample of beer with a known amount of water solution containing solvents of about the same ratio. The solvent ratio can be determined by either mass spectrometry or gas chromatography. If this type of equipment is not available, an older method can be used in which the first step is the determination of acetone in the distillate by changing it to iodoform with hypoiodite. A fresh sample of distillate is salted out to obtain an anhydrous solution of solvents. This anhydrous solution is adjusted to a 25% acetone content by the addition of either acetone or butanol. By titrating the adjusted anhydrous solution to a turbidity point with water and comparing the result with prepared tables, a value for the ethanol content is obtained. The concentration of butanol is obtained by difference (McCutchan and Hickey, 1954).

V. STERILITY AND CONTAMINATION

A. Sterilization and Cleanliness

The butanol–acetone fermentation is less subject to contamination problems than are many others because it operates in the absence of air, thus discouraging aerobic organisms. Steam is the general means of maintaining sterile conditions. To make the steam heat effective, it is necessary to operate without building up organic products in the system.

The primary purpose of cooking the medium is to sterilize it. However, in the case of corn it is also necessary to gelatinize the starch, which makes it necessary to cook under more extreme conditions.

The mash can be cooked in either a batch cooker or a continuous cooker with high temperatures and short residence times. The latter is more adaptable to molasses mashes, since the medium is essentially soluble.

B. Bacterial Contamination

Although the butanol–acetone fermentation is less subject to contamination than are most aerobic processes, it must be operated essentially aseptically. The lactobacilli are particularly troublesome in the corn fermentation and to a lesser extent in the molasses fermenting processes. Samples of the fermenting medium from the laboratory through the final fermentors must be checked for bacterial contamination on aerobically incubated agar plates.

C. Bacteriophage

Problems with bacteriophage have been reported wherever the butanol–acetone process has been operated. It is detected when the fermentation slows as evidenced by low gas rate, nonacceptance of usual ammonia additions in the case of molasses fermentations, and nondisappearance of carbohydrates. A fermentation infected with a phage may stop without apparent cause. Although the fermentation may start again after 24–48 hours, it will probably be slow and give low yields. In order to recover values from a fermentor which has been infected with a phage, some workers have added an inoculum of *Candida utilis* to complete the utilization of carbohydrates (Kinoshita *et al.*, 1954).

The bacteriophage can be separated from its bacterial host by passing through a filter such as Berkefeld or a sintered glass, which retains the bacteria but allows the bacteriophage to pass through. The presence of phage is confirmed when the filtrate interferes with fermentations by its host in laboratory tests.

The culture can be made phage resistant by repeated transfers in the presence of the phagic filtrate. A small amount of the filtrate is added after each transfer. Legg (1928), when using the *C. acetobutylicum* culture, allowed each transfer to incubate to completion and "heat shocked" the culture after each transfer for 3 minutes at 100°C just prior to addition of the phagic filtrate. Legg and Walton (1938), in working with molasses fermenting cultures, found it preferable to make each transfer as soon as the culture recovered, adding fresh phagic filtrate at the same time. McCoy (1946) employed a special medium consisting of liver infusion, glucose, and Bactotryptone for carrying the resistant culture.

The development of a phage-resistant culture is not a complete solution of the problem. The resistant cultures contain particles of phage added to develop the property of resistance, and unless these particles are removed, they will accompany the culture in a new fermentation. The phage will then increase in virulence and in many instances the fermentation will again be affected within a short time (Hongo and Murata, 1966). Physical or chemical methods of inactivation are seldom successful because the tolerance of the phage and bacteria are quite similar.

In 1937 M. T. Walton in hitherto unpublished work discovered a method, based on antiphagic serum, by which he could inactivate the phage particles associated with the resistant bacteria. This method is carried out by injecting 1.0–5.0 ml of the phagic filtrate intravenously in rabbits every other day for 20–30 days. The development of the antiphagic property is confirmed when a sample of phagic filtrate to which the serum is added fails to inhibit the susceptible culture. The serum is judged acceptable when 1 ml inactivates all the phagic particles in 10 ml of filtrate. Two to five milliliters of the serum is then added to the phage-resistant culture immediately after the heat shocking step. The second generation is used to make soil cultures.

The butyl cultures may be attacked by more than one bacteriophage and the culture made resistant to one phage may be susceptible to another. It would, in that event, be necessary to repeat the foregoing process using filtrates of the new phage. Some phages are specific to a given strain of butanol–acetone-producing cultures; others may attack more than one strain, usually to a varying degree.

VI. BY-PRODUCTS

A. Gases

The gas produced from a molasses culture giving solvents with a butanol content of around 70% consists of 67% CO_2 and 33% H_2. Cultures giving a lower butanol percentage produce gases with a higher content of hydrogen. About 4.7 ft³ of gas is produced per pound of sucrose. About 5.9 ft³ of gas is produced per pound of dry corn from the

C. acetobutylicum and the gas composition is about 40% hydrogen and 60% CO_2. The gases can be used to produce methanol or ammonia. When the two gases are separated, the hydrogen can be used in chemical synthesis or burned for energy, and the CO_2 solidified into Dry Ice.

The gases leaving the fermentors carry appreciable amounts of solvents with them. The solvents may be recovered by passing the gases through beds of activated carbon or by a water scrubbing system.

B. Solid Residues

The liquid effluent from the bottom of the beer still contains solids which must be processed because of environmental considerations. The amount of solids and the biological oxygen demand (BOD) will vary depending on the raw material fermented and the amount of slopback, but the solids are normally in the range of 2.0–4.5 gm/100 ml and the BOD runs from 20,000 to 40,000 ppm.

In the late 1930s, Miner (1940) found that the bottoms from the beer still contained riboflavin and other vitamins. The bottoms from the beer still are concentrated in multiple-effect evaporators and the concentrate is dried with either drum or spray driers to give solids which can be incorporated in poultry and swine feeds. The dried residue from the fermentation of high-test molasses contains from 60 to 100 μg riboflavin/gm and from 28 to 30% protein.

Because of its high solids content, blackstrap molasses produces more dried residue than do other commercial raw materials. However, the disposal of the residue from a fermentation of blackstrap molasses is a problem because the high salt content leads to a hygroscopic product and it produces diarrhea in poultry.

The foregoing discovery by Miner became the foundation of riboflavin production by fermentation. Arzberger (1943) showed that the fermentation by *C. acetobutylicum* of corn mashes in vessels made of glass or aluminum gave a solid residue which assayed about 4000 μg of riboflavin per gram of solids. Walton (1945) discovered that the fermentation of a mash containing ground whole rice produced a solid which contained up to 5500 μg of riboflavin per gram of solids.

Meade *et al.* (1945, 1947) reported the fermentation by *C. acetobutylicum* of whey or skim milk to produce a beer with a riboflavin concentration of 25–50 μg/ml, while with a substrate of whey plus xylose he obtained riboflavin concentrations in the range 70–97 μg/ml.

Imai (1955) tested a large number of adjuvants in a study of riboflavin formation by *C. acetobutylicum* but the highest concentration of the vitamin obtained was 15 μg/ml.

Oguni (1955) assayed the stillage from a butanol–acetone plant and found B_{12} in a concentration of 11–27 μg/liter.

VII. PRODUCTION OF BUTANOL-ACETONE IN COUNTRIES OTHER THAN ENGLAND AND THE UNITED STATES

Most of the work on butanol–acetone published in the last 20 years has originated in countries other than England and the United States. A considerable portion of this work is summarized in Table II.

Several reports from Russian microbiologists deal with the production of butanol–acetone by continuous fermentation methods, and it is stated the use of this procedure allows the rate of solvents production from a given fermentor volume to be increased from 20 to 60% over the rate obtainable by the batch method (Yarovenko, 1964a,b). Furthermore, it is claimed the continuous method results in a 5–17% increase in the percentage of butanol in the solvent mixture (Yarovenko et al., 1966).

Yarovenko (1975) has recently published a book on the continuous fermentation procedure applied to butanol–acetone production.

VIII. FUTURE PROSPECTS FOR THE BUTANOL-ACETONE FERMENTATION

As the prices of raw materials based on petroleum increase and as the supplies diminish, the cost of producing butanol–acetone by the synthetic route will increase substantially. At some time in the next several years, this price increase may make the production of butanol–acetone by fermentation the process of choice. Whether cereal grains and molasses will then be available is not certain since there is mounting pressure to use these products for human food. However, there is an abundance of waste carbohydrates which could be transformed into a substrate for the butanol–acetone fermentation. In the United States alone we produce nearly 1 billion tons per year of solid waste, about one-half of which is cellulosic material (Humphrey, 1974). Cellulose (hexosans) and pentosans can be hydrolyzed to their constituent sugars, glucose and xylose, each of which is a suitable raw material for butanol and acetone.

In the next few paragraphs we list a number of problems associated with the use of waste carbohydrates as raw materials for fermentations. If one or more of these problems can be solved, it would greatly enhance the competitive position of the fermentation process for butanol–acetone.

204

TABLE II. Production of Butanol–Acetone in Countries Other Than England and the United States

Culture	Substrate	Time (hours)	Yield (%)	Solvent composition (%) B	A	E	I[a]	References[c]
SES-4	Molasses sugar conc. 5–6%		33	56.8	33.5	9.6		(1)
SES-5	Temperature, 33°C		33	56.2	29.9	13.8		(1)
Culture I (100,000-liter tank)	Molasses [6.37% sugar in mash plus rice bran 0.3%, (NH$_4$)$_2$SO$_4$ 0.3%], and CaCO$_3$ 0.4%; temperature 37.5°–27°C	104	33.6[b]					(2)
SES-5 (100,000-liter tank)	Molasses [6.28% sugar in mash plus rice bran 0.3%, (NH$_4$)$_2$SO$_4$ 0.3%], and CaCO$_3$ 0.4%; temperature 37°–35°C	90	32.0[b]					(2)
Butyl culture	By addition of calcium acetate and acetic acid to fermentation yield of 40% is claimed		40					(3)
C. acetobutylicum	Fermentation of 12% sugar solution in presence of 6% active carbon; carbon could be reused		36	66	33			(4)
Butyl culture	Fermentation of starch and starch plus beet molasses							(5)
	Starch, 100%		37.1					
	Starch, 90%; beet molasses, 10%		36.7					
	Starch, 50%; beet molasses, 50%		38.6					
	Starch, 37%; beet molasses, 63%		37.9					
C. acetobutylicum isolated from sugar cane roots	Mash ingredients in 1 liter: 250 ml sugar cane juice 20°Brix. 1.0 gm ground Vicia sativa, 2.5 gm KH$_2$PO$_4$, 4 gm CaCO$_3$		32	73	19–23	3–4		(6)
C. acetobutylicum	Molasses plus rye or wheat meal		26.5	66	31	3		(7)
C. acetobutylicum	Addition of 1.0% animal charcoal increases yield of solvents							(8)

Organism	Remarks						Reference[c]
C. saccharoperbutyl-acetonicum ATCC 13564	4% sugar solution with $(NH_4)_2SO_4$ 0.2%, Ca super-phosphate 0.1%, and $CaCO_3$ 0.3%	60	28.5				(9)
Butyl culture	In a medium of rye flour, sterile air blown in for 2–3 min every hour; butanol output increased 3.4–9.1%						(10)
C. aurianticum	Organism used glucose, fructose, sucrose and starch; nitrogen sources used: $(NH_4)_2SO_4$, $NaNO_3$, peptone, urea	30–40	60.5	0.5	7.4	25.6	(11)
C. saccharoperbutyl-acetonicum N-I-41	Molasses plus $(NH_4)_2SO_4$, $CaCO_3$, and Ca super-phosphate	60	None given	82.1	15	2.9	(12)
C. saccharoperbutyl-acetonicum	Plant-scale test using blackstrap molasses (5.14 gm sugar/100 ml mash); great deal of trouble with phage	55	34.2	75.4	20.2	4.4	(13)
C. saccharo-butyl-acetonicum-liquefaciens	Molasses (5% sugar in mash), protein nutrients, and an alkaline buffer added to mash	36–48	30	60	30	10	(14)
C. acetobutylicum S$_{25}$	5% Millet		34.4				(1)
	5% Corn		33.9				
	5% Sweet potatoes		33.7				
	7% Corncobs		26.2				
	12% Molasses		24.9				
	6% Corn, 4% molasses		24.3				
	7% Sawdust		22.2				

[a] B, butanol; A, acetone; E, ethanol; I, isopropyl alcohol.

[b] The yields in this reference have been recalculated to conform to the conventional formula:

$$\text{Yield} = \frac{\text{weight of solvents}}{\text{weight of sugar added}} \times 100\%.$$

[c] Key to references: (1) Bah et al. (1950); (2) Lee and Leu (1951); (3) Karsch and Schoedler (1956); (4) Hongo and Nagata (1958); (5) Zalesskaya et al. (1958); (6) Perdomo (1958); (7) Logotkin and Zaritskii (1959); (8) Bahadur and Saroj (1960); (9) Hongo (1960); (10) Nakhmanovich and Koch-kina (1960); (11) Suto et al. (1960); (12) Motoe (1961); (13) Hongo et al. (1965); (14) Srivastava (1968).

If the rate of hydrolysis of hexosans and pentosans by enzymes could be substantially increased and if excessive pretreatment costs were not encountered, this method could be cheaper than acid hydrolysis. The enzyme system would be expected to form fewer products toxic to the butanol–acetone organisms.

At present, acid hydrolysis is the best method of hydrolyzing hexosans and pentosans, but better data are needed on the hydrolysis conditions for each type of waste product so as to get the highest conversion to sugar consistent with a product which does not contain reversion compounds harmful to the bacterium being used for the fermentation.

Some of the waste products, e.g., corncobs and cornstalks, suffer from seasonal output and from the difficulty of collecting and delivering them cheaply to a central location. If these products are to be used, the formidable logistics problems must be solved. There are, however, numerous products such as bagasse, sawdust, other wood by-products, manure, and garbage which are already collected and which could be delivered to a plant.

Fermentation tests are needed for each candidate substrate in order to find the optimum conditions for the highest yield and the best ratio of butanol to acetone.

With the microorganisms now available, fermentation activity is severely depressed when the concentration of butanol in the mash goes above 13 gm/liter. If an organism could be found, either by isolation from natural sources or by genetic manipulation, which would function in mashes of higher butanol concentration, the operating costs of the fermentation process would be reduced considerably.

Hongo and Nagata (1958) report a method for doubling the amount of butanol which can be tolerated by the C. acetobutylicum organism. Their method involved the addition of 6% activated carbon to a 12% sugar solution, which was then fermented in good yields to give a calculated final butanol concentration of 28.5 gm/liter. This method should be tested on the substrates from waste carbohydrates to see whether it gives an economic advantage over the conventional fermentation process.

The continuous fermentation procedure should be tested with the substrates derived from waste carbohydrates to determine whether this principle is applicable to these products. If these tests prove that the increased productivity from a given fermentor volume and the increased ratio of butanol to acetone claimed in the literature can be verified, they would show a significant improvement in the economics of the fermentation processes.

By short hourly periods of aeration in a fermenting mash of rye flour,

Nakhmanovich and Kochkina (1960) were able to increase the amount of butanol in the range of 3.4–9.1%, with a concomitant sharp reduction in the oxidation–reduction potential. This experiment should be repeated with substrates from waste carbohydrates to determine whether aeration increases the ratio of butanol to acetone in the fermentation of mashes containing them.

As indicated earlier, present technology does not permit butanol made by fermentation to compete with butanol made by the oxo process using propylene as a feed stock. If the price of propylene rises to the point where present fermentation technology becomes competitive, the price of butanol may be so high that it will become a specialty product with a limited market. We believe that the future of producing butanol–acetone on a large scale by the fermentation process depends on the successful solution of the foregoing problems relating to the use of waste carbohydrates.

REFERENCES

Arzberger, C. F. (1938). U.S. Patent 2,139,108.
Arzberger, C. F. (1943). U.S. Patent 2,326,425.
Baba, T. (1953). *Bull. Fac. Eng., Hiroshima Univ.* **2,** 19–32. *Chem. Abstr.* **48,** 1624 (1954).
Baba, T. (1954a) a. *Bull. Fac. Eng., Hiroshima Univ.* **3,** 147–152; *Chem. Abstr.* **48,** 10289 (1954).
Baba, T. (1954b) b. *Bull. Fac. Eng., Hiroshima Univ.* **3,** 229–235; *Chem. Abstr.* **49,** 4227 (1955).
Bah, H., Lee, H., and Leu, C. (1950). *Rep. Taiwan Sugar Exp. Stn.* **6,** 193–206; *Chem. Abstr.* **49,** 14263 (1955).
Bahadur, K., and Saroj, K. K. (1960). *Jpn. J. Microbiol.* **4,** 43–51; *Chem. Abstr.* **55,** 8753 (1961).
Carnarius, E. H. (1940). U.S. Patent 2,198,104.
Carnarius, E. H., and McCutchan, W. N. (1938). U.S. Patent 2,139,111.
Dunning, J. W., and Lathrop, E. C. (1945). *Ind. Eng. Chem.* **37,** 24–29.
Gabriel, C. L. (1928). *Ind. Eng. Chem.* **20,** 1063–1067.
Gabriel, C. L., and Crawford, F. M. (1930). *Ind. Eng. Chem.* **22,** 1163–1165.
Hongo, M. (1958). *Nippon Nogei Kagaku Kaishi* **32,** 219–223; *Chem. Abstr.* **57,** 5123 (1962).
Hongo, M. (1960). U.S. Patent 2,945,786; *Chem. Abstr.* **54,** 21632 (1960).
Hongo, M., and Murata, A. (1966). *Hakko Kyokaishi* **24,** 354–358; *Chem. Abstr.* **66,** 1565 (1967).
Hongo, M., and Nagata, K. (1958). *Nippon Nogei Kagaku Kaishi* **32,** 585–594, 679–689, 758–770, and 855–858; *Chem. Abstr.* **55,** 19126 (1961).
Hongo, M., Harada, K., Akahoshi, K., Nagata, K., and Takahashi, S. (1965). *Nippon Nogei Kagaku Kaishi* **39,** 247–256; *Chem. Abstr.* **63,** 15505 (1965).
Horsley, L. H. (1973). *In* "Azeotropic Data-III," p. 24. Am. Chem. Soc., Washington, D.C.
Humphrey, A. E. (1974). *Chem. Eng. (N.Y.)* **81,** 98–112.

Imai, K. (1955). *Mem. Coll. Agric., Kyoto Univ., Chem. Ser.* **29,** 1–31; *Chem. Abstr.* **50,** 13363 (1956).

Karsch, W., and Schoedler, K. (1956). German Patents 941,184 and 941,185; *Chem. Abstr.* **52,** 19014 (1958).

Kinoshita, S., Shiga, A., and Okumura, T. (1954). *Nippon Nogei Kagaku Kaishi* **28,** 83–87; *Chem. Abstr.* **51,** 6937 (1957).

Langlykke, A. F., Van Lanen, J. M., and Fraser, D. R. (1948). *Ind. Eng. Chem.* **40,** 1716–1719.

Lee, H., and Leu, C. (1951). *Rep. Taiwan Sugar Exp. Stn.* **7,** 149–156; *Chem. Abstr.* **50,** 2115 (1956).

Legg, D. A. (1928). U.S. Patent 1,668,814.

Legg, D. A., and Tarvin, N. R. (1938). U.S. Patent 2,107,262.

Legg, D. A., and Walton, M. T. (1938). U.S. Patent 2,132,358.

Leonard, R. H., and Peterson, W. H. (1947). *Ind. Eng. Chem.* **39,** 1443–1445.

Logotkin, I. S., and Zaritskii, I. M. (1959). *Spirt. Prom'st.* **25,** 14–20; *Chem. Abstr.* **53,** 9562 (1959).

McCoy, E. F. (1946). U.S. Patent 2,398,837.

McCutchan, W. N., and Hickey, R. J. (1954). "Industrial Fermentations" (L. A. Underkofler and R. J. Hickey, eds.), Vol. I, pp. 347–352 and 364. Chem. Publ. Co., New York.

Meade, R. E., Pollard, H. L., and Rodgers, N. E. (1945). U.S. Patent 2,369,680.

Meade, R. E., Rodgers, N. E., and Pollard, H. L. (1947). U.S. Patent 2,433,232.

Miner, C. S. (1940). U.S. Patent 2,202,161.

Moldenhauer, O., and Lechner, R. (1951). German Patent 824,932; *Chem. Abstr.* **49,** 1354 (1955).

Motoe, M. (1961). Japanese Patent 61/5797; *Chem. Abstr.* **56,** 7806 (1962).

Müller, J. (1938). U.S. Patent 2,123,078.

Nakhmanovich, B. M., and Kochkina, L. V. (1960). *Tr. Tsentr. Nauchno-Issled. Inst. Spirt. Likero-Vodochn. Prom'sti* **9,** 147–153; *Chem. Abstr.* **58,** 9594 (1963).

Nakhmanovich, B. M., Senkevich, V. V., Shcheblykina, N. A., and Lipshits, V. V. (1960). *Tr. Inst. Lesokhoz. Probl. Khim. Drev., Akad. Nauk Latv. SSR* **21,** 47–53; *Chem. Abstr.* **56,** 15960 (1962).

Nakhmanovich, B. M., Lipshits, V. V., and Pavlovich, L. A. (1965). *Prikl. Biokhim. Mikrobiol.* **1,** 635–639; *Chem. Abstr.* **64,** 13351 (1966).

Oguni, Y. (1955). *Vitamins* **9,** 227–231; *Chem. Abstr.* **50,** 15023 (1956).

Perdomo, E. V. (1958). *Inst. Cubano Invest. Tecnol. Ser. Estud. Trab. Invest.* **3,** 1; *Chem. Abstr.* **52,** 20871 (1958).

Reports of Patent Design and Trademark Cases (1926). Vol. XLIII, No. 7, pp. 185–238. (Great Britain.)

Salle, A. J. (1939). *In* "Fundamental Principles of Bacteriology," pp. 247–251. McGraw-Hill, New York.

Srivastava, S. B. (1968). *Chem. Age India* **19,** 295–296; *Chem. Abstr.* **69,** 42777 (1968).

Suto, T., Namba, Y., Kurashima, K., Matsuki, M., and Furusaka, C. (1960). *Sci. Rep. Res. Inst., Tohoku Univ. Ser. D* **11,** 115–127 and 129–137; *Chem. Abstr.* **55,** 14582 and 14583 (1961).

Taha, S. M., Mahmoud, S. A. Z., Ishac, Y. Z., El-Sawy, M., and El-Demerdash, M. E. (1974). *Egypt. J. Microbiol.* **8,** 15–27; *Chem. Abstr.* **83,** 145, 747 (1975).

Tarvin, N. B. (1941). U.S. Patent 2,260,126.

Tornescher Hefe, G.m.b.H. (1954). German Patent 920,724; *Chem. Abstr.* **53,** 4649 (1959).

Tsuchiya, H. M., Van Lanen, J. M., and Langlykke, A. F. (1949). U.S. Patent 2,481,263.

U.S. Patent Quarterly (1932a). **12,** 47–57.

U.S. Patent Quarterly (1932b). **15,** 237.

Walton, M. T. (1945). U.S. Patent 2,368,074.

Weizmann, C. (1915). British Patent 4845.

Weizmann, C. (1919). U.S. Patent 1,315,585.

Wiley, A. J., Johnson, M. J., McCoy, E., and Peterson, W. H. (1941). *Ind. Eng. Chem.* **33,** 606–610.

Woodruff, J. C., Stiles, H. R., and Legg, D. A. (1937). U.S. Patent 2,089,522.

Yarovenko, V. L. (1964a). *Continuous Cultiv. Microorganisms, Proc. Symp., 2nd, 1962* Vol. 2, pp. 205–219; *Chem. Abstr.* **63,** 1193 (1965).

Yarovenko, V. L. (1964b). *Tr. Vses. Nauchno-Issled. Inst. Fermentn. Spirt. Prom'sti.* **15,** 105–124; *Chem. Abstr.* **63,** 15,504 (1965).

Yarovenko, V. L. (1975). *Pishch. Prom'st. SSSR* pp. 1–102; *Chem. Abstr.* **84,** 134121 (1976).

Yarovenko, V. L., Nakhmanovich, B. M., Shcheblykin, N. F., and Shcheblykina, N. A. (1966). *Prikl. Biokhim. Mikrobiol.* **2,** 344–347; *Chem. Abstr.* **65,** 6257 (1966).

Zalesskaya, M. I., Logotkin, I. S., Marima, A. M., Guskova, N. P., and Chekasina, E. V. (1958). *Tr. Tsentr. Nauchno-Issled. Inst. Spirt. Likero-Vodochn. Prom'sti* **8,** 52–60; *Chem. Abstr.* **55,** 3918 (1961).

Chapter 7

Microbial Production of
Amino Acids

YOSHIO HIROSE
HIROSHI OKADA

I. INTRODUCTION

Konbu (kelplike seaweed) has been widely used as an important traditional seasoning source in Japan. In 1908, the taste of konbu was identified as being due to L-glutamic acid (Ikeda, 1908). Based on this discovery, the industrial production of monosodium L-glutamate was initiated by Ajinomoto Co. in 1909. At that time, L-glutamic acid was produced by the acid hydrolysis of wheat gluten or soybean protein. A half-century after this discovery, it was reported that considerable quantities of L-glutamic acid accumulated in bacterial cultures (Kinoshita *et al.*, 1957a). The research and development, carried out mainly in Japan, resulted in the successful and economical production of L-glutamic acid

MICROBIAL TECHNOLOGY, 2nd ed., VOL. I
Copyright © 1979 by Academic Press, Inc.
All rights of reproduction in any form reserved. ISBN 0-12-551501-4

by the fermentative process. The significance of the establishment of microbial production of L-glutamic acid cannot be overestimated. Such essential metabolites as amino acids or nucleotides were considered not to be accumulated in microbial cultures due to regulatory mechanisms in the cell. The discovery of L-glutamic acid fermentation stimulated a wide variety of research aimed at the isolation of wild strains and the genetic derivation of mutants which could accumulate large amounts of primary metabolites. Fortunately, fundamental knowledge about biosynthetic pathways of amino acids and their regulatory mechanisms had already been elucidated to some extent.

Amino acids are constituents of proteins and play various important roles in living organisms. There have been increased demands for amino acids for use in the areas of food and feed additives and drug manufacturing.

Amino acids are used in foods as flavor enhancers, seasonings, and nutritional additives. They are also used to improve bread quality. Monosodium L-glutamate, the most important commercial amino acid, is widely used as a flavor enhancer. L-Alanine and glycine are used as flavoring agents. The essential amino acids which are often present only in insufficient amount in cereals are L-lysine, L-threonine, and L-methionine. L-Lysine and DL-methionine are now used practically to increase the efficiency of protein utilization of animal feeds. The reductive power of L-cysteine is utilized to improve bread quality in the baking process.

In the field of medicine, amino acids are used for infusions and as therapeutic agents. In the past, based on their nutritional values, essential amino acids were the principal components of the infusion mixtures, only glycine among nonessential amino acids being used to increase nitrogen content. However, it has recently been disclosed that nonessential amino acids also have nutritional significance (Breuer et al., 1964) and the development of amino acid infusions has been stimulated. Infusion formulas now contain various dispensable amino acids as well as essential ones. When the condition of the patients does not allow them to take required amounts of protein in a normal manner, for example, in a postoperative situation, an infusion formula is used.

Based on the specific physiological properties, some amino acids are also being used therapeutically. For example, L-arginine is effective in the alleviation of hyperammonemia and liver disorders by increasing arginase activity in the liver. L-Cysteine is used for the treatment of bronchitis or nasal catarrh, and L-glutamine and L-histidine for gastric ulcers.

Amino acid derivatives are also used in the chemical industry. They are applied in cosmetics, synthetic leathers, surface-active agents, fungicides, and pesticides. One typical example is use of poly-γ-methyl glutamate in the manufacture of synthetic leather.

To meet the diversified demands mentioned above, almost all of the amino acids are now commercially produced. The production methods developed so far may be summarized as follows: (1) protein hydrolysis methods; (2) chemical synthesis methods; (3) microbiological methods, including (a) direct production of amino acids from carbon source such as glucose; (b) precursor addition methods; and (c) enzymatic methods.

Table I presents production methods, uses, annual outputs, and representative manufactures of each amino acid now produced industrially. In the following, the microbial methods of amino acid production are reviewed with special emphasis on L-glutamic acid and L-lysine, which are the ones most abundantly produced. Production of other amino acids is referred to only very briefly.

Amino acid production methods include (1) biosynthesis and excretion of amino acids from carbohydrate, as in the case of L-glutamic acid and L-lysine accumulation from glucose; (2) salvage synthesis of amino acids from intermediate metabolites, for example, L-tryptophan accumulation from anthranilic acid; and (3) enzymatic production of amino acids from their substrates, for example, L-aspartic acid from fumaric acid and ammonia by aspartase of *Escherichia coli*. With selection of a suitable strain, careful determination of culture medium, and optimization of cultural condition, amino acid production can be precisely controlled and enhanced.

Since there have been many excellent reviews on amino acid fermentations exemplified by those of Kinoshita(1963), Yamada *et al.*(1972), Tsunoda and Okumura(1972), and Nakayama(1976), the authors' efforts are focused here on two essential aspects of the microbial production of amino acids: the strains and the process control.

II. MICROBIAL STRAINS EMPLOYED IN AMINO ACID PRODUCTION

A great variety of microorganisms has been isolated or induced for amino acid production. They are classified into wild strains, auxotrophic mutants, and regulatory mutants.

A. Direct Production of Amino Acids from Carbon Sources

The ultimate goal of microbiological amino acid production may be the direct accumulation of amino acid from cheap carbon sources.

TABLE I. Production Methods, Annual Outputs, and Representative Manufacturers of Amino Acids

Amino acids	Industrial production methods[a]	Annual output (tons/year)	Main uses	Representative manufacturers
L-Ala	C-3	10–50	Flavoring	Ajinomoto Kyowa Tanabe
DL-Ala	B C-1	150–200	Flavoring	Ajinomoto Tanabe
L-Arg	C-1 A	200–300	Infusion, therapeutic	Ajinomoto Kyowa
L-Asp	C-3	500–1000	Therapeutic flavoring	Ajinomoto Kyowa Tanabe
L-Asn	B	10–50	Therapeutic	Ajinomoto Tanabe
L-CysH	A	100–200	Improver of bread quality, antioxidant	Ajinomoto Kyowa Kyowa
L-Glu	C-1	200,000	Seasoning	Ajinomoto Kyowa Stauffer, Orsan
L-Gln	C-1	300	Therapeutic	Ajinomoto Kyowa
Gly	B	5000–6000	Sweetner	Ajinomoto Kyowa Tanabe
L-His	C-1	100–200	Therapeutic	Ajinomoto Tanabe
L-Ile	C-1	10–50	Infusion	Ajinomoto Kyowa Tanabe
L-Leu	A	50–100	Infusion	Ajinomoto Kyowa Tanabe
L-Lys	C-1	15,000–20,000	Feed additive	Ajinomoto Kyowa Tanabe
DL-Met	B	70,000	Feed additive	Ajinomoto Sumitomo Chemical Degussa
L-Phe	B	50–100	Infusion	Ajinomoto Tanaba
L-Pro	C-1	10–50	Infusion	Ajinomoto Tanabe
L-Ser	C-2	10–50	Cosmetic	Ajinomoto Kyowa Tanabe
L-Thr	C-1, B	50–100	Food additive	Ajinomoto Kyowa Tanabe

(continued)

TABLE I (*Continued*)

Amino acids	Industrial production methods[a]	Annual output (tons/year)	Main uses	Representative manufacturers
L-Trp	B	50–100	Infusion	Ajinomoto Tanabe
L-Tyr	A	50–100	Infusion	Ajinomoto Kyowa Tanaba
L-Val	C-1	50–100	Infusion	Ajinomoto Tanabe

[a] Key to methods: A, protein hydrolysis method; B, chemical synthesis method; C, methods employing microbial activities: (C-1) direct production from glucose or other carbon sources; (C-2) precursor addition methods; (C-3) enzymatic methods.

Therefore, much effort has been made to isolate wild strains and to derive genetic mutants, auxotrophs and regulatory mutants, which can accumulate amino acids directly from glucose. Wild strains are used in L-glutamic acid and L-glutamine fermentations. Amino acids which are excreted by auxotrophic mutants include L-citrulline, L-leucine, L-lysine, L-ornithine, L-proline, L-threonine, and L-tyrosine. Regulatory mutants are more widely used for the accumulation of L-arginine, L-histidine, L-isoleucine, L-leucine, L-lysine, L-methionine, L-phenylalanine, L-threonine, L-tryptophan, L-tyrosine, and L-valine.

1. Wild Strains

A large number of L-glutamic acid-producing bacteria have been isolated. Representative strains are shown in the following tabulation:

Genus *Corynebacterium:*	*C. glutamicum* (Kinoshita *et al.*, 1957b); *C. lilium* (Lee and Good, 1963); *C. callunae* (Lee and Good, 1963); and *C. herculis* (Distillers Co., Ltd., 1965)
Genus *Brevibacterium:*	*B. divaricatum* (Su and Yamada, 1960); *B. aminogenes* (Ota and Tanaka, 1959); *B. flavum* (Okumura *et al.*, 1962a); *B. lactofermentum* (Okumura *et al.*, 1962a); *B. saccharolyticum* (Okumura *et al.*, 1962a); *B. roseum* (Okumura *et al.*, 1962a); *B. immariophilum* (Okumura *et al.*, 1962a); *B. ammoniagenes* (Oishi and Aida, 1965); *B. alanicum* (Ogawa *et al.*, 1959); and *B. thiogenitalis* (Kanzaki *et al.*, 1967)
Genus *Microbacterium:*	*M. salicinovolum* (Doi and Kaneko, 1960); *M. flavum* var. *glutamicum* (Masuo *et al.*, 1962); and *M. ammoniaphilum* (Miyai *et al.*, 1964)
Genus *Arthrobacter:*	*A. globiformis* (Veldkamp *et al.*, 1963); and *A. aminofaciens* (Mogi *et al.*, 1967)

Most of these L-glutamic acid-producing bacteria are gram-positive, nonspore-forming, and nonmotile and require biotin for growth (Oishi, 1967). Among various components of the culture medium, the concentration of biotin is the most decisive factor in controlling the fermentation.

A great variety of investigations have clarified the role of biotin in the L-glutamic acid production. These works may be summarized as follows.

1. Product formation is greatest when the biotin concentration is suboptimal for the maximum growth. Excess biotin in the medium supports abundant growth but seriously inhibits product formation (Tanaka et al., 1960).

2. Even in the presence of excess biotin, addition of penicillin at the growth phase permits the cells to accumulate L-glutamic acid. Other antibiotics which inhibit cell wall synthesis, such as cephalosporin C, could replace penicillin (Sommerson and Phillips, 1962).

3. The addition of $C_{16}–C_{18}$ saturated fatty acids or their esters with hydrophilic polyalcohols also permits the cells to accumulate L-glutamic acid even in the medium of excess biotin (Takinami et al., 1963, 1964, 1965).

4. Oleic acid could replace biotin (Okumura et al., 1962b). An oleic acid-requiring mutant of B. thiogenitalis could accumulate L-glutamic acid even in the presence of excess biotin. The amount of L-glutamic acid was maximal when the oleic acid supply was limited (Kanzaki et al., 1967).

Discovery of penicillin and saturated fatty acids as antibiotin agents allowed industrial utilization of such biotin-rich raw materials as cane and beet molasses for L-glutamic acid production. It is physiologically and industrially important that these chemical agents be added to the medium during a specific phase of the cell growth.

The excretion of L-glutamic acid is closely related to the cell permeability, which is associated with both chemical and physical constituents of the cell membrane. It is due not to the lack of biosynthesizing activity but to the lack of L-glutamic acid permeability that the biotin-rich cells could not accumulate L-glutamic acid, as demonstrated by Shiio et al. (1963). The glutamic acid-producing cells, e.g., biotin-limited cells, penicillin- or Tween 60-treated cells, and oleic acid-limiting cells, excreted intracellular L-glutamic acid when they were washed with phosphate buffer. On the other hand, biotin-rich cells did not excrete endogeneous L-glutamic acid when washed with the buffer. It was clarified that the cells, whose biotin content was greater than 0.5 ng/mg of cell, biosynthesized sufficient amounts of oleic acid, which resulted in a high content of phospholipid in the cell membrane. These cells excreted endogenous L-glutamic acid poorly, as demonstrated by Takinami *et al.* (1966, 1967). Oleic acid-dependent mutants and C_{16}–C_{18} saturated fatty acid-treated cells, as well as the cells with biotin content less than 0.5 ng/mg of cell, synthesized insufficient amounts of phospholipid. Phospholipid-poor cells excreted endogenous L-glutamic acid. These results are in good agreement with the reports of Izumi *et al.* (1973) and Kamiryo *et al.* (1976) that acetyl-CoA carboxylase, which is a biotin enzyme, is involved in the biosynthesis of oleic acid and other fatty acids and that C_{16}–C_{18} saturated fatty acids inhibited oleic acid synthesis by repressing acetyl-CoA carboxylase (Fig. 1).

The effect of penicillin on L-glutamic acid permeability was of particular interest. The cells grown in the presence of penicillin or C_{16}–C_{18} saturated fatty acids showed enhanced permeability toward L-glutamic acid. Penicillin inhibited cell wall synthesis, resulting in elongated or swelled cells. Saturated C_{16}–C_{18} fatty acids did not have as drastic an effect on the cell structure. The effects of penicillin and C_{16}–C_{18} saturated fatty acids were compared, in connection with osmotic pressures of the

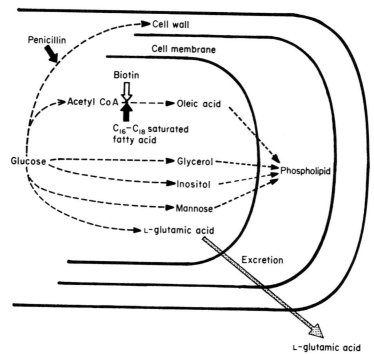

FIGURE 1. L-Glutamic acid permeability in relation to phospholipid biosynthesis. Filled, solid arrows show inhibition; the clear arrow shows promotion.

medium, by Shibukawa *et al.* (1968). The addition of polyoxyethylene monopalmitate enhanced L-glutamic acid permeability in a biotin-rich medium independently of the osmotic pressures. On the other hand, penicillin promoted the excretion of L-glutamic acid only in a medium of low osmotic pressure and was ineffective in a medium of high osmotic pressure. The cell structure was changed by penicillin only at low osmotic pressure. These results led to the conclusion that penicillin had a secondary effect on the membrane function. It primarily inhibited cell wall synthesis leaving the cell membrane unprotected, thus breaking the permeability barrier causing physical damage to the cell membrane.

An example of time course curves of L-glutamic acid fermentation is shown in Fig. 2. L-Glutamic acid-producing bacteria assimilated 100–200 gm/liter of sugar, giving about 50% of the production yield. Usable raw materials included carbon sources, such as a hydrolysate of starch, cane, or beet molasses, and nitrogen sources, such as urea or gaseous ammonia.

Some other amino acids are also extracellularly accumulated by wild strains. *Paracolobacterium coliform* accumulated 15 gm/liter of L-valine from 100 gm/liter of glucose (Udaka and Kinoshita, 1960). DL-Alanine was produced by *B. monoflagelum, B. amylolyticum, C. gelatinosum* (Ozaki *et al.*, 1960), *M. album* (Lafitskaya *et al.*, 1967), *M. ammoni-*

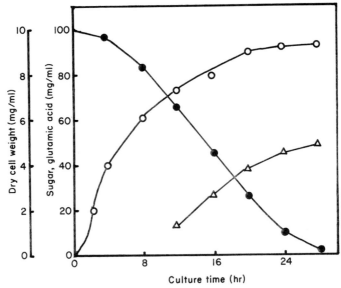

FIGURE 2.   Time course of glutamic acid fermentation by *Brevibacterium lactofermentum*. O, dry cell weight; ●, sugar; △, glutamic acid.

aphilum (Yoshida and Nakayama, 1975), and *Bacillus subtilis* (Matsumoto *et al.*, 1967). L-Glutamine fermentation by *B. flavum* or *C. glutamicum* was of industrial significance. L-Glutamic acid fermentation was converted to L-glutamine fermentation by altering the composition of the culture medium and the pH. In this conversion, N-acetyl-L-glutamine accumulated as a by-product. A high concentration of ammonium chloride, weakly acidic pH, and the presence of zinc ions favored the L-glutamine production at the expense of L-glutamic acid or N-acetyl-L-glutamine accumulation, as was demonstrated by Nakanishi *et al.* (1975). The maximum accumulation of L-glutamine by this wild strain reached 40 gm/liter at the production yield of 30%. However, regulatory mutants were found more potent. A sulfaguanidine-resistant mutant of *B. flavum* accumulated 39 gm/liter of L-glutamine at the production yield of 39% (Yoshinaga *et al.*, 1976).

2. Auxotrophic Mutants

Among auxotrophic mutants derived from *B. flavum* and *C. glutamicum,* several amino acids are produced, including L-citrulline, L-leucine, L-lysine, L-ornithine, L-proline, L-threonine, and L-tyrosine. The application of a homoserine (or L-threonine plus L-methionine)- requiring auxotrophic mutant of *C. glutamicum to* L-lysine production is one example (Kinoshita *et al.*, 1958; Nakayama *et al.*, 1961, 1966). For the effective production of L-lysine, sufficient amounts of biotin must be added to prevent the excretion of L-glutamic acid. Furthermore, the concentration of homoserine must be suboptimal for the growth, because the limitation of the feedback inhibition of aspartate kinase is necessary. Figure 3 shows the biosynthetic pathway of L-lysine and its metabolic controls in *C. glutamicum.* L-Lysine is synthesized through the diaminopimelic acid pathway. The key enzyme, aspartate kinase, is under the influence of the concerted feedback inhibition by L-threonine and L-lysine; excess homoserine increases the L-threonine pool within the cells, and aspartate kinase is inhibited.

Table II shows amino acids which are accumulated by auxotrophic mutants.

3. Regulatory Mutants

Knowledge of regulatory mechanisms of amino acid biosynthesis and the role of amino acid analogs has been applied to the breeding of regulatory mutants for amino acid production. Regulatory mutants possess a feedback-insensitive key enzyme and can be expected to accumulate the related amino acid in large amounts. A typical example of the application of regulatory mutants to amino acid fermentation was the

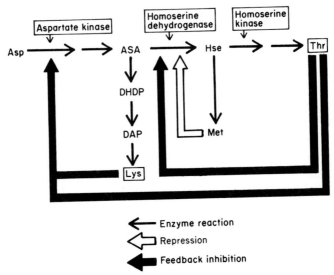

FIGURE 3. Regulation in lysine biosynthesis in *Brevibacterium flavum* and *Corynebacterium glutamicum.* ASA, aspartate semialdehyde; DADP, dihydrodipicolinate; Hse, hemoserine; DAP, diaminopimelate.

derivation by Sano and Shiio (1970, 1971) of a potent L-lysine producer. Unlike the regulatory mechanism in *E. coli* (Fig. 4), the regulation of L-lysine synthesis by *B. flavum* was rather simple: Only aspartate kinase was sensitive to feedback inhibition. This implied that *B. flavum* was a suitable parent strain from which to obtain regulatory mutants for L-lysine production. Regulatory mutants were obtained by the isolation of L-lysine analog-resistant mutants whose growth was not inhibited by an L-lysine analog, which behaved as a feedback inhibitor on aspartate kinase and which did not allow growth of the parent strain. Some of the mutants resistant to the analog were expected to be regulatory mutants with aspartate kinase desensitized to concerted feedback inhibition. Regulatory mutants might also be obtained by the induction of revertants from aspartate kinase-deficient mutants. Various changes in the enzyme structure might take place and some of them were expected to desensitize the regulatory enzyme.

S-(2-Aminoethyl)-L-cysteine (AEC) was employed for L-lysine production.

$$CH_2(NH_2) - CH_2 - CH_2 - CH_2 - CH(NH_2)COOH$$
L-Lysine
$$(NH_2) - CH_2 - CH_2 - S - CH_2 - CH(NH_2)COOH$$
S-(2-Aminoethyl)-L-cysteine (AEC)

TABLE II. Amino Acids Accumulated by Auxotrophic Mutants

Amino acids	Microorganism	Genetic markers	Accumulation and product yield (gm/liter) (yield, %)	References
L-Cit	Bacillus subtilis	Arg⁻	16.5 (12.7)	Okumura et al. (1964)
	Corynebacterium glutamicum	Arg⁻	10.7 (10.0)	Nakayama and Hagino (1966)
L-Lys	Corynebacterium glutamicum	Homoserine⁻	44 (22)	Kinoshita et al. (1958)
L-Orn	Corynebacterium glutamicum	Arg⁻	26 (26)	Kinoshita et al. (1957b)
	Arthrobacter citreus	Thiamine⁻, Arg⁻	27 (27)	Shibuya et al. (1967)
L-Pro	Brevibacterium flavum	Ile⁻, Sulfa-guanidine^r	35 (35)	Yoshinaga et al. (1966, 1976)
L-Tyr	Corynebacterium glutamicum	Phe⁻, Purine⁻	15.1 (7.5)	Hagino et al. (1973)
L-Leu	Corynebacterium glutamicum	Phe⁻, His⁻	16 (13.3)	Araki et al. (1974a)
L-Thr	Escherichia coli	Diaminopimelate⁻ Met⁻, Ile revertant	13.8 (18.4)	Kase et al. (1971)

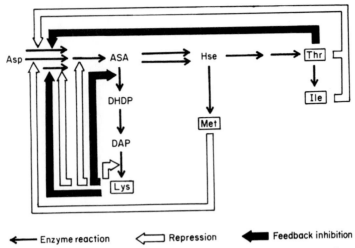

FIGURE 4. Regulation in lysine biosynthesis in *Escherichia coli.*

Figure 5 shows that the combined dose of AEC and L-threonine showed strong inhibitory effect on the cell growth of *B. flavum*. This inhibition was recovered by the addition of L-lysine. Among AEC-resistant mutants which could grow in the presence of AEC and L-threonine, there appeared some potent L-lysine producers. The best producer, FA 1-30, produced 57 gm/liter of L-lysine from 130 gm/liter of glucose. The regulatory mutant exhibited the following characteristics:

1. Its growth was no longer influenced by AEC, which inhibited growth of the parent strain with false feedback inhibition.
2. It produced large amounts of L-lysine.
3. L-Lysine production by the mutant was not restricted by exogenous L-threonine and L-lysine.
4. Activity of its aspartate kinase was not inhibited by L-threonine and L-lysine.

Similar techniques of breeding regulatory mutants were successfully applied to obtain other amino acid producers, examples of which are given in Table III.

B. Precursor Addition Methods

For some amino acids, e.g., L-tryptophan and L-serine, direct fermentation was very difficult to establish, perhaps due to the difficulty in avoiding metabolic control. Instead, the precursor addition method was

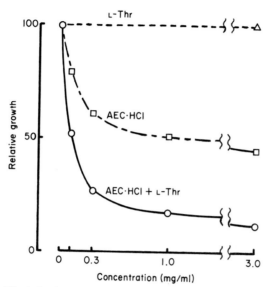

FIGURE 5. Effect of L-threonine and AEC [S-(2-aminoethyl)-L-cysteine] on cell growth of *Brevibacterium flavum.*

developed for these amino acids. Some adequate intermediate metabolites in the biosynthetic pathway of these amino acids were converted to the corresponding amino acids by microorganisms. Amino acid biosynthesis was accompanied by some other side reactions and an ATP supply was required. Therefore, glucose or some other carbon source was also supplied in a medium.

Terui and Niitsu (1969) isolated *Hansenula anomala,* which accumulated only traces of L-tryptophan from glucose; however, significant amounts of the amino acid where accumulated in the medium when anthranilic acid was added as a precursor. Furthermore, indole-resistant mutants of *H. anomala* were derived. They had the following favorable characteristics for L-tryptophan production: (1) Anthranilic acid and L-tryptophan decomposing activities were very weak; and (2) the enzymes concerned with L-tryptophan synthesis were not subject to repression and inhibition by the end product. Glucose, ethanol, or glycerol was used as a carbon source, and anthranilic acid or indole was used as a precursor. With stepwise feeding of the mixed solution of glycerol and anthranilic acid, 3 gm/liter of L-tryptophan was accumulated at the production yield of about 90%.

L-Serine was also produced by *Corynebacterium glycinophilum* (Kubota *et al.,* 1971, 1972). About 20 gm/liter of glycine were converted into 10 gm/liter of L-serine per 100 gm/liter of glucose.

TABLE III. Amino Acids Accumulated by Regulatory Mutants

Amino acids	Microorganisms	Genetic markers[a]	Accumulation and yield (gm/liter) (yield, %)	References
L-Arg	Brevibacterium flavum	TA^r	34.8 (26.7)	Kubota et al. (1973)
	Bacillus subtilis	$ArgHX^r$	4.5 (4.5)	Kisumi et al. (1971)
	Corynebacterium glutamicum	$D\text{-}Ser^r$, $D\text{-}Arg^r$, $ArgHX^r$, Ile revertant	19.6 (13)	Nakayama and Yoshida (1972)
L-His	Brevibacterium flavum	TA^r, SG^r, AHV^r, ETH^r, ABT^r	10 (13)	Mihara et al. (1973)
	Corynebacterium glutamicum	TRA^r, MG^r, AG^r, TU^r, MP^r, $5MT^r$	15 (10)	Araki et al. (1974b)
	Serratia marcescens	Histidase$^-$, TRA^r, MH^r, TA^r	13	Kisumi et al. (1975)
L-Ile	Brevibacterium flavum	AHV^r, OMT^r	14.5 (14.5)	Shiio et al. (1973)
	Serratia marcescens	$IleHX^r$, ABA^r	12 (12)	Komatsubara et al. (1973)
	Corynebacterium glutamicum	Met^-, AHV^r, AEC^r, THI^r, ETH^r, AL^r, ABA^r	10.6 (10.6)	Kato and Nakayama (1976)
L-Leu	Brevibacterium lactofermentum	TA^r, MET^-, Ile^-	28 (22)	Tsuchida et al. (1975a)
	Serratia marcescens	ABA^r, Ile revertant	13.5 (11)	Kisumi et al. (1973)

L-Lys	*Brevibacterium flavum*	AEC^r	57 (44)	Sano and Shiio (1970)
L-Met	*Corynebacterium glutamicum*	Thr^-, ETH^r, SM^r, MetHX^r	2 (2)	Kase and Nakayama (1975)
L-Phe	*Brevibacterium flavum*	MFP^r	2.2 (2.2)	Sugimoto et al. (1973)
	Corynebacterium glutamicum	Tyr^-, PEP^r, PAP^r	9.5 (9.5)	Hagine and Nakayama (1974)
L-Trp	*Brevibacterium flavum*	Phe^-, 5FT^r, 3FP^r	6.2 (6.2)	Shiio et al. (1975)
	Corynebacterium glutamicum	5MT^r, TrpHX^r, 6FT^r, 4MT^r, PEP^r, PAP^r, TyrHX^r, PheHX^r	12 (12)	Hagino and Nakayama (1975)
L-Tyr	*Brevibacterium flavum*	MFP^r	1.9 (1.9)	Sugimoto et al. (1973)
	Corynebacterium glutamicum	AT^r, PAP^r, PEP^r, TyrHX^r, Phe^-	17.6 (17.6)	Hagino and Nakayama (1973)
L-Thr	*Brevibacterium flavum*	AHV^r, Met^-	18 (18)	Nakamora and Shiio (1972)
	Brevibacterium flavum	AHV^r	13 (13)	Shiio and Nakamori (1970)
L-Val	*Corynebacterium glutamicum*	Met^-, AHV^r, AEC^r	14 (14)	Kase and Nakayama (1972)
	Brevibacterium lactofermentum	TA^r	31 (31)	Tsuchida et al. (1975b)

[a] TA, 2-thiazole alanine; ArgHX, arginine hydroxamate; SG, sulfaguanidine; AHV, α-amino-β-hydroxyvaleric acid; ETH, ethionine; ABT, 2-aminobenzothiazole: TRA, 1,2,4-triazole alanine: MG, 6-mercaptoguanine; AG, 8-azaguanine; TU, 4-thiuracil; MP, 6-mercaptopurine; 5MT, 5-methyltryptophan; OMT, o-methylthreonine; ABA, α-aminobutylic acid; AEC, S-(2-aminoethyl)-L-cysteine; SM, selenomethionine; MetHX, methionine hydroxamate; MH, 2-methylhistidine; PEP, p-fluorophenylalanine; PAP, p-aminophenylalanine; MFP, m-fluorophenylalanine; 5FT, 5-fluorotryptophan; 3FP, 3-fluoro-phenylalanine; TrpHX, tryptophan hydroxamate; PheHX, phenylalanine hydroxamate; PEP, p-ethylphenylalanine; TyrHX, tyrosine hydroxamate; AT, 3-aminotyrosine; IleHX, isoleucine hydroxamate; AL, 4-azaleucine; THI, thiaisoleucine.

225

C. Enzymatic Methods

Another microbial method of amino acid production is that of enzymatic conversion of a substrate to the corresponding amino acid. Amino acids which are effectively produced by this method include L-aspartic acid, L-alanine, L-tyrosine, L-cysteine, L-tryptophan, L-phenylalanine, and L-lysine. Table IV lists the amino acids produced by this method and shows substrate, enzymatic reaction, enzyme source, and productivity of each method. Some amino acids are produced with a single enzyme, for example, L-aspartic acid from fumaric acid and ammonia by aspartase of *E. coli*. On the other hand, the enzymatic conversion of chemically synthesized compounds into amino acids often requires several types of enzymes. For example, the conversion of DL-2-amino-thiazoline-4-carboxylic acid (DL-ATC) into L-cysteine requires three enzymes of *Pseudomonas thiazolinophilum:* ATC racemase, L-ATC hydrolase, and S-carbamoyl-L-cysteine (SCC) hydrolase. The reactions are as follows:

Enzymatic methods have been investigated for the production of L-lysine, L-phenylalanine, and L-tryptophan.

Fukumura (1977) established an enzymatic process for producing L-lysine using DL-α-amino-ϵ-caprolactam (DL-aminolactam) as the starting material. This process consisted of a racemization of D-aminolactam and the hydrolyzation of L-aminolactam to form L-lysine, as shown in the reactions below. *Achromobacter obae* nov. sp. and *Cryptococcus laurentti* were selected as the enzyme sources for D-aminolactam racemase and L-aminolactam hydrolase. DL-Aminolactam at a concentration of 100 gm/liter was converted into L-lysine in 25 hours. The production yield was 99.8% (moles product per mole of substrate).

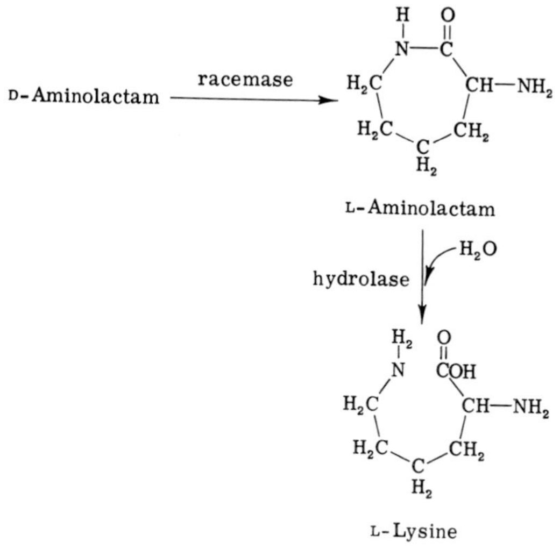

D-Aminolactam $\xrightarrow{\text{racemase}}$

L-Aminolactam

hydrolase

L-Lysine

III. PROCESS CONTROL IN AMINO ACID FERMENTATION

During the course of industrialization of penicillin fermentation, pure-culture technology was developed in large-scale submerged fermentors. It included (1) maintenance of the pure-culture condition (sterilization of equipment, media, and air); (2) control of pH, temperature, and foaming; and (3) aeration and agitation. All these advances in biochemical engineering were successfully applied to industrialization of amino acid fermentation. It would be out of place, however, to attempt to cover all these diverse advances in this chapter. Instead, a few aspects which have particular relation to the large-scale production of amino acids are described.

A. Maintenance of Pure–Culture Conditions

In amino acid fermentation, contamination is controlled according to the procedures which have been established in penicillin fermentation.

TABLE IV. Amino Acids Produced by Enzymatic Methods

Amino acids	Reaction[a]	Enzymes	Enzyme sources	Product concentration and yield (gm/liter (yield, %)	References
L-Asp	Fumaric acid + NH_4^+ → L-Asp	Aspartase	Escherichia coli	560 (99)	Kitahara et al. (1959)
L-Ala	L-Asp → L-Ala + CO_2	Aspartate decarboxylase	Pseudomonas dacunhae	268 (100)	Chibata et al. (1965)
L-Tyr	Phenol + pyruvate + NH_4^+ → L-Tyr + H_2O	Tyrosinephenol-lyase	Erwinia herbicola	61 (94)	Enei et al. (1973)
L-DOPA	Pyrocatechol + pyruvate + NH_4^+ → L-DOPA + H_2O	Tyrosinephenol-lyase	Erwinia herbicola	59 (95)	Enei et al. (1973)
L-Trp	Indole + pyruvate + NH_4^+ → L-Trp + H_2O	Tryptophanase	Proteus rettgeri	63 (98)	Nakazawa et al. (1972)
5-(OH)-L-Trp	5-(OH)-Indole + pyruvate + NH_4^+ → 5-(OH)-L-Trp + H_2O	Tryptophanase	Proteus rettgeri	28 (60)	Nakazawa et al. (1972)
L-Trp	DL-Trp hydantoin + 2 H_2O → L-Trp + CO_2 + NH_3	D-Trp hydantoin racemase, L-Trp hydantoin hydrolase, N-carbamoyl-Trp hydrolase	Flavobacterium aminogenes	50 (100)	Yokozeki et al. (1976)

Amino acid	Reaction	Enzymes	Microorganism	Yield	Reference
L-CysH	DL-ATC + 2 H_2O → L-CysH + CO_2 + NH_3	D-ATC racemase, L-ATC hydrolase, S-carbamoyl-L-CysH hydrolase	*Pseudomonas thiazolinophilum*	30 (95)	Sano et al. (1976)
L-CysH	β-Chloro-L-Ala + Na_2S → L-CysH + NaCl + NaOH	Cysteine desulfhydrase	*Aerobacter aerogenes*	49 (89)	Ogishi et al. (1977)
L-Phe	DL-Phe hydantoin + 2 H_2O → L-Phe + CO_2 + NH_3	D-Phe hydantoin racemase, L-Phe hydantoin hydrolase, N-carbamoyl-L-Phe hydrolase	*Flavobacterium aminogenes*	50 (100)	Yokozeki et al. (1976)
L-Lys	DL-Aminocaprolactam + H_2O → L-Lys	D-Caprolactam racemase	*Achromobacter obae*	100	Fukumura (1974)
		L-Caprolactam hydrolase	*Cryptococcus laurentii*	(100)	

[a] L-DOPA, 3-(3,4-dihydroxphenyl)-L-alanine; DL-Trp hydantoin; DL-5-indolylmethyl-hydantoin; DL-ATC; DL-2-aminothiazoline-4-carboxylic acid; DL-Phe hydantoin; DL-5-benzyl-hydantoin; DL-aminocaprolactam; DL-α-amino-ε-caprolactam.

Unlike penicillin fermentation, in which penicillinase-producing microorganisms often inactivate penicillin, no specific contaminants have been observed in amino acid fermentations. However, amino acids have a nutritional value to many microorganisms, and the culture conditions of amino acid fermentations are very favorable for many undesirable microbes. To ensure the maximum productivity, therefore, the maintenance of pure-culture conditions is essential.

In order to eliminate foreign microbes in the media, heat sterilization is most commonly used. For large volumes of the media in the commercial operation, continuous rather than batchwise sterilization has proved to be more successful. The following advantages, pointed out by Humphrey and Deindoerfer (1959), have been substantiated in amino acid fermentations: (1) energy saving, (2) better quality control, and (3) improved productivity. Figure 6 shows an example of a continuous sterilization system with a plate exchanger employed for L-glutamic acid fermentation.

The problem of producing a large quantity of sterile air is not peculiar (but is no less important) to amino acid fermentation. In laboratories, cotton plugs are satisfactory, and the oxygen transfer characteristics of various types of cotton plugs have been quantitatively elucidated in relation to the productivity of L-glutamic acid (Hirose et al., 1966a). In pilot-plant jar fermentors and large-scale commercial fermentors, air sterilization is performed with glass-wool filters. Developments in the theoretical basis of air filtration were very effective in the design and construction of air filters (Aiba et al., 1973).

B. Automatic Control of Amino Acid Fermentation

Figure 7 shows a typical structure of a commercial fermentor for amino acid fermentation. Although most of the microorganisms for amino acid production are bacteria and less power input may be needed for agitation than is used in antibiotic fermentations, it should be noted that the oxygen requirement and heat evolution per unit volume of fluids and time are large because of the high rate of sugar consumption and respiration.

Utilization of gaseous ammonia is a great advantage in amino acid fermentation. Since the molecules of amino acids contain much nitrogen, the selection of an adequate nitrogen source and its sufficient supply are essential for the performance of this type of fermentation. The use of gaseous ammonia as a nitrogen source and as a pH-controlling agent solves various technological problems. It avoids undesired dilution of the fermentation broths, controls the pH more precisely, and makes the equipment and operation more simple.

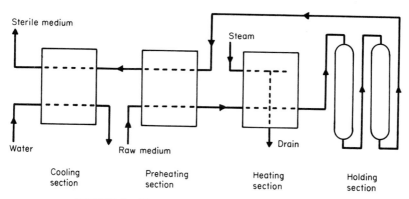

Sterile medium

Steam

Water   Raw medium   Drain

| Cooling | Preheating | Heating | Holding |
| section | section | section | section |

FIGURE 6. Plate–exchanger continuous sterilizer.

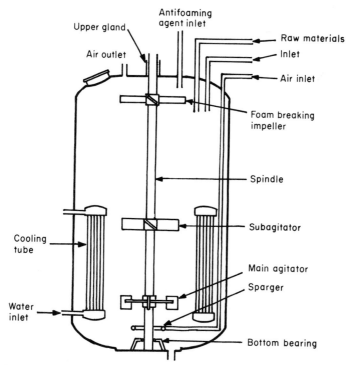

Antifoaming agent inlet

Upper gland

Air outlet

Raw materials

Inlet

Air inlet

Foam breaking impeller

Spindle

Cooling tube

Subagitator

Main agitator

Sparger

Water inlet

Bottom bearing

FIGURE 7. Structure of fermentor for an aerobic fermentation process.

Foaming is usually controlled by the automatic addition of selected antifoaming agents, with an electrode to detect the foaming or with a watt-meter to monitor the load changes on the foam-breaking impeller. A mechanical method may also be employed.

Figure 8 shows a diagrammatic scheme of an automatically controlled fermentation process for amino acid production. The rates of air flow, temperature, pH, and foaming are strictly controlled to achieve the maximum productivity.

A large-scale sequential control system was developed for L-glutamic acid fermentation with the object of reducing labor costs by minimizing manpower requirements (Yamashita et al., 1969). The L-glutamic acid fermentation is a typical batch reaction and therefore requires the sequential operation of many valves to sterilize the system, and then to introduce the medium to the fermentor. During the fermentation, conventional analog control instruments regulate the sterilized air flow, temperature, pressure, and pH. Some of these analog controllers are programmed to execute a desired profile in time.

C. Agitation–Aeration Effectiveness

The biosynthesis of L-glutamic acid is an aerobic process. Experimentally, the oxygen requirement for the L-glutamic acid fermentation is as shown in Eq. (1) (Okada and Tsunoda, 1965).

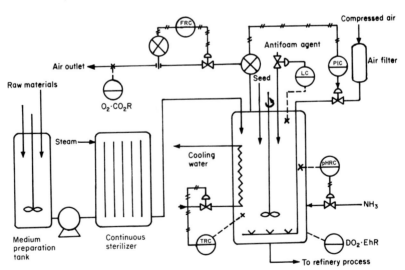

FIGURE 8. An example of an automatically controlled fermentation process. LC, level control; PIC, pressure control; PHRC, pH recording and control; FRC, flow rate recording and control; TRC, temperature recording and control; EhR, redox potentials recording.

$$C_6H_{12}O_6 + 2.33 \; O_2 \longrightarrow 0.82 \; C_5H_9O_4N + 1.94 \; CO_2 \tag{1}$$

Oxygen is consumed in large amounts. Therefore, quantitative discussion on oxygen transfer during the fermentation is indispensable for both the optimization of operation and the scaleup.

The following equations are the simplest expression of the basis for discussing oxygen transfer during the fermentation [Eq. (2)]:

$$
\begin{aligned}
r_{ab} = K_d(P_B - P_L) &= K_r M \; (P_L \geq P_{Lcrit}) \\
&< KM \quad (P_L < P_{Lcrit})
\end{aligned}
\tag{2}
$$

where r_{ab} is the rate of cell respiration (moles of O_2/ml/min); K_d is the oxygen absorption coefficient (mole/ml/min/atm); P_B is the gas phase oxygen tension (atm); P_L is the liquid phase oxygen tension (atm); K_r is the cell's oxygen demand (mole/min/gm of cell); M is the cell density (gm/ml); and P_{Lcrit} is the critical level of liquid phase oxygen for cell respiration (atm). Rate of cell respiration (r_{ab}) equals $K_r M$ at P_L levels above P_{Lcrit}; in other words, cell respiration is satisfied when oxygen supply is sufficient to maintain P_L above P_{Lcrit}.

Values of P_{Lcrit} are usually less than 0.01 atm, too low to be determined with conventional membrane-coated oxygen electrodes. The determination of extremely low P_{Lcrit} values was indispensable; therefore, a new method was developed in which both P_L and the redox potentials of culture medium were measured simultaneously (Shibai et al., 1974). This value was 0.0002 atm for B. flavum (Akashi et al., 1977a).

Lack of quantitative analysis on oxygen transfer sometimes leads to inaccurate conclusions. For example, L-glutamic acid-producing bacteria are known to excrete large amounts of lactic acid in a biotin-rich medium. However, the analysis of oxygen transfer clarified the true factor: It was due not to excess biotin but to oxygen deficiency that lactic acid became the dominant product in a biotin-rich medium (Hirose et al., 1968). In biotin-rich media cell density became high, which resulted in a high oxygen demand of culture broth. Consequently, the amount of oxygen was insufficient. In fact, the cell did not accumulate lactic acid even in biotin-rich media when sufficient oxygen was supplied (Fig. 9).

In L-glutamic acid fermentation, the P_L level is closely related to the product formation, as shown in Fig. 10 (Hirose et al., 1966b). The production yield markedly decreased when the P_L level was zero as measured by conventional membrane-coated oxygen electrodes. This implies that aeration and agitation must be controlled so as to keep P_L slightly above zero for the effective production of L-glutamic acid. In these cultures of high productivity, cell respiration was satisfied. Limited oxygen supply resulted in lactic acid or succinic acid accumulation at the expense of L-glutamic acid formation (Okada et al., 1961). Oxygen tension in the

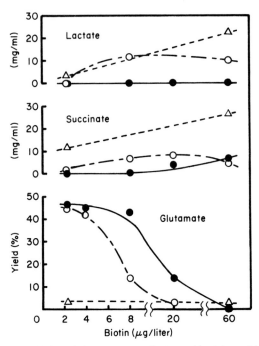

FIGURE 9. Accumulation of glutamate, succinate, and lactate vs. biotin concentration in L-glutamic acid fermentation. ●, P_L controlled at 0.15 atm; ○, sulfite $r_{ab} = 10.5 \times 10^{-7}$ mole/ml/minute; △, sulfite $r_{ab} = 1.2 \times 10^{-7}$ mole/ml/minute.

growth phase also seriously influenced the product formation (Hirose et al., 1967). Oxygen deficiency in the growth phase did not bring about any marked change in L-glutamic acid formation, whereas a high aerating condition seriously reduced the L-glutamic acid-producing capability of the cell.

Although the productivity of other amino acids was also influenced by oxygen supply, the extent or the mode of the influence was diverse according to the characteristics of biosynthetic pathways of each amino acid. Figure 11 shows the relationship between the degree of oxygen satisfaction (r_{ab}/K_rM) and the relative productivity of each amino acid. For example, oxygen shortage caused inhibition of both L-glutamic acid and L-lysine production, but the extent of the inhibition was more serious in L-glutamic acid fermentation than in L-lysine fermentation (Akashi et al., 1977b). Moreover, the maximum production of L-leucine was observed when the cell respiration was slightly repressed. Aerobic microorganisms require oxygen to reoxidize $NADH_2$ to form NAD and to generate ATP for growth and metabolism. L-Glutamic acid is biosynthesized

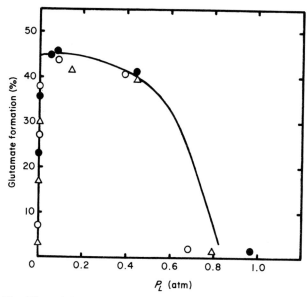

FIGURE 10. Effect of dissolved oxygen level, P_L, on the production of L-glutamic acid. P_L was controlled by changing gas-phase oxygen tension. ●, 20 ml working volume in 500-ml flask; ○, 40 ml working volume in 500-ml flask; △, 80-ml working volume in 500-ml flask.

by way of pyruvic acid, citric acid, and α-ketoglutaric acid. Formation of $5\frac{1}{3}$ moles of $NADH_2$ and generation of 2 moles of ATP are accompanied with assimilation of 1 mole of glucose. On the other hand, L-lysine is biosynthesized by way of pyruvic acid, citric acid, fumaric acid, and L-aspartic acid, forming $2\frac{2}{3}$ moles of $NADH_2$ and $\frac{2}{3}$ moles of ATP from 1 mole of glucose. In case of L-leucine biosynthesis, assimilation of 1 mole of glucose accompanies generation of 2 moles of ATP and $NADH_2$. Table V summarizes ATP consumption and $NADH_2$ formation in relation to the biosynthesis of these amino acids. The amount of $NADH_2$ formed in L-glutamic acid fermentation is more than that in L-lysine fermentation by $2\frac{2}{3}$ moles. This suggests that the inhibition of product formation due to oxygen shortage was more serious in L-glutamic acid fermentation. L-Leucine is biosynthesized not through the TCA cycle but by way of pyruvic acid. Therefore, the maximum production might occur when aerobic degradation of glucose was inhibited due to oxygen shortage. However, extremely deficient condition of oxygen might inhibit the cell to reoxidize 2 moles of $NADH_2$ and lead to the accumulation of lactic acid in place of L-leucine biosynthesis. Therefore, it is reasonable that the

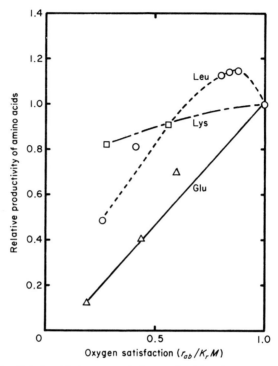

FIGURE 11. Relationship between relative productivity of amino acids and degree of oxygen satisfaction.

maximum production of L-leucine was observed when the cell respiration was slightly inhibited.

IV. CONCLUSION

The demands for amino acids are still increasing in the fields of foods, animal feeds, pharmaceuticals, chemicals, etc. To meet these increas-

TABLE V. ATP Consumption and $NADH_2$ Formation in the Biosynthesis of L-Glutamic acid, L-Lysine, and L-Leucine

Amino acids	Formation of $NADH_2$ (moles)	Consumption of ATP (moles)
L-Glu	$5\frac{1}{3}$	-2
L-Lys	$2\frac{2}{3}$	$-\frac{2}{3}$
L-Leu	2	-2

ing and diversified demands, almost all amino acids are now being produced commercially. Most of them are fermentation products. Large amounts of L-glutamic acid and L-lysine, the most important commercial amino acids, are available at a very low price, but the productivity of some other amino acids, such as L-tryptophan and L-histidine, is still insufficient and, consequently, their selling prices are still high. Every effort to supply each amino acid in sufficient quantities at moderate prices will surely continue. Promising approaches for improving the productivity of amino acid fermentations in the future include: (1) selection and improvement of cultures, (2) selection of raw or starting materials, and (3) physical and metabolic controls of fermentation. Remarkable progress in microbial genetics should accelerate the effective breeding of potent producers of amino acids. In addition to auxotrophs and regulatory mutants, which are now widely employed for industrial production of amino acids, other types of mutants induced by gene engineering technique are expected to be developed. Pursuits of more prospective raw materials are endless. Progress in organic chemistry may supply the fermentation industry with new starting materials. A combination of the synthetic techniques of organic chemistry with enzymatic techniques of microbiology will surely bring about new industrial methods for amino acid production. The productivity of amino acid fermentations will be increased also by better physical and metabolic controls in growing cultures or reaction mixtures. Advances in biochemical engineering will contribute most to successful improvement. An understanding of the microbial physiology and biosynthetic mechanisms involved in the fermentation process, as well as quantitative approaches to the activity of microorganisms, is usually fundamental to the successful control of amino acid fermentations.

REFERENCES

Aiba, S., Humphrey, A. E., and Mills, N. F. (1973). "Biochemical Engineering," 2nd ed., p. 270. Academic Press, New York.

Akashi, K., Shibai, H., and Hirose, Y. (1977a). *Biotechnol. Bioeng.* **20,** 27 (1978).

Akashi, K., Shibai, H., and Hirose, Y. (1977b). *Proc. Annu. Meet. Agric. Chem. Soc. Jpn.* p. 211.

Araki, K., Ueda, H., and Saigusa, S. (1974a). *Agric. Biol. Chem.* **38,** 565.

Araki, K., Shimijo, S., and Nakayama, K. (1974b). *Agric. Biol. Chem.* **38,** 837.

Breuer, L. H., Pond, W. G., Warner, R. G., and Loosli, J. K. (1964). *J. Nutr.* **82,** 499.

Chibata, I., Kamimoto, T., and Kato, J. (1965). *Appl. Microbiol.* **13,** 638.

Distillers Co., Ltd. (1965). British Patent 996,906.

Doi, S., and Kaneko, Y. (1960). *Hakko To Taisha* **2,** 26.

Enei, H., Nakazawa, H., Okumura, S., and Yamada, H. (1973). *Agric. Biol. Chem.* **37,** 725.

Fukumura, T. (1974). *Proc. Int. Congr. IAMS, 1st, 1974, IX, Tokyo.* Abstract Vol. V, p. 495.
Hagino, H., and Nakayama, K. (1973). *Agric. Biol. Chem.* **37,** 2013.
Hagino, H., and Nakayama, K. (1974). *Agric. Biol. Chem.* **38,** 157.
Hagino, H., and Nakayama, K. (1975). *Agric. Biol. Chem.* **39,** 343.
Hagino, H., Yoshida, H., Kato, F., Arai, T., Katsumata, R., and Nakayama, K. (1973). *Agric. Biol. Chem.* **37,** 2001.
Hirose, Y., Sonoda, H., Kinoshita, K., and Okada, H. (1966a). *Agric. Biol. Chem.* **30,** 49.
Hirose, Y., Sonoda, H., Kinoshita, K., and Okada, H. (1966b). *Agric. Biol. Chem.* **30,** 585.
Hirose, Y., Sonoda, H., Kinoshita, K., and Okada, H. (1967). *Agric. Biol. Chem.* **31,** 1210.
Hirose, Y., Sonoda, H., Kinoshita, K., and Okada, H. (1968). *Agric. Biol. Chem.* **32,** 855.
Humphrey, A. E., and Deindoerfer, F. H. (1959). *Appl. Microbiol.* **7,** 256.
Ikeda, K. (1908). *J. Tokyo Chem. Soc.* **30,** 820.
Izumi, Y., Tani, Y., and Ogata, K. (1973). *Biochim. Biophys. Acta* **326,** 485.
Kamiryo, T., Parthasarathy, S., and Numa, S. (1976). *Proc. Natl. Acad. Sci. U.S.A.* **73,** 386.
Kanzaki, T., Isobe, K., Okazaki, H., Mochizuki, K., and Fukuda, H. (1967). *Agric. Biol. Chem.* **31,** 1416.
Kase, H., and Nakayama, K. (1972). *Agric. Biol. Chem.* **36,** 1611.
Kase, H., and Nakayama, K. (1975). *Agric. Biol. Chem.* **39,** 153.
Kase, H., Tanaka, H., and Nakayama, K. (1971). *Agric. Biol. Chem.* **35,** 2089.
Kato, H., and Nakayama, K. (1976). *Proc. Annu. Meet. Agric. Chem. Soc. Jpn.* p. 127.
Kinoshita, S. (1963). *Recent Prog. Microbiol., Symp. Int. Congr. Microbiol., 8th, 1962* p. 317.
Kinoshita, S., Udaka, S., and Shimono, M. (1957a). *J. Gen. Appl. Microbiol.* **3,** 193.
Kinoshita, S., Nakayama, K., and Udaka, S. (1957b). *J. Gen. Appl. Microbiol.* **3,** 276.
Kinoshita, S., Nakayama, K., and Kitada, S. (1958). *J. Gen. Appl. Microbiol.* **4,** 128.
Kisumi, M., Kato, J., Sugiyama, M., and Chibata, I. (1971). *Appl. Microbiol.* **22,** 987.
Kisumi, M., Komatsubara, S., and Chibata, I. (1973). *J. Biochem. (Tokyo)* **73,** 107.
Kisumi, M., Nakazawa, N., Takagi, T., and Chibata, I. (1975). *Proc. Annu. Meet. Agric. Chem. Soc. Jpn.* p. 30.
Kitahara, K., Fukui, S., and Misawa, M. (1959). *J. Gen. Appl. Microbiol.* **5,** 74.
Komatsubara, S., Kisumi, M., and Chibata, I. (1973). *Proc. Annu. Meet. Soc. Ferment. Technol., Jpn* p. 98.
Kubota, K., Kageyama, K., Shiro, T., and Okumura, S. (1971). *J. Gen. Appl. Microbiol.* **17,** 167.
Kubota, K., Kageyama, K., Maeyashiki, I., Yamada, K., and Okumura, S. (1972). *J. Gen. Appl. Microbiol.* **18,** 365.
Kubota, K., Onoda, T., Kamijo, H., Yoshinaga, F., and Okumura, S. (1973). *J. Gen. Appl. Microbiol.* **19,** 339.
Lafitskaya, T. N., and Aseeva, I. V. (1967). *Mikrobiologia* **36,** 568.
Lee, W. H., and Good, R. C. (1963). U. S. Patent 3,087,863.
Masuo, E., Wakisaka, Y., and Hattori, M. (1962). Japanese Patent 37-6342.
Matsumoto, T., Yano, T., and Yamada, K. (1967). *Agric. Biol. Chem.* **31,** 1381.
Mihara, O., Kamijo, H., and Kubota, K. (1973). *Proc. Annu. Meet. Agric. Chem. Soc. Jpn.* p. 112.
Miyai, K., Osawa, T., and Tsuruo, I. (1964). Japanese Patent 39-2988.
Mogi, M., Sugizaki, Z., and Mizusawa, K. (1967). Japanese Patent 42-502.
Nakamori, S., and Shiio, I. (1972). *Agric. Biol. Chem.* **36,** 1675.
Nakanishi, T., Nakajima, J., and Kanda, K. (1975). *J. Ferment. Technol.* **53,** 543.
Nakayama, K. (1976). *Process Biochem.* **12,** 4.
Nakayama, K., and Hagino, H. (1966). *J. Agric. Chem. Soc. Jpn.* **40,** 377.

Nakayama, K., and Yoshida, H. (1972). *Agric. Biol. Chem.* **36,** 1675.

Nakayama, K., Kitada, S., and Kinoshita, S. (1961). *J. Gen. Appl. Microbiol.* **7,** 145.

Nakayama, K., Tanaka, H., Hagino, H., and Kinoshita, S. (1966). *Agric. Biol. Chem.* **30,** 611.

Nakazawa, H., Enei, H., Okumura, S., Yoshida, H., and Yamada, Y. (1972). *FEBS Lett.* **25,** 43.

Ogawa, C., Ode, M., and Midorikawa, Y. (1959). *Hakko To Taisha* **1,** 45.

Ogishi, N., Nishikawa, D., Kumagai, H., and Yamada, H. (1977). *Proc. Annu. Meet. Agric. Chem. Soc. Jpn.* p. 90.

Oishi, K. (1967). *J. Agric. Chem. Soc. Jpn.* **41,** R35.

Oishi, K., and Aida, K. (1965). *Agric. Biol. Chem.* **29,** 83.

Okada, H., and Tsunoda, T. (1965). *Agric. Biol. Chem.* **29,** 923.

Okada, H., Kameyama, I., Okumura, S., and Tsunoda, T. (1961). *J. Gen. Appl. Microbiol.* **7,** 177.

Okumura, S., Tsugawa, R., Tsunoda, T., Kono, K., Matsui, T., and Miyachi, N. (1962a). *J. Agric. Chem. Soc. Jpn* **36,** 141.

Okumura, S., Tsugawa, R., Tsunoda, T., and Motozaki, S. (1962b). *J. Agric. Chem. Soc. Jpn.* **36,** 204.

Okumura, S., Shibuya, M., Konishi, S., Ishida, M., and Shiro, T. (1964). *Agric. Biol. Chem.* **28,** 742.

Ota, S., and Tanaka, M. (1959). *J. Ferment. Technol.* **37,** 261.

Ozaki, S., Kono, K., Okumura, S., Okada, H., and Sakaguchi, K. (1960). *Proc. Symp. Amino Acids Ferment., 2nd, 1960,* p. 4.

Sano, K. (1976). *J. Agric. Chem. Soc. Jpn.* **50,** R201.

Sano, K., and Shiio, I. (1970). *J. Gen. Appl. Microbiol.* **16,** 373.

Sano, K., and Shiio, I. (1971). *J. Gen. Appl. Microbiol.* **17,** 97.

Sano, K., Matsuda, K., Yokazeki, K., Tamura, F., Yasuda, N., Noda, I., and Mitsugi, K. (1976). *Proc. Annu. Meet. Agric. Chem. Soc. Jpn.* p. 238.

Shibai, H., Ishizaki, A., and Hirose, Y. (1974). *Proc. Int. Congr. IAMS, 1st, 1974, IX, Tokyo.* Abstract Vol. V, p. 154.

Shibukawa, M., Kurima, M., Okabe, S., and Osawa, T. (1968). *Hakko To Taisha* **17,** 61.

Shibuya, M., Yoshinaga, F., and Okumura, S. (1967). *Proc. Annu. Meet. Agric. Chem. Soc. Jpn.* p. 178.

Shiio, I., and Nakamori, S. (1970). *Agric. Biol. Chem.* **34,** 448.

Shiio, I., Otsuka, S., and Katsuya, N. (1963). *J. Biochem. (Tokyo)* **53,** 333.

Shiio, I., Nakamori, S., and Sasaki, A. (1973). *Proc. Annu. Meet. Agric. Chem. Soc. Jpn.* p. 113.

Shiio, I., Sugimoto, S., and Nakagawa, M. (1975). *Agric. Biol. Chem.* **39,** 627.

Sommerson, N. L., and Phillips, T. (1962). U. S. Patent 3,080,297.

Su, Y. C., and Yamada, K. (1960). *Bull. Agric. Chem. Soc. Jpn.* **24,** 69.

Sugimoto, S., Nakagawa, M., Tsuchida, T., and Shiio, I. (1973). *Agric. Biol. Chem.* **37,** 2327.

Takinami, K., Okada, H., and Tsunoda, T. (1963). *Agric. Biol. Chem.* **27,** 858.

Takinami, K., Okada, H., and Tsunoda, T. (1964). *Agric. Biol. Chem.* **28,** 114.

Takinami, K., Yoshii, H., Tsuri, H., and Okada, H. (1965). *Agric. Biol. Chem.* **29,** 351.

Takinami, K., Yamada, K., and Okada, H. (1966). *Agric. Biol. Chem.* **30,** 674.

Takinami, K., Yamada, K., and Okada, H. (1967). *Agric. Biol. Chem.* **31,** 223.

Tanaka, K., Iwasaki, T., and Kinoshita, S. (1960). *J. Agric. Chem. Soc. Jpn.* **34,** 593.

Terui, G., and Niitsu, H. (1969). *Biotechnol. Bioeng. Symp.* **1,** 33.

Tsuchida, T., Yoshinaga, F., Kubota, K., Momose, H., and Okumura, S. (1975a). *Agric. Biol. Chem.* **39,** 1149.

Tsuchida, T., Yoshinaga, F., Kubota, K., and Momose, H. (1975b). *Agric. Biol. Chem.* **39,** 1319.

Tsunoda, T., and Okumura, S. (1972). *Proc. Int. Symp. Convers. Manuf. Foodst. Microorg.,* p. 229.

Udaka, S., and Kinoshita, S. (1960). *J. Gen. Appl. Microbiol.* **5,** 159.

Veldkamp, H., Berg, G. V. D., and Zevenhuizen, L. P. T. M. (1963). *Antonie van Leeuwenhoek* **29,** 35.

Yamada, K., Kinoshita, S., Tsunoda, T., and Aida, K. (1972). "Microbial Production of Amino Acids." Kodansha, Tokyo.

Yamashita, S., Hoshi, H., and Inagaki, T. (1969). *In* "Fermentation Advances" (D. Perlman, ed.), p. 444. Academic Press, New York.

Yokozeki, K., Sano, K., Eguchi, T., Yasuda, N., Noda, I., and Mitsugi, K. (1976). *Proc. Annu. Meet. Agric. Chem. Soc. Jpn.* p. 238.

Yoshida, H., and Nakayama, K. (1975). *J. Agric. Chem. Soc. Jpn.* **49,** 527.

Yoshinaga, F., Konishi, S., Okumura, S., and Katsuya, N. (1966). *Proc. Symp. Amino Acids Nucleic Acids Ferment.* p. 4.

Yoshinaga, F., Kikuchi, K., Tsuchida, T., and Okumura, S. (1976). Japanese Patent 51-44196.

Chapter 8

Microbial Production of Antibiotics

D. PERLMAN

I. INTRODUCTION

The production of antibiotics by microorganisms is similar in many respects to the synthesis of other compounds of extreme molecular complexity. In general, the special reactions required for synthesis are superimposed upon the metabolic systems responsible for growth and reproduction. These special steps may be often regarded as distortions or elaborations of normal cellular mechanisms in that they give such structural peculiarities as fused rings, D-amino acid residues, and rare saccharides.

241

MICROBIAL TECHNOLOGY, 2nd ed., VOL. I
Copyright © 1979 by Academic Press, Inc.
All rights of reproduction in any form reserved. ISBN 0-12-551501-4

In modern industrial microbial processes, the products may account for as much as 30% by weight of the nutrients supplied, so that what was previously a minor metabolic route represents a substantial portion of the metabolic activity of the organism. The biosynthesis of antibiotics may be regarded as the result of a series of "inborn errors of metabolism." These errors may be exaggerated by subjecting the original microorganism to mutagenic influences. The high yields of commercially important antibiotics owe much to the selection of such mutant strains. For example, the yield of benzylpenicillin was improved to well over 20 mg/ml (equivalent to 33,000 Oxford units/ml) (Chain, 1964) by the use of strains selected from *Penicillium chrysogenum* NRRL 1951, a culture producing only 120 μg/ml (Backus and Stauffer, 1955; Coghill and Koch, 1945).

In a period of a little more than 35 years, over 4000 antibiotics have been discovered, and the industry producing them has become highly competitive. World production of antibiotics exceeds 100,000 tons annually, and in the United States alone, sales of antibiotics for all purposes are over $1,000,000,000 per year. Of the thousands of antibiotics known, however, only a mere handful—the 90 listed in Table I—are produced on a commercial scale. In addition, 46, including 23 penicillins, 13 cephalosporins, dihydrostreptomycin, clindamycin, 6 tetracyclines, and 2 rifamycins, are prepared by a combination of microbial synthesis and chemical modification. Only chloramphenicol and pyrrolnitrin are manufactured on a commercial scale by chemical synthesis. The total chemical syntheses for penicillins, cephalosporins, polymyxins, gramicidin, tetracyclines, kanamycins, griseofulvin, and cycloheximide have been completed on a laboratory scale.

The cephalosporins group accounts for the largest sales volume (greater than $800,000,000 annually) with the Tetracycline group next in importance and leading in tonnage. Some clinical uses of antibiotics are listed in Table II. A number of antibiotics which were produced on a commercial scale have been withdrawn from large-scale distribution. These include stendomycin, ristocetin, and fumagillin. Those which are currently used for agricultural purposes include the antifungals blasticidin S and cycloheximide; the antibacterial kasugamycin; the coccidiostats monensin, lasalocid, and salinomycin; and the peptides nosiheptin, siomycin, and thiopeptin (all useful as growth promotants in poultry and animals). Antibiotics used only in veterinary medicine include hygromycin B, thiostrepton, and tylosin. Nisin, a polypeptide antibiotic from *Streptococcus cremoris*, is used as a food preservative in England.

II. GENERAL PRODUCTION METHODS

Since all of the antibiotics listed in Table I are made by aerobic fermentations, there are several similarities in the processes used for their production. The industry has been rather reluctant to publish details of the production methodology, but the general outline is fairly well known. Useful general accounts were described by Hastings (1955) and Jackson (1958), and particulars can be pieced together from the patent literature.

The culture vessels may have a capacity of as much as 100,000 gal. They usually measure two or three times as deep as they are wide. Stainless steel or nickel–chrome alloys are normally used in constructing these vessels, though in the past years less expensive grades of steel have been widely used. The vessels are fitted with aerating and agitating devices and with cooling coils and/or jackets for accurate temperature control.

The medium may be sterilized in a separate cooker and then pumped into the sterile culture vessels, or it may be sterilized *in situ*. Strict attention is paid to sterility in the antibiotic industry, and it is claimed that fewer than 2% of the batches are lost as a result of infections with antibiotic-resistant microorganisms or with phages. Since fermentation operations are costly, efforts are usually directed toward running as short a fermentation operation as economically possible. Large inoculum levels, e.g., as much as 10% of the volume of the fermentation, have been reported; careful choice and control of incubation temperatures, pH control, and continuous addition of energy sources and precursors, and other techniques are usually helpful in keeping the fermentation process operating at maximum economic efficiency.

In addition to concern with an efficient economic process from a fermentation viewpoint, there is concern with the problems of extraction and sometimes recovery of the antibiotics. An antibiotic is often present in the culture medium at concentrations of less than 1 gm/liter, and recovery of a pure antibiotic from the fermented medium (containing as much as 40 gm of dissolved solids per liter) is a very difficult assignment. At times a change of fermentation medium composition, e.g., use of invert cane molasses as a replacement for glucose solutions, which is economically attractive, introduces such complications into the recovery process for purifying the antibiotic that the economic advantage is lost.

Each of the 50 commercially operated antibiotic fermentations has its own peculiarities, and space limitations prevent a detailed analysis of

TABLE I. Some Antibiotics Produced on a Commercial Scale

Antibiotic	Microbial source	Antibiotic spectrum[b]						Chemical type	Therapeutic or other use	Some commercial source[c]
		G+	G−	My	AF	AT	Other			
Amphomycin	Streptomyces canus	+						Peptide	Topical	42
Amphotericin B	Streptomyces nodosus			+				Polyene	Oral or parenteral	58
Avoparcin[a]	Streptomyces candidus	+	+	+				Glycopeptide	Animal growth promotant	4
Azalomycin F[a]	Streptomyces hygroscopicus	+	+		+				Topical (AF)	54
Bacitracin	Bacillus subtilis	+						Peptide	Topical; also animal growth promotant	7, 33, 42, 45, 47, 48
Bambermycins	Streptomyces bambergenesis	+						Phosphogly-colipid	Animal growth promotant	31
Bicyclomycin[a]	Streptomyces sapporonensis	+	+						Topical	27
Blasticidin S[a]	Streptomyces griseochromogenes				+			Nucleoside	Agricultural (AF)	35
Bleomycins	Streptomyces verticillus	+	+			+		Peptide	Parenteral (AT)	44a
Cactinomycin[a]	Streptomyces chrysomallus	+				+		Peptide	Parenteral (AT)	22
Candicidin B	Streptomyces griseus			+				Polyene	Topical	42, 47
Candidin[a]	Streptomyces viridoflavus			+				Polyene	Topical	47
Capreomycin	Streptomyces capreolus						+	Peptide	Parenteral	21

Antibiotic	Produced by / Remarks						Rickettsia	Chemical class	Administration	References
Cephalosporins	Cephalosporin C is produced by *Cephalosporium acremonium* and converted to 7-ACA, which is used for prep. of semisynthetic cephalosporins	+	+					Peptide	Oral and parenteral	5, 12, 14, 18, 23, 26, 27, 29, 41, 45, 49, 52, 58, 59, 60
Chloramphenicol	*Streptomyces venezuelae;* commercial manufacture is by chemical synthesis	+	+				Rickettsia		Oral or parenteral	46
Chromomycin A$_3$[a]	*Streptomyces griseus*	+						Peptide	Parenteral (AT)	59
Colistin	*Bacillus colistinus*	+	+					Peptide	Parenteral	8, 37, 39, 52
Cycloheximide	*Streptomyces griseus*			+					Agricultural (AF)	35, 61
Cycloserine	*Streptomyces orchidaceus*				+			Amino acid	Parenteral (TB)	5, 33
Dactinomycin	*Streptomyces antibioticus*		+					Peptide	Parenteral (AT)	44
Daunorubicin	*Streptomyces peucetius*		+						Parenteral (AT)	23, 52
Doxorubicin	*Streptomyces peucetius*		+						Parenteral (AT)	23
Enduracidin[a]	*Streptomyces fungicidus*	+				+		Peptide	Animal growth promotant	59
Erythromycin	*Streptomyces erythreus*	+						Macrolide	Oral and parenteral; animal growth promotant	1, 5, 13, 14, 18, 19, 21, 41, 48, 49, 52, 61
Fortimicins[a]	*Micromonospora olivoasterospora*	+	+					Aminoglycoside	Parenteral	39

(continued)

245

TABLE I (Continued)

Antibiotic	Microbial source	Antibiotic spectrum[b]						Chemical type	Therapeutic or other use	Some commercial source[c]
		G+	G−	My	AF	AT	Other			
Fungimycin[a]	Streptomyces coelicolor var. aminophilus					+		Polyene	Topical	42
Fusidic Acid[a]	Fusidium coelcineum	+	+					Steroid	Parenteral	40
Gentamicins	Micromonospora purpurea	+	+					Aminoglycoside	Parenteral	14, 15, 56
Gramicidin A	Bacillus brevis	+						Peptide	Topical	42, 47, 62
Gramicidin J[a](S)	Bacillus brevis	+						Peptide	Topical	43
Griseofulvin	Penicillium griseofulvum				+			Spirolactone	Oral	14, 29, 34, 40, 44$_n$, 59
Hygromycin B	Streptomyces hygroscopicus	+	+				Helminths	Aminoglycoside	Animal feed suppl.	11, 41, 59
Josamycin[a]	Streptomyces narbonesis	+						Macrolide	Oral and parenteral	55
Kanamycins	Streptomyces kanamyceticus	+	+	+				Aminoglycosides	Parenteral	12, 14, 43, 44a, 52
Kasugamycin[a]	Streptomyces kasugaensis	+	+					Aminoglycoside	Agricultural antibacterial	8, 43, 55
Kitasatamycin[a]	Streptomyces kitasatoensis	+						Macrolide	Oral and parenteral	60
Lasalocid	Streptomyces hazelensis	+					Coccidia	Polyether	Agricultural use as coccidiostat and growth promotant	22
Lincomycin	Streptomyces lincolnensis	+							Oral and parenteral	61
Lividomycin[a]			+	+						36, 52

Antibiotic	Producing organism	Activity	Class	Application	References
Macarbomycins[a]	Streptomyces phaechromogenes	−	Phosphoglycolipid	Animal growth promotant	43
Mepartricin[a]		−	Polyene	Topical	57
Midecamycin[a]		+	Macrolide	Oral and topical	43
Mikamycins[a]	Streptomyces mitakaensis	+	Peptide	Animal growth promotant	8
Mithramycin	Streptomyces species	+ + +		Parenteral (AT)	48
Mitomycin C	Streptomyces caespitosus	+ + +		Parenteral	12, 39
Mocimycin[a]		+		Animal growth promotant	28
Monensin	Streptomyces cinnamonensis	+ (Coccidia)	Polyether	Animal growth promotant	41
Myxin	Chromobacterium iodinum plus chemical modification	+	Phenazine	Topical in veterinary use	32
Neocarzinostatin	Streptomyces carzinostaticus	+ +	Peptide	Parenteral (AT)	
Neomycins	Streptomyces fradiae	+ +	Aminoglycoside	Oral and topical	6, 7, 11, 14, 23, 42, 44a, 47, 48, 52, 53, 58, 59, 61
Nosiheptide[a]	Streptomyces actinosus	+	Peptide	Animal growth promotant	52
Nisin	Streptococcus cremoris	+	Peptide	Food preservative	
Novobiocin	Streptomyces niveus	+		Oral and topical	43, 52, 61
Nystatin	Streptomyces noursei	+	Polyene	Oral and topical	4, 14, 15, 52, 58
Oleandomycin	Streptomyces antibioticus	+	Macrolide	Oral and parenteral	48
Paromomycin	Streptomyces rimosus	+ + (Protozoa)	Aminoglycoside	Oral	46

(continued)

247

TABLE I (Continued)

Antibiotic	Microbial source	Antibiotic spectrum[b]						Chemical type	Therapeutic or other use	Some commercial source[c]
		G+	G−	My	AF	AT	Other			
Penicillin G[a]	Penicillium chrysogenum	+						Peptide	Oral and parenteral; also as animal growth promotant	2, 3, 6, 8, 9, 10, 12, 14, 22, 25, 28, 29, 31, 41, 43, 44, 45, 48, 52, 58, 60, 63
Penicillin V[a]	Penicillium chrysogenum	+						Peptide	Oral	1, 2, 3, 8, 10, 11, 12, 21, 25, 28, 29, 31, 41, 43, 48, 52, 58, 60, 63
Pimaricin	Streptomyces natalensis						+	Polyene	Topical; also used for food preservation	28
Polymyxin B Polyoxins[a]	Bacillus polymyxa Streptomyces cacaoi var. asoensis		+		+			Peptide	Parenteral Agriculture (AF)	45, 48 35, 55
Pristinamycins[a]	Streptomyces pristinaspiralis	+						Peptide	Parenteral	52
Quebemycin[a]	Streptomyces viridans	+						Phosphogly-colipid	Animal growth promotant	39, 52
Ribostamycin[a]	Streptomyces ribosidificus	+	+					Aminoglycoside	Parenteral	43
Rifamycin SV[a]	Nocardia mediterranei	+		+				Anasamycin	Parenteral	30, 26
Ristocetin	Nocardia lurida	+						Glycopeptide	Parenteral	42

Antibiotic	Producing organism	Special organisms	Activity	Chemical class	Use	References
Sagamycin[a]	*Micromonospora sagamiensis*		+ +	Aminoglycoside	Parenteral	39
Salinomycin[a]		Coccidia	+	Polyether	Veterinary use	35
Siccanin[a]	*Streptomyces albus* *Helminthosporium siccans*		+		Veterinary use	54
Siomycin	*Streptomyces sioyaensis*		+	Peptide	Animal growth promotant	56a
Sisomicin	*Micromonospora inyoensis*		+ +	Aminoglycoside	Parenteral	22, 56
Spectinomycin	*Streptomyces spectabilis*		+	Aminocyclitol	Parenteral	1, 35, 61
Spiramycin[a]	*Streptomyces ambofaciens*		+	Macrolide	Parenteral and oral	39, 52
Streptomycin	*Streptomyces griseus*		+ + +	Aminoglycoside	Parenteral; use in agriculture to control bacteria	6, 14, 28, 29, 43, 44, 48, 52, 58, 63
Dihydrostreptomycin	*Streptomyces humidus* (Also chemical reduction of streptomycin)		+ + +	Aminoglycoside	Parenteral	6, 14, 28, 43, 45, 48, 52, 63
Tetracyclines						
Chlortetracycline	*Streptomyces aureofaciens*	Rickettsia	+ +	Tetracycline	Parenteral and oral; animal growth promotant	4, 5, 14, 20, 23, 49, 52, 50, 57,
6-Demethyl-7-chlortetracycline	*Streptomyces aureofaciens*	Rickettsia	+ +	Tetracycline	Parenteral and oral	4, 18, 49, 52,
5-Hydroxytetracycline	*Streptomyces rimosus*	Rickettsia	+ +	Tetracycline	Parenteral and oral animal growth promotant	5, 11, 14, 15, 18, 28, 34, 38, 48, 49, 50
Tetranactin[a]	*Streptomyces flaveolus*	Insects	+	Macrotetralide	Insecticide	17

(continued)

249

TABLE I *(Continued)*

Antibiotic	Microbial source	Antibiotic spectrum[b]						Chemical type	Therapeutic or other use	Some commercial source[c]
		G+	G−	My	AF	AT	Other			
Thiopeptin[a]	Streptomyces tateyamensis	+						Peptide	Animal growth promotant	27, 35
Thiostrepton	Streptomyces azureus	+						Peptide	Animal growth promotant	58
Tobramycin[a]	Streptomyces tenebrarius	+	+					Aminoglycoside	Parenteral	11, 41
Trichomycin[a]	Streptomyces hachijoensis					+	Trichomonas	Polyene	Topical	27
Tylosin	Streptomyces fradiae	+					PPLO	Macrolide	Veterinary; animal growth promotant	21, 41
Tyrothricin	Bacillus brevis	+	+					Peptide	Topical	10, 42, 47, 62
Tyrocidine	Bacillus brevis	+	+					Peptide	Topical	42
Validamycin[a]	Streptomyces hygroscopicus var. limoneus	+	+					Aminoglycoside	Parenteral	59
Vancomycin	Streptomyces orientalis	+						Glycopeptide	Parenteral	41
Variotin[a]	Paecilomyces varioti				+			Peptide	Topical	40, 44a
Viomycin	Streptomyces floridae	+		+				Peptide	Parenteral	48, 60
Virginiamycin	Streptomyces virginiae	+						Peptide	Animal growth promotant	51

[a] Not distributed in the United States (1978).

[b] G+, gram-positive bacteria; G−, gram-negative bacteria; My, mycobacteria; AF, antifungal; AT, antitumor.

c key to sources (from Perlman, 1977):

1. Abbott Laboratories (U.S.A.)
2. Aktiebolaget Astra (Sweden)
3. Aktiebolaget Fermenta (Sweden)
4. American Cyanamid (U.S.A.)
5. Ankerfarm S.p.A. (Italy)
6. Antibioticos S.A. (Spain)
7. Apothekernes Laboratorium fur Specialpraeparater A/S (Norway)
8. Banyu Pharmaceutical Company (Japan)
9. Beecham Pharmaceutical Company (England)
10. Biochemie GmbH (Austria)
11. Biogal (Hungary)
12. Bristol-Myers Company (U.S.A.)
13. Chemibiotic Ltd. (Ireland)
14. China National Chemicals Import and Export Corporation (China)
15. Chinoin (Hungary)
16. Chong-Kun-Dong Corporation (S. Korea)
17. Chugai Pharmaceutical Company (Japan)
18. Companhia Industrial Produtora de Antibioticos S.A.R.L. (CIPAN) (Portugal)
19. Compania Espanola de la Penicillina y Antibioticos S.A. (Spain)
20. Diaspa S.p.A. (Italy)
21. Dista Products Ltd. (England)
22. Farbenfabriken Bayer AG (W. Germany)
23. Farmitalia S.p.A. (Italy)
24. Fermentfarma S.p.A. (Italy)
25. Fermion Oy (Finland)
26. Fervet S.p.A. (Italy)
27. Fujisawa Pharmaceutical Company (Japan)
28. Gist-Brocades n.v. (Holland)
29. Glaxo Laboratories Ltd. (England)
30. Gruppo Lepetit S.p.A. (Italy)
31. Hoechst AG (W. Germany)
32. Hoffmann-LaRoche Inc. (U.S.A.)
33. IMC Chemicals Group, Inc. (U.S.A.)
34. Imperial Chemical Industries Ltd. (England)
35. Kaken Chemical Company (Japan)
36. Kowa Company (Japan)
37. Kayaku Antibiotics Research Company Ltd. (Japan)
38. Krakow Pharmaceutical Works ("Polfa") (Poland)
39. Kyowa Hakko Kogyo Company (Japan)
40. Leo Pharmaceutical Products (Denmark)
41. Eli Lilly Company (U.S.A.)
42. H. Lundbeck and Company (Denmark)
43. Meiji Seika Kaisha Ltd. (Japan)
44. Merck and Company, Inc. (U.S.A.)
44a. Nihon Kayaku Company (Japan)
45. Novo Industri A/S (Denmark)
46. Parke, Davis and Company (U.S.A.)
47. S. B. Penick and Company (U.S.A.)
48. Pfizer, Inc. (U.S.A.)
49. Proter S.p.A. (Italy)
50. Rachelle Laboratories Inc. (U.S.A.)
51. Recherche et Industrie Therapeutique (Belgium)
52. Rhone-Poulenc S.A. (France)
53. Roussel-UCLAF (France)
54. Sankyo Company Ltd. (Japan)
55. Sanraku Ocean Company, Ltd. (Japan)
56. Schering Corporation (U.S.A.)
56a. Shionogi & Co., Ltd. (Japan)
57. Societa Prodotti Antibiotici S.p.A. (Italy)
58. E. R. Squibb and Sons (U.S.A.)
59. Takeda Chemical Industries Ltd. (Japan)
60. Toyo Jozo Company Ltd. (Japan)
61. The Upjohn Company (U.S.A.)
62. Wallerstein Laboratories, Inc. (a division of G. B. Fermentation Industries, Inc., USA)
63. Wyeth Laboratories (U.S.A.)

d Both penicillins G and V are used as sources of 6-aminopenicillanic acid, an intermediate in the synthesis of semisynthetic penicillins. Also, penicillin V is used as a starting material for the chemical synthesis of cephalexin.

TABLE II. Some Antibiotics Useful in Treatment of Infections

Infective microorganism	Disease	Suggested Antibiotics
I. Gram-positive cocci		
Staphylococcus aureus	Abscess *Penicillin G-sensitive:* Bacteremia Endocarditis *Penicillin G-resistant:* Pneumonia Meningitis *Methicillin-resistant:* Osteomyelitis	Penicillin G; gentamicins; erythromycin; lincomycin; cephalosporins; clindamycin Penicillinase resistant penicillin Vancomycin; cephalosporin
Streptococcus pyogenes (groups (A, B, C)	Pharyngitis Scarlet fever Otitis media Pneumonia Cellulitis	Penicillin G and V; cephalosporin; clindamycin; erythromycin; tetracycline
Streptococcus viridans	Dental infections Subacute bacterial endocarditis Urinary tract infections	Penicillin G ± streptomycin; cephalosporin; erythromycin; vancomycin
Streptococcus faecalis (enterococcus)	Endocarditis Urinary tract infections Bacteremia Meningitis Brain abscess	Penicillin + aminoglycoside Ampicillin; vancomycin; penicillin G + aminoglycoside; erthromycin ± aminoglycoside
Streptococcus (anaerobic)	Bacteremia Endocarditis Brain abscess and other abscesses	Penicillin G; chloramphenicol; clindamycin; erythromycin; tetracycline
Streptococcus pneumoniae (diplococcus)	Pneumonia Meningitis Endocarditis Arthritis	Penicillin G; cephalosporin; chloramphenicol; lincomycin; clindamycin; erythromycin
II. Gram-negative cocci		
Neisseria gonorrhoeae (gonococcus)	Genital infections Arthritis Meningitis Endocarditis	Penicillin G; ampicillin; spectinomycin; erythromycin; tetracycline
Neisseria meningitidis	Meningitis Bacteremia Carrier state	Penicillin G; chloramphenicol; erythromycin Rifampin
III. Gram-positive bacilli		
Clostridium perfringens	Gas gangrene	Penicillin G; erythromycin; tetracycline; cephalosporin
Clostridium tetani	Tetanus	Penicillin G; tetracycline; erythromycin

(*continued*)

TABLE II   *(Continued)*

IV. Gram-negative bacilli		
Escherichia coli	Urinary tract infection	Ampicillin; cephalosporin; carbenicillin; tetracycline; chloramphenicol
	Other infections	Ampicillin; gentamicin; cephalosporin; chloramphenicol; tetracycline; colistin; kanamycin; polymyxin B
Enterobacter aerogenes	Urinary tract infection	Gentamicin; carbenicillin; chloramphenicol; kanamycin; tetracycline; colistin
Alcaligenes faecalis	Urinary tract infection	Chloramphenicol; tetracycline; colistin; polymyxin B; kanamycin; gentamicin
Proteus mirabilis	Urinary tract infection	Ampicillin; cephalosporin; gentamicin; kanamycin; chloramphenicol; tetracycline
Pseudomonas aeruginosa	Urinary tract infection	Carbenicillin; gentamicin; colistin; polymyxin B
Salmonella sp.	Typhoid	Ampicillin; chloramphenicol
Shigella sp.	Gastroenteritis	Ampicillin; chloramphenicol; polymyxin B; kanamycin; tetracycline
Vibrio cholerae	Cholera	Tetracycline; chloramphenicol; erthromycin
V. Acid-fast Bacilli		
Mycobacterium tuberculosis	Tuberculosis	Streptomycin + synthetic drugs; rifampin; cycloserine
Atypical *Mycobacterium*	Pulmonary infection	Streptomycin + synthetic drugs; erythromycin; rifampin; cycloserine
VI. Miscellaneous Agents		
Mycoplasma pneumonia	Atypical viral pneumonia	Erythromycin; tetracycline
Rickettsia	Typhus fever	Tetracycline; chloramphenicol
	Rocky Mountain spotted fever	
	Q fever	
VII. Fungi		
Candida albicans	Skin lesions	Amphotericin B; nystatin
Candida albicans	Pneumonia	Amphotericin B
Cryptococcus	Meningitis	
Aspergillus	Bone lesions	
	Disseminated disease	
Microsporon	Skin lesions	Griseofulvin
Trichophyton	Hair lesions	
Epidermophyton	Nail lesions	

the literature on each. However, the techniques used in production of penicillins, tetracyclines, and bacitracin are fairly typical of those used for all antibiotics (the reader is directed to the patent literature for details of processes for producing other antibiotics).

III. PROCESSES USED FOR PENICILLIN PRODUCTION

A. Historical Background

Penicillin is the name applied by Fleming (1929) to the bacteriostatic principle produced by a mold later identified as *Penicillium notatum*. Since Fleming's discovery in 1929, it has been found that penicillin is produced by a variety of molds belonging to other species and genera, and also that there is a series of closely related penicillins, all of which show approximately the same antibiotic characteristics. The naturally occurring penicillins differ from each other in the R group as follows:

$$\underset{\substack{\\ \\}}{\overset{O \quad H \quad H \quad H}{R\text{--}C-N-C-C}} \overset{S}{\underset{O=C-N}{\diagup}} \overset{CH_3}{\underset{\substack{| \\ H}}{\overset{|}{C}}} \overset{CH_3}{\underset{COOH}{\diagdown}}$$

R group	Chemical name	Trivial name
$CH_3CH_2CH{=}CHCH_2-$	6-(2-Hexenamido)penicillin	Penicillin F
$CH_3(CH_2)_3CH_2-$	6-(Hexanamido)penicillin	Dihydropenicillin F
$CH_3(CH_2)_5CH_2-$	6-(Heptamido)penicillin	Penicillin K
⟨phenyl⟩$-CH_2-$	6-(Phenylacetamido)penicillin	Penicillin G
$HO-$⟨phenyl⟩$-CH_2-$	6-(*p*-Hydroxyphenylacetamido) penicillin	Penicillin X

Following Fleming's report, Clutterbuck *et al.* (1932) undertook to study the chemistry of penicillin. Their work showed that penicillin was an organic acid which was extractable into organic solvents from aqueous solutions of low pH, and that it was extremely labile to hydrogen ion and heat. In nearly all of their experiments the bioactivity disappeared upon evaporation of a solution to dryness. In view of this great lability and the very low fermentation yields (less than 0.1% of current practice), work on penicillin was abandoned.

In 1937, stimulated by an interest in antiinfective agents, Chain *et al.*

(1940) undertook a reinvestigation of penicillin. They cultured Fleming's organism in surface culture in what amounted to a small pilot-plant scale. By keeping a low temperature during their extraction procedure, they were able to concentrate the penicillin 1000-fold and produce a dry powder in the form of a salt of penicillin, which had a reasonable stability on storage. This represented the first great advance in penicillin production and furnished material used to demonstrate the phenomenal curative properties of this antibiotic (Abraham et al., 1941).

Because of the virtual impossibility of quickly producing a significant amount of penicillin under wartime conditions in England, Drs. Florey and Heatley came to the United States in July 1941 in order to enlist the aid of the American government and pharmaceutical industry. The story of their success in this direction has been told before and there is no need to repeat it here (Chain et al., 1940; Abraham et al., 1949; Coghill, 1944; Stewart, 1961). Suffice it to say that by the time of the invasion of France in 1944, adequate amounts of penicillin were available on all fronts to effect a tremendous saving of life among the wounded. Penicillin was most certainly one of the few benefits accruing from World War II. Had it not been for the urgent need for such a curative agent, the wartime ease of securing government money for such research projects, and the excess profits tax which made it possible for industry to take a terrific gamble with fifteen-cent dollars, we might not have penicillin even today.

B. Selection of Culture

The mold described by Fleming (1929) and subsequently identified as a strain of *Penicillium notatum* was used in the early developmental stages of the production of penicillin. The yields with this culture were poor, and considerable work was carried out in several laboratories in an attempt to obtain a higher yielding strain. Several strains, including a natural variant of Fleming's culture (given the registry number NRRL 1249B21), were found to produce higher yields of penicillin (Moyer and Coghill, 1946). These new strains, together with other improvements in the process, combined to give yields of penicillin in the range of 200 Oxford units/ml (1 Oxford unit = 0.6 μg) by the surface culture method; none of these cultures performed well in submerged culture.

Many strains of *P. notatum* and the related *P. chrysogenum* were tested in an effort to find one which would produce good yields in submerged culture processes. One of these, *P. chrysogenum* NRRL 1951, was found to be a "superior producer." By plating NRRL 1951, followed by selective isolation, strain NRRL 1951B25 was obtained. This strain was significantly better than the parent. Demerec (at the Carnegie Institution) subjected NRRL 1951B25 to X-ray irradiation, and one of the survivors, X-1612, was found to produce twice as much penicillin (it gave yields of 300 units/ml) (Demerec, 1948). This culture was subjected to ultraviolet irradiation by Backus and Stauffer (Coghill and Koch, 1945) at the University of Wisconsin, and a strain designated as Wisconsin (Wis.) Q176 (producing 1200 units/ml) (Foster et al., 1946) was selected. This strain was adopted by most of the penicillin manufacturers, and monospore isolates from it are the parent cultures of those now in use. The genealogy of this family of strains is given in Fig. 16 (Volume II, Chapter 12). Descendents of these cultures have many properties not found in the parents. They produce 10 times as much penicillin [titers of 30,000 units are reported (Chain, 1964)] and they do not produce the yellow pigment which hampered early efforts at isolating pure, colorless benzylpenicillin. The practice of improvement of antibiotic production by selection of "better" cultures first tested in the penicillin manufacturing program has been widely applied to other antibiotic processes, and the role of the microbial geneticist in the fermentation industry has increased in importance.

The penicillin-producing molds are characterized by unusual variability; in general, the greater the productivity of the strain, the less stable it is. Stock cultures can be maintained on agar slants, in dry soil, in lyophilized form, or as spore or cell suspensions stored in liquid nitrogen. Stocks carried on agar slants are perhaps most liable to variation since successive transfer of the organisms provides greater opportunity for variation to occur. Frequent transfers tend to propagate selectively those portions of the culture population that sporulate most readily. If agar slants are used, transfers should be relatively infrequent. The wide variation in penicillin-producing ability of single-spore isolates from P. chrysogenum Wis. Q176 was described by Reese et al. (1949) and others (Coghill, 1944; Haas et al., 1956). Similar variation has been noted in the cultures producing streptomycin (Perlman et al., 1954).

C. Fermentation Medium and Conditions

Surface culture of P. notatum for penicillin production was first carried out in simple chemically defined media, such as that of Czapek and Dox. Later it was found that the culture grown in this medium supplemented with organic materials, e.g., casein digest, yeast extract, or meat products, resulted in higher antibiotic titers. Lactose was found to be a suitable replacement for the sucrose or glucose of the chemically defined media (probably due to its slow metabolism by the fungus and resultant steady carbohydrate supply over a protracted incubation period). Increased yields were also obtained with more complex nitrogen sources, such as extracted oilseeds, e.g., cottonseed, peanut meal, and soybean meal (Chain, 1966; Moyer and Coghill, 1946; Perlman, 1949).

A great step forward was brought about by the use of corn steep liquor which, together with a change to *P. chrysogenum* as the source organism, resulted in increasing yields from about 20 units/ml to about 100 units/ml (Perlman, 1950). Apart from the supply of utilizable nitrogenous materials in the corn steep liquor, the corn steep contained phenylalanine and phenethylamine which we now know act as precursors for the formation of benzylpenicillin.

At the same time, efforts were made to devise chemically defined media with a similar potential. The first reasonably successful one was that of Jarvis and Johnson (1947) which contained lactose, mineral salts, ammonium lactate and acetate, and phenylacetic acid as a precursor for benzylpenicillin. Another useful medium was that of Calam and Hockenhull (1949) which supported the production of antibiotic by a large number of strains of *P. chrysogenum*. A comparison of this medium with corn steep liquor (Table III) shows a close relationship, chemically speaking. Inclusion of vegetable oils, first used as antifoams, had a beneficial effect on yield (Davey and Johnson, 1953; Perlman, 1950; and E. R. Squibb and Sons, Inc., 1952). The effect of these lipids was to increase both the amount of mycelium formed and the antibiotic yield.

TABLE III. Comparison of Calam and Hockenhull's Medium with Corn Steep Liquor Medium

	Corn steep liquor medium (%)	Calam and Hockenhull's medium (%)
Main carbohydrate	Lactose, 3.0–4.0	Lactose, 3.0
Other carbohydrates	Glucose, 0–0.5 nonreducing, mainly polysaccharides[a]	Glucose, 1.0 Starch, 1.5
Organic acids	Acetic, ca. 0.05[a] Lactic, ca. 0.5[a]	Acetic, 0.25 Citric, 1.0
Special precursors	Phenethylamine and other precursors[a]	Phenylacetic acid, 0.05
Main nitrogen source	Amino acids, peptides, amines[a]	Ammonium sulfate, 0.5
Other nitrogen sources	Ammonia[a]	Ethylamine, 0.3
Total solids	8.0–9.0	8.5
Total N	0.15–0.2	0.2

[a] Concentration varies from one batch of corn steep liquor to next. Phenethylamine, found in many batches of corn steep liquor, was often increased in these media by addition of pure phenethylamine or phenylacetic acid supplements.

Most of the recent developmental work on the penicillin fermentation process has been concerned with increasing production and decreasing manufacturing costs. Replacement of the expensive lactose component of the medium with continuous addition of glucose or sucrose did not change antibiotic production (Davey and Johnson, 1953; Freaney, 1958; Hosler and Johnson, 1953; and Soltero and Johnson, 1953) and resulted in some financial advantages. Increased antibiotic production was also obtained by controlling the pH of the fermentation between pH 6.8 and 7.4 (Brown and Peterson, 1950; Singh and Johnson, 1948). This was accomplished by adding buffering agents to the medium (both $CaCO_3$ and phosphates have been used) and by additions of sterile H_2SO_4 and NaOH as needed. The availability of glass electrodes which withstand repeated autoclavings has made the pH-controlled fermentation possible on a routine basis (Deindoerfer and Wilker, 1957; Gualandi et al., 1960; West et al., 1961).

Aeration and agitation are fundamental to the penicillin process, and the theory of oxygen transfer in large-scale fermentations has been studied in many laboratories. Applications to the penicillin process were studied by Bartholomew et al. (1950a), Brown and Peterson, (1950), Calam et al. (1951), and Chain and Gulandi (1954). Many parameters have been measured, and in general it is agreed that high levels of aeration and agitation are required to maintain the desired dissolved oxygen levels. Temperature of incubation is also very important in the penicillin production process, and most operations are conducted at 25° $\pm$ 0.5°C. Owen and Johnson (1955) showed that slightly higher incubation temperatures resulted in faster mycelium formation, but best antibiotic production took place at 25°C (with their strain of P. chrysogenum).

The first moiety of the penicillin molecule to be chemically identified was the acyl portion, which differs among the naturally occurring penicillins. The most important of the penicillins in current production is penicillin G, where the acyl moiety is phenylacetyl. This was originally derived from the phenethylamine in the corn steep liquor (Mead and Stack, 1948; Moyer and Coghill, 1947; Smith and Bide, 1948), and inclusion of this liquor in the fermentation medium not only increased antibiotic production but also shifted the major penicillin produced from penicillin F to penicillin G. Addition of phenylacetic acid and phenylacetamide to penicillin-producing fermentations was investigated as a means of increasing production of penicillin G (Behrens et al., 1948; Moyer and Coghill, 1947). When large amounts of phenylacetic acid, e.g., 1 gm/liter, were included in the fermentation medium at time of the inoculation, some inhibition of mycelium growth was observed as well as inefficient conversion to penicillin G (Moyer and Coghill, 1946). These difficulties were overcome by supplying the fermentation with compounds which were slowly metabolized to phenylacetic acid by the particular strain of P. chrysogenum used. Among the compounds found useful for this purpose were N-phenacetylethanolamine (Behrens et al., 1948), N-methylphenylacetamide (Behrens et al., 1948), esters of phenylacetic acid (Tabenkin et al., 1952), and esters of phenethanol (Perlman et al., 1954). As the bioengineering technology became more sophisticated, continuous addition of phenylacetic acid to the growing cultures during the incubation period was evaluated and found to be economically desirable (Pan and Perlman, 1954; Singh and Johnson, 1948).

A flow sheet of a typical fermentation process for production of procaine penicillin G is shown in Fig. 1. Spores of P. chrysogenum strain are used to inoculate 100 ml of medium in a 500-ml Erlenmeyer flask, and the inoculated flasks are placed on a rotary shaker (250 rpm, 2-inch displacement) located in a 25°C incubator. After 4 days' incubation, the contents of the flask are transferred to 2 liters of medium (in a 4-liter flask). This second flask fermentation is incubated on the shaker for 2 days and the contents are transferred to 500 liters of medium in an 800-liter stainless steel tank. This tank is equipped with air spargers, agitators, cooling coils for temperature control, and antifoam addition devices. After 3 days' incubation, the contents of the tank are used to inoculate 180,000 liters of fermentation medium in a 250,000-liter fermentor. This larger fermentor is equipped with devices for continuous addition of sterile glucose syrups, pH control (automatic addition of NaOH and H_2SO_4 to keep the pH at a preset value), foam sensing devices to activate automatic addition of antifoams (e.g., animal or vegetable oils), and metering pumps for continuous addition of sterile

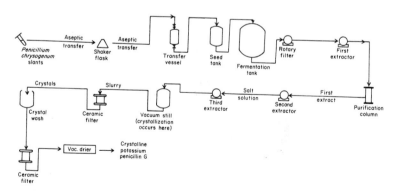

FIGURE 1. Penicillin fermentation process flow chart.

phenylacetic acid. After 5–6 days' incubation, the contents of the fermentor are filtered (all the penicillin is in the cell-free liquid), and the filtrate is passed through a series of Podbelniak extractors where the penicillin is extracted into amyl or butyl acetate. The penicillin in the amyl acetate extract is transferred back into an aqueous solvent by extraction with phosphate buffer, and the potassium penicillin G is crystallized from a butanol–water mixture. The potassium penicillin G may be further purified and used as such, or it can be converted to procaine penicillin G for clinical use.

Most penicillin-producing fungi are capable of incorporating a variety of compounds into the acyl portion of the penicillin molecule when the appropriate compounds are added to the growing culture. More than 1000 compounds have been tested to determine whether they would be incorporated and nearly 100 penicillins have been prepared in relatively pure form (Behrens *et al.*, 1948; Soper *et al.*, 1948). Among the objectives of this research program have been the preparation of penicillins with all of the "good" attributes of penicillin G and additional advantages. These latter include: (1) acid stability so that the penicillin can be administered orally, (2) lowered allergenicity since nearly 8% of the population is allergic to penicillin G, and (3) resistance to the penicillin-inactivating enzyme, penicillinase (Abraham and Chain, 1940; Pollock, 1965; Pollock *et al.*, 1956) found in many bacteria. Phenoxymethyl-penicillin (also known as penicillin V), prepared by addition of phenoxyacetic acid to fermentations (Behrens *et al.*, 1948), has the desired acid stability characteristics and has found a place in clinical practice as an orally absorbed penicillin. Patients allergic to penicillin G were found to be less sensitive to allylmercaptomethylpenicillin (also known as penicillin O), a penicillin produced biosynthetically by addition of allylmercaptoacetic acid to *P. chrysogenum* cultures (Ford *et al.*, 1953). None of the penicillins prepared biosynthetically appeared to be penicillinase resistant.

D. Semisynthetic Penicillins

Workers in the penicillin process development laboratories had noted over a period of years that chemical assays for penicillin in fermented media were frequently higher than bioassays. Kato (1953a,b) reported finding in his fermented media a considerable amount of the substance responsible for the discrepancy and that more of this material was found in fermented media which had not been supplemented with phenylacetic acid. Kato postulated that this material was the penicillin molecule "without" the R group. This hypothesis was confirmed in part by Demain (1953), who found that the substance did not form penicillin in the presence of washed, resting *P. chrysogenum* cells and phenylacetic acid precursor. The identification of the substance as 6-aminopenicillanic acid was reported by Batchelor *et al.* (1959), who showed that it was sensitive to the enzyme pencillinase (which opened the β-lactam ring of penicillin) and also that it could be converted to penicillin G by treatment with phenacetyl chloride.

Since the isolation of the acid from fermented media was different than for penicillin G (because of its hydrophilic nature), other methods were sought for the production. Rolinson *et al.* (1960) noted that certain strains of gram-negative bacteria and streptomycetes converted penicillin G to 6-aminopenicillanic acid (a reaction first proposed by Sakaguchi and Murao (1950; Murao, 1955). Similar results were obtained by Claridge *et al.* (1960) and Huang *et al.* (1960, 1963). The distribution of this hydrolytic enzyme (called penicillin amidase or benzylpenicillin acylase) was studied in many laboratories (Batchelor *et al.*, 1961; Chain, 1964; Huang *et al.*, 1963; Kaufmann and Bauer, 1960) with the results shown in Table IV.

Since 6-aminopenicillanic acid can be readily acylated by chemical means, it has been possible to prepare a new series of semisynthetic penicillins. Tens of thousands of compounds have been prepared (Price, 1969, 1977), and a number have been found to be markedly superior to penicillin G in oral absorption, acid stability, and resistance to penicillin-inactivating enzymes. The structures of some of these which have been of clinical importance are shown in Fig. 2 and their special properties summarized in Table V.

IV. CEPHALOSPORINS

The production of antibiotic activity by a particular strain of *Cephalosporium* was first noted by Brotzu (1948). Several antibiotics with differing

TABLE IV. Distribution of Penicillin Amidase

Group of microorganisms	Genera in which penicillin amidase was reported
Bacteria	*Pseudomonas, Xanthomonas, Alcaligenes, Flavobacterium, Escherichia, Aerobacter, Erwinia, Serratia, Proteus, Bordetella, Micrococcus, Sarcina, Corynebacterium, Cellomonas, Nocardia, Bacillus*
Molds	*Alternaria, Aspergillus, Epicoccum, Fusarium, Mucor, Penicillium, Phoma, Trichoderma*
Yeasts	*Cryptococcus, Saccharomyces, Trichosporon*
Actinomycetes	*Streptomyces*

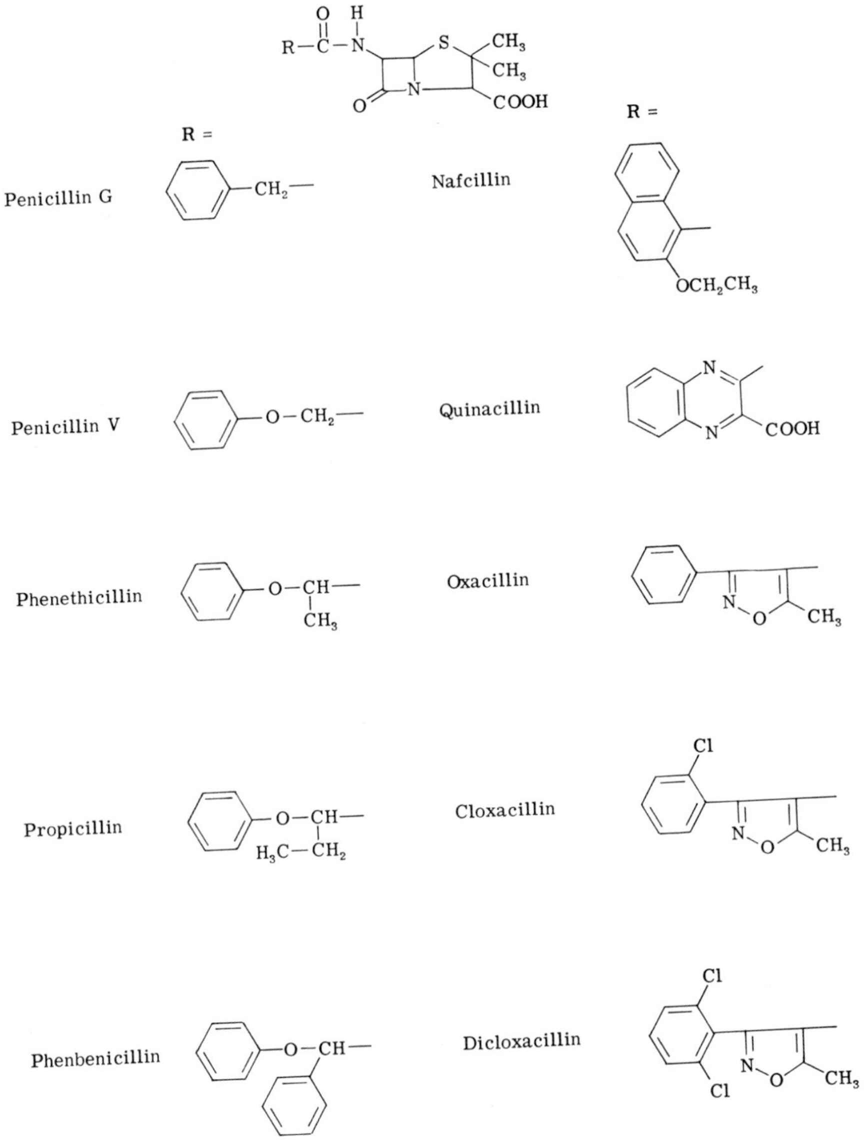

FIGURE 2. Structures of some clinically useful penicillins.

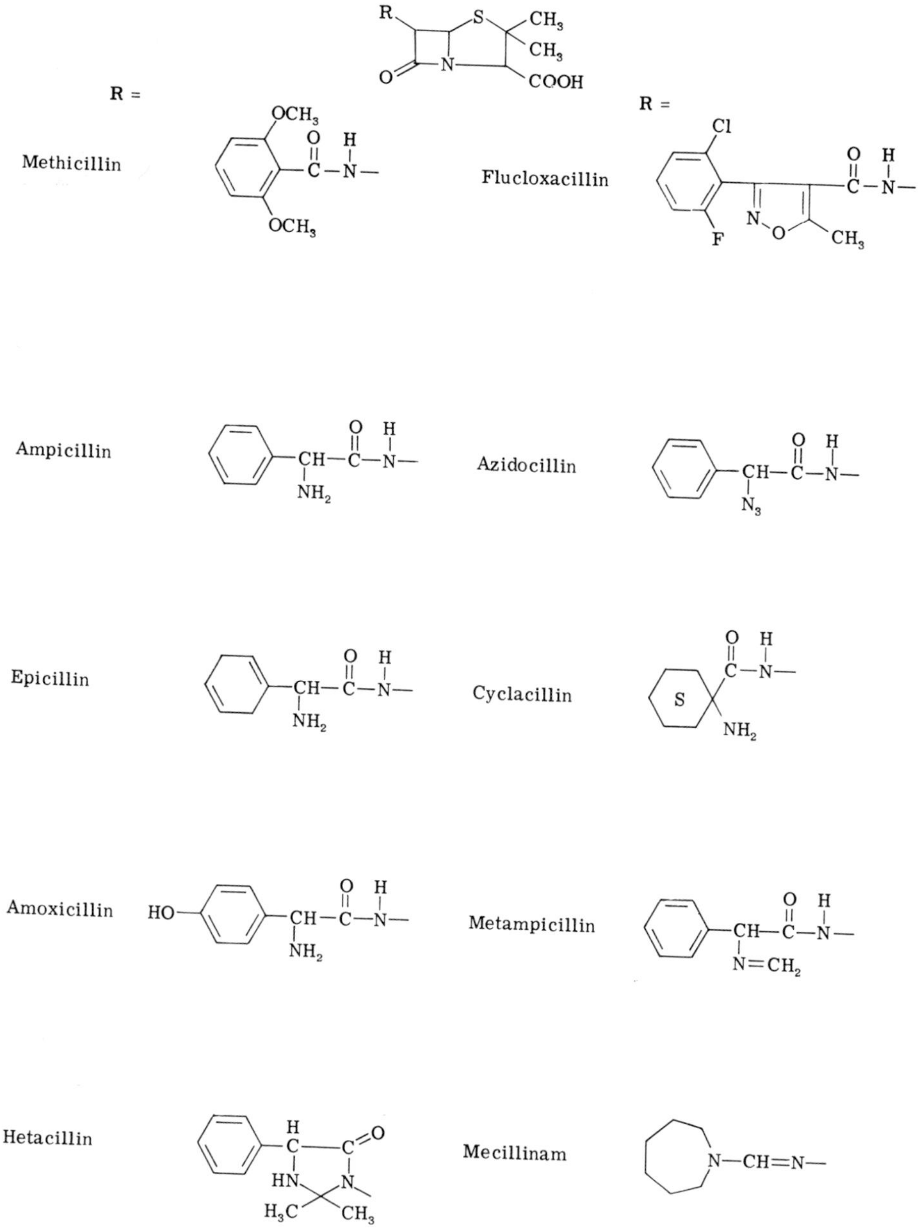

FIGURE 2. (Continued.)

TABLE V. Some Characteristics of Clinically Used Penicillins

Name	Method of preparation	Gram+ cocci	Gram− rods	Acid stability	Resistance to staphyloccal penicillinase
Amoxicillin	Semisynthetic	+	+	+	Low
Ampicillin	Semisynthetic	+	+	+	Low
Azidocillin[a]	Semisynthetic	+	+	+	Low
Azolocillin[a]	Semisynthetic	+	Some	+	Low
Bacampicillin[a, b]	Semisynthetic	+	+	+	Low
Carbenicillin	Semisynthetic	+	Some	−	Low
Cloxacillin	Semisynthetic	+	Nil	+	High
Cyclacillin[a]	Semisynthetic	+	+	+	High
Dicloxacillin	Semisynthetic	+	Nil	+	High
Epicillin[a, b]	Semisynthetic	+	+	+	Low
Flucloxacillin[a]	Semisynthetic	+	Nil	+	High
Hetacillin[b]	Semisynthetic	+	+	+	Low
Mecillinam	Semisynthetic	+	+	+	High
Metamacillin	Semisynthetic	+	+	+	Low
Methicillin	Semisynthetic	+	−	−	High
Mezlocillin[a]	Semisynthetic	+	+	+	Low
Nafcillin	Semisynthetic	+	Nil	+	High
Oxacillin	Semisynthetic	+	Nil	+	High
Penicillin G	Fermentation	+	Nil	−	Nil
Penicillin V	Fermentation	+	Nil	+	Nil
Phenethicillin	Semisynthetic	+	Nil	+	Nil
Pirbenicillin[a]	Semisynthetic	+	+	+	Medium
Pivampicillin[a, b]	Semisynthetic	+	+	+	Low
Pivmecillinam[a]	Semisynthetic	+	+	+	High
Propicillin[a]	Semisynthetic	+	Nil	+	Low
Sulbenicillin[a]	Semisynthetic	+	Some	+	Low
Talampicillin[a, b]	Semisynthetic	+	+	+	Low
Ticarcillin	Semisynthetic	+	Some	+	Low

[a] Not available in the United States (1978).

[b] Hetacillin, bacampicillin, epicillin, pivampicillin, and talampicillin are converted to ampicillin in the body.

properties were isolated by Abraham and Newton (1961) and named cephalosporin C, N, and P. Cephalosporin C resembled penicillin in possessing a fused β-lactam ring, but it contained two oxygen and two carbon atoms more, and had D-α-aminoadipic acid in the side chain (R group). The full structure shown in Fig. 3 was elucidated (Hodgkin and Maslen, 1961) and the nucleus shown to consist of a six-membered dihydrothiazine ring (in place of the five-membered thiazolidine ring of the penicillins) fused to a β-lactam ring. This nucleus (7-amino-

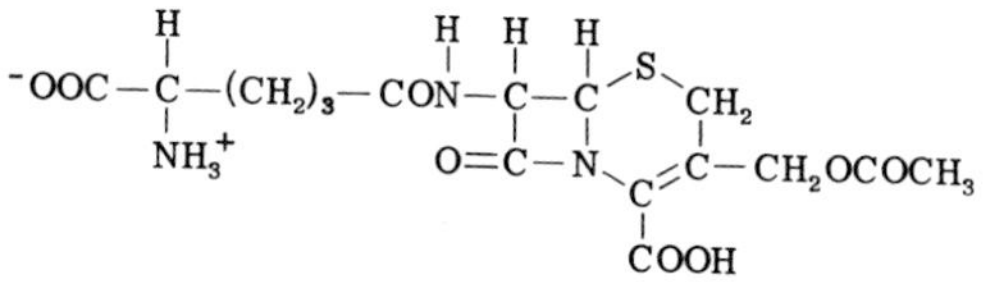

$$^-OOC - \underset{\underset{NH_3^+}{|}}{\overset{\overset{H}{|}}{C}} - (CH_2)_3 - CON - \underset{\underset{O=C-N}{|}}{\overset{\overset{H}{|}}{C}} - \overset{\overset{H}{|}}{\underset{\underset{C}{\diagup}}{C}} \overset{S}{\diagdown} CH_2 \quad \overset{}{\underset{COOH}{|}} C - CH_2OCOCH_3$$

FIGURE 3. Structure of cephalosporin C, an antibacterial agent produced by *Cephalosporium acremonium.*

cephalosporanic acid) contains structures which are present in 6-aminopenicillanic acid, including a peptide linkage (CO : NH) vulnerable to hydrolysis by amidase (Holt and Stewart, 1964; Walton, 1964a,b) and the β-lactam grouping susceptible to enzymatic hydrolysis by lactamase (Fleming et al., 1963; Sabath and Abraham, 1966). Cephalosporins are in general less susceptible than penicillin G or ampicillin to hydrolysis by β-lactamase from staphylococci or *Bacillus cereus,* but they are readily hydrolyzed by enzymes from gram-negative bacteria. The aminoadipic acid side chain (Fig. 3) lessens susceptibility to β-lactamase from certain bacteria.

The antistaphylococcus activity of cephalosporin C is relatively low, but it has some action against streptococci, and also against certain gram-negative bacilli, and is remarkably nontoxic in experimental animals. In view of similarity in structure to penicillin, it is not surprising to find sensitivity in subjects who are hypersensitive to penicillins.

Microbiological processes for production of cephalosporin C resemble in many respects those used for penicillin production. Special strains of *Cephalosporium* have been selected which produce more cephalosporin C and less cephalosporin N than the parent culture, and the growth of these in certain special fermentation media has resulted in higher antibiotic titers. Even with these modifications in processing, the antibiotic yields are much lower than those reported for the penicillins. Manufacturing operations on a commercial scale are carried out by many companies (see Table I).

The possibilities of preparing derivatives of cephalosporin C with new attributes were investigated by Abraham and Newton, who found that the pyridine-treated material had higher antibacterial activity especially against staphylococci (Abraham, 1962; Loder et al., 1961). Profiting from the experience with semisynthetic penicillins, scientists at the Eli Lilly Company laboratories, and at other companies, used the 7-aminocephalosporanic acid prepared by chemical degradation of cephalosporin C (Morin et al., 1962) to prepare thousands of semisynthetic cephalosporins (Webber and Ott, 1977). The structures of some of the clinically useful group are shown in Fig. 4 and some of their therapeutic uses are mentioned in Table VI.

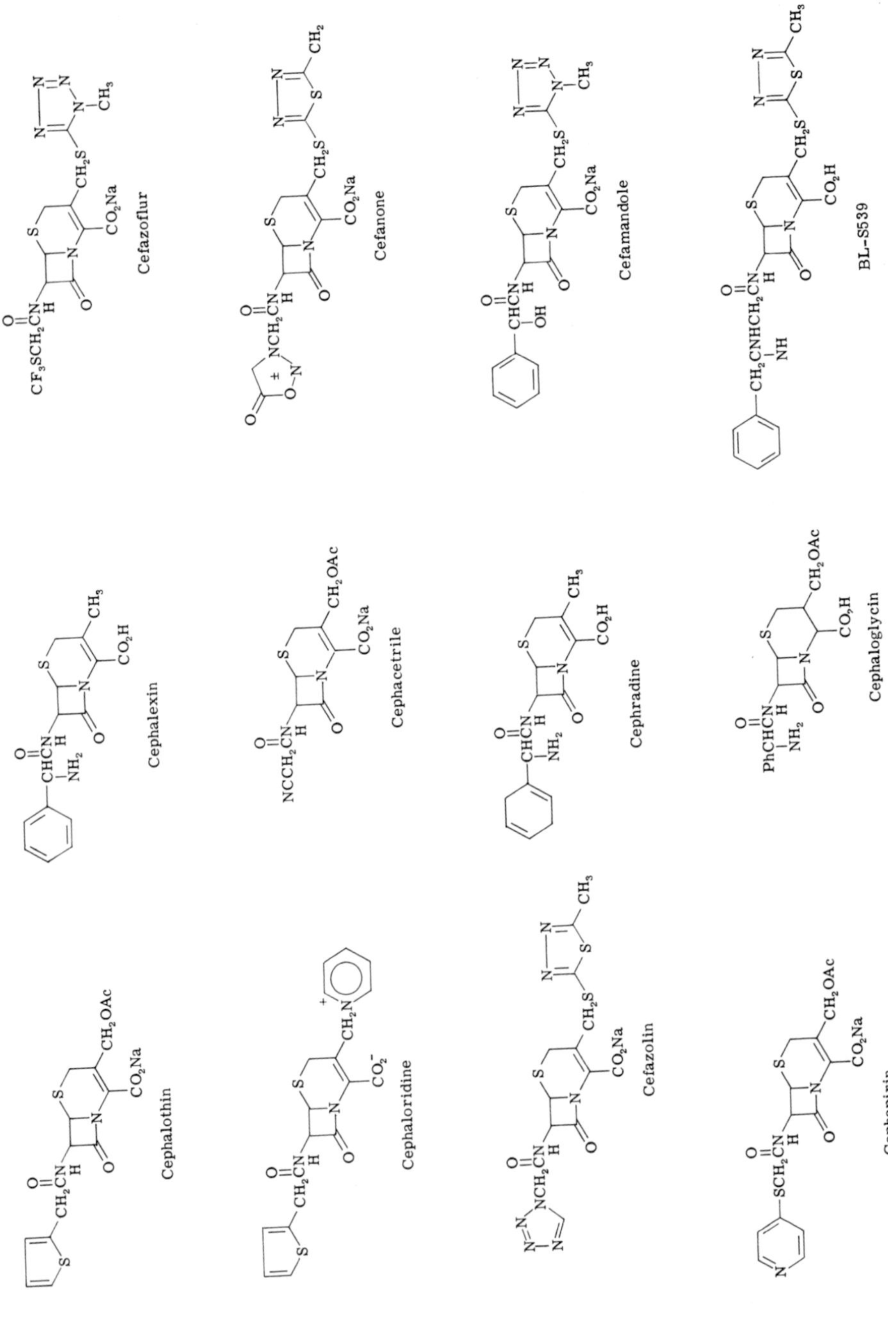

FIGURE 4. Structures of some clinically useful cephalosporins.

TABLE VI. Some Characteristics of Cephalosporins of Clinical Interest

Name	Method of Administration		Specific activity (minimum inhibitory concentration, ug/ml) against:		
	Injection	Oral	S. aureus	Pseudomonas	E. coli
Cefanone[a]	+	−	0.5	0.7	>200
Cefamandole	+	−	0.9	0.8	>200
Cefazolin	+	−	0.8	2	>200
Cefoxitin	+	−	2	6	>200
Cefuroxime[a]	+	−	0.7	8	>200
Cefazaflur	+	−	0.6	0.7	>200
Cefaparole[a]	+	−	4	4	>200
Cefatrizine[a]	+	−	3	3	>200
Ceftezol[a]	+	−	0.6	3	>200
Cephalexin	−	+	5	9	>200
Cephaloglycin	−	+	2	2	>200
Cephaloridine	+	−	4	5	>200
Cephalothin	+	−	0.4	17	>200
Cephapirin	+	−	0.6	13	>200
Cephacetrile	+	−	2	8	>200
Cephradine	−	+	5	20	>200
BL-S339[a]	+	−	0.6	3	>200
BL-S217[a]	+	−	0.7	6	>200
SCE-129[a]	+	−	6	58	0.9

[a] Not approved for use in the United States (1978).

V. PRODUCTION OF TETRACYCLINES

The tetracyclines are a group of crystalline antibiotics possessing a common hydronaphthacene skeleton. Tetracycline can be considered the parent compound of the group and is 4-dimethylamin-1, 4, 4α, 5, 5α, 6, 11, 12α-octahydro-3, 6, 10, 12, 12α-pentahydroxy-6-methyl-1, 11-dioxo-2-naphthacenecarboxamide (Stephens et al., 1954; Waller et al., 1952) (see Fig. 5 for a diagrammatic presentation). Chlortetracycline (Aureomycin) is 7-chloro, and oxytetracycline (Terramycin) the 5-hydroxy derivative. 7-Bromotetracycline (Lepetit S.p.A. 1957a,b) and 6-demethyl-7-chlortetracycline (Declomycin) have also been of clinical interest. All of these antibiotics are manufactured by fermentation processes using species of Streptomyces and, in addition, tetracycline is also manufactured by hydrogenolysis of chlortetracycline (Boothe et al., 1953; Forbath, 1957).

The tetracyclines are active in vivo against numerous gram-positive and gram-negative organisms, and some of the pathogenic rickettsiae

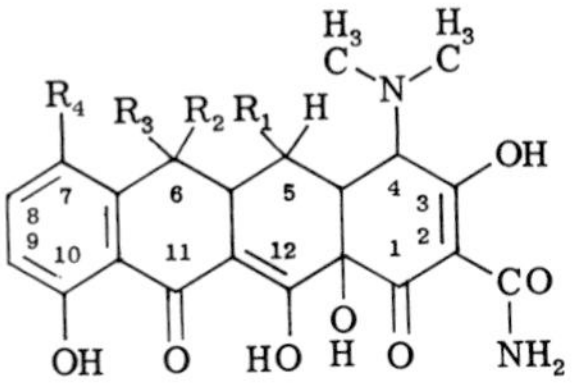

FIGURE 5. Structures of clinically important tetracyclines.

	R_1	R_2	R_3	R_4	Manufactured by
Tetracycline	H	OH	CH_3	H	Fermentation
7-Chlortetracycline	H	OH	CH_3	Cl	Fermentation
5-Oxytetracycline	OH	OH	CH_3	H	Fermentation
6-Demethyl-7-chlortetracycline	H	OH	H	Cl	Fermentation
6-Deoxy-5-oxytetracycline	OH	H	CH_3	H	Semisynthetic
7-Dimethylamino-6-deoxy-6-demethyl-tetracycline	H	H	H	$N(CH_3)_2$	Semisynthetic

and large viruses. Some infections where tetracyclines are therapeutic agents of choice are listed in Table II. The antibiotics are primarily bacteriostatic in action and in higher concentrations do have a bactericidal effect.

The systemic administration of tetracyclines may be carried out utilizing either oral or intravenous dosage forms; the intramuscular route is not feasible due to considerable pain and irritation at the site of injection. For veterinary use, tetracyclines are given by intravenous injection in bovines and equines, except that oral routes are used for young non-ruminating calves.

In connection with work on the "animal protein factor," it was found that crude chlortetracycline fermentation mash (containing some vitamin B_{12}) gave growth responses in animals well above those obtained with supraoptimal levels of vitamin B_{12}. At the present time, both chlortetracycline and oxytetracycline are being used extensively for growth stimulation and improvement of feed efficiency in poultry and hogs, and for the reduction of losses from certain disease conditions. Purified antibiotics as well as dried fermentation residues and the mycelium of *Streptomyces aureofaciens* grown on shredded barley (Belik *et al.*, 1957; Herold *et al.*, 1958; Herold and Necasek, 1959) are used for these purposes. The action of antibiotics in increasing poultry growth is apparently confined to their effect on the bacteria within the intestinal tract. Chlortetracycline and oxytetracycline have also been used in the preservation of fish and poultry.

A. Selection of Culture

Chlortetracycline, the first of the group to be investigated (Broschard *et al.*, 1949), is prepared from *S. aureofaciens* fermentations. The original strain, isolated by Duggar from a soil sample from Missouri (Duggar, 1948, 1949), shows extreme variability in microscopic, macroscopic, and physiological characteristics, and a variety of natural and artificially induced mutants have been examined (Backus *et al.*, 1954; Duggar *et al.*, 1954). Van Dyck and deSomer (1952), in attempts to correlate mycelium pigmentation and chlortetracycline production of their mutant strains, concluded that the best producers were the highly yellow pigmented strains.

Many investigators have observed that in the absence of special precautions, degeneration of antibiotic-producing ability took place in selected strains (Haas *et al.*, 1956). Storage of spores in lyophilized form has been used in some laboratories, while in others the spores have

TABLE VII. Streptomycetes Species Producing Tetracyclines

Name	Products[a]	References
S. alboflavus	CTC, TC, OTC, actinomycin	Villax (1963)
S. antibioticus	OTC, TC	Villax (1963)
S. aureofaciens	CTC, TC	Duggar (1948, 1949)
S. aureus	TC	Minieri *et al.* (1954, 1956)
		Villax 1963)
S. californicus	TC, actinomycin	Villax (1963)
S. cellulosae	OTC, actinomycin	Villax (1963)
S. feofaciens	TC	Lepetit S.p.A. (1957a)
S. flaveolus	OTC, TC, actinomycin	Villax (1963)
S. flavus	CTC, TC, OTC, actinomycin	Villax (1963)
S. fuscofaciens	OTC, quinocycline	Chas. Pfizer and Co. (1958)
S. lusitanus	CTC, TC	Villax (1963)
S. parvus	OTC, TC, actinomycin	Villas (1963)
S. platensis	OTC	McGuire (1954)
S. rimosus	OTC, TC, rimocidin	Finlay *et al.* (1950);
		Perlman *et al.* (1960);
		Sobin *et al.* (1950)
S. sayamaensis	TC, CTC	Arishima and Sekizawa, (1959)
S. vendargensis	OTC, vengacide	N.V. Koniklijke Nederlandsche
		Gist-en Spiritus-Fabriek (1956)
S. viridofaciens	CTC, TC	Lepetit S.p.A. (1957a)

[a] CTC, 7-chlortetracycline; TC, tetracycline; OTC, oxytetracycline.

been stored in sterile, desiccated soil in a refrigerator. Among the complicating factors in the preservation and maintenance of high antibiotic-producing strains is the possible contamination with actinophages. Although phage-resistant strains can be obtained without too much difficulty, some of these may be lysogenic and potentially dangerous.

In the 20 years since Duggar's isolation of S. aureofaciens, a number of other streptomycetes producing various tetracyclines, together with other antibiotics, have been described. The names given to some of these and the fermentation products are presented in Table VII. In a number of instances, the nature of the antibiotics produced by these cultures was determined by paper chromatographic methods (Martin et al., 1955), while in others (e.g., S. aureofaciens, S. feofaciens, S. fuscofaciens, S. lusitanus, S. platensis, S. rimosus, S. sayamaensis, S. vendargensis, and S. viridofaciens) the products were isolated in pure form.

As with penicillin production, certain cultures of the tetracycline-producing streptomycetes produce much more antibiotic activity than others. Intensive screening of S. aureofaciens mutants (Backus et al., 1954; Duggar et al., 1954) showed wide variation in antibiotic production, and some were obtained which produced as much as 50 times the amount of chlortetracycline as the original culture. The original S. aureofaciens produced both chlortetracycline and tetracycline when grown in media where chloride ion was limiting (Demain, 1953; Federal Trade Commission, 1958; Minieri et al., 1954, 1956).

Mutants with impaired halogen metabolism derived from this culture are now available and produce only tetracycline when grown in media containing large amounts of chloride (Doerschuk et al., 1956, 1959). Certain other cultures are nonselective in choice of halogen, accepting bromide in place of chloride, and thus producing 7-bromotetracycline (Lepetit S.p.A., 1957a,b). Streptomyces rimosus, the oxytetracycline-producing culture, synthesizes traces of tetracycline (Perlman et al., 1960) but apparently will not synthesize the chloro analog of 5-oxytetracycline. Mutants from S. aureofaciens with impaired methylation systems synthesize 6-demethyltetracyclines (McCormick et al., 1957). These antibiotics, which are chemically more stable than the tetracycline analogs, are widely used clinically.

B. Fermentation Medium and Conditions

Most of the available information on processes for the fermentation production of chlortetracycline (Duggar, 1949; Niedercorn, 1952), oxytetracycline (Sobin et al., 1950), and tetracycline (Minieri et al., 1956; Lepetit S.p.A., 1957a,b; Gourevitch and Lein, 1955; Hatch et al., 1956; McGhee and Megna, 1957) is in the patent literature. Although different media are used for each of the fermentation processes, the general operations appear to be the same for all: Spores of the streptomycete taken from stock slants or soil stocks are used to inoculate agar plates. The plates are incubated for several days, and the second crop of spores is used to inoculate flasks of liquid media. These flasks are placed on shakers, and when vegetative growth is at a maximum, they are used to inoculate aerated bottles or small fermentors. Several small fermentors are sometimes used to inoculate larger tanks, and eventually the final fermentation stage (20,000–100,000 liters) is seeded with about 5% of inoculum. Among the media used for the agar slants is Duggar's agar (Duggar, 1948; Duggar et al., 1954) which contains 2% beef extract, 0.05% asparagine, 1% glucose, 0.5% K_2HPO_4, and 1.3% agar. The shaken flasks and aerated bottles may contain a variety of media. A typical one contains 2% corn steep liquor (50% solids), 3% sucrose, and 0.5% $CaCO_3$.

The medium described by Duggar (1949) for the fermentation phase contains 1% sucrose, 1% corn steep liquor, 0.2% $(NH_4)_2HPO_4$, 0.2% KH_2PO_4, 0.1% $CaCO_3$, 0.025% $MgSO_4 \cdot 7 H_2O$, 0.005% $ZnSO_4 \cdot 7 H_2O$, 0.00033% $CuSO_4 \cdot 5 H_2O$, and 0.00033% $MnCl_2 \cdot 4 H_2O$. Chlortetracycline yields of 170 $\mu g/ml$ were obtained by Duggar (1949) with this medium. Van Dyck and deSomer (1952) used a medium containing sucrose, peanut oil meal, corn steep liquor, molasses, sucrose, $(NH_4)_2SO_4$, $CaCO_3$, and NaCl and, with selected cultures of S. aureofaciens, obtained yields of 1250 $\mu g/ml$ of chlortetracycline. A yield of 2000 $\mu g/ml$ was reported by McCormick et al. (1959) when they grew selected cultures in chemically defined media, showing that no special "stimulating materials" found in seed meals and other nitrogenous substances are needed for chlortetracycline production. However, they did find (McCormick et al., 1959, 1960, 1961) that a cosynthetic factor produced by certain streptomycetes stimulated production of the tetracyclines by mutants of S. aureofaciens. Yields obtained in the other tetracycline-producing fermentations range from 1000 $\mu g/ml$ for oxytetracycline (Regna et al., 1951; Sobin et al., 1950) to 2000–9500 $\mu g/ml$ for tetracycline (Laboratories Pro-Ter, 1964; Rolland and Sensi, 1955; Sensi et al., 1955; Villax, 1965).

The direct microbial production of tetracycline is the practical alternative to the chemical dechlorination of chlortetracycline for the manufacture of tetracycline (Forbath, 1957). When S. aureofaciens (strain NRRL 2209, the Duggar culture) is grown on Duggar's medium (Duggar, 1949) a mixture of chlortetracycline and tetracycline is produced (Federal Trade Commission, 1958). Growth of this and other chlortetracycline-producing organisms in media containing biologically insignificant amounts of chloride results in biosynthesis of tetracycline instead of chlortetracycline (Federal Trade Commission, 1958; Gourevitch and Lein, 1955; McGhee and Megna, 1957; Minieri et al., 1954, 1956; Petty and Matrishin, 1950; Rolland and Sensi, 1955; Sensi et al., 1955). Since corn steep liquor and other nitrogenous materials of natural origin contain significant amounts of inorganic chloride, this must be removed before these substances can be used in media for a tetracycline-producing fermentation. Among the methods used for removal of the chloride is absorption by ion-exchange resins (applied to corn steep liquor) (Hatch et al., 1956) and precipitation of the chloride as the insoluble AgCl (Gourevitch and Lein, 1955). Chemically defined media, where only chloride-free reagents are used, have also been successfully prepared (Darken et al., 1960).

A biochemical alternative to the use of chloride-free media for the production of tetracycline by S. aureofaciens is the inclusion in the media of compounds which inhibit the incorporation of chloride into the antibiotic. Among the more useful are the halides, bromide (Gourevitch et al., 1956; Lepetit S.p.A., 1957a,b) and fluoride (Szumski, 1959), and a variety of organic compounds, including 2-mercapto-2-thiazoline, 2-mercapto-4,5-dimethylthiazole, 2-benzoxazolethiol, 2-mercaptobenzimidazole (Lein et al., 1959), thiadiazoles, oxadiazoles, triazoles, and benzthiazoles (Goodman et al., 1959). The use of bromide, which blocks chloride utilization in certain cultures (Gourevitch et al., 1956), results in biosynthesis of bromotetracycline with others (Doerschuk et al., 1956, 1959; Lepetit S.p.A., 1957a,b).

The addition of antimetabolites to tetracycline-producing fermentations has resulted in modifying the biosynthetic pathways, with "new" antibiotic entities being produced. For example, addition of sulfonamides (Goodman and Miller, 1963; Perlman et al., 1961; Perlman and Heuser, 1962) to chlortetracycline-producing fermentations interfered with some of the methylation steps in the biosynthesis, resulting in biosynthesis of 7-chlor-6-demethyltetracycline. A similar result was obtained when aminopterin and ethionine were added (Neidleman, 1962; Neidleman et al., 1963), while addition of ethionine to an oxytetracycline fermentation resulted in biosynthesis of 4-N-ethylmethyl-5-oxytetracycline (Goodman and Miller, 1963). There is some variation from strain to strain in response to these antimetabolites, and what is one strain's "poison" is the next strain's nutrient.

VI. PRODUCTION OF BACITRACIN

Bacitracin is a polypeptide antibiotic active against many gram-positive and a few gram-negative species of bacteria. It is not assimilated significantly into circulating body fluids following oral ingestion, but it exerts its activity locally in the gastrointestinal tract. Although it is highly active parenterally, its systemic use has so far been considerably restricted because of the appearance of kidney damage when it is administered under some conditions. In addition to its considerable success in topical applications, bacitracin is used effectively in the treatment of surgical wounds. Among the important infectious problems of animals in which bacitracin has been used with reasonable success are bovine mastitis, infected wounds, hemorrhagic septicemia, otitis, enteritis (necrotic), infectious keratitis, dermatitis, and some types of dysentery. Feed supplements containing bacitracin in various forms have proved to be quite effective and economical in animal and poultry nutrition, increasing feed efficiency and reducing infectious diseases.

The original bacitracin-producing organism was obtained from debrided tissue taken from a compound fracture of the tibia of 7-year-old Margaret Tracy (Johnson et al., 1945; Meleney et al., 1949). The culture, first identified as *Bacillus subtilis*, was later classified as *Bacillus licheniformis*. Several cultures of *B. licheniformis* were later found to produce antibiotics, and one of these produced the antibiotic ayfivin, a peptide antibiotic shown to be chemically in the bacitracin group of antibiotics (Arrigada et al., 1949; Sharp et al., 1949). Bacitracin has also been identified in *B. subtilis* cultures (Aida, 1962; Freaney and Allen, 1958; Keko et al., 1953; Snoke, 1960).

The various products generally referred to as "bacitracin" are mixtures of related polypeptides. They are generally known as bacitracins A, A', B, C, D, E, F, F_1, F_2, F_3, and G. Bacitracin A is of primary importance and is the most potent, antibiotically speaking. Bacitracins B and C are less potent antibiotics than A, and the rest have very little antibacterial activity. Commercial products contain mostly bacitracin A, but some preparations have had as much as 20% bacitracin F (Craig and Konigsberg, 1957).

A unit of bacitracin is the amount which, in a dilution of 1 : 1024 based on serial twofold dilutions in beef infusion broth, inhibits the growth of a particular group A hemolytic streptococcus under defined conditions (Jackson, 1958). An international standard has been set up and has an activity of 55 units/mg (Craig and Konigsberg, 1957). Products with potencies of the order of 80 units/mg have been reported in the literature (Hickey, 1964).

The amino acid sequence of bacitracin A is:

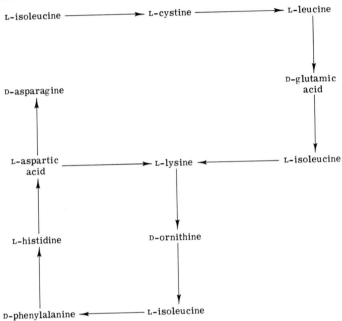

Methods for biosynthesis of bacitracin evolved from static procedures employing flasks and bottles and commercially impractical laboratory media to deep tank aerated operations employing low-cost media for industrial production. Flask and bottle surface cultures were used on an industrial scale in 1946 at the Ben Venue Laboratories to produce enough bacitracin for clinical and pharmacological trials. As the medical potential developed and more material was required to meet the demand, the submerged culture process was developed. Some of the characteristics of processes mentioned in the patent literature (and probably used on an industrial scale) are given in Table VIII.

A flow sheet describing the large-scale manufacture of bacitracin at Terre Haute, Indiana, plant of IMC Chemicals Group, Inc. (formerly Commercial Solvents Corporation) is shown in Fig. 6 (Inskeep *et al.*, 1953). The bacitracin-producing strains of *B. licheniformis* were carried on agar slants or in the spore form dispersed in dry, sterile soil. For culture development, a suitable small amount of soil culture was used to inoculate a 4-liter Erlenmeyer flask of tryptone or peptone broth. This was shaken on a reciprocating shaker located in a 37°C incubator. After

TABLE VIII. Characteristics of Processes Used in the Production of Bacitracin

Medium composition	Aeration method	Incubation period	Bacitracin yield (units/ml)	References
Tryptone, meat infusion, other broths	Surface culture	3–5 days	10–26	Jackson (1958)
Soybean meal, starch, calcium lactate	Surface culture	7 days	88	Darker (1951)
Soybean meal, calcium lactate, $CaCO_3$, starch	Submerged culture	26 hours	80–90	Keko et al. (1953)
Cottonseed and soybean meals, $CaCO_3$, dextrin	Submerged culture	24 hours	125	Cohen (1957)
Soybean meal, starch, $CaCO_3$	Submerged culture	24 hours	325	Freaney and Allen (1958)
Soybean grits, sucrose	Submerged culture	30 hours	400	Flickinger and Perlman (1979)

18–24 hours' incubation, the contents of the flask were used to inoculate 150 gal of medium contained in a 200-gal stainless steel tank. The culture was aerated vigorously for 6 hours (temperature maintained at 37°C), and this cell suspension was used to inoculate 800 gal of medium in a seed tank. The seed tank is made of stainless steel and is operated in a manner similar to the 150-gal stage, except that 800 gal of medium are prepared in a separate 1200-gal carbon steel cooker. This medium is blown hot to the seed tank, where it is cooled. A typical seed medium contained 4% soybean meal, 0.5% $CaCO_3$, and 0.5% starch. The seed culture is aerated and agitated after inoculation, and when the bacterial culture is in the log phase of growth, this inoculum is blown with sterile air via sterile pipe lines to a fermentor of cooled, sterile medium.

The fermentors are vertical, cylindrical, stainless steel vessels of 24,000-gal total capacity, having somewhat dished bottoms. The vessels are equipped with sterile air spargers, stainless steel tubing acting as cooling coils, and antifoam addition devices.

At the end of the incubation period, the bacitracin is recovered from the fermented medium by extraction with n-butanol. The bacitracin in the rich butanol is extracted with buffers, and the aqueous phase is concentrated (see Fig. 6). Ion-exchange resins have been used to effect further purifications (Gollaher and Honohan, 1956). Zinc bacitracin is used as a high-potency product for certain pharmaceutical preparations since its

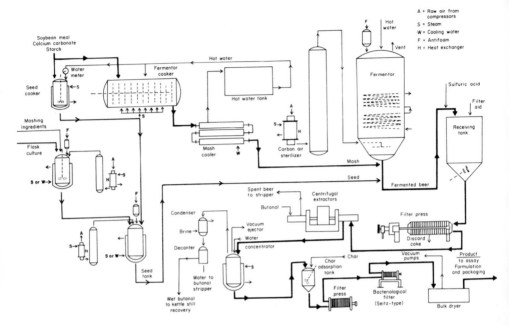

FIGURE 6. Bacitracin fermentation process flow chart.

storage stability is better than that of bacitracin itself. For animal feed purposes, the processing is quite simple, involving evaporation of the whole fermentation broth, drying, and blending with other animal feed supplements. Zinc bacitracin and bacitracin–methylene–disalicylate are two forms used for some feed formulations.

REFERENCES

Abraham, E. P. (1962). *Pharmacol. Rev.* **14,** 473.

Abraham, E. P., and Chain, E. (1940). *Nature (London)* **146,** 837.

Abraham, E. P., and Newton, G. G. F. (1961). *Biochem. J.* **79,** 393.

Abraham, E. P., Chain, E., Fletcher, C. M., Florey, H. W., Gardner, A. D., Heatley, N. G., and Jennings, M. A. (1941). *Lancet* **2,** 177.

Abraham, E. P., Chain, E., Florey, H. W., Florey, M. E., Heatley, N. G., Jennings, M. H., and Sanders, A. G. (1949). *In* "The Antibiotics" (H. W. Florey, ed.), p. 631. Oxford Univ. Press, London and New York.

Aida, T. (1962). *J. Agric. Chem. Soc. Jpn.* **36,** 724.

Arishima, M., and Sekizawa, Y. (1959). Japanese Patent 9,497.

Arrigada, A., Florey, H. W., Florey, M. E., Jennings, M. A., and Wallmark, I. G. (1949). *Br. J. Exp. Pathol.* **30,** 458.

Backus, E. J., Duggar, B. M., and Campbell, T. H. (1954). *Ann. N.Y. Acad. Sci.* **60,** 86.

Backus, M. P., and Stauffer, J. F. (1955). *Mycologia* **47,** 429.

Bartholomew, W. H., Karow, E. O., Stat, M. R., and Wilhelm, R. H. (1950a) *Ind. Eng. Chem.* **42,** 1801.

Bartholomew, W. H., Karow, E. O., Sfat, M. R., and Wilhelm, R. H. (1950b). *Ind. Eng.Chem.* **42,** 1810.

Batchelor, F. R., Doyle, F. P., Naylor, J. H. C., and Rolinson, G. N. (1959). *Nature (London)* **183,** 257.

Batchelor, F. R., Chain, E. B., Richards, M., and Rolinson, G. N. (1961). *Proc. R. Soc. London, Ser. B* **154,** 522.

Behrens, O. K., Corse, J., Edwards, J. P., Garrison, L., Jones, R. G., Soper, Q. F., van Abeele, F. R., and Whitehead, C. W. (1948). *J. Biol. Chem.* **175,** 793.

Belik, E., Herold, M., and Doskocil, J. (1957). *Folia Biol. (Prague)* **3,** 229.

Boothe, J. H., Morton, J., Petisi, J. P., Wilkinson, R. G., and Williams, J. H. (1953). *J. Am. Chem. Soc.* **75,** 4621.

Broschard, R. W., Dornbush, A. C., Gordon, S., Hutchings, B. L., Kohler, A. R., Krupa, G., Kushner, S., Lefemine, D. W., and Pidacks, C. (1949). *Science* **109,** 199.

Brotzu, G. (1948). *Lav. Ist. Ig. Cagliari*

Brown, W. E., and Peterson, W. H. (1950). *Ind. Eng. Chem.* **42,** 1769.

Calam, C. T., and Hockenhull, D. J. D. (1949). *J. Gen. Microbiol.* **3,** 19.

Calam, C. T., Driver, N., and Bowers, R. H. (1951). *J. Appl. Chem.* **1,** 209.

Chain, E. B. (1964). *In* "New Perspectives in Biology" (M. Sela, ed.), p. 205. Elsevier, Amsterdam.

Chain, E. B. (1966). *Antimicrob. Agents Chemother.* p. 1.

Chain, E., and Gulandi, G. (1954). *Rend. Ist. Super. Sanita* **17,** 5.

Chain, E., Florey, M. W., Gardner, A. D., Heatley, N. G., Jennings, M. A., Orr-Ewing, J., and Sanders, A. G. (1940). *Lancet* **2,** 226.

Chas. Pfizer and Co. (1958). German Patent 1,028,744.

Claridge, C. A., Gourvitch, A., and Lein, J. (1960). *Nature (London)* **187,** 237.

Clutterbuck, P. W., Lovell, R., and Raistrick, H. (1932). *Biochem. J.* **26,** 1907.

Coghill, R. D. (1944). *Chem. & Eng. News* **22,** 588.

Coghill, R. D., and Koch, R. D. (1945). *Chem. & Eng. News* **23,** 2310.

Cohen, I. R. (1957). U.S. Patent 2,789,941.

Craig, L. C., and Konigsberg, W. (1957). *J. Org. Chem.* **22,** 1345.

Darken, M. A., Berenson, H., Shirk, R. J., and Sjölander, N. O. (1960). *Appl. Microbiol.* **8,** 46.

Darker, G. D. (1951). U.S. Patent 2,567,698.

Davey, V. F., and Johnson, M. J. (1953). *Appl. Microbiol* **1,** 208.

Deindoerfer, F. H., and Wilker. B. L. (1957). *Ind. Eng. Chem.* **49,** 1223.

Demain, A. L. (1953). *Adv. Appl. Microbiol.* **1,** 23.

Demerec, M. (1948) U.S. Patent 2,445,748.

Doerschuk, A. P., McCormick, J. R. D., Goodman, J. J., Szumski, S. A., Growich, J. A., Miller, P. A., Bitler, B. A., Jensen, E. R., Petty, M. A., and Phelps, A. S. (1956). *J. Am. Chem. Soc.* **81,** 1058.

Doerschuk, A. P., McCormick, J. R. S., Goodman, J. J., Szumski, S. A., Growich, J. A., Miller, P. A., Bitler, B. A., Jensen, E. R., Matrishin, M., Petty, M. A., and Phelps, A. S. (1959). *J. Am. Chem. Soc.* **81,** 3069.

Duggar, B. M. (1948). *Ann. N.Y. Acad. Sci.* **51,** 177.

Duggar, B. M. (1949). U.S. Patent 2,482,055.

Duggar, B. M., Backus, E. J., and Campbell, T. H. (1954). *Ann. N.Y. Acad. Sci.* **60,** 71.

E. R. Squibb and Sons, Inc. (1952). British Patent 679,087.

Federal Trade Commission (1958). "Economic Report on Antibiotic Manufacture," US Gov. Printing Office, Washington, D.C.

Finlay, A. C., Hobby, G. L., P'an, S. Y., Regna, P. P., Routien, J. B., Seeley, D. B., Shull, G. M., Sobin, B. A., Solomons, I. A., Vinson, J. W., and Kane, J. H. (1950). *Science* **111,** 85.

Fleming, A. (1929). *Br. J. Exp. Pathol.* **10,** 226.

Fleming, P. C., Goldner, H., and Glass, D. G. (1963). *Lancet* **1,** 1399.

Flickinger, M. C., and Perlman, D. (1979). *Appl. Environ. Microbiol.* **15,** 222.

Forbath, T. P. (1957). *Chem. Eng. (N.Y.)* **64,** 228.

Ford, J. H., Churchill, B. W., and Colingsworth, D. R. (1953). *Antibiot. Chemother. (Washington, D.C.)* **3,** 1149.

Foster, J. W., Woodruff, M. B., Perlman, D., McDaniel, L. E., Wilker, B. L., and Hendlin, D. (1946). *J. Bacteriol.* **51,** 695.

Freaney, T. E. (1958). U.S. Patent 2,830,934.

Freaney, T. E., and Allen, L. P. (1958). U.S. Patent 2,828,246.

Gollaher, M. G., and Honohan, E. J. (1956). U.S. Patent 2,763,590.

Goodman, J. J., and Miller, P. A. (1963). *Biotechnol. Bioeng.* **4,** 391.

Goodman, J. J., Matrishin, M. M., Young, R. W., and McCormick, J. R. D. (1959). *J. Bacteriol.* **78,** 492.

Gourevitch, A., and Lein, J. (1955). U.S. Patent 2,712,517.

Gourevitch, A., Misiek, M., and Lein, J. (1956). *Antibiot. Chemother. (Washington, D.C.)* **5,** 448.

Gualandi, G., Caldarola, E., and Chain, E. B. (1960). *Rend. Ist. Super. Sanita* **23,** 869.

Haas, F. L., Puglisi, T. A., Moses, A. J., and Lein, J. (1956). *Appl. Microbiol.* **4,** 187.

Hastings, J. J. H. (1955). *Ind. Chem. Manuf.* **31,** 494.

Hatch, A. B., Hunt, G. A., and Lein, J. (1956). U.S. Patent 2,763,591.

Herold, M., and Necasek, J. (1959). *Adv. Appl. Microbiol* **1,** 1.

Herold, M., Matelova, V., Necasek, J., and Hostalek, Z. (1958). *Cesk. Mikrobiol.* **3,** 313.

Hickey, R. J., (1964). *Prog. Ind. Microbiol.* **5,** 93.

Hodgkin, D., and Maslen, E. N. (1961). *Biochem. J.* **79,** 393.

Holt, R. J., and Stewart, G. T. (1964). *J. Gen. Microbiol.* **36,** 203.

Hosler, P., and Johnson, M. J. (1953). *Ind. Eng. Chem.* **45,** 871.

Huang, H. T., English, A. R., Seto, T. A., Shull, G. M., and Sobin, B. A. (1960). *J. Am. Chem. Soc.* **82,** 3790.

Huang, H. T., Seto, T. A., and Shull, G. M. (1963). *Appl. Microbiol.* **11,** 1.

Inskeep, G. C., Bennett, R. E., Dudley, J. F., and Shepard, M. W. (1953). *Ind. Eng. Chem.* **43,** 1488.

Jackson, T. (1958). In "Biochemical Engineering" (R. Steel, ed.), p. 185. Macmillan, New York.

Jarvis, F. G., and Johnson, M. J. (1947). *J. Am. Chem. Soc.* **69,** 3010.

Johnson, B. A., Anker, H., and Meleney, F. (1945). *Science* **102,** 376.

Kato, K. (1953a). *J. Antibiot. Ser. A* **6,** 130.

Kato, K. (1953b). *J. Antibiot. Ser. A* **6,** 184.

Kaufmann, W., and Bauer, K. (1960). *Naturwissenschaften* **47,** 474.

Keko, W. L., Bennett, R. E., and Arzberger, C. F. (1953). U.S. Patent 2,627,494.

Laboratories Pro-Ter (1964). Belgian Patent 649,537.

Lein, J., Sawmiller, L. F., and Cheney, L. E. (1959). *Appl. Microbiol.* **7,** 149.

Lepetit S.p.A. (1957a). British Patent 772,149.

Lepetit S.p.A. (1957b). British Patent 775,139.

Loder, B., Newton, G. G. F., and Abraham, E. P. (1961). *Biochem. J.* **79,** 408.

McCormick, J. R. D., Sjölander, N. O., Hirsch, U., Jensen, E. R., and Doerschuk, A. P. (1957). *J. Am. Chem. Soc.* **74,** 4561.

McCormick, J. R. D., Sjölander, N. O., Johnson, S., and Doerschuk, A. P. (1959). *J. Bacteriol.* **77,** 475.

McCormick, J. R. D., Hirsch, U., Sjölander, N. O., and Doerschuk, A. P. (1960). *J. Am. Chem. Soc.* **82,** 5006.

McCormick, J. R., Sjölander, N. O., and Hirsch, U. (1961). U.S. Patent 2,998,352.

McGhee, W. J., and Megna, J. C. (1957). U.S. Patent 2,776,243.

McGuire, J. M. (1954). British Patent 713,795.

Martin, J. H., Shay, A. J., Pruess, L. M., Porter, J. N., Mowat, J. H., and Bohonos, N. (1955). *Antibiot. Annu.* p. 1020.

Mead, T. M., and Stack, M. V. (1948). *Biochem. J.* **42,** XVIII.

Meleney, F., and Johnson, B. A. (1949). *Am. J. Med.* **7,** 794.

Minieri, P. P., Firman, M. C., Mistretta, A. G., Abbey, A., Bricker, C. E., Rigler, N. E., and Sokol, H. (1954). *Antibiot. Annu.* p. 81.

Minieri, P. P., Sokol, H., and Firman, M. C. (1956). U.S. Patent 2,734,018.

Morin, R. B., Jackson, B. G., Flynn, E. H., and Roeske, R. W. (1962). *J. Am. Chem. Soc.* **84,** 3400.

Moyer, A. J., and Coghill, R. D. (1946). *J. Bacteriol.* **51,** 57.

Moyer, A. J., and Coghill, R. D. (1947). *J. Bacteriol.* **53,** 329.

Murao, S. (1955). *J. Agric. Chem. Soc. Jpn.* **29,** 400.

Neidleman, S. L. (1962). U.S. Patent 3,061,522.

Neidleman, S. L., Bienstock, E., and Bennett, R. E. (1963). *Biochim. Biophys. Acta* **71,** 199.

Niedercorn, J. G. (1952). U.S. Patent 2,609,329.

N. V. Koninklijke Nederlandsche Gist-en Spiritus-Fabriek (1956). British Patent 764,193.

Owen, S. P., and Johnson, M. J. (1955). *Appl. Microbiol.* **3,** 375.

Pan, S. C., and Perlman, D. (1954). *Anal. Chem.* **26,** 1432.

Perlman, D. (1949). *Bull. Torrey Bot. Club.* **76,** 79.

Perlman, D. (1950). British Patent 700,316.

Perlman, D. (1977). *CHEM TECH* **7,** 434.

Perlman, D., and Heuser, L. J. (1962). U.S. Patent 3,028,311.

Perlman, D., and O'Brien, E. (1954). *Arch. Biochem. Biophys.* **51,** 266.

Perlman, D., Greenfield, R. B., and O'Brien, E. (1954). *Appl. Microbiol.* **2,** 199.

Perlman, D., Heuser, L. J., Dutcher, J. D., Barrett, J. M., and Boska, J. A. (1960). *J. Bacteriol.* **80,** 419.

Perlman, D., Heuser, L. J., Semar, J. B., Frazier, W. R., and Boska, J. A. (1961). *J. Am. Chem. Soc.* **83,** 4481.

Petty, M. A., and Matrishin, M. (1950). *Abstr. 118th Meet., Am. Chem. Soc.* p. 18A.

Pollock, M. R. (1965). *Biochem. J.* **94,** 666.

Pollock, M. R., Torriani, A. M., and Tridgell, E. J. (1956). *Biochem. J.* **62,** 387.

Price, K. E. (1969). *Adv. Appl. Microbiol.* **11,** 17.

Price, K. E. (1977). *In* "Structure-Activity Relationships among the Semisynthetic Antibiotics" (D. Perlman, ed.), p. 61. Academic Press, New York.

Reese, E. T., Sanderson, K., Woodward, R., and Eisenberg, G. M. (1949). *J. Bacteriol.* **57,** 15.

Regna, P. P., Solomons, I. A., Murai, K., Timreck, A. E., Brunnings, K. J., and Lazier, W. A. (1951). *J. Am. Chem. Soc.* **73,** 4211.

Rolinson, G. R., Batchelor, F. R., Butterworth, D., Cameron-Wood, J., Cole, M., Eustale, G. C., Hart, M. W., Richards, M., and Chain, E. B. (1960). *Nature (London)* **187,** 236.

Rolland, G., and Sensi, P. (1955). *Farmaco, Ed. Sci.* **10,** 37.

Sabath, L. D., and Abraham, E. P. (1966). *Antimicrob. Agents Chemother.* p. 392.

Sakaguchi, K., and Murao, S. (1950). *J. Agric. Chem. Soc. Jpn.* **23,** 411.

Sensi, P., de Ferrari, G. A., Gallo, G. G., and Rolland, G. (1955). *Il Farmaco, Ed. Sci.* **10,** 337.

Sharp. V. E., Arrigada, A., Newton, G. G. F., and Abraham, E. P. (1949). *Br. J. Exp. Pathol.* **30,** 444.

Singh, K., and Johnson, M. J. (1948). *J. Bacteriol.* **50,** 339.

Smith, E. L., and Bide, A. E. (1948). *Biochem. J.* **42,** XVII.

Snoke, J. E. (1960). *J. Bacteriol.* **80,** 552.

Sobin, B. A., Finlay, A. C., and Kane, J. H. (1950). U.S. Patent 2,516,080.

Soltero, F. V., and Johnson, M. J. (1953). *Appl. Microbiol.* **1,** 52.

Soper, Q. F., Whitehead, C. W., Behrens, O. K., Corse, J. J., and Jones, R. G. (1948). *J. Am. Chem. Soc.* **70,** 2849.

Stephens, C. R., Conover, L. H., Hochstein, F. A., Moreland, W. T., Regna, P. P., Pilgrim, F. J., Brunings, K. J., and Woodward, R. B. (1954). *J. Am. Chem. Soc.* **76,** 3568.

Stewart, G. T. (1961). *Lancet* **1,** 509.

Szumski, S. A. (1959). U.S. Patent 2,871,167.

Tabenkin, B., Lehr, H., Wayman, A. C., and Goldberg, M. W. (1952). *Arch. Biochem. Biophys.* **38,** 43.

Van Dyck, P., and deSomer, P. (1952). *Antibiot. Chemother. (Washington, D.C.)* **2,** 184.

Villax, I. (1963). *Antimicrob. Agents Chemother.* p. 661.

Villax, I. (1965). Eire Patent 49565.

Waller, C. W., Hutchings, B. L., Wolf, C. F., Goldman, A. A., Broschard, R. W., and Williams, J. H. (1952). *J. Am. Chem. Soc.* **74,** 4981.

Walton, R. B. (1964a). *Dev. Ind. Microbiol.* **5,** 349.

Walton, R. B. (1964b) French Patent 1,357,977.

Webber, J. A., and Ott, J. L. (1977). *In* "Structure-Activity Relationships among the Semisynthetic Antibiotics" (D. Perlman, ed.), p. 161. Academic Press, New York.

West, J. M., Stickle, G. P., Walter, K. D., and Brown, W. E. (1961). *J. Biochem. Microbiol. Technol. Eng.* **3,** 125.

Chapter 9

Production of Microbial Enzymes

KNUD AUNSTRUP
OTTO ANDRESEN
EDVARD A. FALCH
TAGE KJAER NIELSEN

MICROBIAL TECHNOLOGY, 2nd ed., VOL. I
Copyright © 1979 by Academic Press, Inc.
All rights of reproduction in any form reserved. ISBN 0-12-551501-4

I. INTRODUCTION

Production and use of microbial enzymes is an ancient art. By experience and empirical methods it has developed to a highly sophisticated state; the preparation of fermented oriental food is an example of this.

The combination of microbiological and biochemical science with modern technology has made possible the development of a large enzyme industry. This industry was founded well over 50 years ago, but the economically important developments have taken place only within the last decade.

A. Recent Developments

The regrowth of the enzyme industry began in the early 1960s with the introduction of glucoamylase in starch hydrolysis. The enzyme method was substituted for the traditional acid hydrolysis and resulted in an increased dextrose yield.

Enzyme-containing detergents have been known since 1913, but their use was limited because the enzymes available were unstable in detergent formulations. Around 1965 a new, stable protease produced by *Bacillus licheniformis* was introduced, which was very successful and found wide application for use in detergents.

In 1968–1969, it was discovered that some workers handling enzyme concentrates were experiencing allergic reactions. This caused a strong public reaction against the use of enzymes. Consequently, enzymes were removed from most detergents used in the United States. An official investigation showed that there was no risk involved for the detergent user. After the introduction of dust-free, encapsulated enzyme products, the risk involved in handling enzyme concentrates was also eliminated. Since that time the use of detergent enzymes has grown steadily.

The substitution of microbial enzymes for calf rennet is an old dream, and numerous microbial proteases have been tested for this purpose. In the period 1965–1970 three microbial rennets were successfully introduced, and they are now being widely used in the rennet market.

Due to the high cost of sucrose and the availability of large amounts of inexpensive dextrose, the development of an enzymatic process which would isomerize glucose to fructose has been very attractive. The first commercial process was in use around 1970, and since then it has grown to become one of the most important enzyme applications. The use of immobilized enzymes for this purpose has made it possible to keep the conversion costs to a low level.

The brewing industry has traditionally used malt as an enzyme source.

The partial substitution of barley for malt and the inclusion of microbial amylase, protease, and β-glucanase has avoided some of the difficulties caused by variation of the malt quality. Furthermore, this change has been economically advantageous. The distilling industry, another traditional malt consumer, has in many countries substituted the microbial enzymes, α-amylase and glucoamylase, for malt.

In the pharmaceutical industry, the use of enzymes has been limited. The most important application is for digestive aids, in which a mixture of pancreatic and microbial enzymes is used. Another important application is the use of penicillin acylase in the manufacture of semisynthetic penicillins.

Apart from these large-scale industrial applications of enzymes, there has been a substantial growth in the preparation of microbial enzymes for diagnostic, scientific, and analytical purposes. These enzymes are usually prepared under laboratory-type conditions and are not dealt with here.

B. Present Position

A rough estimate is that the present world market for industrial microbial enzymes represents a sales value of $150–175 million. The economically most important enzymes are listed in Table I, together with an estimate of the relative value of the enzymes and the amount of enzyme produced per year, calculated as pure enzyme protein.

The major producers of industrial enzymes are listed in Table II. NOVO and Gist-Brocades have by far the largest share of the market.

TABLE I. Production of Industrial Enzymes

Enzyme	Amounts produced per year (tons of pure enzyme protein)	Relative sales value (%)
Bacillus protease	500	40
Glucoamylase	300	14
Bacillus amylase	300	12
Glucose isomerase	50	12
Microbial rennet	10	7
Fungal amylase	10	3
Pectinases	10	10
Fungal protease	10	1
Other	—	1

TABLE II. Companies Producing Microbial Enzymes

Denmark	Grindstedvaerket
	NOVO Industri
France	Sté. Rapidase
Germany	Boehringer Ingelheim
	Hoechst
	Miles-Kalichemie
	Röhm
Holland	Gist Brocades (KNGS)
Japan	Amano
	Daiwa Kasei
	Godo Shusei
	Kyowa Hakko Kogyo
	Meito Sangyo
	Nagase
	Sankyo
	Shin Nihon
Switzerland	Swiss Ferment
United Kingdom	ABM
	Glaxo
	ICI
United States	CPC
	G. B. Fermentation Industries, Inc.
	Miles
	Novo Biochemical Industries
	Pfizer
	Rohm & Haas
	Standard Brands

II. DEVELOPMENT OF NEW ENZYMES

Fewer than 50 microbial enzymes are of industrial importance today, but patents have been applied for on more than a thousand different enzymes. This reflects the increasing interest in developing new enzyme products and shows that it is easier to find a new enzyme than to find a profitable application.

A new enzyme product becomes a commercial success only if a demand exists and if the product possesses properties which satisfy the technical and economic requirements of the process. Most new enzymes fail in at least one of these respects. Demands for new enzymes arise from the development of new processes or from the unsatisfactory performance of known enzymes in established processes.

The properties of enzymes can to some extent be changed by chemical modification of the molecule. The scientific literature on the subject is extensive, but the methods have so far not been used on technical enzymes, presumably because the results have not been economically or technically attractive. The search for new enzymes has, therefore, been a search for new microorganisms. The basic principles which are guidelines in this search are described below.

A. Strain Selection

1. Enzyme Properties and Taxonomy

It is generally held that macromolecules, such as nucleic acids and proteins, are specific for the species and that the phylogenetic development which has given rise to the microbiological variation basically has been caused by variation in these molecules. As a consequence, it is to be expected that enzyme types, such as protease, α-amylase, and lactase, which are found in several species, will have properties which vary as much as the other properties of the organism. One usually finds that closely related organisms have enzymes with closely related properties, while unrelated organisms have enzyme systems which differ widely.

The protease subtilisin Carlsberg from *Bacillus licheniformis* is closely related to the protease subtilisin NOVO from *Bacillus amyloliquefaciens,* just as the species are closely related. Yet, the differences are distinct, and it appears that these enzymes are specific for the species. Unfortunately, many incorrect classifications of microorganisms in the biochemical literature make the establishment of such rules as this difficult.

2. Enzyme Properties and the Environment

Extracellular enzymes must work in the environment of the microorganism. One may therefore rightly assume that they have their optimum activity and stability close to the optimum conditions for growth of the microorganism. Numerous examples of this may be cited.

Proteases from alkalophilic *Bacillus* species have pH optima several units higher than that of the protease from the "neutralophilic" *B. licheniformis.* The α-amylase from the thermophile *Bacillus coagulans* can be used at 10°C higher temperatures than that of the mesophile *B. amyloliquefaciens.* There are known exceptions. The α-amylase of the mesophile *B. licheniformis* is more thermostable than the amylase of *B.*

coagulans. Several *Aspergillus* species, which have growth optima at about pH 4, produce an alkaline protease, with pH optimum about 10, in addition to an acid protease.

The properties of intracellular enzymes are usually not too dependent on environmental conditions. The internal pH of the cells is close to neutrality and appears to be little influenced by external conditions. Thermostable enzymes, however, are usually found in thermophilic organisms. The thermostable lactase of *Bacillus stearothermophilus* is such an example. Thermostability of an isolated enzyme may be lower than the growth optimum of the strain would indicate, because the enzyme is especially stabilized in the intact cell.

3. Enzyme Properties and the Presence of Substrate

Microorganisms are responsible for the majority of the mineralizations of organic material in nature. It is claimed that all organic material can be microbiologically decomposed. Microorganisms which will produce enzymes for the degradation of certain materials are, therefore, usually found where these substances are abundant. For example, producers of lignolytic enzymes, pentosanases, and cellulases are found in forest soil, and uric acid decomposers in chicken pens. Microorganisms capable of metabolizing a new compound are often quickly established in sewage plants and effluent streams.

4. Methods for the Isolation of Enzyme-Producing Organisms

Traditional microbiological methods are most often used for isolation. Especially important is the use of enrichment cultures and selective media. In the ideal system, the substrate for the enzyme serves as the sole source of one or more vital elements. For example, a uric acid–salts medium is used in the isolation of urate oxidase-producing microorganisms.

Agar diffusion tests are used whenever possible for the detection of enzyme activity. The principle is valuable as a qualitative or very rough quantitative test but the results are often misleading, particularly if the enzyme substrate is incorporated in the growth medium for the microorganism.

In the design of isolation methods, the imagination of the researcher is essential for success. A combination of a good source of microorganisms and a specific isolation method, often nontraditional, will lead to a useful result. An example is the search for alkali-stable proteases, where enrichment and selection on media with 0.1 M sodium sesquicarbonate and with casein as a protease indicator led to the discovery of a large group of *Bacillus* species that produce alkaline proteases.

5. Safety Measures

Microorganisms may be pathogenic or producers of toxic materials, and this must constantly be borne in mind when working with unknown organisms. Until they are identified, therefore, cultures and products should be handled with the proper precautions. This is especially important for fungal isolates since fungal spores spread easily.

B. Strain Development

1. Desired Properties of Production Strains

The organism should grow on an inexpensive medium and give a constant, high yield of enzyme in a short time. Secondary enzyme activities and the content of metabolites in the fermented broth should be minimal. Furthermore, recovery of the enzyme should be simple and inexpensive and lead to a stable product which can be handled safely and which has an acceptable appearance. Last but not least, the process must be safe to the personnel in the production plant, and the effluents from the plant should not disturb the environment.

The fulfillment of these objectives requires a combined optimization of the strain properties and process parameters. Optimization of strain properties is very attractive because this, as a rule, offers an inexpensive and permanent solution to the problem. An example is the development of a constitutive mutant which eliminates the need for an expensive inducer, a problem which has been relevant in the production of the xylose-induced glucose isomerase.

Health and safety problems can often be alleviated by strain development. The viable counts of the products, for example, can be reduced by using asporogenic mutants of *Bacillus* strains. Objectionable by-products, such as antibiotics, can be eliminated by using mutants.

Wild strains will often have undesirable enzymatic side activities which have to be removed in the purification process. This is the case with the transglucosidase in glucoamylase. A better method is to develop a mutant which does not produce the side activity.

2. Methods for Strain Development

The exciting progress in microbial genetics offers many attractive methods to the industrial researcher. The use of most of these methods, however, is hampered by the facts that the enzyme-producing organism usually is genetically unknown and that the efforts and especially the time which can be spent on a single organism is limited. The method of choice, therefore, will, as a rule, be straightforward mutagenization and selection. The mutagens used are those commonly applied in microbiology. In the development of selection methods, experience and personal skill are important.

The success of the mutation work is of great economic value to the enzyme industry. The methods used are therefore carefully guarded and very little information about them is published.

3. Strain Maintenance

The highly developed production strains must be protected against degeneration, contamination, and loss of viability. The most convenient way to secure this is to store the strains lyophilized or at the temperature of liquid nitrogen. It is possible to store the strains almost indefinitely without any changes using these methods.

III. THE FERMENTATION PROCESS

Traditionally, microbial enzymes were produced by surface cultures, i.e., cultures of microorganisms in thin layers of liquid or moist, solid media. This technique is still used for a few products, primarily of fungal origin, such as *Aspergillus* amylase, proteases from *Aspergillus* and *Mucor* species, pectinases from *Penicillium* and *Aspergillus* species, and also cellulases. Originally, the cultures grew in trays that were handled manually, but today mechanical systems for cleaning, filling, and emptying the trays are used. Systems where the semisolid culture is tumbled in a rotating drum also have been developed. Control of infections and also uniform control of temperature, humidity, and aeration present difficulties. Few developments of this technique have been published in the last decade.

Submerged culture methods today dominate in the production of enzymes because handling costs and the risk of infection are reduced, and because modern methods of control are more easily adapted to these processes. Yields are also generally higher. The production methods discussed below refer to submerged fermentation.

A. Inoculum Preparation

Highly mutated strains are increasingly used for enzyme production. An important requirement for a propagation technique, therefore, is that the production capability of the strain be preserved. The technique should also minimize the risk of contamination. Numerous recultivations of laboratory cultures and multiple propagation steps in the factory should be avoided. It is generally possible to develop a method where the seed flask is inoculated directly from a lyophilized culture and where only one seed tank is used.

In a few cases, the enzyme yield or the mycelial form of the culture depends on a specific propagation method (Meyrath et al., 1973); but this is the exception. In general, the process pattern is unaffected by rather wide variations in the propagation technique. Agar media may often be used in the seed flask. These cultures are easier to handle and have a longer period of applicability than shake-flask cultures. Cells or spores from the agar culture are suspended in a sterile solution before being transferred to the seed tank.

The medium in the seed tank often resembles the production medium. Excessive heat sterilization of the medium retards the growth of the inoculum. The volume of the seed tank usually constitutes 3–10% of the volume of the production fermentor.

The propagation time in the seed fermentor varies from 10 to 80 hours, depending on the process. Examples of parameters used as applicability criteria of the seed culture are change of pH, relative mycelial volume, development of carbon dioxide, or an easily detectable enzyme activity.

B. Medium Formulation and Preparation

The medium should provide the energy source for the process. Furthermore, it should include nutrients providing carbon and nitrogen sources and also special growth requirements, such as essential amino acids. Stimulating factors may be added to reduce the lag time or to increase the growth rate.

However, enzyme production may be negligible in a medium designed for good growth. The enzyme production often depends on the presence of an inducer in the medium, or it may be repressed by a component of the medium. Catabolite repression often controls enzyme production (Paigen and Williams, 1970). The strongest repression is seen in media containing glucose. Repression may be avoided by replacing glucose by slowly fermentable carbohydrates or by partly hydrolyzed starch. In a process for pullulanase production, the use of liquefied starch with a dextrose equivalent less than 42 is recommended (Heady, 1971).

In order to increase the productivity, highly concentrated media are used. In a patented process for the production of glucoamylase, concentrations of liquefied starch as high as 25% on a dry weight basis are used (Smith and Frankiewicz, 1978). Often a concentrated medium will result in catabolite repression, in retardation caused by the high osmotic pressure, or in a culture which is difficult to aerate. These difficulties may be overcome by incremental or continuous feeding of a concentrated medium to the culture (Kalaboklas, 1971; Hulme, 1973). Typical protein and carbohydrate components of industrial media for enzyme production are listed in Table III. Salts are often added as supplementary sources of nitrogen, phosphorus, sulfur, or calcium. Trace metals are normally present in sufficient amounts in the tap water or in the main raw materials. Strong pH buffers, mostly phosphate buffers, are still widely used to reduce pH fluctuations during the fermentation. Proper balance of carbon and nitrogen sources is also important to the pH pattern of the process if pH control is not applied.

Economy is very important in medium formulation. Typically, raw materials account for 60–80% of the variable costs of an enzyme fermentation process. Much development work is directed toward the replacement of costly ingredients with components available in large quantities at low cost. An enzyme fermentation medium normally has a high content of solids, such as ground whole grains or flakes.

Certain raw materials must be avoided in enzyme fermentations because the quality is too variable. Either a food quality is not available or they have an adverse effect on enzyme recovery, product quality, or waste disposal.

Most media are still sterilized batchwise in the fermentor. Continuous sterilization methods, however, are gaining wider application, particularly for the feeding medium. In some cases the feeding medium may be sterilized by filtration. Continuous heat sterilization is performed as a high-temperature/short-time process which results in improved preservation of growth factors and less development of color. Recovery of heat may be an additional advantage. Fouling of heat-exchange surfaces and clogging of pipes by grainy material in the medium should be considered when the sterilization system is designed. Direct steam injection may, in some cases, solve this problem.

TABLE III. Typical Medium Constituents for Enzyme Fermentation

Carbohydrates	Starch hydrolysate, molasses, saccharose, corn, barley, wheat
Proteins	Soybean meal, cotton seed meal, peanut meal, corn steep liquor, yeast hydrolysate, whey, gluten

C. Process Conditions and Equipment

Since microbial enzymes are relatively low-volume products, it has been difficult to justify the development and construction of specialized equipment for submerged fermentation. Equipment and techniques are most often adapted from antibiotics fermentations. Tall cylindrical fermentors of stainless steel with capacities of 10–100 tons and furnished with strong mechanical agitators and air spargers are typical. The advantage of this traditional setup is flexibility. It is easy to switch between products.

A schematic diagram of a typical enzyme fermentation process is shown in Fig. 1. The diagram indicates the extent of auxiliary equipment and control systems required for the process. As in many other fermentation processes, control of an enzyme fermentation is hampered by the fact that neither growth nor product formation may be determined rapidly enough to be of value for control. Primary physical variables, such as temperature, air flow, and pressure, are controlled within narrow limits. Other measurable variables, such as pH, oxygen tension, or oxygen consumption, are often also applied in process control. Foaming is normally controlled by automatic oil addition.

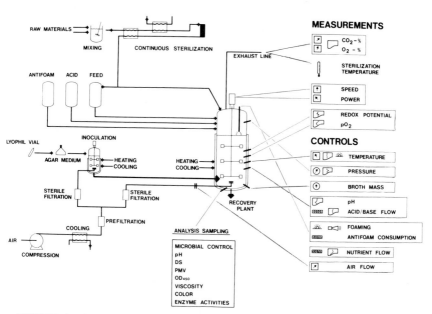

FIGURE 1. Schematic diagram of a typical enzyme fermentation process. Examples of useful measurements and controls are indicated.

Enzyme fermentations are especially vulnerable to microbial contamination. Rich media with a neutral pH value are typical, and the protection afforded by antibiotic activity is normally lacking. Infection generally means loss of a batch. Strict attention must be given to the contamination risk in the design and construction of fermentor, pipes, and auxiliary equipment. A fully welded system is recommended. Ports and valves should be steam-sealed, and all transfers of cultures and media should be done by compressed sterile air. A positive pressure must be maintained in the aseptic system. Mechanical agitator seals are normally used.

As a rule, extracellular enzymes are produced by batch processes which last from 30 to 150 hours. The optimal harvesting time falls between the point of maximum productivity and the point of maximum enzyme activity. Relative costs of raw materials, utilities, and recovery, as well as utilization of the plant capacity, determine the optimum. Often the process is terminated before or carried on beyond the optimum in order to obtain a broth with properties which facilitate product recovery. Finally, the scheduling in the plant restricts the choice of fermentation time. The optimal harvesting time is often signaled by a characteristic change in pH, oxygen tension, or some other parameter which is easy to determine.

Continuous culture techniques have been applied in several studies of enzyme production on a laboratory scale. Jensen (1972) and Fabian (1969), investigating *Bacillus* proteases, and Mitra and Wilke (1975), investigating *Trichoderma* cellulase, point out the advantages of a two-stage system. The first stage is operated under optimal conditions for growth, while conditions in the second stage are optimal for enzyme production. On an industrial scale, continuous methods have been reported in use only for the production of glucose isomerase (Diers, 1976). One reason for the limited application of continuous culture by the industry is the instability of the production strains (Heineken and O'Connor, 1972).

While continuous methods are rarely applied, the batch process is often extended, and the enzyme production favored by continuous feeding of carbohydrate or protein. One feeding strategy is to maintain a low reducing sugar level (Kalaboklas, 1971). The feed rate may also follow an empirical program or be controlled by pH, redox potential, oxygen consumption, or some other measured or calculated variable.

Most enzyme fermentations have a high oxygen demand, requiring aeration and agitation rates similar to those used in antibiotic fermentations. Very viscous media with non-Newtonian behavior are often employed; in other cases heavy mycelial growth complicating the oxygen transfer develops. In the glucoamylase process the productivity is limited by the oxygen transfer rate (Aunstrup, 1978). However, high oxygen tension may, in certain processes, inhibit enzyme formation. Diers (1976) reports that glucose isomerase production by an atypical variety of *Bacillus coagulans* is optimal during oxygen limitation and simultaneous addition of glucose at such a rate that the concentration in the broth is infinitesimal.

During process development much attention must be given to the properties of the broth in product recovery and to the quality of the final product. Since enzymes used for technical purposes are marketed as rather crude protein solutions or precipitates, the quality is strongly influenced by the fermentation method. Choice of raw materials, sterilization method, foam-control method, aeration and agitation rates, and fermentation time affect not only the product recovery yield and cost but also color, smell, stability, powder properties, and similar quality parameters.

A general kinetic model of enzyme synthesis has not been developed, but the regulation of the formation of many specific enzymes has been studied. A few enzymes used commercially are formed during exponential growth, but most are formed in the postexponential growth phase. As mentioned in the preceding sections, many factors other than growth control enzyme formation. A stoichiometric relationship for the process cannot be expressed. The yield of useful enzyme protein may reach 1–5% of the initial medium dry substance. The cell yield in a typical enzyme fermentation may be 2–10% on a similar basis. Residual nutrients and metabolites usually constitute 5–10% of the broth at the end of a fermentation.

IV. RECOVERY AND FINISHING

Methods for the recovery of enzymes for use commercially are simple unit operations such as centrifugation, filtration, vacuum evaporation, and precipitation of proteins. The complications in the processes arise from the character of the fermented broth. The broth has a variable, unspecified composition, a high content of colloidal material, and often a high viscosity.

The most significant development in recent years has taken place in the finishing processes for solid enzymes. Ten years ago, commercial products were dusty powders, but today the majority of solid enzymes are supplied as dustless granulates or as immobilized enzymes.

A. Cooling and Pretreatment of the Broth

After termination of the fermentation, the broth is rapidly cooled from the fermentation temperature of 30°–50°C to about 5°C. The culture broth is an excellent medium for a wide variety of microorganisms. Thorough cooling of the broth and strict hygienic measures in the plant are the most important methods to control infections during the recovery process. The addition of preservatives is normally unacceptable because many enzyme products are used in food processing. Also, many preservatives may effect the biological treatment of the wastewater from the recovery plant.

The refrigerated broth is normally pretreated before the separation processes—filtration or centrifugation. The character of pretreatment depends on the type of broth and on the equipment used for recovery.

Broth fermented by filamentous organisms can often be filtered or centrifuged directly after adjustment of pH to the stability optimum of the enzyme.

Broth from bacterial fermentations is more difficult to process. Fermentations on rich media and fermentations that leave large amounts of colloidal particles in the broth present the largest problem to the recovery plant. In these cases pretreatment with a coagulating or flocculating agent is needed. Inorganic salts, e.g., calcium phosphate precipitated in the broth from soluble phosphate and calcium salts, may be used to enclose cells and colloids in a precipitate. More efficient is a flocculation using synthetic polyelectrolytes which are available in several types, ranging from strongly cationic over nonionic to anionic. Each type can be used alone or in combination with another type or with inorganic chemicals. The synthetic polyelectrolytes are large molecules often based on polymerized ethyleneimine, acrylates, or acrylamide.

A special pretreatment step prior to a filtration may, for instance, be the addition of 2–4% diatomaceous earth to the broth as a body feed.

B. Separation of Solids

Most enzymes of industrial interest are extracellular, i.e., excreted from the cells into the liquid. The first task in recovery of an extracellular enzyme is to remove cells and other suspended material from the broth.

During the past decade there have been few developments in equipment used for this purpose. The main alternatives are still vacuum drum filters and centrifuges. The disk-type centrifuge with a self-cleaning bowl is the one mainly used. The use of vacuum drum filters precoated with diatomaceous earth or expanded perlite are particularly widespread in the industry.

C. Purification

The purification of the enzyme after the initial separation step may be more or less extensive, depending on the intended use of the product.

Figure 2 shows examples of purification steps leading to commercial-grade enzyme products.

After a polishing filtration, the solution is concentrated by vacuum evaporation at low temperature or by ultrafiltration. Ultrafiltration, although a relatively new technique, has been developed to a degree which makes it suitable for large-scale application. Compared with vacuum evaporation, ultrafiltration has the advantage that it removes smaller molecules, e.g., molecules of a MW below 10,000, from the concentrate. Consequently it is often possible to make a more concentrated enzyme solution by ultrafiltration than by vacuum evaporation. One drawback, however, is that the membranes are easily clogged by precipitates which may form during the concentration of the solution.

The concentrated enzyme solution can be clarified by a polishing filtration, and the remaining germs can be removed by a sterile filtration on cellulose-containing filter pads.

Stabilizers or preservatives may be added before or after the sterile filtration. Typical stabilizers are calcium salts, proteins, starch hydrolyzates, and sugar alcohols. Microbial stability of a liquid product may be secured by addition of 18–20% NaCl or by the use of a food-grade preservative, such as benzoate, parabens, or sorbate.

There is a preference of liquid rather than solid enzyme products, because the liquid products are cheaper to produce and are also safer and more convenient to apply. In some cases, however, only a solid preparation can be used. Solid enzyme concentrates can be obtained by precipitation or by direct spray-drying of the enzyme solution. It is often impossible to spray-dry a solution concentrated by evaporation, but it can usually be done on an ultrafiltered solution.

A higher degree of purity is obtained when the enzyme is precipitated with acetone, alcohols, or organic salts, such as ammonium sulfate or sodium sulfate. Fractional precipitation yields a higher purity than a one-step precipitation. In large-scale operations salts are preferable to organic solvents to eliminate the possibility of explosion. The precipitate may be dried by freeze-drying, vacuum-drying, or spray-drying depending on the heat stability of the enzyme.

D. Intracellular Enzymes

Recovery of intracellular enzymes is more difficult. The yield of enzyme on a total broth basis is often less than for extracellular enzymes because the cell mass is limiting.

To harvest the cells, centrifugation is usually preferred to filtration to avoid mixing the cells with diatomaceous earth. In industry the cells are often used without extraction of the enzyme, thereby saving activity and

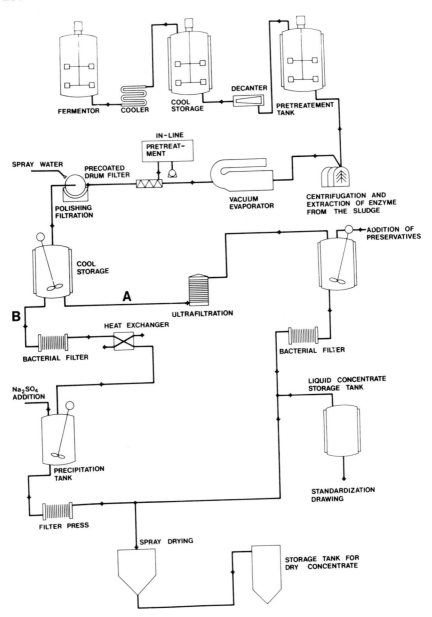

FIGURE 2. Examples of enzyme recovery processes.

expensive purification steps. If the enzymes must be extracted, autolysis may be used, but on an industrial scale physical methods are preferred, e.g., disruption of the cells by means of a homogenizer or a bead mill (Melling and Phillips, 1975). When the cells have been broken, the enzyme can be purified as are extracellular enzymes, but the process is usually more difficult due to the content of cell debris and nucleic acids from the broken cells.

E. Finishing of Solid Enzymes

A few years ago, all solid enzymes were sold as dusty powders with a small particle size. A significant proportion of the particles were smaller than 10 μm. This type of product requires very careful handling in order to avoid exposure to enzyme dust. Problems in handling and admixture of the powders, especially in the detergent industry, led to the development of methods to convert the enzyme concentrate to dustless granulates. An additional advantage of the granulates is often improved storage stability.

In the field of detergent enzyme granulation several methods have been developed, but on an industrial scale only a few are applied. A common method is to embed the enzyme into spheres of a waxy material consisting of a nonionic surfactant by means of a spray-cooling or prilling process. The "marumerizer" process is another important method by which the enzyme is mixed with a filler, a binder, and water; then extruded; and subsequently formed into spheres in a so-called marumerizer. After drying the spheres are coated with a layer of waxy material (see Fig. 3). Enzyme granules produced by these methods are rigid and have a very low dust level. The particles have an average diameter of about 500 μm, and practically no particles are smaller than 200 μm. Furthermore, the granules may be coated with an inert film. Today most enzymes used in the detergent industry are granulated.

In other applications, where the enzymes normally are used in industrial processes, the handling properties are less critical, and simple powders are still used. One reason is that the additives used in the granulates for the detergent industry are unacceptable in many other applications. A significant improvement in handling properties may be obtained, however, by a relatively simple granulation step, such as a fluid bed agglomeration.

F. Immobilization of Enzymes

Methods for the immobilization of enzymes have been known for several years, but the technique has gained industrial importance only

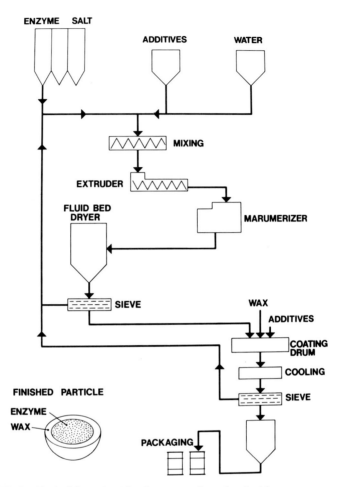

FIGURE 3. Typical flow sheet for the preparation of a dust-free enzyme product.

within the last few years. By far the most significant immobilized enzyme product at present is glucose isomerase, an intracellular enzyme produced by various microorganisms. Several manufacturers use the enzyme-containing cells for immobilization without further purification.

Numerous immobilization methods have been published (Barker and Kay, 1975). In some methods of industrial importance, glutaraldehyde is used to cross-link the enzyme or the supporting protein in which the cells are suspended. According to a method described by Van Velsen (1971), the enzymes or the cells are suspended in an aqueous solution of a gelling protein. The protein is suspended in an organic water-immiscible liquid to produce enzyme-containing droplets. The gelling protein is forced to gel. Subsequently, the particles are cross-linked with glutaraldehyde. Amotz et al. (1976) describe a method in which the cells are harvested, concentrated, and homogenized. Glutaraldehyde is added, and the resulting gel is shaped into granules which are dried.

Another attractive immobilization method is the use of polyelectrolytes to flocculate the enzyme-containing cells. After filtration the filter cake is shaped to particles of the desired size and dried (Long, 1974). Other methods are based on the fixation of the extracted enzyme through chemical reaction or adsorption to solid supports, such as ion exchangers, glass beads, or cellulose fibers.

The particles of an immobilized enzyme must be porous, uniform in size, and physically strong in order to perform satisfactorily in a column over a long period of time. Glucose isomerase, for example, is usually used more than 1000 hours at temperatures between 60° and 65°C.

V. REGULATIONS AND SPECIFICATIONS

Because microbial enzymes are natural products, and in view of their traditional use for food purposes in the orient, they were regarded as inherently harmless substances until about 15 years ago. The increasing knowledge of mycotoxins and the allergenic properties of enzymes has made everybody concerned more cautious. Microbial enzyme products must now meet strict specifications with regard to toxicity and other safety aspects.

A. Toxicity

As a rule, enzymes are completely nontoxic. The only recorded exception is phospholipase C, which is formed by several pathogenic bacteria. Toxic materials may, however, be present in the raw materials, and theoretically they may be formed during the manufacturing process by the metabolism of the enzyme-generating organism or contaminants. Since most enzyme products undergo a relatively crude purification process, they contain all components of the fermentation broth in small or large quantities. It is necessary, therefore, to ensure that no toxic products are formed at any stage of the process. This is done by prescribing food- or feed-grade raw materials throughout the process, by regularly testing for microbial contamination, and by a thorough investigation of the enzyme-generating organism and of the enzyme product. This investigation also includes a complete survey of the scientific literature on the organism and closely related organisms, followed by extensive feeding studies on several species of animals. The feeding studies are usually made with a mixture of several batches, e.g., ten batches of enzyme concentrate prepared according to the actual production process. The amounts used are chosen to ensure the highest possible safety factor. Usually the limit is set by the amount of solids that the animals are able to ingest. In some cases a high content of salts, such as NaCl, Na_2SO_4, and phosphates, will set an unacceptably low limit, necessitating desalting of the enzyme preparation before it is fed to the animals.

The preparations will always be tested for aflatoxin. They will be tested also for other mycotoxins if their presence is theoretically possible.

B. Allergenicity

Like all proteins, enzymes are antigenic and may thus cause allergy. The antigenicity varies from one enzyme to another, but no general rules can be given. The antigenicity of the enzyme as such cannot be changed, but the formulation can be made so that exposure of the user of the enzyme is minimized. The most important factor is the level of dust formed by handling the enzyme. To keep it low, the enzyme may be encapsulated in inert material. An encapsulated product may contain less than 0.5 μg of pure enzyme dust per kilogram as determined according to Harris and Rose (1972). Whenever possible, liquid products are preferred because they can easily be handled without exposing personnel to enzyme dust.

C. Microbial Safety

The enzyme-generating strain must be a harmless saprophyte. This is ensured by a literature survey and by animal pathogenicity tests, in which the organism must be administered both orally and by injection. In addition, the organism, as a rule, will be completely removed from the preparations, a precaution that serves the additional purpose of preventing competitors from obtaining the strain.

During the fermentation process contamination tests are regularly performed, and contaminated batches discarded. During recovery and handling of the finished product the hygienic standard must be as high as in the food industry. Nevertheless a certain degree of contamination is unavoidable. Regular controls of the microbial standard of the finished product are therefore necessary.

D. Regulations

Responsibility for the safety of an enzyme product remains with the manufacturer. In most countries, though, a new enzyme product for use in food requires approval by the appropriate authorities. In the United States, approval of the Federal Drug Administration is required. National and international bodies have worked out recommendations and specifications for a number of known enzyme products (Food Chemicals Codex, 1975; World Health Organization, 1972, 1975).

E. Specifications

All technical enzyme preparations are sold on an activity basis. This means that an analytical method must be specified by the manufacturer. Although recommendations for enzyme analysis have been given by the Commission on Enzyme Nomenclature, the different enzyme manufacturers generally use their own method of analysis. Even when activities are given in so-called international units, they should be interpreted with caution, because small differences in reaction conditions may have an important effect on the result. The manufacturer will do his best to reduce the deviation of the analytical method. For most preparations it is possible to give the activity with a standard deviation better than 5%.

Usually the manufacturer will guarantee a certain storage stability under specified conditions. For example, less than 10% loss per 6 months at 25°C for liquid products or less than 10% loss per year for solid preparations. In addition, the gross chemical composition may be given as well as the kind and amount of preservatives added. For special enzymes, such as detergent proteases, the specifications usually include limits for enzyme dust, particle size, bulk density, etc.

Immobilized enzymes present special problems, since performance cannot be judged on activity alone. The activity test must be performed in a column operating over a considerable period of time, and other parameters, such as particle size, particle strength, and color formation, become important.

The customer expects uniform, reliable, and safe enzyme products, and the manufacturer must prepare specifications which ensure that these expectations are met.

VI. SURVEY OF ENZYMES AND APPLICATIONS

A list of the enzymes used commercially is given in Table IV All the enzymes with the exception of two are hydrolases. It can thus be said

TABLE IV. Microbial Enzymes for Industrial Use

Enzyme	Microbial source
α-Amylase	*Aspergillus oryzae*
	Bacillus amyloliquefaciens
	Bacillus licheniformis
β-Amylase	*Bacillus cereus*
	Bacillus megaterium
	Bacillus polymyxa
β-Glucanase	*Aspergillus niger*
	Bacillus amyloliquefaciens
Cellulase	*Aspergillus niger*
	Trichoderma reesei
Dextranase	*Penicillium* sp.
	Trichoderma sp.
Glucoamylase	*Aspergillus awamori*
	Aspergillus niger
	Rhizopus sp.
Glucose isomerase	*Actinoplanes* sp.
	Arthrobacter sp.
	Bacillus sp. (thermophilic)
	Streptomyces sp.
Glucose oxidase	*Aspergillus niger*
Invertase	*Saccharomyces* sp.
Lactase	*Aspergillus niger*
	Kluyveromyces fragilis
	Kluyveromyces lactis
Lipase	*Aspergillus* sp.
	Candida lipolytica
	Geotrichum candidum
	Rhizopus sp.
Pectinase	*Aspergillus niger*
	Aspergillus sp.
Penicillin acylase	*Bacillus megaterium*
	Erwinia carotovorum
	Escherichia coli
Pentosanase	*Aspergillus* sp.
	Bacillus amyloliquefaciens
Protease, acid	*Aspergillus* sp.
Protease, neutral	*Aspergillus oryzae*
	Bacillus amyloliquefaciens
	Bacillus thermoproteolyticus
Protease, alkaline	*Aspergillus oryzae*
	Bacillus amyloliquefaciens
	Bacillus licheniformis
	Bacillus sp. (alkalophilic)
	Streptomyces griseus
Pullulanase	*Klebsiella aerogenes*
Rennet	*Endothia parasitica*
	Mucor miehei
	Mucor pusillus

that enzyme technology so far has been concentrated on the relatively simple hydrolytic processes, whereas the more complicated enzymatic reactions have not been possible on a technical scale. Most of the more complicated systems require co-enzymes, and this has deterred their use since an economical way of applying them has not been found. The explanation may be that there is no demand for a large-scale application of such complicated systems.

A. Microbial Sources of Technical Enzymes

The majority of industrially produced enzymes are derived from microorganisms of two genera, *Aspergillus* and *Bacillus*. It is characteristic of these microorganisms that they are common, harmless saprophytes; they grow rapidly, are metabolically very active, and they are highly variable. They secrete a number of extracellular enzymes in relatively large quantities. Usually several enzymes are produced at the same time.

Bacillus amyloliquefaciens produces α-amylase, β-glucanase, and neutral and alkaline protease. *Bacillus licheniformis* normally produces an almost pure serine protease with minimal amounts of α-amylase. However, special strains that produce high amounts of α-amylase together with the protease have been developed. Although the taxonomic distinction between the two species seems pretty clear, many manufacturers and scientists unfortunately still prefer to name both strains *B. subtilis*.

Aspergillus oryzae, when grown in semisolid culture, will produce a large number of enzymes, primarily α-amylase, lactase, glucoamylase, and protease. In submerged culture the α-amylase formation is increased, whereas the formation of other enzymes becomes minimal. *Aspergillus niger* strains are used in the production of glucoamylase, pectinase and protease. Special strains are developed for each of these purposes, but it is characteristic that all strains produce most of the enzymes in varying amounts. The nomenclature of the *Aspergilli* is complicated and unclear. *Aspergillus niger* and *A. oryzae* should here be understood as representing the groups of black and green *Aspergillus* species, respectively.

Members of the family *Mucoraceae* have frequently been used for enzyme production, e.g., the use of *Rhizopus* sp. for the manufacture of α-amylase, glucoamylase, protease, and lipase. These organisms are still important enzyme sources in the manufacture of oriental foods but their use in the production of enzymes for technical use is limited. The two thermophilic *Mucor* species, *M. miehei* and *M. pusillus,* are important in the manufacture of microbial rennet.

The *Actinomycetales* have become important sources of glucose isomerase which is produced commercially using several *Streptomyces* species and one *Actinoplanes* species. Few other microbial species are used for commercial enzyme production (Table IV).

B. Applications

Enzymes offer an attractive solution to many catalytic problems, especially within the food industry, because of their specificity, nontoxicity, and mild reaction conditions. Only the most important processes are described here. A list of selected publications on the application of enzymes is given at the end of this chapter.

1. Enzymes for the Detergent Industry

Proteinaceous dirt will often precipitate on soiled clothes, or it may coagulate during the normal washing process. Furthermore, proteins make dirt adhere to the textile fibers. Such stains, which are otherwise difficult to remove, can be dissolved easily by addition of proteolytic enzymes to the detergent.

The alkaline serine protease of *B. licheniformis,* otherwise known as subtilisin Carlsberg, is the preferred protease for this purpose. The enzyme is well suited for this application. It will attack many peptide bonds and therefore easily dissolve proteins. It may be used in the presence of most nonionic and anionic detergents at temperatures up to 65°C, and its pH optimum is close to 9, the pH normally used in washing fluids. Furthermore, extensive tests have proved the enzyme harmless to the user.

A few other proteases are used in detergents. The serine protease of *B. amyloliquefaciens* has found some application, presumably because the preparations usually have a substantial content of α-amylase. For some applications this may be an advantage. Proteases with improved stability and protein solubilizing properties under washing conditions have been found in alkalophilic bacilli (Aunstrup et al., 1972). These enzymes are particularly advantageous in detergents with low phosphate content.

The success of the detergent proteases has inspired numerous attempts to use enzymes for removal of other stubborn stains, such as fruit colors and lipids, especially in low-temperature washing. So far no satisfactory solution to these problems has been found.

2. Enzymes for the Starch Industry

Until around 1960, dextrose and glucose syrups were prepared from starch exclusively by acid hydrolysis. The process was corrosive, and the dextrose yield low. It was, therefore, a great step forward when pure glucoamylase in combination with bacterial α-amylase made possible a complete enzymatic hydrolysis of starch to dextrose. Today several enzymatic processes are used in the industry. A schematic summary is given in Table V. Most of the processes are applied in combination, starting with a liquefaction followed by a more or less extensive saccharification depending on the desired end product.

Prior to liquefaction the starch should be gelatinized by heat treatment. For the widely used maize starch, temperatures above 100°C are normally applied in order to secure complete gelatinization. It is consequently an advantage to use the extremely heat-stable amylase of B. licheniformis which can function at temperatures up to 110°C. Normally high concentrations (30–50% dry substance) are used in the saccharification process. The equilibrium concentrations of the saccharides formed by resynthesis limit the maximum degree of hydrolysis obtainable, e.g., 97% dextrose at 30% dry matter. Since the activity of glucoamylase toward branching points (α-1,6 bonds) is low, it may be an advantage to use a debranching enzyme, such as pullulanase, early in the hydrolysis process. In the preparation of fructose syrup, a fructose concentration of minimum 42% (on solids) is normally sought. Since this is close to the equilibrium concentration (around 50%), it is important that the initial glucose syrup be as pure as possible, i.e., that the hydrolysis be complete. Glucose isomerase is used in immobilized form, and a major share of the fructose syrup production is made by continuous column processes (Hilmer Nielsen et al., 1976).

Since the processes described here are directly dependent on the enzyme activities, it is important that the enzymes be pure, i.e., free from undesired enzyme activities, such as transglucosidase or protease, and that the activity be accurately standardized. Liquid preparations of amylase and glucoamylase are used, whereas the immobilized glucose isomerase comes in various forms, from frozen cells to dried granules. For column operation, mechanical strength, flow properties, and active lifetime are important parameters.

3. Enzymes for the Dairy Industry

In the dairy industry microbial rennet has to a large extent replaced the expensive rennet from calves. The microbial rennets are acid aspartate

TABLE V. Examples of Enzymatic Processes in the Starch Industry

| | | | Reaction conditions | | |
			pH	Temperature °C	Product
Process	Enzyme	Source			
Liquefaction	α-Amylase	B. amyloliquefaciens	5.5–7	90	Maltodextrins DE 10–20
		B. licheniformis	5.5–9	110	
Debranching	Pullulanase	K. aerogenes	6–7	50–60	Intermediate process in the manufacture of dextrose
Saccharification	α-Amylase	A. oryzae	5–7	50–55	High-maltose syrup, high-DE syrup,
Saccharification	Glucoamylase	A. niger (Rhizopus sp.)	4–5	55–60	High-DE syrup, cryst. dextrose
Isomerization	Glucose isomerase	(Streptomyces sp.), B. coagulans, Actinoplanes sp.	6.5–8.5	60–65	Fructose syrup

proteases. The reaction mechanism with casein is closely related to that of calf rennet. However, minor differences in the enzyme properties, such as dependency on Ca^{2+}, temperature, and pH, necessitate small adjustments of the cheese-making procedure according to the enzyme used. When the proper procedures are applied, the quality of cheese prepared with microbial rennet equals that of cheese prepared with traditional rennet.

Lactase for the hydrolysis of lactose in whey or milk is gaining increasing importance. Lactase from *Kluyveromyces fragilis* or *Kluyveromyces lactis* is usually used, although the properties of these enzymes are not ideal because of a narrow pH optimum and a low maximum temperature.

Lipase for flavor development in special cheeses (romano, provolone) is an interesting application, and microbial esterases for this purpose are presently being developed.

4. Other Applications

In the textile industry, α-amylase has been used for many years to remove starch sizes. This application is still an important outlet for bacterial α-amylase. An improved, rapid, high-temperature process is possible with the thermostable amylase from *B. licheniformis*.

Bacterial α-amylase is also used in the brewing industry for the liquefaction of brewing adjuncts. The substitution of barley for malt requires three enzymes, α-amylase, β-glucanase, and protease. All are produced by *B. amyloliquefaciens*. It is noteworthy that only the neutral protease is of importance because the alkaline protease is inhibited by an inhibitor in barley. β-Glucanase is often used alone to alleviate filtration problems due to poor malt quality or precipitation of glucans during the fermentation.

Substitution of amylase and glucoamylase for malt in the distilling industry makes high-temperature liquefaction possible and renders the process more reproducible.

The use of pectic enzymes in the wine and juice industry improves the yield and the quality of the products. Empirical methods and experience are especially important in the production and use of pectinase products. They contain the enzymes pectinase, pectin esterase, transeliminase, and polygalacturonidase. In addition, several nonpectolytic enzymes, such as protease, pentosanase, and glucoamylase, are present in small quantities. The combined action of several of these enzymes often decides its application.

Alkaline protease from alkalophilic bacteria may be used for dehairing of hides. The process functions well and is considerably more pleasant, both to workers and to the environment, than the usual sulfide process.

The method is little used, partly because it requires strict control of the process conditions, and partly for economic reasons.

VII. CONCLUSION

The industrial use of enzymes is insignificant and the processes are unimportant when compared with the overwhelming magnitude and importance of the enzymatic processes performed in nature. Some of these processes, such as lignocellulose hydrolysis, are relatively simple and seem to be technologically possible within a few years. Others, such as nitrogen fixation, are extremely complex, and their industrial realization is unthinkable with present technology.

Solution to such important problems as these will require systems of a higher degree of complexity than are so far known in the enzyme technology. The development will include multienzyme systems, free or immobilized, and in most cases coupled to cofactors. The possibilities are legion but the research effort necessary is some orders of magnitude larger than that for the simple problems so far attempted by the industry.

There is no doubt that industrial evolution will favor the use of enzymes. Their specificity, mild reaction conditions, and nontoxicity make them ideal catalysts in a world which becomes more and more conscious of pollution and energy waste.

REFERENCES

Amotz, S., Nielsen, T. K., and Thiesen, N. O. (1976). U.S. Patent 3,980,521.

Aunstrup, K. (1978). *In* "Biotechnology and Fungal Differentiation" (J. Meyrath and J. D. Bu'lock, eds), p. 157. Academic Press, New York.

Aunstrup, K., Outtrup, H., Andresen, O., and Dambmann, C. (1972). *Ferment. Technol. Today, Proc. Int. Ferment. Symp., 4th, 1972* pp. 299–305.

Barker, S. A., and Kay, I. (1975). *In* "Handbook of Biotechnology" (A. Wiseman, ed.), pp. 89–95. Halsted Press, New York.

Diers, I. (1976). *In* "Continuous Culture 6: Applications and New Fields" (A. C. R. Dean *et al.*, eds.), pp. 208–225. Halsted Press, New York.

Fabian, J. (1969). *In* "Continuous Cultivation of Microorganisms" (I. Málek, ed), pp. 489–495. Academia, Prague.

Food Chemicals Codex (1975). 2nd ed. Natl. Acad. Sci., Washington, D.C.

Harris, R. G., and Rose, T. J. (1972). British Patent 1,343,963.

Heady, R. E. (1971). British Patient 1,232,130.

Heineken, F. G., and O'Connor, R. J. (1972). *J. Gen. Microbiol.* **73,** 35–44.

Hilmer Nielsen, M., Zittan, L., and Hemmingsen, S. H. (1976). *In* "Chemical Engineering in a Changing World" (W. T. Koetsier, ed.), pp. 183–198. Elsevier, Amsterdam.

Hulme, M. A. (1973). U.S. Patent 3,734,831.

Jensen, D. E. (1972). *Biotechnol. Bioeng.* **14,** 647–662.

Kalaboklas, G. (1971). U.S. Patent 3,623,956.

Long, M. E. (1974). U.S. Patent 3,935,069.

Melling, J., and Phillips, B. W. (1975). *In* "Handbook of Enzyme Biotechnology" (A. Wiseman, ed.), pp. 63–64. Halsted Press, New York.

Meyrath, J., Volavsek, G., and Stahl, U. (1973). *Biotechnol. Bioeng. Symp.* **4,** 257.

Mitra, G., and Wilke, C. R. (1975). *Biotechnol. Bioeng.* **17,** 1–13.

Paigen, K., and Williams, B. (1970). *Adv. Microb. Physiol.* **4,** 251–324.

Smith, J. A., and Frankiewicz, J. R. (1978). British Patent 1,526,237.

Van Velsen, A. G. (1971). U.S. Patent 3,838,007.

World Health Organization (1972). "Food Additives Series," No. 2. WHO, Geneva.

World Health Organization (1975). "Food Additives Series," No. 6, WHO, Geneva.

GENERAL REFERENCES

de Becze, G. I. (1970). *Crit. Rev. Food Technol.* **1,** 479–518.

Meltzer, Y. L. (1973). "Encyclopedia of Enzyme Technology." Future Stochastic Dynamics Inc., New York.

Reed, G. (1975). "Enzymes in Food Processing," 2nd ed. Academic Press, New York.

Spencer, B., ed. (1974). "Industrial Aspects of Biochemistry." Vol. 30, Part I. Am. Elsevier, New York.

Whitaker, J. R. (1974). "Food related Enzymes," Adv. Chem. Ser. No. 136. Am. Chem. Soc., Washington, D.C.

Wiseman, A., ed. (1975). "Handbook of Enzyme Biotechnology." Halsted Press, New York.

Chapter 10

Microbial Production of Nucleosides and Nucleotides

YOSHIO NAKAO

MICROBIAL TECHNOLOGY, 2nd ed., VOL. I
Copyright © 1979 by Academic Press, Inc.
All rights of reproduction in any form reserved. ISBN 0-12-551501-4

I. INTRODUCTION

In Japan, *konbu* (kelplike seaweed) and *katsuobushi* (dried bonito) have been widely used as food seasoning substances from the early seventeenth century. The flavor-enhancing material in konbu was identified as sodium glutamate by Ikeda in 1908, and that of katsuobushi, as histidine salt of inosinic acid (IMP) by Kodama in 1913. The industrial manufacturing method for sodium glutamate was established immediately following the discovery by Ikeda. First, sodium glutamate was produced by the acid hydrolysis of plant proteins. In 1957, a process for the fermentative production of L-glutamic acid was developed in Japan by Kinoshita *et al*. L-Glutamic acid is now industrially produced, mainly by the fermentation process, worldwide.

In the 1950s, a wide range of research was initiated for the development of methods for industrial production of 5'-IMP when knowledge of the biosynthesis of nucleic acids, its regulation, and enzymatic degradation of nucleic acids was accumulating and analytical methods for nucleosides, nucleotides, and nucleic acids were becoming established. At that time applied microbiologists were stimulated by the success in glutamic acid fermentation and were aware that they could contribute to the development of processes for the fermentative production of other useful substances.

Techniques for the derivation of various mutants were being developed. Mutants useful for the industrial production of primary metabolites had to possess the following properties: high productivity of metabolites, lack of specific enzyme in the biosynthetic pathway, release from metabolic regulation, and altered membrane permeability.

Attempts to produce 5'-IMP and 5'-guanylic acid (GMP) were borne on the background of growing concepts and techniques in applied microbiology and increasing knowledge and methods in nucleic acid biochemistry and biochemical genetics.

In 1960, the manufacturing method of 5'-IMP and 5'-GMP by the enzymatic hydrolysis of ribonucleic acid (RNA) was established in Japan. Several years later, a process for fermentative production of 5'-IMP and 5'-GMP was developed. The industrial production of 5'-IMP and 5'-GMP is at present carried out by both hydrolysis of yeast RNA and direct fermentation.

Observation of the synergy of flavor between sodium glutamate and the sodium salt of 5'-IMP or 5'-GMP has resulted in a rapid enlargement of the market size of complex food seasonings (a mixture of sodium glutamate, 5'-IMP and 5'-GMP) in which 5'-IMP and 5'-GMP are usually added at levels of 8–12% or 1.5–2.0%.

Microbial technology established for the production of 5'-IMP and 5'-GMP has been applied to develop processes for the production of nucleic acid-related substances used for medicinal purposes. Amounts and uses of nucleic acid-related substances produced in Japan are shown in Table I (Yamada, 1976).

In this chapter, production methods of nucleosides and nucleotides are described with particular emphasis on the industrial manufacturing of 5'-IMP and 5'-GMP. Preparation of adenosine triphosphate (ATP), nicotinamide adenine dinucleotide (NAD), flavin adenine dinucleotide (FAD), coenzyme A (CoA), adenosine 3',5'-cyclic phosphate (cAMP), and guanosine 5'-triphosphate 3'-diphosphate (pppGpp) is also described.

II. CLASSIFICATION OF METHODS FOR PRODUCTION OF 5'-IMP AND 5'-GMP

Nucleotides are the building units of polymerized nucleic acids which are present in all living organisms: viruses, microorganisms, and plant and animal cells. They also occur in the nucleotide pool as free or coenzyme forms. Therefore, nucleotides can be produced by the methods listed below:

1. Enzymatic or chemical degradation of nucleic acids contained in organisms
2. Direct fermentation using mutants which are genetically blocked in

TABLE I. Outputs and Utilization of Nucleic Acid-Related Compounds[a]

Compound	Amount (tons/year)	Use
5'-IMP	2000	Seasonings
5'-GMP	1000	Seasonings
Inosine	25	Drug (coronary insufficiency, myocardial infarction)
ATP	6	Drug (metabolic disorders, muscular dystrophy)
CDP-choline	2	Drug (clouding of consciousness, accompanying head trauma or cerebral surgery)
FAD	Trace	Drug (hepatic or renal diseases)
NAD	Trace	Drug (hepatic or renal diseases)
5'-AMP	?	Drug (circulatory disorders, rheumatism)
Orotic acid	?	Drug (hepatic diseases)

[a] From Yamada (1976).

the biosynthetic pathway of nucleic acids and which are released from feedback regulation by end products

3. Extraction of free nucleotides existing in the pool

Method (1) is broken down into three procedures: (a) Hydrolysis of yeast RNA by microbial enzymes; (b) degradation of microbial endogenous RNA by endogenous enzymes; and (c) chemical hydrolysis of yeast RNA into nucleosides and its chemical phosphorylation.

Method (2) is broken down into four procedures: (a) Fermentative production of nucleosides and their phosphorylation; (b) fermentative production of 5'-nucleotides; (c) salvage synthesis of nucleosides from bases and phosphorylation of the nucleosides; and (d) salvage synthesis of 5'-nucleotides from bases.

Methods (1a), (1c), (2a), and (2b) are now industrially used for the production of 5'-IMP and 5'-GMP and are described in detail in the sections below.

III. PRODUCTION OF 5'-IMP AND 5'-GMP BY HYDROLYSIS OF RNA

A. Production of 5'-IMP and 5'-GMP by Hydrolysis of RNA with Microbial Enzymes

Industrial production of 5'-IMP and 5'-GMP started in Japan in 1961 using the method of hydrolysis of RNA by microbial enzymes. The manufacturing process includes (1) production of yeast cells with high RNA content, (2) extraction of RNA from yeast cells, (3) production of RNA-hydrolyzing enzymes, (4) enzymatic hydrolysis of RNA, and (5) recovery and purification of 5'-IMP and 5'-GMP from the enzymatic digest. Discovery of microbial nucleolytic enzymes which yield 5'-mononucleotides as a degradation product of RNA was an important factor which stimulated the establishment of this manufacturing method. The properties of nucleolytic enzymes and their producing organisms are described in the next section.

1. Nucleolytic Enzymes and Their Producing Microorganisms

As shown in Table II nucleolytic enzymes are classified on the basis of the substrates and degradation products of the enzymes. Figure 1 shows the scheme of production of 5'-IMP and 5'-GMP from RNA. Enzymatic formation of 5'-IMP and 5'-GMP requires two reactions: Hydrolysis of RNA into 5'-mononucleotides and deamination of 5'-adenylic acid

TABLE II. Classification of Nucleolytic Enzymes

Enzyme				Substrate	Product
Nucleolytic enzyme	(A) Nuclease	Endonuclease	RNase	RNA	Oligonucleotide with 3'-terminal phosphate / Oligonucleotide with 5'-terminal phosphate
			Nuclease	RNA, DNA	Oligonucleotide with 3'-terminal phosphate / Oligonucleotide with 5'-terminal phosphate
			DNase	DNA	Oligonucleotide with 3'-terminal phosphate / Oligonucleotide with 5'-terminal phosphate
		Exonuclease		RNA, DNA	3'-Mononucleotide / 5'-Mononucleotide
	(B) Polynucleotide phosphorylase			DNA	5'-Mononucleotide
				RNA	Nucleoside diphosphate

315

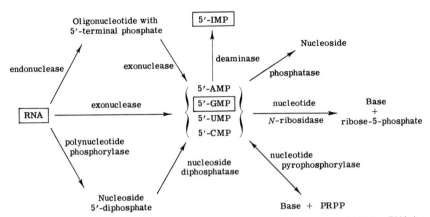

FIGURE 1. Enzymes participating in the production of IMP and GMP by RNA hydrolysis.

(AMP) to 5'-IMP. Therefore, microorganisms capable of producing these two activities should be screened first. In the 1950s, phosphodiesterases of snake venom (Cohn and Volkin, 1953) and calf intestinal mucosa (Cohn and Volkin, 1951) were known to form 5'-mononucleotides from RNA. These enzyme sources, of course, were not applicable for industrial purposes. Then, Kuninaka and his co-workers (1959, 1961) and Ogata and his co-workers (1963) searched for microbial enzymes which hydrolyze RNA to 5'-mononucleotides. Fungi, such as *Penicillium citrinum*; actinomycetes, such as *Streptomyces aureus;* bacteria; and yeasts were found to produce enzymes fitting their purposes. Table III lists those microorganisms capable of hydrolyzing RNA to 5'-nucleotides. The enzymological properties of these enzymes have been extensively studied and the enzymes have been classified according to the type of hydrolysis (Table IV).

2. Production of RNA

Economical production of RNA has been pursued since 1960, when the demand for RNA as raw material for 5'-mononucleotide production increased. RNA is contained in all living organisms and microorganisms would be considered to be the most suitable source of RNA. The RNA content of microorganisms is 5–25% in bacteria, 2.5–15% in yeasts, and 0.7–28% in fungi (Wade and Morgan, 1957; Ogata and Imada, 1962; Belozersky and Spirin, 1960; Belozersky, 1947; Di Carlo and Schultz, 1948; Vendrely, 1955; Kuroiwa and Horie, 1955; Miwa, 1959). Among microorganisms, yeasts have been considered to be the best RNA

TABLE III. Microorganisms Capable of Hydrolyzing RNA to 5′-Nucleotides

Microorganism	References
Mold	
Ascomycetes	
Aspergillus	Ogata et al. (1963);
	Nakao and Ogata (1963)
Chaetomidium	Ogata et al. (1963)
Glomerella	Ogata et al. (1963)
Monascus	Saruno et al. (1964);
	Soeda et al. (1968)
Neurospora	Ogata et al. (1963)
Penicillium	Kuninaka et al. (1959, 1961)
Sordaria	Ogata et al. (1963)
Fungi imperfecti	
Acrocylindrium	Ogata et al. (1963)
Fusarium	Ogata et al. (1963)
Gliomastix	Ogata et al. (1963)
Helminthosporium	Ogata et al. (1963)
Phoma	Tone et al. (1964);
	Tone and Ozaki (1968)
Verticillium	Ogata et al. (1963)
Basidiomycetes	
Pellicularia	Hasegawa et al. (1964);
	Fujimura et al. (1964, 1967)
Actinomycetaceae	
Streptomyces	Ogata et al. (1963);
	Sugimoto et al. (1964a,b)
Bacteria	
Bacillus	Ogata et al. (1963)
Micrococcus	Berry and Campbell (1967a,b)
Yeast	
Rhodotorula	Nakao et al. (1963)

source because their nucleic acids consist mainly of RNA, with very limited deoxyribonucleic acid (DNA) content.

Moreover, yeast cells are easily collected by yeast separators and many strains of the yeasts can grow on cheap carbon sources, such as molasses or sulfite waste liquor. *Candida utilis,* among the yeasts, is most widely used as an RNA source on the industrial scale. The RNA content of yeast cells can be increased by optimization of culture condition. Mutants with high RNA content were also isolated (Akiyama et al., 1975). As in bacteria, the RNA content of yeasts is high at their logarithmic growth phase with active protein biosynthesis (Katchman and Fetty, 1955; Chayen et al., 1955). RNA content in bakers' yeasts increased in media containing higher ammonium and phosphate con-

TABLE IV. Microorganisms Producing 5'-P-Forming Nucleolytic Enzyme[a]

Nucleolytic enzyme				
Enzyme	Substrate	Product	Microorganism	References
RNase	RNA	Oligonucleotides with 5'-terminal phosphate	Bacillus AU2	Jacobsen and Rodwell (1972)
			Escherichia coli	Spahr and Schlessinger (1963); Spahr (1964)
			Lactobacillus plantarum	Logan and Singer (1968)
			Lactobacillus casei	Keir et al. (1964)
			Physarum polycephalum	Hiramaru et al. (1969a,b)
Endonuclease	RNA	Oligonucleotides with 5'-terminal phosphate	Azotobacter agilis	Stevens and Hilmoe 1960a,b)
	DNA		Serratia marcescens	Nestles and Roberts (1969a,b)
			Streptomyces aureus	Yoneda (1964,b)
			Saccharomyces fragilis	Nakao et al. (1968); Lee et al. (1968)
			Acrocylindrium sp.	Suhara and Yoneda (1973); Suhara (1973)
			Aspergillus oryzae	Uozumi et al. (1968a,b,c, 1969a,b,c, 1972); Hino et al. (1971); Uozumi and Arima (1974)
Exonuclease	RNA	5'-Mononucleotide	Neurospora crassa	Linn and Lehman (1965a,b, 1966)
	Oligonucleotide		Micrococcus sodonensis	Berry and Campbell (1967a,b)
			Streptomyces aureus	Nakao (1976)
			Streptomyces sp. No. 41	Sugimoto et al. (1964a,b)
			Aspergillus oryzae	Ando (1966) Shishido and Ikeda (1970, 1971a,b) Shishido and Ando (1972)
			Aspergillus quercinus	Ohta and Ueda (1968)
			Acrocylindrium sp.	Suhara (1974)
			Monascus purpureus	Saruno et al. (1964); Soeda et al. (1968, 1969)
			Penicillium citrinum	Fujimoto et al. (1969, 1974a—d, 1975a,b); Kuninaka et al. (1975a,b)
			Phoma cucurbitacearum	Tone et al. (1964); Tone and Ozaki (1968)
			Physarum polycephalum	Hiramaru et al. (1969a,b)

[a] Enzymes hydrolyzing nucleic acid to mono- and/or oligonucleotides terminated by 5'-phosphate.

centrations (Di Carlo and Schultz, 1948). Addition of copper, zinc, and ferrous ions to the culture medium greatly stimulated RNA formation in *C. utilis*; the RNA content reached about 11% in media containing more than 0.25 ppm zinc ion (Yamashita *et al.*, 1974). Several antibiotics, such as blasticidin S (Hashimoto and Okauchi, 1965) and anisomycin (Watanabe, 1976), were observed to increase RNA content in *C. utilis* and *Saccharomyces cerevisiae*. The Waldhof-type fermentor was devised in Germany during World War II in order to economically produce fodder yeast on a large scale. The Waldhof-type fermentor has been used for cultivation of yeasts as an RNA source. Recently, the air-lift-type fermentor (Kanazawa, 1974) was found to be effective for cultivation of yeasts. In Japan, yeasts are now industrially cultivated on sulfite waste liquor or molasses by continuous culture systems. RNA content is about 10 and 15% on the respective carbon sources. Yeasts are separated by yeast separators and dried, and then the RNA is extracted with hot dilute alkaline saline. After yeast cells are removed, RNA in the extract is precipitated by the addition of acid or ethanol. Crude RNA powder is obtained by dehydration of the precipitate. For the production of 5′-IMP and 5′-GMP, such RNA powder is subjected to enzymatic hydrolysis. Fukuda *et al.* (1963) devised the method of direct enzymatic hydrolysis of dried yeast without RNA extraction.

3. Production of Nucleolytic Enzymes

Penicillium citrinum and *S. aureus* are industrially used as the enzyme producers. Each organism has been improved by repeated mutation and selection. The improved mutant of *P. citrinum*, which has lost the ability to form pigment and has acquired the ability to form abundant nuclease, is cultivated on wet wheat bran at 30°C for 5 days. Wheat bran with grown mycelia is extracted with water. The extract contains at least two enzymes: exonuclease, which hydrolyzes RNA into 5′-mononucleotides (Fujimoto *et al.*, 1974a–d), and phosphatase, which dephosphorylates 5′-mononucleotides to nucleosides. Phosphatase should be selectively inactivated by heat treatment before the extract is subjected to hydrolysis of RNA.

In the case of *S. aureus*, improved mutants which contain higher endonuclease and exonuclease and lower phosphatase activities have been isolated. The selected mutant is cultivated at 28°C for 30 hours in a liquid medium with agitation and aeration. The liquid medium contains (%): enzymatic hydrolyzate of starch, 3.0; soybean meal, 2.0; corn steep liquor, 1.0; $(NH_4)_2SO_4$, 0.1; $MgSO_4 \cdot 7 H_2O$, 0.05; $CaCO_3$, 0.5; soybean oil, 0.005. The culture filtrate contains endonuclease (Yoneda, 1964a,b), exonuclease, adenylic deaminase, and phosphatase (Nakao, 1976).

4. Enzymatic Hydrolysis of RNA and Recovery of 5'-IMP and 5'-GMP

Phosphatase contained in enzyme solutions dephosphorylates 5'-mononucleotides into nucleosides. Therefore, it is highly important to destroy or suppress the activity of phosphatase. Heat treatment of enzyme solutions greatly reduces the phosphatase activity, but complete selective inactivation of phosphatase is difficult. Enzyme solution obtained from wheat bran culture of *P. citrinum* contains exonuclease, which acts optimally at 65°C, and phosphatase, which acts optimally at 45°C. Digestion of RNA with *P. citrinum* enzyme solution at 65°C for about 4 hours at pH 5.0 yields 5'-mononucleotides with little concomitant formation of nucleosides. Since this enzyme solution does not contain adenylic deaminase, the degradation products of RNA are 5'-AMP, 5'-GMP, 5'-cytidylic acid (CMP), and 5'-uridylic acid (UMP). The 5'-AMP in the digest is deaminated into 5'-IMP with adenylic deaminase of *Aspergillus oryzae*.

In the case of hydrolysis with the enzyme solution of *S. aureus*, RNA is hydrolyzed into 5'-mononucleotides by the cooperative action of endo- and exonucleases. The 5'-AMP, once formed, is further deaminated to 5'-IMP by deaminase present in the same enzyme solution. Therefore, as the hydrolysis reaction with a multienzyme system proceeds, suitable conditions for enzyme reaction should be set to produce 5'-IMP and 5'-GMP in maximum yield. Industrial production of 5'-IMP and 5'-GMP by hydrolysis of RNA with the enzyme solution of *S. aureus* is carried out at 0.5–1.0% of RNA and at 42°–65°C for about 10 hours in a pH region of 7.0–8.0.

The 5'-mononucleotides in the digest are recovered by adsorption on activated charcoal at acidic pH, followed by elution with dilute methanol–ammonia solution. 5'-IMP and 5'-GMP can be separated from 5'-CMP and 5'-UMP by activated charcoal column (Tanaka *et al.*, 1964a,b). Each nucleotide can be fractionated with anion- or cation-exchange resins (Tanaka *et al.*, 1963; Ueno *et al.*, 1965). Sanno *et al.* (1964) and Ishibashi *et al.* (1965) proposed a fractional precipitation method for the separation of 5'-purine nucleotides from 5'-pyrimidine nucleotides. Industrially, 5'-IMP and 5'-GMP are recovered by the treatment of the enzymatic hydrolysate with ion-exchange resin and charcoal, followed by the fractional precipitation with a lower alcohol.

B. Formation of 5'-Nucleotides by Degradation of Cellular RNA by Endogenous Enzymes

In the processes described in the previous section, yeast RNA was hydrolyzed with enzymes obtained from other microorganisms. An idea to simplify the process is to digest microbial RNA with endogenous nucleolytic enzymes. Several bacteria and yeasts were found to digest their cellular RNA by their endogenous enzymes and excrete 5'-mononucleotides (Imada, 1976). Such a process has not been adopted for industrial production of 5'-IMP and 5'-GMP. If economical mass production of single-cell protein (SCP) is to be practiced to cover the worldwide shortage of protein sources, it would be necessary to reduce nucleic acid content in single-cell protein. In such a case, the autodigestive process would be useful to achieve both the reduction of nucleic acid content and the production of 5'-nucleotides.

C. Production of Nucleosides by Chemical Hydrolysis of RNA and Their Chemical Phosphorylation

It is well known that 3'(2')-nucleotides are formed by alkaline hydrolysis of RNA, but 5'-mononucleotides have never been directly formed from RNA by chemical methods. As a result of extensive investigation, production of nucleosides from RNA by chemical processes followed by their phosphorylation have been applied for the industrial manufacturing of 5'-IMP and 5'-GMP.

Production of nucleosides from RNA is composed of two reactions: Formation of 3'(2')-nucleotides by alkaline hydrolysis of RNA and hydrolysis of the phosphomonoester linkage. These two reactions, however, can be accomplished by a single process. RNA is quantitatively converted into nucleosides by heating RNA at 130°C for 3–4 hours in calcium hydroxide solution (Koshiro et al., 1972). Guanosine in the hydrolysate is easily crystallized due to its low solubility in water. Adenosine, cytidine, and uridine are separated with anion-exchange resin (Numata et al., 1969), and each nucleoside is crystallized by concentration of each fractionated eluate. Adenosine is converted into inosine by deamination with nitrous acid. Inosine and guanosine are phosphorylated to 5'-IMP and 5'-GMP by the chemical method described in later sections.

IV. PRODUCTION OF 5'-IMP AND 5'-GMP BY FERMENTATION

After establishment of the industrial production of 5'-IMP and 5'-GMP by enzymatic hydrolysis of RNA, applied microbiologists in Japan actively explored the fermentative production of these 5'-nucleotides. In order to produce purine nucleotides by direct fermentation, the biosynthetic pathway of purine nucleotides and the regulation involved in it should be taken into account.

A. Biosynthesis of Purine Nucleotides and Its Regulation

Many reviews have appeared on the biosynthesis of purine nucleotides (Buchanan and Hartman, 1959; Moat and Friedman, 1960; Magasanik, 1962; Hartman, 1970; Blakley and Vitols, 1968; Momose, 1968; Gots, 1971). Therefore only a summary is presented here.

1. De novo Biosynthesis of Purine Nucleotides and Its Regulation

a. Precursors of Purine Nucleus. Isotope incorporation studies (Buchanan and Hartman, 1959) proved that nine atoms of purine nucleus come from five precursor molecules, as shown in Fig. 2. For the synthesis of 1 mole of 5'-IMP, 1 mole each of phosphoribosyl pyrophosphate (PRPP), aspartic acid, glycine, and CO_2 and 2 moles each of formic acid and glutamine are required as substrate. Eight moles of ATP are necessary as energy donor.

b. Biosynthetic Pathway. The biosynthetic pathway of purine nucleotides was first elucidated in pigeon liver homogenate by Buchanan and Hartman (1959). The pathway in microorganisms was later found to be identical with that in pigeon liver. As shown in Fig. 3, 5'-IMP is synthesized through 11 reactions starting from ribose 5-phosphate. 5'-AMP and 5'-GMP are synthesized through branch pathways from 5'-IMP. The 5'-AMP and 5'-GMP are interconvertible through reactions of the so-called purine nucleotide cycle. The 5'-AMP and 5'-GMP thus formed are used as building units of RNA, DNA, and various coenzymes.

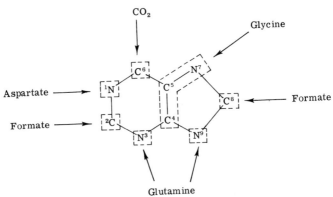

FIGURE 2. Precursors of the purine ring system.

c. Mutants of Purine Nucleotide Biosynthesis.   Magasanik (1957) classified purine-requiring mutants into several classes according to their requirements for purine bases and intermediates accumulated by them (Fig. 4). Class 1 mutants required for their growth either adenine, hypoxanthine, xanthine, or guanine. Class 1 mutants, therefore, had an intact purine nucleotide cycle but lacked some enzymes for the synthesis of 5'-IMP. These mutants were classified into Classes 1a, 1b, and 1c: Class 1b mutants accumulated aminoimidazole riboside (AIR) and Class 1c mutants 5-amino-4-imidazole-carboxamide riboside (AICAR).

Class 2 mutants lacked 5'-IMP dehydrogenase and required xanthine or guanine for growth. Class 3 mutants lacked 5'-XMP aminase, required guanine for growth, and accumulated xanthosine in the culture medium. Class 4 mutants showed specific requirement for adenine and were subdivided into two groups: mutants lacking adenylosuccinate (SAMP) lyase and accumulating 5-amino-4-imidazole-N-succinocarboxamide riboside (SAICAR), and mutants lacking SAMP synthetase and accumulating inosine. Mutants being industrially used for fermentative production of nucleic acid-related substances have been isolated and improved from various bacterial species, based on the findings described above.

d. Regulation of Biosynthetic Pathway.   Complex and fine control of purine biosynthetic pathway has been extensively studied. In this section, information on the control mechanisms which are indispensable for the development of the fermentation process for production of 5'-IMP and 5'-GMP will be summarized. Details have been reviewed by Blakley and Vitols (1968), Momose (1968), and Gots (1971).

(1) Metabolic Regulation of 5'-IMP Biosynthesis.   PRPP amidotransferase catalyzes the first reaction of purine nucleotide biosynthesis: formation of PRA from PRPP and glutamine. Wyngaarden and Ashton (1959) and Caskey *et al.* (1964) demonstrated that this enzyme purified from pigeon liver was inhibited by AMP, ADP, ATP, GMP, GDP, or GTP. The inhibition was due to competition of these nucleotides for PRPP. Combination of 6-aminopurine and 6-hydroxypurine nucleotides, such as GMP plus AMP or IMP plus AMP, showed stronger inhibition to the reaction than the combination of two 6-aminopurine nucleotides, such as AMP plus ADP, or two 6-hydroxypurine nucleotides, such as GMP plus IMP. PRPP amidotransferase from *Aerobacter aerogenes* showed properties similar to that from pigeon liver (Nierlich and Magasanik, 1965).

Repression and derepression of PRPP amidotransferase in *A. aerogenes* was also studied by Nierlich and Magasanik (1963). The enzyme synthesis was completely repressed by the simultaneous addition of

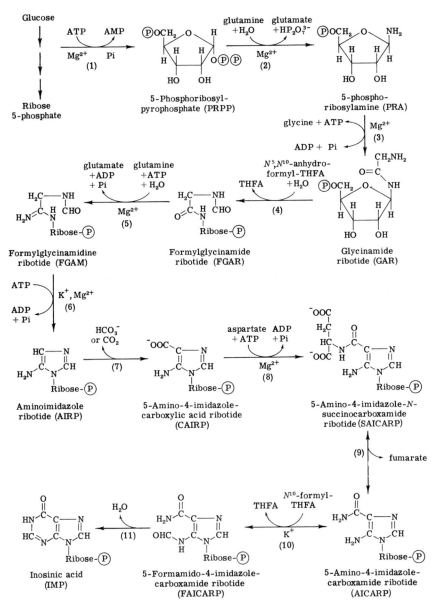

FIGURE 3. Synthesis of IMP. (1) PRPP synthetase, (2) PRPP amidotransferase, (3) phosphoribosylglycinamide synthetase, (4) phosphoribosylglycinamide formyltransferase, (5) phosphoribosylformylglycinamidine synthetase, (6) phosphoribosylaminoimidazole synthetase, (7) phosphoribosylaminoimidazole carboxylase, (8) phosphoribosylamino-imidazole succinocarboxamide synthetase, (9) adenylosuccinate lyase, (10) phosphoribo-

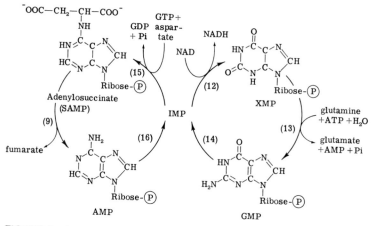

FIGURE 3 (continued)
sylaminoimidazole carboxamide formyltransferase, (11) IMP cyclohydrolase, (12) IMP dehydrogenase, (13) XMP aminase, (14) GMP reductase, (15) Adenylosuccinate synthetase, (16) AMP deaminase. (Microbial production of nucleic acid-related substances (1976), p. 26, Fig. 2.2, Kodansha Ltd. Halsted Press.)

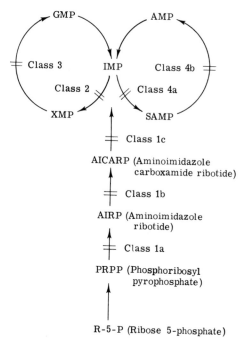

FIGURE 4. Classification of mutants blocked in the purine nucleotide biosynthetic pathway.

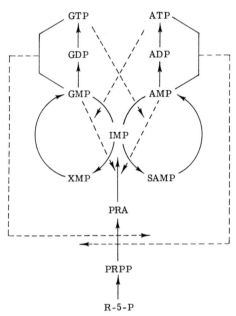

FIGURE 5. Inhibition of enzymes participating in the biosynthesis of purine nucleotides.

adenine and guanine to the medium. The repression was released by limitation of the addition of both adenine and guanine to the medium, but it was partially released in the presence of limited guanine and excess adenine.

(2) Metabolic Regulation of Purine Nucleotide Cycle. The purine nucleotide cycle consists of the AMP cycle and the GMP cycle which are connected at 5′-IMP and are regulated as shown in Fig. 5. The reaction from IMP to XMP is inhibited by GMP and that from GMP to IMP, by ATP. The reaction from IMP to SAMP is inhibited by AMP and that from AMP to IMP, by GTP.

On the other hand, repression of the synthesis of enzymes involved in the purine nucleotide cycle is also demonstrated. In the case of *Bacillus subtilis,* IMP dehydrogenase, which converts IMP to XMP, is repressed by GMP and SAMP synthetase which converts IMP to SAMP by AMP (Momose *et al.,* 1965, 1966; Momose, 1967; Nishikawa *et al.,* 1967).

2. Salvage Synthesis of Purine Nucleotides and Its Regulation

a. Salvage Reactions. Microbial cells synthesize purine nucleotides from purine bases present in a medium. Such a reaction is called a salvage synthesis, in contrast to a *de novo* synthesis. Salvage synthesis takes place by either of the two reactions shown in Fig. 6. One is direct conversion of bases to nucleotides catalyzed by nucleotide pyrophosphorylase. The other is a two-step reaction in which bases are first converted to nucleoside by nucleoside phosphorylase, followed by phosphorylation of nucleosides by nucleoside kinase or nucleoside phosphotransferase.

b. Regulation of Salvage Synthesis. Nucleotide pyrophosphorylase of *B. subtilis* is inhibited by purine nucleotides (Berlin and Stadtman, 1966). Nucleotide formation from adenine and guanine was specifically inhibited by the respective 5'-nucleotides formed as the products. On the other hand, conversion of hypoxanthine and xanthine to IMP was inhibited more strongly by GMP and GTP than by IMP and XMP.

B. Production of 5'-IMP

The following methods have been reported for the production of 5'-IMP.

1. Fermentative production and chemical phosphorylation of inosine
2. Fermentative production of 5'-IMP
3. Fermentative production of adenosine or 5'-AMP followed by chemical or enzymatic conversion to 5'-IMP
4. Salvage synthesis by microorganisms of 5'-IMP from chemically synthesized hypoxanthine

Methods (1) and (2) have proved to be useful for economical production of 5'-IMP and are described in detail in this section.

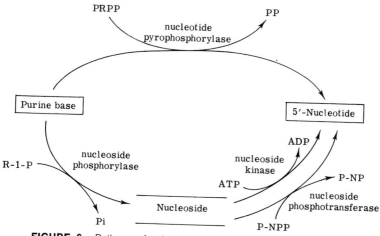

FIGURE 6. Pathway of salvage biosynthesis of purine nucleotide.

1. Production of Inosine by Fermentation and Its Chemical Phosphorylation

a. Fermentative Production of Inosine. *(1) Producing Microorganisms.* Inosine producers should be endowed with the following characteristics: (i) lack of SAMP synthetase, (ii) release from regulation operating in 5'-IMP biosynthesis, and (iii) lack of inosine-degrading enzymes, such as nucleoside phosphorylase and hydrolase.

In 1957, Gots and Gollub and Partridge and Giles (1957) first reported that hypoxanthine was accumulated in the culture broth of an adenine auxotroph of *Escherichia coli* and *Neurospora crassa*. Since the establishment of the manufacturing process for 5'-IMP by enzymatic hydrolysis of RNA, applied microbiologists actively have explored fermentative methods for production of inosine using adenine auxotrophs of species of the genera *Bacillus, Brevibacterium, Corynebacterium, Streptomyces,* and *Saccharomyces.* Mutants of the genera *Bacillus* and *Brevibacterium* have been found to accumulate large amounts of inosine.

In 1961, Uchida *et al.* first reported accumulation of inosine by mutants obtained from species of *Bacillus.* Aoki and co-workers (Aoki, 1963; Aoki *et al.*, 1963a,b, 1968; Yamanoi *et al.*, 1967; Yamanoi and Shiro, 1968) mutated *B. subtilis* and isolated mutant No. 2 which accumulated hypoxanthine with a small amount of inosine. Then mutants requiring various amino acids were isolated from mutant No. 2. One of the isolates, mutant C-30 ($ad^-his^-try^-$), was found to have low inosine-degrading activity and to accumulate 6.3 mg/ml inosine.

Nogami *et al.* (1968) derived a mutant with low nucleoside phosphorylase activity from an inosine-guanosine-producing mutant of *Bacillus* sp. No. 102. They found that concomitant formation of purine bases was negligible with the mutant of low nucleoside phosphorylase activity (Nogami and Yoneda, 1969).

Shiio and Ishii (1971) and Ishii and Shiio (1972) derived mutants resistant to low levels of 8-azaguanine from *B. subtilis* RDA-16 which required adenine and lacked adenine-deaminating activity. They found that one of the mutants, which had PRPP amidotransferase not repressed by adenine, accumulated 18 mg/ml inosine.

Accumulation of inosine by mutants of *Brevibacterium ammoniagenes* was studied by Furuya *et al.* (1970a, 1975). They first derived an adenine auxotroph KY 13714 which was resistant to low levels of 6-mercaptoguanine. Mutant KY 13714 accumulated 9.3 mg/ml of inosine. Then KY 13714 was further mutated to guanine auxotrophy and 6-methylthiopurine resistance. One of such mutants, KY 13761, accumulated as high as 30 mg/ml inosine under well-controlled culture conditions (Kotani *et al.*, 1977).

Industrial production of inosine is mainly carried out by using the $ad^-his^-try^-$ series of mutants derived from *B. subtilis* by Aoki *et al.* described above.

(2) Culture Conditions. Representative fermentation media suitable for inosine production are shown in Table V.

Fermentative production of inosine by mutants of *B. subtilis* is industrially carried out under the following conditions (Aoki *et al.*, 1968; Shibai *et al.*, 1973): carbon source, starch hydrolyzate; adenine source, dried yeast or crude RNA; pH, 6.0–6.2; temperature, 30°–34°C; aeration, more than 0.01 atm dissolved oxygen. The aeration rate for maximum inosine production was determined from two parameters: (i) fulfillment of oxygen absorption by the producing microorganism which requires more than 5 $\times$ 10^{-7} mole/ml/min and (ii) avoidance of inhibition of inosine formation by generated carbon dioxide.

Figure 7 shows an example of inosine fermentation with a *B. subtilis* mutant under such manipulated conditions. Figure 8 shows an example of inosine fermentation by *B. ammoniagenes* KY 13761 derived by Kotani *et al.* (1977).

(3) Isolation of Inosine. Culture filtrate is adjusted to pH 11 with sodium hydroxide, and then concentrated *in vacuo* until the concentration of inosine becomes greater than 30 mg/ml. The concentrate is again adjusted to pH 11 and is further concentrated. Crystals of inosine sodium salt are collected and then suspended in water. The suspension is adjusted to pH 5.2 with hydrochloric acid and concentrated *in vacuo*. Crystalline inosine, separated out by concentration, is collected by centrifugation.

b. Chemical Phosphorylation of Inosine. In order to phosphorylate inosine, enzymatic methods (Brawerman and Chargaff, 1955; Mitsugi, 1964; Mitsugi *et al.*, 1964, 1965) as well as chemical methods (Muramatsu and Takenishi, 1965; Yoshikawa and Kato, 1967; Yoshikawa *et al.*, 1967, 1969, 1970; Honjo *et al.*, 1966) were widely tested. Industrially, chemical phosphorylation methods are employed. Yoshikawa *et al.* (1967, 1969) could chemically convert nucleosides to 5′-nucleotides in 68–90% yield by suspending unprotected nucleosides in trimethyl phosphate or triethyl phosphate followed by a reaction with phosphoryl chloride (Table VI). In the reaction system, 2′(3′),5′-diphosphoryl ester was concomitantly formed, but this disadvantageous by-production of diphosphoryl ester could be suppressed by addition of about equimolar water to nucleoside in the reaction mixture. Conversion of inosine or guanosine to 5′-IMP or 5′-GMP reached more than 90% in the improved

TABLE V. Composition of Media Employed for Inosine Accumulation

	Microorganisms					
Compounds	B. subtilis[a,g] ad⁻, his⁻, tyr⁻	B. subtilis[b,g] ad⁻, try⁻, red⁻, dea⁻, 8AGʳ	Bacillus sp.[c,g] ad⁻, his⁻, red⁻, 8AGʳ	Bacillus sp.[d,g] ad⁻, his⁻, thr⁻, dea⁻	Brev. ammoniagenes[e,g] ad^L, 6MGʳ, gua⁻	Brev. ammoniagenes[f,g] ad^L, 6MGʳ, 6MTPʳ, gua⁻
Glucose	60–70 gm/liter	70 gm/liter	100 gm/liter	100 gm/liter	130 gm/liter	
Maltose						75 (0 hours), 100 (22 hours) gm/liter
Inverted molasses						
NH_4Cl	20 gm/liter	15 gm/liter	20 gm/liter	20 gm/liter		
$(NH_4)_2SO_4$					4 gm/liter	2 gm/liter
$(NH_2)_2CO$					10 gm/liter	1 gm/liter
KH_2PO_4	1 gm/liter	1 gm/liter	2 gm/liter	2 gm/liter	10 gm/liter	1 gm/liter
K_2HPO_4			5 gm/liter			
$CaHPO_4$			5 gm/liter			
$Ca_3(PO_4)_2$			2 gm/liter			
$MgSO_4 \cdot 7 H_2O$	0.4 gm/liter	0.4 gm/liter	2 gm/liter	0.5 gm/liter	10 gm/liter	1 gm/liter
$FeSO_4 \cdot 7 H_2O$	2 ppm (Fe^{2+})	2 ppm (Fe^{2+})			10 mg/liter	
$ZnSO_4 \cdot 7H_2O$					10 mg/liter	
$MnSO_4 \cdot xH_2O$	2 ppm (Mn^{2+})	2 ppm (Mn^{2+})			10 mg/liter	
$MnCl_2 \cdot 4h_2O$				0.01 gm/liter		
Cysteine					20 mg/liter	

	[a]	[b]	[c]	[d]	[e]	[f]
Tryptophan		300 mg/liter			5 mg/liter	
Thiamin					10 mg/liter	
Pantothenate					30 μg/liter	
Biotin					5 mg/liter	
Nicotinic acid			0.2 mg/liter			
Mieki	4 gm/liter	2 gm/liter				
Casamino acid				1 gm/liter		
Yeast extract				10 gm/liter		
Meat extract					10 mg/liter	
Dry yeast	14 gm/liters		10 gm/liters			
Adenine	100 mg/liter	100 mg/liter			100 mg/liter	
Guanine					100 mg/liter	
$CaCO_3$	2%	2.5%	2%	5%		
pH	6.0	7.0	7.6		8.3	
Max. inosine accumulated	10.5 gm/liter	18 gm/liter	14.1 gm/liter	10.8 gm/liter	13.6 gm/liter	30 gm/liter

[a] From Aoki et al. (1968).
[b] From Ishii and Shiio (1972).
[c] From Nogami et al. (1968).
[d] From Komatsu et al. (1972).
[e] From Furuya et al. (1975).
[f] From Kotani et al. (1977).
[g] Key to Abbreviations: ad^-, adenine-requiring; gua^-, guanine-requiring; his^-, histidine-requiring; tyr^-, tyrosine-requiring; thr^-, threonine-requiring; try^-, tryptophan-requiring; ad^L, leaky adenine mutant; red^-, GMP reductase-deficient; dea^-, AMP deaminase-deficient; $8AG^r$, 8-azaguanine-resistant; $6MG^r$, 6-mercaptoguanine-resistant; $6MTP^r$, 6-methylthiopurine-resistant.

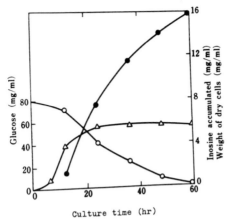

FIGURE 7. Time course of inosine fermentation by a mutant of *B. subtilus* K. (From Shibai *et al.*, 1973.) ●—● Inosine, ○—○ glucose, △—△ weight of dry cells.

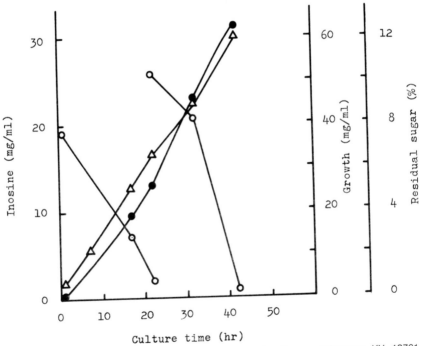

FIGURE 8. Time course of inosine accumulation by *B. ammoniagenes* KY 13761. (From Kotani *et al.*, 1977.) ●—●, Inosine, ○—○, residual sugar, △—△, growth.

TABLE VI. Direct Phosphorylation of Unprotected Nucleosides in Trialkyl Phosphate[a,b]

Nucleoside	R group in $(RO)_3PO$	$POCL_3$ (mmole)	H_2O (mmole)	Time (hours)	Yield of nucleoside-phosphate (mole %)	
					5'-mono	2'(3') 5'-di
Inosine	CH_3CH_2	6	0	2	68	18
Inosine	CH_3CH_2	6	2	2	91	8
Guanosine	CH_3	6	0	6	85	9
Guanosine	CH_3	6	2	6	90	5
Adenosine	CH_3CH_2	4	0	6	84	11
Xanthosine	CH_3	4	0	9	80	5
Uridine	CH_3	4	0	12	89	8
Cytidine	CH_3CH_2	4	0	1	88	8
Deoxyinosine	CH_3	4	0	5	73	6
Deoxycytidine	CH_3	4	0.5	6	64	19
AICAR	CH_3	4	0	4	91	5

[a] From Yoshikawa et al. (1967, 1969).
[b] Conditions: nucleoside, 2 mmole; trialkyl phosphate, 5 ml; at 0°C.

system. The 5'-IMP was neutralized with sodium hydroxide to 5'-IMP ·Na_2· 7.5 H_2O and supplied commercially.

2. Production of 5'-IMP by Fermentation

a. Producing Microorganisms. Phosphate esters, including 5'-nucleotides, are generally impermeable through cytoplasmic membranes. The following four characteristics are considered to be necessary for a 5'-IMP-producing strain: (1) lack of SAMP synthetase, (2) release from regulation operating in the biosynthesis of 5'-IMP, (3) lack of 5'-IMP-degrading enzymes such as 5'-nucleotidase, and (4) cytoplasmic membrane permeable to 5'-IMP.

In order to obtain mutants lacking nucleotide dephosphorylating enzymes, Momose et al. (1964), Fujimoto and Uchida (1965), and Fujiwara et al. (1967) isolated mutants which can grow on adenine but not on 5'-AMP from inosine-producing B. subtilis. The mutants obtained showed reduced nucleotide-degrading activity but, unexpectedly, did not accumulate considerable amounts of 5'-IMP.

On the assumption that trace levels of nucleotide-degrading activity remained, interfering with the accumulation of 5'-IMP, Fujimoto et al. (1965) tried to decrease the nucleotide-degrading activity further and obtained mutant A-1-25. They found that about 0.6 mg/ml of 5'-IMP was accumulated by mutant A-1-25. Akiya et al. (1972) examined culture conditions for 5'-IMP formation with mutant A-1-25Z, which is a derivative of the above A-1-25, and succeeded in accumulating about 3.4 mg/ml 5'-IMP.

Nakayama et al. (1964, 1965) and Satō et al. (1965) mutated a glutamic acid-producing Corynebacterium glutamicus to adenine auxotrophy and 6-mercaptopurine resistance. The mutant accumulated 2 mg/ml 5'-IMP. Demain and co-workers (1965, 1966; Demain, 1968) also derived an adenine-requiring and IMP dehydrogenase-lacking mutant of C. glutamicus and found that the mutant accumulated 0.8–0.9 mg/ml 5'-IMP. Nara et al. (1967) derived an adenine auxotroph, KY 7208, from B. ammoniagenes ATCC 6872. They succeeded in accumulating 5 mg/ml 5'-IMP under elaborated culture conditions. Furuya et al. (1968) derived a leaky mutant, KY 13102, the growth of which was stimulated by adenine, from B. ammoniagenes ATCC 6872. The mutant KY 13102 accumulated 12.8 mg/ml 5'-IMP.

Accumulation of 5'-IMP by B. ammoniagenes mutants obtained by Nara et al. (1968b), Furuya et al. (1968, 1970b), and Komuro et al. (1969) was greatly influenced by Mn^{2+} concentration as well as by adenine concentration in the medium. From the above-described KY 13102, Furuya et al. (1969) isolated mutant KY 13105 and KY 13171 (Kato et al., 1971) which could accumulate 5'-IMP in the presence of excess Mn^{2+}.

b. Culture Conditions. Nara et al. (1967, 1968b, 1969a,b), Furuya et al. (1968, 1970b) , and Komuro et al. (1969) examined culture conditions for 5'-IMP production by B. ammoniagenes mutants. Their results are summarized as follows: (1) optimum concentration of phosphoric acid was as high as 2%; (2) that of $MgSO_4$, about 1%; (3) that of adenine, about 25 mg/liter; (4) that of biotin, 20–30 μg/liter; and (5) that of Mn^{2+}, 10–20 μg/liter. In addition, (6) thiamin and calcium pantothenate were necessary for growth of producer microorganisms and accumulation of IMP and (7) a mixture of amino acids, e.g., histidine plus lysine or casamino acids, stimulated IMP accumulation.

Figure 9 shows the time course of 5'-IMP production by B. ammoniagenes KY 13102 under optimal culture conditions (Furuya et al., 1968). A characteristic feature of the fermentation process is that hypoxanthine is first accumulated at an early stage and 5'-IMP is formed in later stages on consumption of hypoxanthine. Nara et al. (1969a) proposed a mechanism of IMP accumulation by B. ammoniagenes mutants as diagramatically shown in Fig. 10. Later, Furuya et al. (1970b) and Kato et al. (1971) demonstrated that de novo formed 5'-IMP leaked out directly from producer cells in later fermentation stages.

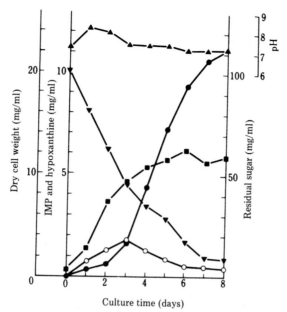

FIGURE 9. Time course of IMP accumulation by *B. ammoniagenes* KY 13102. (From Furuya *et al.*, 1968.) ●—● IMP, ○—○ hypoxanthine, ▲—▲ pH, ▼—▼ residual sugar, ■—■ dry cell weight.

3. Production of Adenosine by Fermentation

Production of 5'-IMP from fermentatively produced adenosine does not seem very practical for economical manufacturing. Adenosine fermentation was first studied for the purpose of producing 5'-IMP, but later studies were made not for 5'-IMP production but for the supply of adenosine or 5'-AMP, which are important as raw materials for synthesis of several valuable medicinary compounds.

a. Producing Microorganisms. Konishi *et al.* (1968) derived an isoleucine auxotroph, which accumulated 2.5 mg/ml adenosine, from *B. subtilis* K. The correlation between isoleucine auxotrophy and adenosine productivity is not clear. Haneda *et al.* (1971) serially mutated *Bacillus* sp. No. 1043 and obtained mutant P53-18, capable of producing 16 mg/ml adenosine. Genetical characteristics of the mutant P53-18 were $his^-thr^-xan^-8AX^r$ and lack of adenosine deaminase and GMP reductase.

The following characteristics of the producing strain favor abundant accumulation of adenosine: (1) lack of IMP dehydrogenase, (2) release

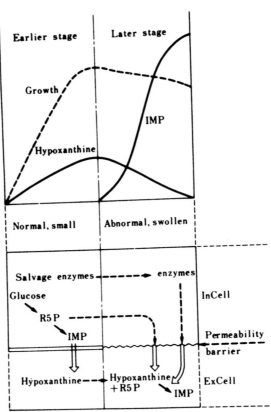

FIGURE 10. Hypothetical diagram for direct IMP fermentation with an adenineless mutant of B. ammoniagenes. (From Nara et al., 1969a.)

from regulation operating in AMP synthesis, (3) lack of adenosine deaminase, and (4) lack of nucleoside phosphorylase and hydrolase.

In order to suppress reverse mutation of the xan⁻ marker, a further genetic defect in XMP aminase was effective (Haneda et al., 1972).

b. Culture Conditions. Haneda et al. (1971) evaluated culture conditions for adenosine production by the mutant P53-18 and found that concentration of yeast extract, calcium carbonate, and aeration rate were important factors for adenosine production. A process of adenosine fermentation by mutant P53-18 under optimum conditions is shown in Fig. 11.

C. Production of 5'-GMP

Accumulation of large amounts of 5'-GMP in the microbial culture fluid appeared to be very difficult since there exist regulation mechanisms to control the overproduction of 5'-GMP.

It was also presumed that 5'-GMP, once formed, is converted to other nucleotides by the purine nucleotide cycle and that 5'-GMP is degraded into guanosine or guanine by phosphatases or nucleoside phosphorylases, which are active in most microbial cells. Based on these presumptions, various indirect methods have been attempted in order to produce 5'-GMP beside trials to accumulate 5'-GMP directly in the fermentation broth. These methods include: (1) fermentative production of AICAR followed by chemical conversion of it to 5'-GMP; (2) fermentative production of guanosine followed by chemical or biochemical phosphorylation of it; (3) mixed culture of two microorganisms, one accumulating 5'-XMP (or xanthosine) and the other converting 5'-XMP (or xanthosine) to 5'-GMP; and (4) direct fermentation of 5'-GMP. Among these methods, (1) and (2) are now industrially employed.

1. Production of AICAR by Fermentation and Its Chemical Conversion to 5'-GMP

a. Fermentative Production of AICAR. *(1) Producing Microorganisms.* Accumulation of AICA in culture broth of sulfonamide-inhibited *E. coli* was first observed in 1942 (Fox, 1942; Stetten and Fox, 1945; Shive *et al.*, 1947). Since then, the accumulation of AICAR was studied using mainly *E. coli* (i) wildtype strains, (ii) mutants genetically blocked in purine nucleotide biosynthesis pathway, and (iii) wildtype strains metabolically blocked by drugs which inhibits purine biosynthesis. However, the amount of AICAR accumulated by *E. coli* was low. Shiro *et al.* (1962) isolated a purine auxotroph, D-421, from X-ray treated *B. subtilis* No. 2093. The mutant accumulated about 2.6 mg/ml of AICAR in its culture broth. This finding led to the establishment of industrial production of AICAR by fermentation. Later, Kinoshita *et al.* (1967) isolated many auxotrophic mutants of *B. subtilis* IAM 1523, *Bacillus megaterium* IAM 1245, and *Brevibacterium flavum* No. 2247 and examined their productivity of AICAR. AICAR was produced only by purine auxotrophs. They found that a purine auxotroph No. 336 of *B. megaterium* accumulated 15 mg/ml AICAR. Shirafuji *et al.* (1968) derived a mutant requiring adenine plus hypoxanthine or guanine from nonexacting purine auxotroph of *Bacillus pumilus* which formed detectable AICAR. The derived mutant accumulated as high as 15 mg/ml of AICAR. For industrial pro-

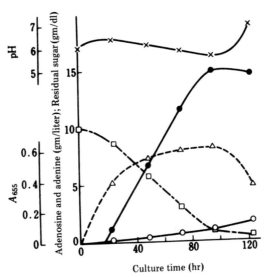

FIGURE 11. Time course of adenosine fermentation by *Bacillus* sp. P53-18. (From Haneda *et al.*, 1971.) ●—● Adenosine, ○—○ adenine, □- — -□ residual sugar, △- - -△ A_{655}, ×—× pH.

duction of AICAR, a purine auxotroph of *B. megaterium* is now being used.

(2) Culture Conditions. Kinoshita *et al.* (1967, 1968, 1969a,b,c, 1973) examined production conditions for AICAR by *B. megaterium* No. 336 and found that suppression of spore formation and back-mutation during fermentation were important factors for the production of AICAR in high yield. The sporulation could be suppressed by the regulation of oxygen supply (Kinoshita *et al.*, 1973) or by the addition of sporulation inhibitors, such as butyric acid (Kinoshita *et al.*, 1969b). Reversion of prototrophy could be controlled by selection of mutants with low reversion rates (Kinoshita *et al.*, 1967) or by the addition of erythromycin, which preferentially inhibits the growth of revertants (Kinoshita *et al.*, 1968). The fermentation medium should be supplemented with purine sources, such as RNA, adenine, or inosine, at a suboptimal level for growth and with suitable amounts of inorganic ions, such as phosphate, potassium, and sulfate, in order to accumulate large amounts of AICAR. Figure 12 shows a typical time course of AICAR fermentation (Kinoshita *et al.*, 1969a).

b. Chemical Synthesis of 5′-GMP from AICAR. AICAR in the fermented broth is recovered in more than 90% yield through several

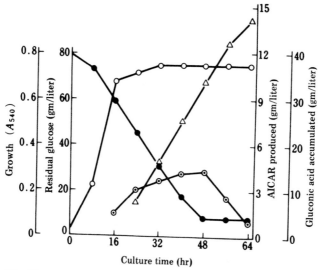

FIGURE 12. Time course of AICAR fermentation by *B. megaterium* MA-336. (From Kinoshita *et al.*, 1969a.) △—△ AICAR, ⊙—⊙ gluconic acid, ○—○ growth, ●—● glucose.

purification steps: adsorption on and elution from cation-exchange resin, concentration, crystallization, and drying. AICAR is chemically converted in 70% yield to guanosine as shown in Fig. 13 and then chemically phosphorylated to 5'-GMP in about 90% yield (Kinoshita *et al.*, 1967; Yamazaki *et al.*, 1967; Kumashiro *et al.*, 1968). The 5'-GMP formed is neutralized with sodium hydroxide and crystalline 5'-GMP · Na$_2$ · 7 H$_2$O is prepared as the final product.

2. Production of Guanosine by Fermentation and Its Phosphorylation

a. Fermentative Production of Guanosine. Investigations to develop a process for the direct fermentative production of guanosine have been attempted since the process appeared to be more useful than the above two-step method of guanosine production from AICAR. Accumulation of guanosine in the culture fluid of *B. subtilis*, *B. pumilus*, *Bacillus licheniformis*, *Corynebacterium petrophilum*, *Corynebacterium guanofaciens*, and *Streptomyces griseus* have been reported. The yield of guanosine production is high with mutants of *Bacillus* sp. and these mutants are now being used for industrial guanosine production.

Producing Organisms. For the accumulation of guanosine, the producing microorganisms should have the following properties: (i) lack of

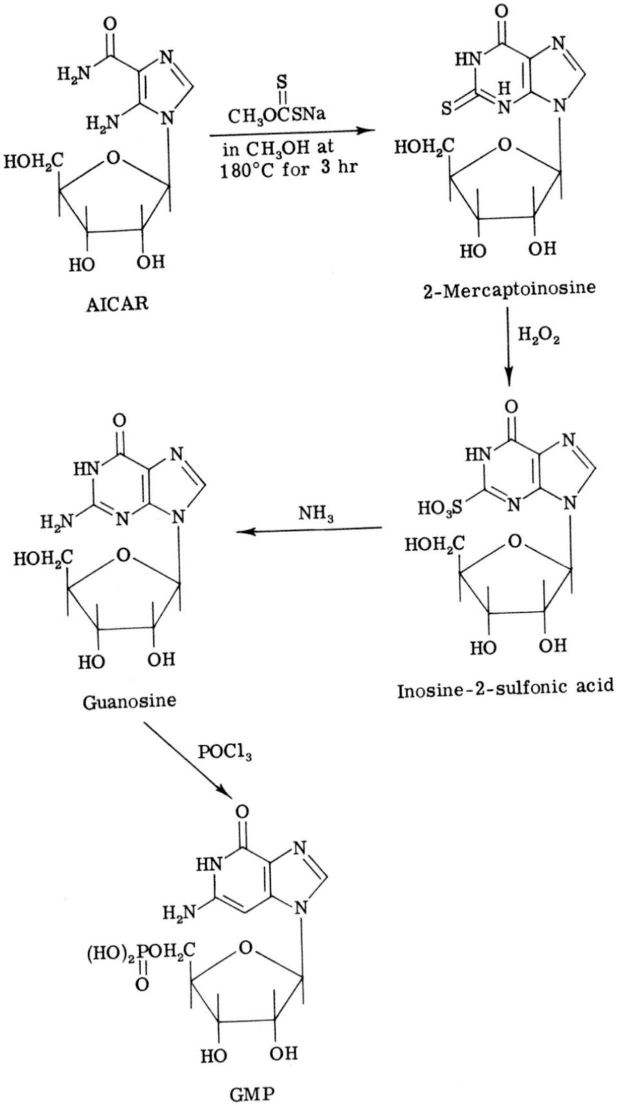

FIGURE 13. Synthesis of GMP from AICAR. (From Yamazaki *et al.*, 1967; and Kumashiro *et al.*, 1968.)

SAMP synthetase and GMP reductase; (ii) lack of guanosine-degrading enzymes, such as purine nucleoside phosphorylase or nucleoside hydrolase; (iii) release of regulation by GMP of GMP-synthesizing enzymes; and (iv) faster formation of 5'-GMP from 5'-IMP than dephosphorylation of 5'-IMP.

Mutants endowed with such favorable characteristics for guanosine formation have been derived from B. subtilis. The results are summarized in Table VII.

Konishi and Shiro (1968) obtained mutants resistant to 8-azaguanine from an inosine-producing mutant, AJ-1987, of B. subtilis. One of such mutants, AJ-1993, accumulated 4.3 mg/ml guanosine. Strain AJ-1993 possessed a higher level of 5'-IMP dehydrogenase than the original AJ-1987 and showed weak GMP-reductase activity.

Momose and Shiio (1969) isolated many 8-azaxanthine-resistant mutants from two kinds of mutants of B. subtilis: an inosine-producing adenine auxotroph K38-3, and a GMP-reductase-lacking adenine and tryptophan double auxotroph No. 30-12, which was derived from mutant K38-3. On examining nucleoside-forming capacities, 57% of 8-azaxanthine-resistant mutants derived from K38-3, which possessed GMP reductase, were found to accumulate inosine and hypoxanthine and the remainder (47%) only accumulated inosine. On the other hand, 70% of 8-azaxanthine resistant mutants derived from GMP reductase-lacking mutant No. 30-12 accumulated guanosine together with inosine, while the remainder (30%) only accumulated inosine. The yield of guanosine by one of the latter mutants, No. 30-12-3, was 5 mg/ml.

Nogami et al. (1968) derived a guanosine-producing mutant, designated T-780, from a Bacillus sp. No. 102 after multistep mutation (Nogami and Yoneda, 1969). They showed that adenine auxotrophy, lack of GMP reductase, and resistance to adenine or adenosine are essential characteristics for guanosine production. Other genetic characteristics, such as high resistance to streptomycin and lack of nucleoside phosphorylase, were effective for guanosine production.

Komatsu et al. (1972) and Komatsu and Kodaira (1973), based on a similar concept, derived a guanosine-producing mutant No. 20, from a Bacillus sp. No. 1043. It produced guanine, hypoxanthine, and xanthine together with guanosine. They further derived a mutant GnR-176, which had weak nucleoside-degrading enzyme activity. The yield of guanosine by GnR-176 increased to 10.6 mg/ml.

Matsui et al. (1974) isolated a guanosine-producing mutant, No. 14119, which became resistant to methionine sulfoxide, from B. subtilis No. 1411, capable of producing inosine. Slight increase in IMP dehydrogenase activity and marked decrease in nucleotidase activity were ob-

TABLE VII. Production of Guanosine by Microorganisms[a]

Microorganism	Genetic character	Guanosine accumulated (gm/liter)	References
Bacillus subtilis	*Ad⁻, red⁻*, 8AG^r	4.3	Konishi and Shiro (1968)
Bacillus subtilis	*Ad⁻, red⁻*, 8AX^r, *try⁻*	5.0	Momose and Shiio (1969)
Bacillus sp.	*Ad⁻, red⁻*, 8AG^r, AAR^r SM^r, NP⁻	5.8	Nogami *et al.* (1968); Nogami and Yoneda (1969)
Bacillus sp.	*Ad⁻, red⁻*, 8AX^r, *his⁻ thr⁻, adenase⁻, NS^pd*	10.6	Komatsu *et al.* (1972); Komatsu and Kodaira (1973)
Bacillus subtilis	*Ad⁻, red⁻*, MSO^r, *his⁻ Psicofuranine^r, Decoyinine^r*	16	Enei *et al.* (1976)

[a] Key to abbreviations; *ad⁻*, adenine-requiring; *his⁻*, histidine-requiring; *try⁻*, tryptophan-requiring; 8AG^r, 8-azaguanine-resistant; 8AX^r, 8-azaxanthine-resistant; AAR^r, adenine or adenosine resistant; SM^r, streptomycin resistant; MSO^r, methionine sulfoxide resistant; *psicofuranine^r*, psicofuranine resistant; *decoyinine^r*, decoyinine-resistant; *red⁻*, GMP reductase deficient; NP⁻, nucleoside phosphorylase deficient; *adenase⁻*, adenase deficient; NS^pd, partially deficient purine nucleoside-hydrolyzing activity.

served with the mutant No. 14119. Enei *et al.* (1976) further derived a psicofuranine-resistant mutant Gp-1 from strain 14119. Since Gp-1 formed xanthosine besides guanosine, they mutated it to be decoyinine resistant and obtained mutant MG-1. GMP synthetase in MG-1 was not inhibited by 5'-GMP. The yield of guanosine by MG-1 was as high as 16 mg/ml. Figure 14 shows an example of guanosine fermentation by mutant MG-1.

b. Isolation and Phosphorylation of Guanosine. To recover guanosine, the fermented broth is first adjusted to pH 12 with sodium hydroxide. After cells are removed by centrifugation, the supernatant fluid is concentrated under reduced pressure. Guanosine, precipitated by neutralizing the concentrate, is phosphorylated and crystalline 5'-GMP · Na_2 · 7 H_2O is prepared as described previously.

3. Production of Xanthosine and 5'-XMP by Fermentation and Their Conversion to 5'-GMP

The accumulation of xanthosine or 5'-XMP, an intermediate of 5'-GMP biosynthesis, in microbial culture fluid appears easier than accumulation of guanosine or 5'-GMP as an end product. Methods for production of xanthine derivatives have been investigated together with the conversion of those to 5'-GMP.

Accumulation of xanthosine or 5'-XMP would take place with microorganisms fulfilling the following two characters: (1) lack of XMP aminase and (2) lack of SAMP synthetase. High producers of xanthosine have been found in mutants of *B. subtilis*, and those of 5'-XMP, in mutants of *B. ammoniagenes*.

Ishii and Shiio (1973) mutated an inosine-producing strain of *B. subtilis*, which requires adenine, lacks GMP reductase and AMP deaminase, and is resistant to 8-azaguanine, to guanine auxotrophy. The obtained guanine auxotroph accumulated 17–18 mg/ml xanthosine under elaborated culture conditions. Xanthosine is phosphorylated to 5'-XMP by the chemical method described previously. One-step conversion of xanthosine to 5'-GMP, however, has not yet been reported.

Misawa *et al.* (1969) found that 6.5 mg/ml 5'-XMP was accumulated in the culture fluid of a guanine-requiring mutant of *B. ammoniagenes*. They concluded that the accumulated 5'-XMP was synthesized *de novo* and excreted through the cytoplasmic membrane of the producing microorganism.

In order to convert the accumulated 5'-XMP to 5'-GMP, Furuya *et al.* (1971) used a mutant of *B. ammoniagenes*, which is resistant to psicofranine or decoyinine and has increased XMP aminase activity. By cultivation of the mutant in a medium containing 13 mg/ml 5'-XMP at 30°C for

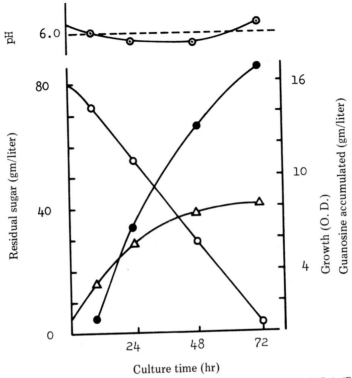

FIGURE 14. Time course of guanosine fermentation by *B. subtilus* MG-1. (From Enei *et al.,* 1976). ●—● Guanosine, ○—○ residual sugar, △—△ growth, ☉—☉ pH.

48 hours, about 10 mg/ml of 5'-GMP was produced in the culture fluid. Recently, they succeeded in the development of a single-step fermentative process for 5'-GMP formation by mixed culture of the above two mutants: One accumulates 5'-XMP and the other converts 5'-XMP to 5'-GMP (Furuya *et al.,* 1973). The yield of 5'-GMP was also about 10 mg/ml. Although they demonstrated a possibility that 5'-GMP is produced by fermentation, the processes have not yet been employed in industrial scale due to insufficient yield of 5'-GMP and formation of GDP and GTP as by-products.

4. Production of 5'-GMP by Fermentation

It seems necessary that 5'-GMP producers have the following characteristics: (1) lack of SAMP synthetase, (2) lack of 5'-GMP-degrading activity, (3) insensitivity of GMP-synthesizing enzymes to the product,

5'-GMP, and (4) membrane permeability from the interior to the exterior for 5'-GMP.

Demain *et al.* (1966) mutated a 5'-IMP-forming *Micrococcus glutamicus* ATCC 13761 and derived a mutant MB-1806 which is released from feedback inhibition by GMP with respect to biosynthesis of 5'-GMP. It accumulated 0.5 mg/ml GMP in its culture broth. Many trials followed in order to derive 5'-GMP producers from microorganisms such as *Pseudomonas aeruginosa, B. subtilis, Brevibacterium helvolum, E. coli, B. ammoniagenes,* and *Candida tropicalis.* Among these mutants, an adenine auxotroph derived from *B. ammoniagenes* ATCC 6871 showed the highest yield: 5.1 mg/ml 5'-GMP in broth (Abe and Udagawa, 1966a,b,c). This yield, however, is not sufficient for economical production of 5'-GMP on an industrial scale.

V. PRODUCTION OF NUCLEIC ACID-RELATED SUBSTANCES BY FERMENTATION

Investigation to develop a process for fermentative production of nucleic acid-related substances other than those related to 5'-IMP or 5'-GMP production has also been carried out. In this section, production by microorganisms of nucleic acid-related substances which are or will be useful for medicinal purposes is described.

A. ATP

Based on an old finding by Lutwal-Mann and Mann (1935), Tochikura *et al.* (1967) developed an improved process to produce ATP. ATP was obtained in a molar yield of 72% by phosphorylation of adenosine or 5'-AMP with bakers' yeast dried with acetone. Such a method was improved for industrial ATP production with several modifications: treatment of yeast with anionic detergents (Kawai *et al.*, 1974) and mild drying of yeasts at 60°C with a liquid-phase dryer to give 8.5% water content, followed by mild rehydration (30°C, more than 10 minutes) (Saeki *et al.*, 1976). In the latter case, the reaction mixture containing 15 mg/ml yeast cells, 36 mg/ml glucose, 2.03 mg/ml $MgCl_2$, and 3.83 mg/ml AMP in phosphate buffer of pH 7.5 was incubated at 30°C with shaking. The molar yield of ATP from AMP was about 70%.

Direct fermentative production of ATP was also attempted by Tanaka *et al.* (1968) and Nara *et al.* (1968a). Addition of 2 mg/ml adenine to the culture of *B. ammoniagenes* ATCC 6872 followed by an additional 1 day incubation yielded 1.57 mg/ml ATP, 1.59 mg/ml ADP, and 2.16 mg/ml AMP. Addition of guanine instead of adenine yielded GTP, GDP, and GMP.

B. NAD

Takebe and Kitahara (1963) demonstrated that the NAD content is highest in lactic acid bacteria among various microorganisms. Yeast cells grown on a medium supplemented with adenine and nicotinamide, however, accumulated NAD in the cells as high as 40 mg/gm (Sakai *et al.*, 1973b). NAD can be economically extracted from such yeast cells.

Nakayama *et al.* (1968) cultivated *B. ammoniagenes* for 2 days on a usual medium and for an additional 2–3 days on a medium containing adenine and nicotinamide. NAD in the culture fluid reached 2–3 mg/ml.

NAD is generally phosphorylated to NADP by NAD kinase. Tochikura *et al.* (1969) found a unique phosphate-transferring activity in *Proteus mirabilis*. In that transferring reaction, *p*-nitrophenylphosphate or nucleoside monophosphates are used as phosphate donor. The enzyme catalyzing the transferring reaction was also found in species of the genera *Aerobacter*, *Bacterium*, *Staphylococcus*, and *Streptomyces* (Kuwahara *et al.*, 1972). Another unique reaction, in which only metaphosphate served as a phosphate donor, was found in *B. ammoniagenes* and several species of the genus *Micrococcus* (Murata *et al.*, 1977). Application of such microorganisms for NADP production has not yet been reported.

C. FAD

Eremothecium ashbyii accumulates much FAD in its mycelia (Kuwata *et al.*, 1954; Kuwata, 1955). Based on that finding, a process to produce FAD was established by extracting it from *E. ashbyii* mycelia. A process to accumulate FAD extracellularly, however, seems to be more favorable than that described above.

Nakamura *et al.* (1968) found that the addition of adenine to the culture of *B. ammoniagenes* resulted in the accumulation of FAD in its culture filtrate. There have been several attempts to accumulate FAD by fermentation (Sakai *et al.*, 1973a). Among them, *Sarcina lutea* showed the highest yield and conditions for FAD production were elaborated. Alteration of cell wall structure by the addition of cycloserine to a culture medium was effective for FAD production. Cultivation of *S. lutea* in a medium containing 100 mg/liter cycloserine for 24 hours, followed by cultivation for 3–4 days more in the presence of adenine and riboflavin resulted in the accumulation of 780 mg/liter FAD (Watanabe *et al.*, 1974). This method may be the most economical way to manufacture FAD.

D. Coenzyme A

Coenzyme A has been industrially produced by extraction from yeast cells (Beinert *et al.*, 1953; Reece *et al.*, 1959). The yield is 8–9 mg/kg of bakers' or brewery yeast. Ogata *et al.* (1972a,b) and Shimizu *et al.* (1972) succeeded in accumulating 2–3 mg/ml CoA in the culture broth by incubating *B. ammoniagenes* IFO 12071 in a medium containing pantothenic acid, cysteine, and ATP as precursors in the presence of anionic detergents. They also improved the method by immobilizing bacterial cells in polyacrylamide gel (Shimizu *et al.*, 1975).

Accumulation of CoA in the culture filtrate is also possible. *B. ammoniagenes* was first cultivated in a medium containing AMP; pantothenic acid, cysteine, and a cationic detergent were then added to the medium. Upon further cultivation, 3–5 mg/ml CoA (as disulfide) were accumulated (Shimizu *et al.*, 1973a). Nishimura *et al.* (1974) demonstrated that *S. lutea* IAM 1099 showed strong activity to synthesize CoA. Shimizu *et al.* (1973b, 1974) have reviewed CoA biosynthesis and its regulation.

E. cAMP

Okabayashi *et al.* (1964) demonstrated the direct fermentative production of cAMP without addition of precursors: *Brevibacterium liquefaciens* ATCC 14929 accumulated 1.2 mg/ml cAMP in an alanine-containing medium. A mutant of *Microbacterium* sp. No. 205, which required biotin and was resistant to 6-mercaptopurine, 8-azaguanine, and methylsulfoxide, accumulated 2 mg/ml cAMP (Ishiyama *et al.*, 1973). It is also known that *Corynebacterium murisepticum* No. 7 and *Microbacterium* sp. No. 205 accumulated 10–100 μg/ml cGMP (Ishiyama, 1975).

However, cAMP can be most effectively produced by salvage synthesis. Ishiyama *et al.* (1974) cultivated mutant ATCC 21376 which was derived from *Microbacterium* sp. No. 205 in nutrient broth and transferred the culture to a medium containing 0.3% hypoxanthine (or 0.5% inosine), glucose, phosphate, peptone, and yeast extract. After cultivation for 40 hours at 28°C, 3–6 mg/ml cAMP was accumulated.

The 3′,5′-cyclic nucleotides can be also prepared by hydrolyzing RNA with a specific RNase followed by incubation in ⅓ M phosphate buffer (pH 7.0–8.7) (Yamasaki and Arima, 1967; Mantani *et al.*, 1972).

F. Guanosine 5'-Triphosphate 3'-Diphosphate

Murao and Nishino (1973, 1974) first demonstrated ATP: nucleotide pyrophosphotransferase in microbial culture broth. This enzyme formed magic spot I (ppGpp) or II (pppGpp) (Cashel and Gallant, 1969; Lazzarini et al., 1972) from ATP and GDP or GTP in the presence of Mg^{2+}. Purine nucleoside 5'-mono-, di-, and triphosphate accept pyrophosphate from ATP and the respective 3'-pyrophosphates are formed (Hamagishi et al., 1975). Distribution of this enzyme is widespread: Streptomyces adephosphoryticus (Murao and Nishino, 1973, 1974), Streptomyces hydrogenans (Betz et al., 1977), Streptomyces morookaensis, Streptomyces aspergilloides, Streptomyces hachijoensis, Actinomyces violasceus, Streptoverticillium septatum (Oki et al., 1976), and mutant KY 13510 (decoyinine resistant) derived from B. ammoniagenes ATCC 6872 (Sato and Furuya, 1976, 1977). Incubation in a reaction mixture containing 50–150 mg/ml of freeze-dried cells of mutant KY 13510, 30 mg/ml 5'-GMP·Na_2·7 H_2O, glucose, potassium phosphate, $MgSO_4$, and polyoxyethylene stearylamine (pH 7.4) gave 12 mg/ml ppGpp and 5 mg/ml pppGpp.

REFERENCES

Abe, S., and Udagawa, K. (1966a). Japanese Patent 41-12,673.
Abe, S., and Udagawa, K. (1966b). Japanese Patent 41-13,796.
Abe, S., and Udagawa, K. (1966c). Japanese Patent 41-13,797.
Akiya, T., Midorikawa, Y., Kuninaka, A., Yoshino, H., and Ikeda, Y. (1972). Agric. Biol. Chem. **36,** 227–233.
Akiyama, S., Doi, M., Arai, Y., Nakao, Y., and Fukuda, H. (1975). U.S. Patent 3,909,352.
Ando, T. (1966). Biochim. Biophys. Acta **114,** 158–168.
Aoki, R. (1963). J. Gen. Appl. Microbiol. **9,** 397–402.
Aoki, R., Momose, H., Kondō, Y., Muramatsu, N., and Tsuchiya, Y. (1963a). J. Gen. Appl. Microbiol. **9,** 387–396.
Aoki, R., Kondō, Y., and Momose, H. (1963b). J. Gen. Appl. Microbiol. **9,** 403–408.
Aoki, R., Kondō, Y., Hirose, Y., and Okada, H. (1968). J. Gen. Appl. Microbiol. **14,** 411–416.
Beinert, H., Korff, R. W., Green, D. E., Buyske, D. A., Handschumacher, R. E., Higgins, H., and Strong, F. M. (1953). J. Biol. Chem. **200,** 385–400.
Belozersky, A. N. (1947). Cold Spring Harbor Symp. Quant. Biol. **12,** 1–6.
Belozersky, A. N., and Spirin, A. S. (1960). In "The Nucleic Acids" (E. Chargaff and J. N. Davidson, eds.), Vol. 3, pp. 147–185. Academic Press, New York.
Berlin, R. D., and Stadtman, E. R. (1966). J. Biol. Chem. **241,** 2679–2686.
Berry, S. A., and Campbell, J. N. (1967a). Biochim. Biophys. Acta **132,** 78–83.
Berry, S. A., and Campbell, J. N. (1967b). Biochim. Biophys. Acta **132,** 84–93.
Betz, J. W., Schneider, B. H., and Träger, L. (1977). Hoppe-Seyler's Z. Physiol. Chem. **358,** 353–359.
Blakley, R. L., and Vitols, E. (1968). Annu. Rev. Biochem. **37,** 201–224.
Brawerman, G., and Chargaff, E. (1955). Biochim. Biophys. Acta **16,** 524–532.

Buchanan, J. M., and Hartman, S. C. (1959). *Adv. Enzymol.* **21,** 199–261.
Cashel, M., and Gallant, J. (1969). *Nature (London)* **221,** 838–841.
Caskey, C. T., Ashton, D. M., and Wyngaarden, J. B. (1964). *J. Biol. Chem.* **239,** 2570–2579.
Chayen, R., Chayen, S., and Roberts, E. R. (1955). *Biochim. Biophys. Acta* **16,** 117–126.
Cohn, W. E., and Volkin, E. (1951). *Nature (London)* **167,** 483–484.
Cohn, W. E., and Volkin, E. (1953). *J. Biol. Chem.* **203,** 319–332.
Demain, A. L. (1968). *Biotechnol. Bioeng.* **10,** 291–302.
Demain, A. L., Jackson, M., Vitali, R. A., Hendlin, D., and Jacob, T. A. (1965). *Appl. Microbiol.* **13,** 757–761.
Demain, A. L., Jackson, M., Vitali, R. A., Hendlin, D., and Jacob, T. A. (1966). *Appl. Microbiol.* **14,** 821–825.
Di Carlo, F. J., and Schultz, A. S. (1948). *Arch. Biochem.* **17,** 293–300.
Enei, H., Sato, K., Matsui, H., and Hirose, Y. (1976). *Abstr. Annu. Meet. Agric. Chem. Soc. Jpn.* p. 119.
Fox, C. L., Jr. (1942). *Proc. Soc. Exp. Biol. Med.* **51,** 102–104.
Fujimoto, M., and Uchida, K. (1965). *Agric. Biol. Chem.* **29,** 249–259.
Fujimoto, M., Morozumi, M., Midorikawa, Y., Miyakawa, S., and Uchida, K. (1965). *Agric. Biol. Chem.* **29,** 918–922.
Fujimoto, M., Kuninaka, A., and Yoshino, H. (1969). *Agric. Biol. Chem.* **33,** 1517–1518.
Fujimoto, M., Kuninaka, A., and Yoshino, H. (1974a). *Agric. Biol. Chem.* **38,** 777–783.
Fujimoto, M., Kuninaka, A., and Yoshino, H. (1974b). *Agric. Biol. Chem.* **38,** 785–790.
Fujimoto, M., Kuninaka, A., and Yoshino, H. (1974c). *Agric. Biol. Chem.* **38,** 1555–1561.
Fujimoto, M., Fujiyama, K., Kuninaka, A., and Yoshino, H. (1974d). *Agric. Biol. Chem.* **38,** 2141–2147.
Fujimoto, M. Kuninaka, A., and Yoshino, H. (1975a). *Agric. Biol. Chem.* **39,** 1991–1997.
Fujimoto, M., Kuninaka, A., and Yoshino, H. (1975b). *Agric. Biol. Chem.* **39,** 2145–2148.
Fujimura, Y., Hasegawa, Y., Kaneko, Y., and Doi, S. (1964). *Nippon Nogei Kagaku Kaishi* **38,** 467–471.
Fujimura, Y., Hasegawa, Y., Kaneko, Y., and Doi, S. (1967). *Agric. Biol. Chem.* **31,** 92–100.
Fujiwara, M., Byhovsky, V., Nakamura, H., Tamura, G., and Arima, K. (1967). *J. Gen. Appl. Microbiol.* **13,** 1–14.
Fukuda, H., Yashima, S., Akiyama, S., Fugono, T., Nakazawa, B., and Tada, T. (1963). British Patent 933,829.
Furuya, A., Abe, S., and Kinoshita, S. (1968). *Appl. Microbiol.* **16,** 981–987.
Furuya, A., Abe, S., and Kinoshita, S. (1969). *Appl. Microbiol.* **18,** 977–984.
Furuya, A., Abe, S., and Kinoshita, S. (1970a). *Appl. Microbiol.* **20,** 263–270.
Furuya, A., Abe, S., and Kinoshita, S. (1970b). *Agric. Biol. Chem.* **34,** 210–221.
Furuya, A., Abe, S., and Kinoshita, S. (1971). *Biotechnol. Bioeng.* **13,** 229–240.
Furuya, A., Okachi, R., Takayama, K., and Abe, S. (1973). *Biotechnol. Bioeng.* **15,** 795–803.
Furuya, A., Kato, F., and Nakayama, K. (1975). *Agric. Biol. Chem.* **39,** 767–771.
Gots, J. S. (1971). *Metab. Pathways, 3rd Ed.* **5,** 225–255.
Gots, J. S., and Gollub, E. G. (1957). *Proc. Natl. Acad. Sci. U.S.A.* **43,** 826–834.
Hamagishi, Y., Nishino, T., and Murao, S. (1975). *Agric. Biol. Chem.* **39,** 1015–1023.
Haneda, K., Hirano, A., Kodaira, R., and Ohuchi, S. (1971). *Agric. Biol. Chem.* **35,** 1906–1912.
Haneda, K., Komatsu, K., Kodaira, R., and Ohsawa, H. (1972). *Agric. Biol. Chem.* **36,** 1453–1460.
Hartman, S. C. (1970). *Metab. Pathways, 3rd Ed.* **4,** 1–68.
Hasegawa, Y., Nakai, T., Fujimura, Y., Kaneko, Y., and Doi, S. (1964). *Nippon Nogei Kagaku Kaishi* **38,** 461–466.
Hashimoto, K., and Okauchi, M. (1965). Japanese Patent 40-15,946.

Hino, T., Uozumi, T., Tamura, G., and Arima, K. (1971). *Agric. Biol. Chem.* **35,** 1109–1115.

Hiramaru, M., Uchida, T., and Egami, F. (1969a). *J. Biochem. (Tokyo)* **65,** 693–700.

Hiramaru, M., Uchida, T., and Egami, F. (1969b). *J. Biochem. (Tokyo)* **65,** 701–708.

Honjo, M., Marumoto, R., Kobayashi, K., and Yoshioka, Y. (1966). *Tetrahedron Lett.* **32,** 3851–3856.

Imada, A. (1976). *In* "Microbial Production of Nucleic Acid-related Substances" (K. Ogata *et al.*, eds.), pp. 101–111. Halsted Press, New York.

Ishibashi, J., Kamio, H., and Yoneda, M. (1965). Japanese Patent 40-12,914.

Ishii, K., and Shiio, I. (1972). *Agric. Biol. Chem.* **36,** 1511–1522.

Ishii, K., and Shiio, I. (1973). *Agric. Biol. Chem.* **37,** 287–300.

Ishiyama, J. (1975). *Agric. Biol. Chem.* **39,** 1331–1332.

Ishiyama, J., Yokotsuka, T., Saito, N., Kato, M., and Yoshida, F. (1973). *Abstr. Annu. Meet. Agric. Chem. Soc. Jpn.* p. 306.

Ishiyama, J., Yokotsuka, T., and Saito, N. (1974). *Agric. Biol. Chem.* **38,** 507–514.

Jacobsen, G. B., and Rodwell, V. W. (1972). *J. Biol. Chem.* **247,** 5811–5817.

Kanazawa, M. (1974). *Sekiyu To Biseibutsu (Petroleum and Microorganisms)* **12,** 33–36.

Katchman, B. J., and Fetty, W. O. (1955). *J. Bacteriol.* **69,** 607–615.

Kato, F., Furuya, A., and Abe, S. (1971). *Agric. Biol. Chem.* **35,** 1061–1067.

Kawai, H., Sako, F., and Endo, K. (1974). *Amino Acid Nucleic Acid (Tokyo)* **30,** 62–73.

Keir, H. M., Mathog, R. H., and Carter, C. E. (1964). *Biochemistry* **3,** 1188–1193.

Kinoshita, K., Shiro, T., Yamazaki, A., Kumashiro, I., Takenishi, T., and Tsunoda, T. (1967). *Biotechnol. Bioeng.* **9,** 329–342.

Kinoshita, K., Aoki, M., Yamanoi, A., and Shiro, T. (1968). *Nippon Nogei Kagaku Kaishi* **42,** 529–535.

Kinoshita, K., Tsuri, H., Sakai, S., Yasunaga, M., Okada, H., and Shiro, T. (1969a). *Nippon Nogei Kagaku Kaishi* **43,** 400–403.

Kinoshita, K., Sakai, S., Yasunaga, M., Sasaki, H., and Shiro, T. (1969b). *Nippon Nogei Kagaku Kaishi* **43,** 404–409.

Kinoshita, K., Yasunaga, M., Sakai, S., and Shiro, T. (1969c). *Nippon Nogei Kagaku Kaishi* **43,** 473–477.

Kinoshita, K., Niwa, K., Sasaki, H., Sakai, S., and Hirose, Y. (1973). *Nippon Nogei Kagaku Kaishi* **47,** 793–798.

Komatsu, K., and Kodaira, R. (1973). *J. Gen. Appl. Microbiol.* **19,** 263–271.

Komatsu, K., Haneda, K., Hirano, A., Kodaira, R., and Ohsawa, H. (1972). *J. Gen. Appl. Microbiol.* **18,** 19–27.

Komuro, T., Nara, T., Misawa, M., and Kinoshita, S. (1969). *Agric. Biol. Chem.* **33,** 1018–1029.

Konishi, S., and Shiro, T. (1968). *Agric. Biol. Chem.* **32,** 396–398.

Konishi, S., Kubota, K., Aoki, R., and Shiro, T. (1968). *Amino Acid Nucleic Acid (Tokyo)* **18,** 15–20.

Koshiro, M., Fukuhara, S., Sowa, T., Ohuchi, S. (1972). Japanese Patent 47-23,296.

Kotani, Y., Yamaguchi, T., Kato, T., and Furuya, A. (1977). *Abstr. Annu. Meet. Agric. Chem. Soc. Jpn.* p. 397.

Kumashiro, I., Yamazaki, A., Meguro, T., Takenishi, T., and Tsunoda, T. (1968). *Biotechnol. Bioeng.* **10,** 303–320.

Kuninaka, A., Otsuka, S., Kobayashi, Y., and Sakaguchi, K. (1959). *Bull. Agric. Chem. Soc. Jpn.* **23,** 239–243.

Kuninaka, A., Kibi, M., Yoshino, H., and Sakaguchi, K. (1961). *Agric. Biol. Chem.* **25,** 693–701.

Kuninaka, A., Fujimoto, M., and Yoshino, H. (1975a). *Agric. Biol. Chem.* **39,** 597–601.

Kuninaka, A., Fujimoto, M., and Yoshino, H. (1975b). *Agric. Biol. Chem.* **39,** 603–610.
Kuroiwa, Y., and Horie, Y. (1955). *Bull Agric. Chem. Soc. Jpn.* **19,** 35–42.
Kuwahara, H., Tachiki, T., Tochikura, T., and Ogata, K. (1972). *Agric. Biol. Chem.* **36,** 745–754.
Kuwata, S. (1955). *Vitamins* **9,** 453–462.
Kuwata, S., Masuda, T., Sawa, Y., and Asai, M. (1954). *Vitamins* **7,** 990.
Lazzarini, R. A., Cashel, M., and Gallant, J. (1972). *J. Biol. Chem.* **246,** 4381–4385.
Lee, S. Y., Nakao, Y., and Bock, R. M. (1968). *Biochim. Biophys. Acta* **151,** 126–136.
Linn, S., and Lehman, I. R. (1965a). *J. Biol. Chem.* **240,** 1287–1293.
Linn, S., and Lehman, I. R. (1965b). *J. Biol. Chem.* **240,** 1294–1304.
Linn, S., and Lehman, I. R. (1966). *J. Biol. Chem.* **241,** 2694–2699.
Logan, D. M., and Singer, M. F. (1968). *J. Biol. Chem.* **243,** 6161–6166.
Lutwak-Mann, C., and Mann, T. (1935). *Biochem. Z.* **281,** 140–156.
Magasanik, B. (1957). *Annu. Rev. Microbiol.* **11,** 221–252.
Magasanik, B. (1962). *In* "The Bacteria" (I. C. Gunsalus and R. Y. Stanier, eds.), Vol. 3, pp. 295–334. Academic Press, New York.
Mantani, S., Fukumoto, J., and Yamamoto, T. (1972). *Agric. Biol. Chem.* **36,** 242–248.
Matsui, H., Satou, K., Anzai, Y., Enei, H., and Hirose, Y. (1974). *Proc. Symp. Amino Acid Nucleic Acid 1974,* p. 20.
Misawa, M., Nara, T., Udagawa, K., Abe, S., and Kinoshita, S. (1969). *Agric. Biol. Chem.* **33,** 370–376.
Mitsugi, K. (1964). *Agric. Biol. Chem.* **28,** 669–677.
Mitsugi, K., Komagata, K., Takahashi, M., Iizuka, H., and Katagiri, H. (1964). *Agric. Biol. Chem.* **28,** 586–600.
Mitsugi, K., Nakazawa, E., Okumura, S., Takahashi, M., and Yamada, H. (1965). *Agric. Biol. Chem.* **29,** 1051–1058.
Miwa, M. (1959). *Hakko Kyokaishi* **17,** 535–540.
Moat, A. G., and Friedman, H. (1960). *Bacteriol. Rev.* **24,** 309–339.
Momose, H. (1967). *J. Gen. Appl. Microbiol.* **13,** 39–51.
Momose, H. (1968). *Tanpakushitsu Kakusan Koso* **13,** 781–797.
Momose, H., and Shiio, I. (1969). *J. Gen. Appl. Microbiol.* **15,** 399–411.
Momose, H., Nishikawa, H., and Katsuya, N. (1964). *J. Gen. Appl. Microbiol.* **10,** 343–358.
Momose, H., Nishikawa, H., and Katsuya, N. (1965). *J. Gen. Appl. Microbiol.* **11,** 211–220.
Momose, H., Nishikawa, H., and Shiio, I. (1966). *J. Biochem. (Tokyo)* **59,** 325–331.
Muramatsu, N., and Takenishi, T. (1965). *J. Org. Chem.* **30,** 3211–3212.
Murao, S., and Nishino, T. (1973). *Agric. Biol. Chem.* **37,** 2929–2930.
Murao, S., and Nishino, T. (1974). *Agric. Biol. Chem.* **38,** 2483–2489.
Murata, K., Uchida, T., Kato, J., and Chibata, I. (1977). *Abstr. Annu. Meet. Agric. Chem. Soc. Jpn.* p. 287.
Nakamura, N., Takasawa, S., and Tanaka, M. (1968). *Vitamins* **37,** 622.
Nakao, Y. (1976). *In* "Microbial Production of Nucleic Acid-related Substances" (K. Ogata et al., eds.), pp. 87–100. Halsted Press, New York.
Nakao, Y., and Ogata, K. (1963). *Agric. Biol. Chem.* **27,** 291–301.
Nakao, Y., Nogami, I.,.and Ogata, K. (1963). *Agric. Biol. Chem.* **27,** 507–517.
Nakao, Y., Lee, S. Y., Halvorson, H. O., and Bock, R. M. (1968). *Biochim. Biophys. Acta* **151,** 114–125.
Nakayama, K., Suzuki, T., Satō, Z., and Kinoshita, S. (1964). *J. Gen. Appl. Microbiol.* **10,** 133–142.
Nakayama, K., Nara, T., Tanaka, H., Satō, Z., Misawa, M., and Kinoshita, S. (1965). *Agric. Biol. Chem.* **29,** 234–238.

Nakayama, K., Satō, Z., Tanaka, H., and Kinoshita, S. (1968). *Agric. Biol. Chem.* **32,** 1331–1336.

Nara, T., Misawa, M., and Kinoshita, S. (1967). *Agric. Biol. Chem.* **31,** 1351–1356.

Nara, T., Misawa, M., and Kinoshita, S. (1968a). *Agric. Biol. Chem.* **32,** 561–567.

Nara, T., Misawa, M., and Kinoshita, S. (1968b). *Agric. Biol. Chem.* **32,** 1153–1161.

Nara, T., Misawa, M., Komuro, T., and Kinoshita, S. (1969a). *Agric. Biol. Chem.* **33,** 358–369.

Nara, T., Komuro, T., Misawa, M., and Kinoshita, S. (1969b). *Agric. Biol. Chem.* **33,** 1030–1036.

Nestle, M., and Roberts, W. K. (1969a). *J. Biol. Chem.* **244,** 5213–5218.

Nestle, M., and Roberts, W. K. (1969b). *J. Biol. Chem.* **244,** 5219–5225.

Nierlich, D. P., and Magasanik, B. (1963). *Fed. Proc., Fed. Am. Soc. Exp. Biol.* **22,** 476.

Nierlich, D. P. and Magasanik, B. (1965). *J. Biol. Chem.* **240,** 358–365.

Nishikawa, H., Momose, H., and Shiio, I. (1967). *J. Biochem. (Tokyo)* **62,** 92–98.

Nishimura, N., Shibatani, T., Kakimoto, T., and Chibata, I. (1974). *Appl. Microbiol.* **28,** 117–123.

Nogami, I., and Yoneda, M. (1969). *Kagaku To Seibutsu* **7,** 371–377.

Nogami, I., Kida, M., Iijima, T., and Yoneda, M. (1968). *Agric. Biol. Chem.* **32,** 144–152.

Numata, T., Kishi, M., Sato, T., Yano, N., Senoo, S., and Hiro, H. (1969). Japanese Patent 44-25,587.

Ogata, K., and Imada, A. (1962). *Annu. Rep. Takeda Res. Lab.* **21,** 31–53.

Ogata, K., Nakao, Y., Igarashi, S., Ohmura, E., Sugino, Y., Yoneda, M., and Suhara, I. (1963). *Agric. Biol. Chem.* **27,** 110–115.

Ogata, K., Shimizu, S., and Tani, Y. (1972a). *Agric. Biol. Chem.* **36,** 84–92.

Ogata, K., Tani, Y., Shimizu, S., and Uno, K. (1972b). *Agric. Biol. Chem.* **36,** 93–100.

Ohta, Y., and Ueda, S. (1968). *Appl. Microbiol.* **16,** 1293–1299.

Okabayashi, T., Ide, M., and Yoshimoto, A. (1964). *Amino Acid Nucleic Acid (Tokyo)* **10,** 117–123.

Oki, T., Yoshimoto, A., Ogasawara, T., Sato, S., and Takamatsu, A. (1976). *Amino Acid Nucleic Acid (Tokyo)* **34,** 52–62.

Partridge, C. W. H., and Giles, N. H. (1957). *Arch. Biochem. Biophys.* **67,** 237–238.

Reece, M. C., Donald, M. B., and Crook, E. M. (1959). *J. Biochem. Microbiol. Technol. Eng.* **1,** 217–228.

Saeki, N., Watanabe, T., Hoda, Y., and Oka, O. (1976). *Abstr. Annu. Meet. Agric. Chem. Soc. Jpn.* p. 117.

Sakai, T., Watanabe, T., and Chibata, I. (1973a). *Agric. Biol. Chem.* **37,** 849–856.

Sakai, T., Uchida, T., and Chibata, I. (1973b). *Agric. Biol. Chem.* **37,** 1041–1048.

Sanno, Y., Nara, K., Minato, S., and Hirose, U. (1964). Japanese Patent 39-15,846.

Saruno, R., Takahira, H., and Fujimoto, M. (1964). *J. Ferment. Technol.* **42,** 475–480.

Sato, A., and Furuya, A. (1976). *Agric. Biol. Chem.* **40,** 465–474.

Sato, A., and Furuya, A. (1977). *Agric. Biol. Chem.* **41,** 641–646.

Satō, Z., Nakayama, K., Tanaka, H., and Kinoshita, S. (1965). *Agric. Biol. Chem.* **29,** 412–418.

Shibai, H., Ishizaki, A., Mizuno, H., and Hirose, Y. (1973). *Agric. Biol. Chem.* **37,** 91–97.

Shiio, I., and Ishii, K. (1971). *J. Biochem. (Tokyo)* **69,** 339–347.

Shimizu, S., Tani, Y., and Ogata, K. (1972). *Agric. Biol. Chem.* **36,** 370–377.

Shimizu, S., Miyata, K., Tani, Y., and Ogata, K. (1973a). *Agric. Biol. Chem.* **37,** 607–613.

Shimizu, S., Kubo, K., Tani, Y., and Ogata, K. (1973b). *Agric. Biol. Chem.* **37,** 2863–2870.

Shimizu, S., Kubo, K., Morioka, H., Tani, Y., and Ogata, K. (1974). *Agric. Biol. Chem.* **38,** 1015–1021.

Shimizu, S., Morioka, H., Tani, Y., and Ogata, K. (1975). *J. Ferment. Technol.* **53,** 77–83.

Shirafuji, H., Imada, A., Yashima, S., and Yoneda, M. (1968). *Agric. Biol. Chem.* **32,** 69–75.

Shiro, T., Yamanoi, A., Konishi, S., Okumura, S., and Takahashi, M. (1962). *Agric. Biol. Chem.* **26,** 785–786.

Shishido, K., and Ando, T. (1972). *Biochim. Biophys. Acta* **287,** 477–484.

Shishido, K., and Ikeda, Y. (1970). *J. Biochem. (Tokyo)* **67,** 759–765.

Shishido, K., and Ikeda, Y. (1971a). *J. Mol. Biol.* **55,** 287–291.

Shishido, K., and Ikeda, Y. (1971b). *Biochem. Biophys. Res. Commun.* **42,** 482–489.

Shive, W., Ackermann, W. W., Gordon, M., Getzendaner, M. E., and Eakin, R. E. (1947). *J. Am. Chem. Soc.* **69,** 725–726.

Soeda, E., Murata, A., Utsu, M., and Saruno, R. (1968). *Seikagaku* **40,** 688.

Soeda, E., Abe, S., Murata, A., and Saruno, R. (1969). *Nippon Nogei Kagaku Kaishi* **43,** N86.

Spahr, P. F. (1964). *J. Biol. Chem.* **239,** 3716–3726.

Spahr, P. F., and Schlessinger, D. (1963). *J. Biol. Chem.* **238,** PC 2251–2253.

Stetten, M. R., and Fox, C. L., Jr. (1945). *J. Biol. Chem.* **161,** 333–349.

Stevens, A., and Hilmoe, R. J. (1960a). *J. Biol. Chem.* **235,** 3016–3022.

Stevens, A., and Hilmoe, R. J. (1960b). *J. Biol. Chem.* **235,** 3023–3027.

Sugimoto, H., Iwasa, T., and Yokotsuka, T. (1964a). *Nippon Nogei Kagaku Kaishi* **38,** 135–143.

Sugimoto, H., Iwasa, T., and Yokotsuka, T. (1964b). *Nippon Nogei Kagaku Kaishi* **38,** 567–575.

Suhara, I. (1973). *J. Biochem. (Tokyo)* **73,** 1023–1032.

Suhara, I. (1974). *J. Biochem. (Tokyo)* **75,** 1135–1141.

Suhara, I., and Yoneda, M. (1973). *J. Biochem. (Tokyo)* **73,** 647–654.

Takebe, I., and Kitahara, K. (1963). *J. Gen. Appl. Microbiol.* **9,** 31–40.

Tanaka, H., Satō, Z., Nakayama, K., and Kinoshita, S. (1968). *Agric. Biol. Chem.* **32,** 721–726.

Tanaka, K., Mizuno, K., Sanno, Y., and Hamuro, Y. (1963). Japanese Patent 38-16,892.

Tanaka, K., Ohmura, E., Honjo, M., Sanno, Y., and Sugino, Y. (1964a). Japanese Patent 39-7779.

Tanaka, K., Ohmura, E., Ogata, K., Sanno, Y., Yoneda, M., and Suhara, I. (1964b). Japanese Patent 39-12,342.

Tochikura, T., Kuwahara, M., Yagi, S., Okamoto, H., Tominaga, Y., Kano, T., and Ogata, K. (1967). *J. Ferment. Technol.* **45,** 511–529.

Tochikura, T., Kuwahara, M., Komatsubara, S., Fujisaki, M., Suga, A., and Ogata, K. (1969). *Agric. Biol. Chem.* **33,** 840–847.

Tone, H., and Ozaki, A. (1968). *Enzymologia* **34,** 101–128.

Tone, H., Umeda, M., Ishikura, T., and Miyachi, N. (1964). *Amino Acid Nucleic Acid (Tokyo)* **10,** 142–149.

Uchida, K., Kuninaka, A., Yoshino, H., and Kibi, M. (1961). *Agric. Biol. Chem.* **25,** 804–805.

Ueno, Y., Wada, S., Imada, I., and Tomoda, K. (1965). Japanese Patent 40-12,515.

Uozumi, T., and Arima, K. (1974). *Agric. Biol. Chem.* **38,** 1739–1740.

Uozumi, T., Tamura, G., and Arima, K. (1968a). *Agric. Biol. Chem.* **32,** 963–968.

Uozumi, T., Tamura, G., and Arima, K. (1968b). *Agric. Biol. Chem.* **32,** 969–974.

Uozumi, T., Tamura, G., and Arima, K. (1968c). *Agric. Biol. Chem.* **32,** 1409–1413.

Uozumi, T., Tamura, G., and Arima, K. (1969a). *Agric. Biol. Chem.* **33,** 25–30.

Uozumi, T., Tamura, G., and Arima, K. (1969b). *Agric. Biol. Chem.* **33,** 636–644.

Uozumi, T., Tamura, G., and Arima, K. (1969c). *Agric. Biol. Chem.* **33,** 645–652.

Uozumi, T., Hino, T., Tamura, G., and Arima, K. (1972). *Agric. Biol. Chem.* **36,** 434–441.

Vendrely, R. (1955). *Int. Rev. Cytol.* **4,** 115–142.

Wade, H. E., and Morgan, D. M. (1957). *Biochem. J.* **65,** 321–331.

Watanabe, K. (1976). *In* "Microbial Production of Nucleic Acid-related Substances" (K. Ogata *et al.*, eds.), pp. 55–65. Halsted Press, New York.

Watanabe, T., Uchida, T., Kato, J., and Chibata, I. (1974). *Appl. Microbiol.* **27,** 531–536.

Wyngaarden, J. B., and Ashton, D. M. (1959). *J. Biol. Chem.* **234,** 1492–1496.

Yamada, K. (1976). *Hakko To Kogyo* **34,** 658–694.

Yamanoi, A., and Shiro, T. (1968). *J. Gen. Appl. Microbiol.* **14,** 1–17.

Yamanoi, A., Konishi, S., and Shiro, T. (1967). *J. Gen. Appl. Microbiol.* **13,** 365–380.

Yamasaki, M., and Arima, K. (1967). *Biochim. Biophys. Acta* **139,** 202–204.

Yamashita, T., Hidaka, T., and Watanabe, K. (1974). *Agric. Biol. Chem.* **38,** 727–734.

Yamazaki, A., Kumashiro, I., and Takenishi, T. (1967). *J. Org. Chem.* **32,** 1825–1828.

Yoneda, M. (1964a). *J. Biochem. (Tokyo)* **55,** 475–480.

Yoneda, M. (1964b). *J. Biochem. (Tokyo)* **55,** 481–485.

Yoshikawa, M., and Kato, T. (1967). *Bull. Chem. Soc. Jpn.* **40,** 2849–2853.

Yoshikawa, M., Kato, T., and Takenishi, T. (1967). *Tetrahedron Lett.* **50,** 5065–5068.

Yoshikawa, M., Kato, T., and Takenishi, T. (1969). *Bull. Chem. Soc. Jpn.* **42,** 3505–3508.

Yoshikawa, M., Sakuraba, M., and Kusashio, K. (1970). *Bull. Chem. Soc. Jpn.* **43,** 456–461.

Chapter 11

Production of Organic Acids by Fermentation

LEWIS B. LOCKWOOD

MICROBIAL TECHNOLOGY, 2nd ed., VOL. I
Copyright © 1979 by Academic Press, Inc.
All rights of reproduction in any form reserved. ISBN 0-12-551501-4

I. INTRODUCTION

Organic acids discussed here are those which are manufactured in large volume or which offer potential for future development. They are marketed as relatively pure chemicals or their salts. These processes have been developed within the last hundred years, in contrast to some of the major food and beverage fermentations, the origin of which is found in antiquity. Operation of the first plant for the bulk manufacture of a fermentation chemical was started in 1881 in Massachusetts for the production of calcium lactate by bacterial fermentation. Wehmer in 1893 reported that certain fungi of the genus *Citromyces* could accumulate citric acid in liquid cultures. The genus *Citromyces* is now included in the monoverticillate group of the genus *Penicillium*. A plant to produce citric acid using Wehmer's culture was established in Alsace, France and operated for about 10 years (Mial, 1975). These plants provided the starting point from which much modern fermentation technology was developed.

II. CITRIC ACID

Citric acid was recovered in 1869 in England from calcium citrate imported from Italy. Calcium citrate was marketed through an Italian government-controlled cartel, which held the price very high and thus retarded the development of new markets for citric acid. Introduction of citric acid made by a new fermentation process in 1923 broke the power of the cartel, and today's prices are considerably lower than those prior to 1923. The price of citric acid dihydrate in the United States in 1976 was $0.495 per pound in 250-lb drums, and the price of anhydrous citric acid was $0.555–0.605 per pound. The price of calcium citrate was $0.69 per pound. The four citric acid plants of two manufacturers in the United States are in the eastern part of the country at Southport, North Carolina and Groton, Connecticut (Charles Pfizer, Inc.), Dayton, Ohio and Elkhart, Indiana (Miles Laboratories). Prices are about $0.045 per lb higher west of the Rocky Mountains.

The fermentation processes now account for almost all the citric acid marketed in the world today, although a small amount is recovered from citrus fruit in Mexico and in South America. This is estimated to amount to less than 1% of total world production. Citric acid production in the United States and Europe probably amounts to 210,000,000 lb annually. In addition, Japan produces about 10,000,000 lb, and about 20,000,000 lb are produced in Brazil, Israel, Mexico, Colombia, and Argentina.

Three processes are used in the manufacture of citric acid. These are the Japanese koji process, in which special strains of *Aspergillus niger* are used; the liquid surface culture fermentation, in which *A. niger* floats on the surface of a solution; and the submerged fermentation process, in which the fungal mycelium grows throughout a solution in a deep tank.

Scheele (1789) first reported the isolation and crystallization of the sour constituent of lemon juice in 1784. Grimaux and Adams (1880) synthesized citric acid from glycerol.

Wehmer (1893) established the occurrence of citric acid as a microbial metabolite. The next noteworthy step in the development of the citric acid production process was the recognition by Molliard (1922) that citric acid accumulates in cultures of *A. niger* under conditions of nutrient deficiency. He attributed his results to phosphate deficiency in his culture media. Yields were too small to permit commercial exploitation, but his publication lead to further development of the concept of nutrient deficiency especially in the development of the submerged fermentation process. In a culture solution adequate to permit the maximum growth of mycelium or to permit sporulation, little or no citric acid accumulates.

A. Koji Fermentation Process

The Japanese wheat bran process accounts for about one-fifth of Japanese citric acid production annually. Originally the solid residues left after sweet potato starch recovery were the fermentation medium. Later, wheat bran was substituted for the sweet potato material, but data are not available on whether sweet potato wastes are still used to any extent. In terms of process operations, the conditions of fermentation are almost identical with those when fungal amylase and protease of *Aspergillus oryzae* are produced. Prior to sterilization, the pH of the bran is adjusted to between 4 and 5, and additional moisture is picked up during steaming, so that the final water concentration of the mash is 70–80%. When the bran has cooled to 30°–36°C the mass is inoculated with a koji which was made by a special strain of *A. niger* which is probably not as responsive to the presence of ions of iron as the culture strains used in other processes. Hisanaga and Nakamura (1966) and Yamada (1965) reported that the temperature of the mash during fermentation should not exceed 28°C. Addition of 3–7% of filter cake from a glutamic acid fermentation resulted in improvement in yield. Starch originally present in the bran is saccharified by the amylase of *A. niger*. Addition of α-amylase to the bran after cooling was also beneficial. The inoculated bran is spread in trays to a depth of 3–5 cm or spread in windrows on a floor. After 5–8 days, the koji is harvested and placed in percolators, and the citric acid is extracted with water. Purification and recovery follow essentially the same process steps used in the recovery of the acid when either the liquid surface culture process or the submerged fermentation processes is used.

B. Liquid Surface Culture Process

The liquid surface culture process is the one which first supplied citric acid at a low price. It is still used by many manufacturers, but operations are highly secret, and only one detailed description of the process has been published (Mallea, 1950). The inoculated fermentations are dispersed in shallow pans made of high-purity aluminum or of stainless steel. Humidified air is blown over the surface of the solution for 5 or 6 days, after which dry air is used. If the CO_2 content of the air rises as high as 10%, a marked reduction of yield occurs. Spores germinate within 24 hours, and a white, wrinkled mycelium covers the surface of the solution. Eight or 10 days after inoculation, the original sugar concentration of 20–25% has been reduced to the range of 1–3%. At the end of the fermentation, the liquid can be drained off and replaced by fresh culture solution if care is taken to introduce the replacement solution under the mycelium to ensure that the mycelium still floats on the surface. Any portion of the mycelial mat which becomes submerged is inactivated. The average length of the fermentation cycle is reduced since the initial growth period of about 3 days is bypassed. Little citric acid is produced during the growth phase. Such reuse of mycelium has been done in Argentina and in Russia, but it is doubtful if it is done in any large plant today.

The preferred sugar substrate for the fermentation is sugar beet molasses which has not undergone the Steffens process for sucrose recovery. This molasses contains sucrose as the principal carbohydrate, and some glucose as well as proteins, peptides, amino acids, and inorganic ions. It has been subjected to heat during the concentration of the beet extract and sucrose crystallization, so doubtless it contains saccharinic acids and related compounds at least in traces. These may serve to counteract in part the effect of traces of undesirable metallic ions.

Mallea (1950) appreciated the importance of metallic ions in the control of the accumulation of citric acid in solutions for the surface culture of A. niger. He developed on a commercial scale a system for removal of metallic ions, or reduction in quantity of undesirable ions in sucrose syrup by adsorption with a combination of $CaCO_3$, colloidal silica, tricalcium phosphate, and starch. Steinberg (1945) had previously removed trace metallic ions from sucrose solutions by heating with $Ca(OH)_2$. It appears probable that some activated carbons might also be effective in trace element removal from liquid surface culture media. Some manufacturers precipitate iron by addition of calcium ferrocyanide.

The pH of the culture solution for citric acid accumulation is usually initially in the range of 5–6 but, on spore germination, rapidly approaches the range of pH 1.5–2.0 as ammonium ions are removed from the solution. If it rises to about 3.5 after the initial change or is adjusted to 3.5 or above, some oxalic acid is formed along with citric acid. The presence of iron or sodium salts also favors the accumulation of oxalic acid, and of yellow or yellow-green pigments in the mycelium. These pigments are sometimes excreted into the culture solution and are difficult to remove during product recovery and purification.

The final yield of citric acid in the liquid surface culture process is in the range of 80–85% of the weight of the carbohydrate initially supplied.

C. Submerged Culture Processes

The advantages of the submerged fermentation process over the liquid surface fermentation processes in terms of investment and operating costs induced investigators to attempt to develop a submerged fermentation process for citric acid manufacture. Amelung (1930), using a black *Aspergillus* which he called *A. niger japonicus* (probably *Aspergillus japonicus* Saito) slowly bubbled a stream of air through a culture solution of 15-cm depth. He obtained subsurface growth and found citric acid in the culture solution. Yields were inferior to those commonly found in liquid surface culture fermentation.

Kluyver and Perquin (1933) described the use of cultures on shakers, a technique for the study of submerged fermentation in small scale. Up to this time, the study of submerged fermentation required large equipment and an air compressor. Perquin (1938) used shaker cultures for the study of the production of organic acids. He published in his doctoral dissertation a laboratory-scale method for the production of citric acid by *A. niger* in which he extended Molliard's phosphate deficiency concept to submerged fermentations. Neither Molliard nor Perquin appears to have recognized the role of trace metallic ions which are commonly occurring impurities in phosphate salts. Following the Molliard–Perquin concepts, Szücs (1944) obtained a patent for the submerged culture production of citric acid by the method of growing *A. niger* on a culture solution which contained KH_2PO_4 and transferring the mycelium to a new culture solution to which he had not added phosphate, or by the method of growing mycelium in a culture solution initially inadequately supplied with phosphate. He assumed that the growing mycelium would remove all the phosphate from the solution, leaving it free of phosphate.

Moyer (1953) found that addition of methanol to his shaker cultures of *A. niger* resulted in improvement in yield of citric acid. Neither this nor the application of the Molliard–Perquin concepts has led to a commercially successful process. Studies conducted by Miles Laboratories have failed to demonstrate any relationship between assimilable phosphorus in the culture solution and citric acid accumulation after the initial mycelial growth period (Lockwood and Snell, 1976).

Shu and Johnson (1947, 1948a,b) demonstrated that when ions of manganese or iron were present in sufficient concentration little or no citric acid accumulates in the culture solutions, but *A. niger* grew much better than when these ions were inadequately supplied. Removal from the culture solution of undesirable metallic ions by ion exchange was demonstrated by Woodward *et al.* (1949) and the use of copper ion as an antagonist to iron, discovered by Schweiger (1957), made possible the present commercial process. Tables I and II show the response of the citric fermentation to ferric ions, and the counteraction of this response by copper ions. Bruchmann (1961) found that in the presence of copper ions, the enzyme aconitase is inhibited. He believed that the action of the zinc ion was at some later point in the destruction of citric acid.

Culture solutions are flash sterilized by passage through the pipes of a

TABLE I. Effect of Iron on the Production of Citric Acid from Decationized Sucrose by *A. niger* in Submerged Culture[a]

Iron[b] (mg/liter)	Weight yield citric acid[c] (%)
0.0	67.0
0.05	71.0
0.5	88.0
0.75	79.0
1.00	76.0
2.00	71.0
5.00	57.0
10.0	39.0

[a] From Schweiger (1957). Medium composition: sucrose purified by mixed resin bed ion exchange from 3,600,000 ohms resistance at 40% concentration, diluted with deionized water at 18,000,000 ohms resistance to 14.2% sugar content, and recrystallized nutrient salts added: KH_2PO_4, 0.014%; $MgSO_4 \cdot 7 H_2O$, 0.1%; $(NH_4)_2CO_3$, 0.2%; HCl to pH 2.6.

[b] Supplied as $FeCl_3$.

[c] (Grams citric acid produced/ Equivalent grams hexose moiety supplied) $\times$ 100.

TABLE II. Effect of Ions of Copper and Iron on the Production of Citric Acid from Glucose by *A. niger* in Submerged Culture[a]

Fe^{3+} (mg/liter)	Cu^{2+} (mg/liter)	Citric acid yield[b] (%)
10	50	77.8
50	50	69.1
100	50	50.7
150	50	14.2
10	100	77.2
50	100	65.4
100	100	53.9
150	100	29.8
10	500	74.0
50	500	65.4
100	500	60.6
150	500	27.6

[a] From Schweiger (1957). Medium composition: glucose (commercial grade), approx. 14%, $(NH_4)_2CO_3$, 0.2%; KH_2PO_4, 0.014%; $MgSO_4 \cdot 7 H_2O$, 0.1% Fe and Cu supplied as sulfates.
[b] (Grams citric acid produced/grams glucose supplied) × 100.

steam-jacketed heat exchanger and immediately cooled to about 30°C in another heat exchanger. To operate efficiently, it may be necessary to induce turbulent flow through the first heat exchanger. For decationized solutions No. 316 stainless steel is necessary in the construction of the heat exchanger and lines to the fermentor in order to avoid corrosion by the acidic solutions.

Inocula are usually spores of a suitable strain of *A. niger* grown on solid nutrient media. Stock cultures are kept refrigerated in tubes of a good loam soil, or as lyophilized spores. *Aspergillus niger* is an unstable organism, and it is desirable to minimize the number of culture transfer generations from a culture which has proved satisfactory.

The range of copper ion required to be added for process control varies from 0.1 to 50.0 mg/liter, depending on the amount of iron in the culture solution (Table II). Other additives to the culture solution are KH_2PO_4, 0.01–0.3% (but may be as great as 2%), and $MgSO_4 \cdot 7 H_2O$ to give up to 0.25%. If the content of iron is low enough, iron must be added to give 0.1–0.2 mg/liter. Copper ion may be required to counteract any excess of iron. The pH of the decationized solutions is between 1.5 and 2.0. Before inoculation of decationized solutions, the pH value is adjusted with ammonia to about pH 4.0. In both fermentations, the pH rapidly changes to the range of 1.5–2.0 by rapid uptake of ammonium ion, leaving inorganic acid residues. Little citric acid is formed in young cultures before the pH of 2.0 is reached.

Aeration of the submerged fermentation must be continuous. Air is bubbled through the solution at a rate of 0.5–1.5 v/v solution per minute. Mechanical stirring is not necessary. Batti (1966) has shown that even a brief failure of air supply stops the fermentation and that it will not be reactivated for several days unless the pH is adjusted to permit new growth. At the same time, additional copper ion must be added to ensure that the new growth is of the right biochemical type. The combination of aeration, low pH, and the extreme sensitivity to iron and copper make the use of ordinary iron fermentors impossible unless some impervious coating is applied to them. Glass and plastic coatings have been used, but the common blue glass linings are unsatisfactory because cobalt leaches out of the glass. Antifoam agents are necessary to prevent loss of material due to foamover at any time during the fermentation. Such agents must be free of iron, cobalt, or nickel.

Process controls in the submerged fermentation depend primarily on microscopic examination of mycelia. Schweiger (1957) has described the short, thick, stubby branches and abundant chlamydospores of active mycelia. An experienced technician can estimate the amount of iron or copper to be added to control the rate and efficiency of the fermentation. Figure 1 is a flow sheet of the citric fermentation process.

A series of organic compounds which counteract the effect of iron on

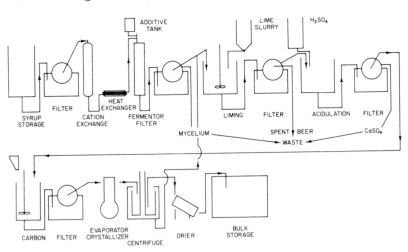

FIGURE 1. Flow sheet of the manufacture of citric acid. (From Lockwood and Schweiger, 1967; reprinted by permission of the Van Nostrand-Reinhold Co., New York.)

citric acid accumulation has been patented (Lockwood and Batti, 1965). Their action appears to be similar to the action of copper ion in that growth and sporulation are inhibited and mycelium produced has the proper enzymatic makeup for citric acid accumulation. The organic compound antagonists are probably not used in the commercial manufacture of citric acid at this time since the copper ion is effective at lower cost.

Clark (1962) described a process in which ferrocyanide was used to remove or complex iron in beet or cane molasses. The initial pH before dilution of the molasses was pH 6 before sterilization but was adjusted to pH 6.5 after sterilization. Ferrocyanide concentrations were in the range of 20–150 mg/liter. Optimal yields were obtained when the initial concentrations of ferrocyanide were 20 mg/liter and were increased to 80 mg/liter after the initial growth period. The removal of trace metallic ions by ion exchange or precipitation by ferrocyanides are probably the only practical methods of purifying culture solutions for use in submerged culture production of citric acid. It is not known how many plants use each of these methods.

Solutions of high-test cane syrup, glucose, or sucrose are suitable carbohydrate substrates for the production of citric acid by either process. High-test syrup is concentrated cane juice in which about 30–35% of the sucrose has been inverted by yeast invertase. When the ferrocyanide treatment is used the substrate can be a solution of any of the above, or beet molasses which has not been treated by the Steffens process. Ion exchange may be too costly for use with the cruder molasses solutions. Yields obtained from decationized refined syrups are generally superior to those obtained from the less pure sugar sources. The initial sugar concentration is about 20–25% and yields of citric acid approximate those obtained in surface culture.

Meyrath (1967) has discussed the energy requirements for growth and citric acid excretion in relation to the acid production by nonproliferating cells. A significant amount of energy is required since the acid is excreted into a solution of much greater concentration. The culture solution concentration of citric acid may be as great as 1 M, while that within the cell probably never exceeds 0.01 M.

Continuous culture techniques are not considered suitable for use in citric acid production. A multitank system would be required for continuous fermentation in any process in which cell growth and metabolic product formation occur primarily at different times. This would require a significantly larger capital expenditure and would probably not be economically competitive.

D. Product Recovery

The filtered culture solution is subjected to a polishing filtration if the solution is hazy due to the presence of residual antifoam agents, mycelium, or oxalate. Calcium citrate is precipitated from the clear solution by addition of a slurry of $Ca(OH)_2$. This must be very low in magnesium content if losses due to the relatively soluble magnesium citrate are to be avoided. Calcium citrate is filtered off on any conventional type of industrial filter, and the filter cake is transferred to a tank where it is treated with sulfuric acid to precipitate calcium sulfate. The dilute filtrate containing the citric acid is purified by treatment with carbon which has been activated by treatment with heat or an hydrochloric acid wash. It is demineralized by successive passages through ion-exchange beds, and the purified solution is evaporated in a circulating granulator or in a circulating evaporator–crystallizer. Crystals are removed by centrifugation, and a predetermined fraction of the mother liquor is returned to the evaporator. The remaining mother liquor is returned to the recovery stream at a point prior to liming and decolorization. Citric acid so prepared may require recrystallization from water in order to meet the standards of the United States Pharmacopoeia. Mother liquors from recrystallization are recycled to a point prior to liming in primary product recovery. Direct crystallization of citric acid from fermented beers has been studied, but quality problems have always precluded the use of this method of recovery.

In the Usines de Melle patented recovery process (Anonymous, 1963), the filtered beer is extracted countercurrently at 10°–30°C with a mixture of 100 parts tri-n-butyl phosphate and 5–30 parts n-butyl acetate or methyl isobutylketone. The solvent is then extracted with water at 70°–95°C. Citric acid is further concentrated in the hot-water extract, decolorized, and crystallized in the usual manner.

E. Chemistry of the Citric Acid Fermentation

Citric acid is excreted into and accumulates in culture solutions of pH about 1.8–2.0, an acidity too great to be tolerated by many organisms. Data are not available as to the pH within the mitochondrion or even within the cell. Data are available which clarify some aspects of the problem. Blumenthal (1965) has reported on the relative importance of different metabolic pathways in fungi. In A. niger 78% of the sugar consumed passes through the Embden–Meyerhof–Parnas (EMP) pathway. Under the special conditions of the citric acid fermentation, this amount may be either reduced or increased, but such data are not available for cultures under the conditions required for citric acid accumulation. Both EMP and hexose monophosphate (HMP) pathways of glycolysis were reported to be active at all times, the greatest activity of the EMP pathway occurred during the vegetative stage, and that of the HMP pathway occurring during conidiation. Relatively little citric acid is found in cultures during the growth period, and most of the citric acid is produced by nonproliferating cells. Since cultures which sporulate accumulate almost no citric acid, it appears that the EMP pathway may play the major role in glycolysis in the citric acid fermentation.

Acetyl-CoA derived by the EMP pathway condenses with oxalacetate to form citrate, the point of entry into the oxidative sequence of an abortive citric acid cycle (Fig. 2). The cycle is aborted by inhibition of the enzymes aconitase and isocitric dehydrogenase either by copper ions, by H_2O_2 (Bruchmann, 1961), or possibly by the action of some constituent of beet molasses. Iron is an essential cofactor for these enzymes in the conversion of either citrate to aconitase or isocitrate to aconitate by isocitrate dehydrogenase. In noninhibited cultures supplied with iron, citric acid is an excellent substrate for growth. La Nauze (1966) has compared activity of aconitase, and NAD- and NADH-linked isocitrate dehydrogenases of culture strains which gave widely different citric acid yields. All three enzymes were present in both culture strains when grown on media in which citric acid did or did not accumulate. Copper ion was supplied to the culture solutions at 0.3 mg/liter and ferrous ion at 2 mg/liter. The ratio of copper ion to iron ion would not permit activity of the three enzymes at pH 2.0, an acidity approximating that of the commercial fermentations. She also found some citric acid inside the mycelium, an observation which I have confirmed. This would indicate that the intracellular pH was in the neighborhood of 2, a value much too

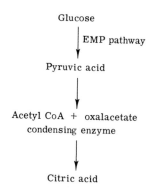

FIGURE 2. Biochemical sequence in citric acid fermentation.

acidic for these enzymes to be active. Thus, citric acid accumulation would result not from failure to produce the citric acid cycle enzymes, but from the inability of the enzymes to function under the physiologically adverse conditions of the fermentation. *Aspergillus niger* and *Penicillium javanicum* convert several intermediates of the citric acid cycle to citric and oxalic acids. However, Cleland and Johnson (1954), in radioactive carbon studies, found no evidence of recycling of carbon in the citric fermentation. Commercial yields of citric acid from D-glucose are great enough to preclude the possibility of significant recycling of carbon.

F. Marketing: Market Forms and Control Procedures

Citric acid is marketed as an anhydrous crystalline chemical, as the crystalline monohydrate, or as the crystalline sodium salt. Most of the free acid meets the specifications of the United States Pharmacopoeia, which sets maxima and minima for ash, heavy metals, sugar, other acids, and materials which cause a dark color when a sample of citric acid is heated in concentrated sulfuric acid (90°C for 1 hour). Anhydrous and monohydrate crystals are available in granular (15–50 mesh), fine, and powder form (35–100 mesh for anhydrous and 60–200 mesh for the monohydrate).

G. Packaging and Stability

Both anhydrous and monohydrate forms of citric acid are supplied in 100-lb multiwall moisture-barrier paper bags and in 100- and 250-lb drums. Caking may occur if citric acid is stored at high temperature and humidity. Storage at 21°C and 50–70% relative humidity is satisfactory. Citric acid does not oxidize under these conditions. Its solutions are only mildly corrosive.

H. Uses and Competition

Citric acid is the principal food acid used in the preparation of soft drinks, desserts, jams, jellies, candies, wines, and frozen fruits. In this area, malic acid offers a small degree of competition. Citric acid is also used in gelatin food products and artificial flavors of dry compounded materials, such as soft drink tablets and powders. Here, fumaric acid offers competition.

In pharmacy, citrates are used in blood transfusion, and the free acid is used in effervescent products. It is rapidly metabolized almost completely in the human body and can serve as a source of energy. Cosmetic uses of citric acid include those in astringent lotions, where it is used to adjust pH and act as a sequestrant and in hair rinses and hair setting fluids.

Citric acid is a superior sequestering agent in the neutral and low pH range, exhibiting some activity at pH values as low as 3.0. This complexing activity permits its wide industrial use in electroplating, in leather tanning, and in the reactivation of old oil wells, where the pores of the sand face have become clogged with iron.

III. ITACONIC ACID

Itaconic acid (methylene succinic acid) accumulates in cultures of *Aspergillus terreus* and to a lesser extent in those of *Aspergillus itaconicus*. The compound is a structurally substituted metharyllic acid.

$$CH_2{=}\underset{\underset{\displaystyle COOH}{\overset{\displaystyle |}{\underset{\displaystyle |}{CH_2}}}}{\overset{\displaystyle |}{C}}{-}COOH$$

Itaconic acid

Consequently its principal use would be expected to be as a copolymer in acryllic of methacryllic resins. In these resins, inclusion of about 5% itaconate in the polymer imparts the ability to take and hold printing inks. Resins derived by polymerization of methyl itaconate tend to be rather brittle. Copolymerization with other esters makes a less brittle resin. Itaconic acid may also find some use in detergents.

Itaconic acid is marketed in two grades: refined, which is a pale tan to white crystalline solid and sells for $0.83 per pound in car lots, and the industrial grade, which is darker in color and sells for $0.22 per pound (U.S. Tariff Commission, 1974). The Charles Pfizer Co. is the sole manufacturer of this substance in the United States at the present time.

Itaconic acid was formerly obtained by pyrolysis of citric acid. Water is lost to produce aconitic acid. On loss of CO_2 from aconitic acid, two isomeric compounds are produced, the anhydrides of itaconic and citraconic acids. The process was not commercially successful. Aconitic acid is present in sugar cane juice and was believed to interfere in sucrose crystallization. Removal of calcium aconitate by sugar refiners, and its conversion to itaconic acid by heating of its solutions, was the principal source of itaconic acid until the fermentation process was developed.

In fermentation solutions, itaconic acid is accompanied by succinic acid and itatartaric acid. These interfere in analysis of the solutions when itaconic acid is measured by titration with alkali but do not interfere in the bromination method of Friedkin (1945).

A. Technological Development

Itaconic acid was first reported as a fungal metabolite by Kinoshita (1931, 1937) who described a new species, *Aspergillus itaconicus*, which produced as its principal water-soluble metabolites mannitol and itaconic acid. Calam *et al.* (1939) found that the compound accumulated in cultures of *A. terreus*. Moyer made a survey of isolates of *A. terreus* and found one strain, NRRL 1960, which was superior to his other strains in either surface or submerged culture. His strain or a mutant derived from it is believed to be the one now used industrially.

In early studies it was observed that sugar concentrations greater than 7% were not fermented efficiently. This value was confirmed by Tsao *et al.* (1962; Tsao and Su, 1964) and by Pfeifer *et al.* (1953, 1957). Kobayashi (1967) has questioned the economics of the process developed by Pfeifer *et al.* (1953, 1957) and considered his continuous fermentation to be preferable. More recently, it was observed that if beet molasses was used in the spore germination medium, it was possible to ferment cane molasses at 15% sugar concentration, with an itaconic acid concentration in the beer of 8.5%, a yield of 85% of the theoretical maximum.

B. Current Manufacturing Process

The current manufacturing process is the one based on the use of molasses. Detailed information on this process has not been published; however, the manufacturer owns the patent of Neubel and Ratajak (1962) which probably provides the best index of current usage.

Stock cultures are maintained on Czapek agar slants or as spores scraped off these slants and suspended in garden soil. Spores for inoculum are grown on either a liquid or solid vegetable medium. Spores are germinated in a solution of the composition shown in the tabulation below.

Component	Amount
	Sugar-beet molasses to give 15% sugar
$ZnSO_4$	1.5 gm
$MgSO_4 \cdot 7 H_2O$	5.0 gm
$CuSO_4 \cdot 5 H_2O$	0.02 gm
Soybean oil	0.25 ml
Water to	1 liter

Aeration at one-fourth v/v solution and high-speed agitation are maintained for about 18 hours; the temperature is held at 33°–37°C. Under these conditions spore germination results in the formation of mycelia of a suitable biochemical type. The pH changes from about 7.5 to 4.0.

The fermentation solution is inoculated with one-fifth volume of the suspension of germinated spores. This is an excessively large volume of inoculum but is probably desirable on account of the high temperature of the fermentation (39°–42°C), and to obtain enough beet molasses into the solution to give the benefits of beet molasses. The composition of the fermentation solution is shown below:

Component	Amount
	Cane molasses to give 150 gm sugar
$ZnSO_4$	1 gm
$MgSO_4 \cdot 7 H_2O$	3.0 gm
$CuSO_4 \cdot 5 H_2O$	0.01 gm
Water to	1 liter

Air is bubbled through the culture solution, and vigorous agitation is applied. During the first 24 hours, the pH changes from about 5.1 to 3.1. Lime or ammonia is added to readjust the pH to about 3.8, and the fermentation is continued for 2 days more. The final itaconate concentration is about 85 gm/liter. Use of ammonia is preferable to use of lime because calcium itaconate has low solubility in the solution, and much of it adheres to the mycelium during filtration. On the other hand, if ammonia is the neutralizing agent, ammonium itaconate crystallizes out with the itaconic acid on evaporation of the solution. This is not too serious because the major market is for the esters which are distilled off on heating with sulfuric acid.

In another process (Batti and Schweiger, 1964) spore germination and growth of *A. terreus* is restricted by adding $CuSO_4 \cdot 5 H_2O$ to the decationized syrups. Sugar concentration as great as 30% can be used with final beer concentrations of itaconic acid reaching 180 gm/liter. Ten to fifteen percent of the acidity may be succinic and itatartaric acids. Alkali earth metal salts added to the initial culture solution restrict the accumulation of these acids to trace amounts and result in improvement in itaconic acid yield (Table III).

In several characteristics both processes are similar and resemble the citric fermentation. Copper ion restricts growth and product destruction in both processes. The concentrations of copper ions in the two processes are similar to those used in the citric fermentation. In both processes excessive concentration of iron in the culture solution results in reduction of the accumulation of itaconic acid (Table IV); a similar

TABLE III. Effect of Addition of Alkaline Earth Metals on Itaconic Acid Accumulation by *Aspergillus terreus*[a]

Metallic ion added		
(%) Type	Amount (mg/liter)	Weight yield[b]
None		9
Calcium	337	43
Calcium	2700	59
Magnesium	337	20
Magnesium	2700	48

[a] From Batti and Schweiger (1964).
[b] (Grams itaconic acid produced/grams sugar supplied) × 100. Itaconic acid determined by bromination.

effect is observed in the manufacture of citric acid. Both itaconic fermentations and the citric fermentation require continuous aeration, and stoppages of air of very brief duration may be enough to stop the fermentation. Fermentation will not be rapidly resumed until new growth of mycelium in the presence of copper ion in inhibitory concentration has occurred.

Saccharinic acid-type compounds present in molasses may in part substitute for the copper ion requirement in the present manufacturing process. A similar affect has been observed in the citric fermentation.

In either process whenever the concentration of itaconic acid exceeds 7% the fermentation rate is much reduced. This is not an effect of pH as no significant change occurs at this time. If the excess titratable acidity is neutralized with ammonia or NaOH, the fermentation rate remains constant, and total itaconate and itaconic acid concentrations may reach 15–18%.

TABLE IV. Effect of Iron on the Production of Itaconic Acid by *Aspergillus terreus*[a]

Iron (mg/liter)	Weight yield[b] (%)
0	57
1	25
2	17
4	17

[a] From Batti and Schweiger (1964).
[b] (Grams itaconic acid produced/grams sugar supplied) × 100

C. Product Recovery

If part of the acid has been neutralized with NH_4OH, NaOH, or $Ca(OH)_2$ during the fermentation, the solution must be acidified with an inorganic acid prior to evaporation. Mycelium and any suspended solids are removed by filtration on any conventional industrial precoat filter with a filter aid precoat. If a refined-grade free acid product is made, the hot solution is treated with activated carbon and filtered. This filtrate is then evaporated and cooled to crystallize. A second crop of crystals may be obtained on further evaporation of the mother liquor, but these are industrial grade. When only the industrial grade is made, the carbon treatment may be omitted since this product will be esterified and the ester distilled. The mother liquor obtained from the crystallization steps may be extracted with a solvent, such as n-amyl alcohol, which has a favorable partition coefficient. The solvent extracts too much color, however, and further carbon treatment may be required. However, carbon treatment of the solvent may be much less effective than carbon treatment of the aqueous solution. Alternatively the mother liquor may pass through an anion-exchange resin bed, and the itaconic acid eluted with mineral acid. Crude itaconate is suitable for esterification without further purification or may be precipitated from fermentation filtrates by liming and the calcium precipitated with sulfuric acid. The itaconic acid crystallizes on concentration. Crude itaconic acid obtained from mother liquor of recrystallization may be esterified directly or, if a higher purity acid is desired, recrystallized from hot water after decolorization with carbon and filtration. This requires considerable care in temperature control to avoid large losses on the filter. A flow sheet is given in Fig. 3.

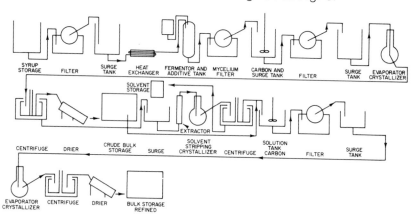

FIGURE 3. Flow sheet for itaconic acid manufacture. (From Lockwood and Schweiger, 1967; reprinted by permission of the Van Nostrand-Reinhold Co., New York.)

D. Chemistry of Itaconic Acid Fermentation

As in the citric acid fermentation, it is very probable that the EMP glycolytic pathway is followed as far as pyruvic acid. Here a variation occurs. Instead of acetyl-CoA combining with oxalacetate to form citric acid, acetyl-CoA condenses with pyruvic acid to form citramalic acid (Jakubowska and Metodiewa, 1974) (Fig. 4). This is then converted to itatartaric acid which is reduced to itaconic acid. It was formerly believed that itaconic acid was formed by loss of CO_2 from *cis*-aconitic acid of the citric acid cycle (Bentley and Theisen, 1957; Casida, 1968; Berry, 1975). *Aspergillus terreus* has the enzyme system for this sequence of reactions when grown in the absence of copper ion, and at a higher pH than that of the itaconic fermentation. If the conversion of citric acid to *cis*-aconitic acid is inhibited by copper ion which inhibits the action of aconitase, it should also result in reduction in itaconic acid yield. Actually addition of copper ion results in an increase in yield. Although either mechanism can result in the accumulation of itaconic acid, it is doubtful if the decarboxylation of *cis*-aconitic acid contributes significantly to itaconic acid accumulation. The role of copper ion must be the inhibition of some enzyme involved in the catabolism of itaconic acid.

As in the citric acid fermentation, aeration requirements are small, but aeration must be continuous. Cessation of aeration for as little as 15

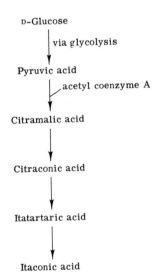

FIGURE 4. Proposed metabolic sequence in the production of itaconic acid.

seconds stops the fermentation and it will not be resumed until new mycelial growth of suitable biochemical type has occurred.

IV. LACTIC ACID

Lactic acid was first isolated by Scheele (1789) from sour milk. Many studies on the physical and chemical properties of the compound established its structure and applications and found that the compound occurred in two isomeric forms and as a racemic mixture of these. It was also found that the acid is produced in muscle when work is done and that the acid is produced by many microorganisms by the fermentation of sugars.

$$
\begin{array}{c}
COOH \\
| \\
CHOH \\
| \\
CH_3
\end{array}
$$

Lactic acid

A. Technological Development

Lactic acid was the first organic acid to be manufactured industrially by fermentation. In 1880 a plant was built in Littleton, Massachusetts, to manufacture calcium lactate by fermentation. Little is known about the process beyond the fact that calcium lactate was manufactured and marketed. The Clinton Processing Co., Clinton, Iowa is the only manufacturer using the fermentation process in the United States.

Numerous types of microorganisms have been isolated which accumulate lactic acid or lactates in their culture solutions. Homofermentative lactobacilli, the ones which produce only cells and lactic acid in significant amounts, are used commercially, although algae, yeasts, and phycomycetous fungi are also known to produce the acid. Yields comparable to those obtained by homofermentative lactic acid bacteria have been obtained only in glucose cultures of the fungus *Rhizopus oryzae*. The homofermentative lactic acid fermentation produces almost 2 moles lactic acid per mole of glucose supplied. The common heterofermentative types have not been used on account of their much lower yields although they occur not infrequently as contaminants.

B. Isomers

Lactic acid of commerce is the L(+)-isomer, D(−)-isomer, or any possible mixture of the two. The entire range of isomers has been found.

The capital letters prefixed to the names of the isomers indicate configuration in relation to the isomers of glyceraldehyde, and the symbols (+) and (−) indicate the direction of rotation of a plane of polarized light. The mixture of isomers is now called DL-lactic acid. Most lactic acid-producing organisms produce only one isomer but a few bacteria, such as *Lactobacillus plantarum,* which are frequent contaminants produce enzymes which convert one isomer to a mixture of the two. These enzymes, called racemases, are now known to be lactic dehydrogenases which maintain an equilibrium between the lactic acid isomers and pyruvic acid (Dennis and Kaplan, 1960). When both dehydrogenase enzymes are present, racemization occurs.

Identification of the isomers by polarimetric methods is made difficult by the low optical activity of the pure isomers and by their spontaneous polymerization in aqueous solution. The dimers, polymers, salts, and esters have rotation reversed to that of the free acid solutions. This is due to the occurrence of an epoxide form (Lockwood et al., 1965). Observed rotations of the free acids are of direction opposite to that of the monomers on account of the presence of the epoxide.

C. Current Manufacturing Process

Thirty years ago five lactic acid fermentation plants operated in the United States. Today only one remains. Much lactic acid is made by hydrolysis of lactonitrile, a by-product of another process.

Peckham (1944) and Schopmeyer (1954) have described the lactic acid fermentation process. Little change has occurred since these publications appeared. Lactic acid is very corrosive to metals. Consequently, wooden fermentors have been used in most plants. These were uncovered or were covered with loosely fitting wooden lids. They were steamed empty before charging, and fermentation solutions were pasteurized or sterilized by passage through steam-jacketed heat exchangers. Contamination of culture solutions by thermophilic clostridia resulted in the production of some butanol and butyric acid contaminating the lactic acid. The product could be sold only to leather tanners for deliming of hides. Such contamination actually forced one manufacturer out of the lactic acid business.

Inoculum cultures of *Lactobacillus delbrueckii* are transferred from test tubes through successively larger culture vessels held at 45°–55°C. Each stage of culture buildup requires 16–18 hours. A slight excess of $CaCO_3$ is present in each stage. Inoculum volume is usually about 5% of fermentation solution volume.

The fermentation solution is usually hydrolyzed starch or dextrose syrups, although D-glucose, maltose, lactose, or sucrose can be fermented. The sugar concentration does not exceed 12%. At a greater sugar concentration, calcium lactate may crystallize in the fermentor as a solid mass which is difficult to handle.

Lactic acid bacteria have complex requirements for B vitamins. This ordinarily is met by enrichment of the culture solution with crude vegetable materials. Malt sprouts, the rootlets produced by germinating barley grains in the malting process, are the most commonly used vegetable material. If they have been overheated in drying, they lose some of their value as a nutrient for lactobacilli. The culture solution contains $CaCO_3$ to neutralize 2 moles lactic acid per mole hexose supplied, or a slight excess. After inoculation the solution is slowly stirred to keep the $CaCO_3$ suspended, but aeration of the solution is not necessary.

Plant fermentation time is usually 5–10 days, varying from batch to batch with uncontrolled conditions. An excess of $CaCO_3$ in the culture solution keeps the pH in the range of 5.5–6.5. The pH necessary varies somewhat with the composition of the culture solution. Control of pH between 6.3 and 6.5 by continuous neutralization with a slurry of $Ca(OH)_2$ will permit the complete fermentation of 12–13% glucose in 72 hours. It is very important that residual sugar be reduced to 0.1% or less during the fermentation because residual sugar makes recovery of the better quality lactic acid more difficult. Commercial fermentation yields are 93–95% of the weight of the glucose supplied. Recovery yields vary with the various recovery processes and product grades.

D. Product Recovery and Grades

The suspended solids and most of the bacteria are removed from the solution on any conventional industrial precoat filter. A polishing filtration may also be required. For technical grade product, calcium is precipitated as $CaSO_4 \cdot 2 H_2O$ which is filtered off. On evaporation of the filtrate to 35–40% lactic acid content, more $CaSO_4 \cdot 2 H_2O$ precipitates and is removed by filtration. Further evaporation to 44–45% total acidity gives the technical-grade product.

Food-grade lactic acid is a pale yellow, straw-colored solution of about 50% total acidity. The acid is made by fermenting a higher grade of sugar, with minimal protein content of the culture solution. Calcium is precipitated as $CaSO_4 \cdot 2 H_2O$ and washed, and the filtrate and wash water are combined. The solution is bleached with activated carbon, evaporated to about 25% solids, then bleached again, further evaporated to about 50% total acidity, and further bleached. If the final product is off-color, it may be treated with ferrocyanide to precipitate iron or copper ions and filtered. An alternative procedure is the evaporation of the fermented beer to recover calcium lactate which is washed with cold water in a centrifuge before decomposing with H_2SO_4. Further processing is as above.

Plastics-grade lactic acid is a colorless product which can be obtained by esterification with methanol after concentration. Another procedure is solvent extraction with isopropyl ether followed by reextraction of the isopropyl ether with water (Anonymous, 1959).

Lactic acid, USP, is a colorless solution of 85% total acidity, 76–78% lactic acid concentration. It contains 2–3.5% volatile acids and 0.5–1.0% ash. Approximately 30% of the acid occurs as polymers, either lactide, linear polymers, or both.

Entry into the market of DL-lactic acid obtained as a by-product altered the economic value of the fermentation process very unfavorably, especially in the paler grades. In the event that a shortage of hydrocarbons occurs, lactonitrile will be in short supply or available only at higher cost. In this instance, fermentation lactic acid will again be economically competitive. However, it appears probable that the production of L(+)-lactic acid by *Rhizopus* will be preferred on account of the much simpler recovery process and shorter fermentation time. Yields of lactic acid obtained by use of homofermentative lactobacilli and *Rhizopus* are identical but the 48-hour fermentation by *Rhizopus* gives an almost colorless fermented solution. The lactic acid on acidulation of the solution with H_2SO_4 and filtration to remove $CaSO_2 \cdot 2\,H_2O$ will crystallize on evaporation. Crystals so obtained are colorless and are not hygroscopic but tend to liquefy by loss of water due to polymerization.

V. GLUCONIC ACID AND GLUCONO-δ-LACTONE

D-Glucono-δ-lactone, the simplest of the direct dehydrogenation products of D-glucose, and its free acid form, gluconic acid, are produced by a large variety of bacteria and fungi. Both types of microorganisms are used commercially. Bacterial accumulation of gluconic acid was observed by Alsburg (1911) in cultures of a bacterial parasite of olive trees now known as *Pseudomonas savastanoi*. Molliard (1922) described its occurrence in cultures of *A. niger*. Fungi which produce D-glucono-δ-lactone and D-gluconic acid are found in the genera *Aspergillus, Penicillium, Gliocladium, Scopulariopsis, Gonatobotrys,* and *Endomycopsis*. Doubtless many other fungi will be found to produce these compounds. Bacteria which produce D-glycono-δ-lactone and D-gluconic acid are found among the genera *Pseudomonas, Vibrio,* and

Gluconobacter. The organisms commonly used in this fermentation are *A. niger* or *Gluconobacter suboxydans.* The larger volume production uses the fungal process. Most of the *Gluconobacter* production is marketed as glucono-δ-lactone for use in baking powders and effervescent products.

The equilibrium of the lactone and the free acid in solution is readily controlled by pH and temperature. The relative amounts of lactone and free acid can be determined approximately by titration as rapidly as possible up to the first end point with phenolphthalein as indicator with standard alkali. An excess of standard alkali is added to the solution heated to at least 53°C and then back-titrated with standard acid. The difference between initial and total alkali titer represents the lactone, while the initial titer represents the free acid. The lactone may be determined more precisely and independently by the hydroxamate method of Hestrin (1949). Some *Pseudomonas* strains produce a lactonase which hydrates the lactone and the lactone does not accumulate in their culture solutions. The lactone may also arise spontaneously in solutions of the free acid, but the fermentation product is the δ-lactone.

A. Fermentation Processes

Only submerged process technology has been used for many years, although at one time at least one manufacturer in the United States made gluconates by a surface liquid culture process. The manufacturers of gluconic acid and gluconates in America use the fungal process. Only one chemical process may be competitive. Although numerous chemically mild oxidizing agents convert D-glucose to D-gluconic acid, they are usually too costly, even though a lesser initial capital investment may be required. Manufacturers of gluconic acid or its salts in the United States are Bristol-Meyers Co., Syracuse, New York; Pfizer Inc., New York, New York; and Premier Malt Products Inc., Milwaukee, Wisconsin.

B. Calcium Gluconate Fermentation

Calcium gluconate is manufactured primarily to meet the pharmaceutical market. The solubility of calcium gluconate in water is about 4% at 30°C (Ward, 1967). Supersaturation of this salt commonly occurs in fermentation solutions up to about 15% or more glucose concentration. Consequently this imposes a limitation on initial glucose concentration to about 13–15% (Moyer *et al.*, 1940). The calcium source must also be sterilized separately from the glucose solution to avoid the Lobry–du Bryn–Van Ekenstein reaction, which alters the conformation of about

30% of the glucose, thus reducing the yield. This reaction occurs slowly at pH 5.5 and at values above pH 7 is almost instantaneous. The neutralizing agent must be added aseptically to the sterile glucose solution at intervals after inoculation. If added before spore germination, spores will not germinate; if added to mycelium in large enough quantity to make the solution alkaline the mycelium will be killed. Addition of $CaCO_3$ slurry must be delayed until enough glucono-δ-lactone has been formed to buffer the solution against the high alkalinity of a heat sterilized slurry of $CaCO_3$. On sterilization of an aqueous suspension of $CaCO_3$ with steam under pressure, considerable $Ca(OH)_2$ is formed. Dry sterilization of $CaCO_3$ in large scale has not proved practical.

Moyer et al. (1940) tried to increase the solubility of gluconate by adding borates or boric acid to the fermentation solutions. Calcium borogluconate is formed. Although one company used this procedure for a short time, the borogluconate was found to be deleterious to blood vessels of animals, and the product was withdrawn from the market.

Culture solutions used for this fermentation are D-glucose solutions to which several inorganic salts and corn steep liquor have been added. The solution has not been deionized. Much variation in mineral salt concentration and corn steep liquor concentration can be tolerated with little effect on yield. Most if not all the mineral salt nutrient requirements are supplied by the corn steep liquor.

The beneficial effects of growing the mold under increased air pressure were demonstrated by May et al. (1934), who studied the fermentation in submerged aerated cultures at various pressures above atmospheric. Pressures up to 3 atm above sea level pressure gave more rapid fermentations than cultures at 1 atm pressure. Most of the advantage obtained was found at 2 atm gauge pressure, and no further improvement was found between 45 and 105 lb gauge pressure.

C. Sodium Gluconate Fermentation

The major differences between the production of calcium and sodium gluconate are in the concentrations of initial glucose and final concentration of gluconate (Fig. 5) and in pH control. In the calcium gluconate fermentation pH control results from the $CaCO_3$ slurry added to the solution during the fermentation, and in the sodium gluconate fermentation pH control is accomplished by automatic addition of NaOH solution. Sodium gluconate is much more soluble than is calcium gluconate (Table V). Consequently, the controlled automatic addition of NaOH provides an easy and precise means of neutralizing the acid as it is formed. Again, it is important that the pH of the fermentation solution

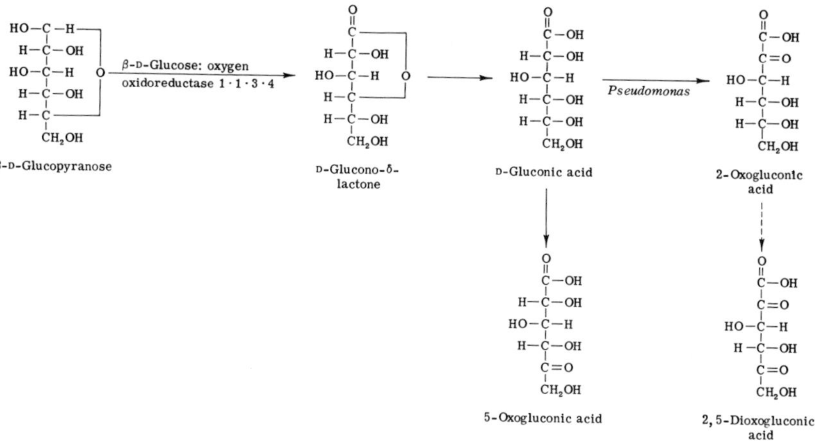

FIGURE 5. Gluconate and oxogluconate production.

should never be less acid than pH 6.5 after inoculation. The method described by Blom et al. (1952), developed at the Northern Regional Research Laboratory of the United States Department of Agriculture, appears to be typical of the industrial production of sodium gluconate. Efficient oxygenation of the fermentation solution is ensured by the use of 1.0–1.5 volumes air per volume solution in baffled fermentors equipped with turbomixers. The air is introduced directly beneath the turbomixer blade. A pressure of 30 psig is maintained. The fermentation solution must be cooled by circulation of water through coils inside the fermentor or in the jacket of the tank.

A typical fermentation solution contains, per liter, glucose (anhydrous basis), 250–350 gm; corn steep liquor, 3.7 gm; $MgSO_4 \cdot 7 H_2O$, 3.5–4.0

TABLE V. Solubility of Sodium Gluconate in Water at Various Temperatures[a]

Temperature (°C)	Sodium gluconate in saturated solution (wt. %)
15	34.6
20	36.7
30	39.6
50	45.9
60	49.9
70	53.0
80	56.9

[a] From Ward (1967).

gm; KH_2PO_4, 0.2–0.3 gm; $(NH_4)_2HPO_4$, 0.4 gm; urea, 0.1 gm; and H_2SO_4 to pH 4.5. Sterilization may be done by direct steam injection, steam in the jacket, or steam around the tubes of an heat exchanger through which the solution flows. After sterilization, the pH is automatically adjusted and maintained throughout the fermentation at pH 6.5. The complete conversion of 300 gm glucose per liter to gluconate requires about 36 hours when the inoculum is germinated spores.

Many strains of A. niger have been examined, and nearly all produce a mixture of gluconic acid, citric acid, and oxalic acid. Citric acid contamination of the product may be eliminated by the manganese content of the corn steep liquor, but the presence of sodium ion favors oxalic acid production by many strains of A. niger. Strain NRRL 3, the one used by Blom et al. (1952), appears to be satisfactory in that only glucono-δ-lactone and gluconic acid are produced under the conditions described above.

Glucono-δ-Lactone Fermentation

In the Gluconobacter suboxydans fermentation, a 10% concentration of glucose is converted to glucono-δ-lactone and free gluconic acid in about 3 days in the absence of pH control. Little is known about this fermentation, but it appears that either corn steep liquor or cooked yeast cells provide all the essential nutrients. Approximately 40% of the gluconic acid is in the δ-lactone form. Most of the gluconic acid is converted to the δ-lactone during recovery operations.

It has been informally reported to the writer that one major manufacturer in the United States now recovers the lactone directly and converts it to the desired salt by heating with the hydroxide of the desired cation. No details of the fermentation were reported but it has been known for more than 50 years that such fermentation is possible with A. niger or with one of several penicillia as well as by Gluconobacter.

D. Product Recovery

1. Calcium Gluconate

Calcium gluconate may be recovered by removal of the mycelium and any other suspended solids on any conventional industrial filter, carbon treatment of the filtrate to decolorize it, filtration to remove carbon, and evaporation to a 15–20% calcium gluconate concentration. On cooling to a temperature just above 0°C, most of the calcium gluconate in the highly supersaturated solution crystallizes. Seeding with calcium gluconate crystals may be necessary in order to obtain rapid crystallization.

The needle crystals tend to form a spongy mass so pressure filtration or centrifugation in a perforate basket centrifuge is necessary to remove most of the mother liquor. Washing should be done with cold water.

2. Sodium Gluconate

Sodium gluconate in the commercial grades may be prepared from filtrates of the sodium gluconate fermentation by concentrating to 42–45% solids, adjusting to pH 7.5 with NaOH, and drum-drying (Ward, 1967). A more refined product may be made by carbon treatment of the hot solution before drying, or by recrystallization from the crude drum-dried product after carbon treatment.

3. Crystalline Lactones of Gluconic Acid and the Crystalline Acid

Aqueous solutions of gluconic acid are an equilibrium mixture of free gluconic acid, glucono-δ-lactone, and glucono-γ-lactone. Crystals separating out of a supersaturated solution at a temperature below 30°C (preferably near 0°C) will be predominantly free gluconic acid; from 30° to 70°C the crystals will be principally glucono-δ-lactone, and at temperatures greater than 70°C they will be principally the γ-lactone. Seeding with the desired compound and close control of conditions are required to obtain crystals of high purity (Ward, 1967).

4. Free Gluconic Acid Solution

The standard 50% solution of gluconic acid may be prepared by treatment of the calcium gluconate solutions with H_2SO_4 and heating to precipitate calcium sulfate, filtration, and concentration of the filtrate to about 50% solids. If any residual sugar or fungal polysaccharide is present, carbon treatment may be necessary.

E. Market Data

Gluconic acid is usually marketed as a 50% technical-grade aqueous solution (w/w). It finds use as a mild acidulant in metal pickling and in foods. Calcium gluconate is a readily available form of soluble calcium widely used for oral and intravenous therapy. Sodium gluconate is an excellent sequestering agent in neutral or alkaline solutions, where it finds use in the cleansing of glassware, such as returnable bottles and dishes, and as a constituent of compounds for washing walls.

Prices of gluconic acid, calcium gluconate, and sodium gluconate in March 1976 varied considerably with market form (U.S. Tariff Commission, 1976). Technical-grade sodium gluconate sold for $0.80 per pound in 100-lb bags. It was sometimes blended with NaOH to make the alkaline product used in bottle washing solutions and was sold at a slightly lower price. Calcium gluconate USP grade in ton lots packed in 100-lb drums brought $0.75 per pound. The price of the free acid sold as a 50% solution and delivered in the United States was $0.33–0.355 per pound. The sodium salt is the principal market form, amounting in 1974 in the United States to 8,169,000 lb (U.S. Tariff Commission, 1974).

1. Specifications

The specifications shown in Table VI are typical of those now prevailing. Calcium gluconate must meet USP standards.

TABLE VI. Specifications[a]

Substance	USP Standard
Sodium gluconate	
Appearance	White to tan powder
Purity	98.8–99.8%
Moisture	0.1–0.2%
Reducing substances	0.1–0.5%
Solubility, 20°C	
In water	Very soluble
In ethanol	Slightly soluble
In ether	Insoluble
Aqueous solution	pH 5.3–6.3
Stability, 0°–100°C	Stable
Glucono-δ-lactone	
Specific rotation	
(freshly prepared) $[\alpha]20$	+64°
Decomposition point	158°–163°C
Solubility, 20°C	
In water	Very soluble
In ethanol	Slightly soluble
In ether	Insoluble
Appearance	White crystalline powder
Taste	Sweet
Gluconic acid (technical 50%)	
Color	Light yellow
Gluconic acid content	49–51%
Ash	Not over 0.4%
Specific gravity	1.23–1.24
Sulfate	Trace
Chloride	Trace
Reducing substances	Not more than 3% on 100% acid basis
	Nontoxic

[a] From Ward (1967).

2. Toxicity

Storage above 18°C is recommended to prevent crystallization of the acid or the lactone.

3. Uses

The principal market for gluconic acid and its derivatives is for the sodium salt. Hydroxyl groups of the gluconate molecule bind di- and trivalent metallic ions in soluble form, thus preventing precipitation of calcium, magnesium, or iron compounds from solutions used in bottle and dish washing. Alkaline bottle washing compounds which contain 2.5–10% sodium gluconate, depending on the hardness of the water, are used extensively in automatic cleansing equipment. Metal carbonate precipitates are removed by sodium gluconate containing products for the washing of painted walls or metals without inducing appreciable corrosion. When added to cement, gluconates control setting time and increase the strength and water resistance of the cement.

In pharmaceutical use, calcium gluconate has long been used as a preferred source of calcium in cases of calcium deficiency due to diet or pregnancy, and for rapid relief of symptoms of allergies. Many drugs can be administered more effectively when combined with gluconates.

Glucono-δ-lactone finds use as a constituent of baking powders, bread mixes, and other foods where its latent acidogenic properties may be useful.

VI. OXOGLUCONATES

Three oxogluconates are accumulated in glucose culture by bacteria. *Gluconobacter suboxydans* converts D-glucose into a mixture of 2-oxogluconate and 5-oxogluconate. 5-Oxogluconate predominates, and 2-oxogluconate may be found in only traces. Many species of *Pseudomonas* accumulate 2-oxogluconate in D-glucose cultures. Weight yields as great as 90% are not uncommon. Katznelson *et al.* (1953) produced 2,5-dioxogluconate from glucose using *Gluconobacter melanogenum*.

The metabolic scheme (Fig. 5) showing the biochemical mechanism is probably correct. D-Glucono-δ-lactone is an intermediate in the production of D-gluconates by these organisms. D-Gluconate salts are a good substrate for *Pseudomonas* growth and conversion to 2-oxogluconate.

2-Oxogluconate is the only one of the oxogluconates that has been manufactured commercially. Operating conditions are the same as those used in the fungal production of calcium gluconate except that 5 ml of corn steep liquor or 0.3 gm of yeast extract may be added per liter of culture solution. 2-Oxogluconate formation follows quickly on the initial formation of gluconate. Figure 6 illustrates the simultaneous kinetic pattern of the fermentation of 10% glucose medium under 30 lb gauge pressure of air in the presence of an excess of CaCO$_3$ (Stubbs *et al.*, 1940). Yields of calcium 2-oxogluconate equivalent to 90% of theory have been obtained but industrial yields are believed to be slightly less than this. Commercial usage calls for the fermentation of 20% glucose solutions by *Pseudomonas fluorescens* strains acclimatized to high osmotic pressure. In order to maintain acclimatization, stock cultures are maintained on 20% sodium gluconate agar slants, and seed buildup plant stages all require 20% glucose solutions.

In a few individual culture strains, the fermentation does not proceed beyond the gluconate stage. In others, although 2-oxogluconate accumulates as the major metabolite, if the fermentations are prolonged δ-oxoglutaric acid, succinic acid, and pyruvic acid accumulate in considerable quantity (Lockwood and Stodola, 1946).

2-Oxogluconic acid is usually recovered and shipped in the free acid state after centrifugation or filtration to remove cells. Calcium is removed

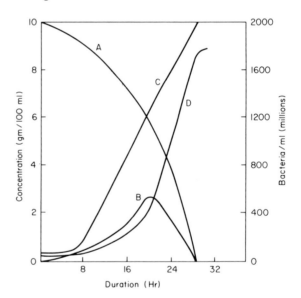

FIGURE 6. Composition of fermenting culture solution at various ages for 2-oxogluconate production based on data of Stubbs *et al.* (1940). Curve A: glucose in solution (gm/100 ml); curve B: glucono-δ-lactone and gluconate in solution (gm/100 ml); curve C: 2-oxogluconate in solution (gm/100 ml); curve D: *Pseudomonas* cells per milliliter (millions).

by precipitation with H_2SO_4. The filtered acid may be shipped as a syrup or as a crystalline material after evaporation under reduced pressure at a temperature which must never exceed 50°C. The material may be shipped as esters since esterification is necessary to conversion to erythorbic acid.

Productions of erythorbic acid, its esters, and its salts are the major markets for 2-oxogluconic acid. They are used as water- or fat-soluble antioxidants to retain color, flavor, and nutritive values in canned fruits and vegetables and in meats and meat products.

Simultaneous analysis of fermenting solutions which contain calcium salts of two acids, gluconic and 2-oxogluconic, and two compounds which reduce alkaline copper sulfate is done by a method developed by G. E. Ward (Stubbs *et al.*, 1940) which distributes the copper-reducing activity on the basis of the optical rotation of the plane of polarized light on passing through a standard layer of the solution.

5-Oxogluconate has not been manufactured on a commercial scale but is of interest as a potential source of L(+)-tartaric acid.

VII. MISCELLANEOUS ACIDS FORMERLY MADE BY FERMENTATION

Examination of the history of lactic acid production in the United States suggests the possibility that lactic acid may someday be transferred to this category. However, only two organic acids of any volume now belong to this group.

The manufacture of fumaric acid, formerly done by fermentation, is now entirely done by oxidation of aromatic hydrocarbons. When the time comes that hydrocarbons are in very short supply, or at too high a price, further consideration of fermentation methods for the production of fumaric and lactic acids is to be expected. Fumaric acid was made by the aerobic fermentation of glucose by selected strains of *Rhizopus* in the presence of $CaCO_3$. It is also found in cultures of *Mucor* and *Cunninghamella*. Most strains of *Rhizopus* produce L(+)-lactic acid under the ordinary fermentation conditions, or a mixture of lactic and fumaric acids. Strain selection offers the promise of improvement in yield. Only inorganic nutrients other than the carbon source are required. Calcium fumarate has low solubility in cold water, so it precipitates among the filaments of mycelium. It can be extracted with hot water by heating the culture solution before filtration.

The other organic acid made by fermentation in the United States about 1916 was gallic acid. Tannase of *Aspergillus wentii* was produced in deep tanks with agitators by growing the fungus on an extract of gallotannins. Aeration was discontinued after the third day, more tannins were added, and agitation was stopped. Excreted tannase hydrolyzed the gallotannins. The gallic acid so obtained was used in blue pigments.

REFERENCES

Alsburg, C. L. (1911). *J. Biol. Chem.* **9,** 1–7.
Amelung, H. (1930). *Chem.-Ztg.* **44,** 118.
Anonymous (1959). *Chem. & Eng. News,* June 15.
Anonymous (1963). *Chem. Week,* Sept. 21.
Batti, M. A. (1966). U.S. Patent 3,290,227.
Batti, M. A., and Schweiger, L. B. (1964). U.S. Patent 3,078,217.
Bentley, R. C., and Theisen, C. (1957). *J. Biol. Chem.* **226,** 703–720.
Berry, D. R. (1975). *In* "The Filamentous Fungi, Industrial Mycology" (J. E. Smith and D. R. Berry, eds.), Vol. I. Arnold, London.
Blom, R. H., Pfeifer, V. F., Moyer, A. J., Traufler, D. H., Conway, D. H., Crocker, H. F., Farison, R., and Hannibal, D. B. (1952). *Ind. Eng. Chem.* **44,** 435.
Blumenthal, H. (1965). *In* "The Fungi: An Advanced Treatise" (G. C. Ainsworth and A. S. Sussman, eds.), Vol. I. Academic Press, New York.
Bruchmann, E. E. (1961). *Biochem. Z.* **335,** 199–211.
Calam, C. J., Offord, A. C., and Raistrick, H. (1939). *Biochem. J.* **33,** 1488–1495.
Casida, L. E. (1968). "Industrial Microbiology." Wiley, New York.
Clark, D. S. (1962). *Biotechnol. Bioeng.* **4,** 17–21.
Cleland, W. W., and Johnson, M. J. (1954). *J. Biol. Chem.* **208,** 679–689.
Dennis, D., and Kaplan, N. (1960). *J. Biol. Chem.* **235,** 810–818.
Friedkin, M. (1945). *Ind. Eng. Chem., Anal. Ed.* **17,** 637–638.
Grimaux, P., and Adams, P. (1880). *C. R. Hebd. Seances Acad. Sci.* **90,** 1252.
Hestrin, S. (1949). *J. Biol. Chem.* **180,** 249–261.
Hisanga, W., and Nakamura, S. (1966). Japanese Patent 16555/66.
Jakubowska, J., and Metodiewa, S. (1974). Final Report, Grant No. FG-P1-244. U.S. Dept. of Agriculture, Lodz, Poland.
Katznelson, H., Tannenbaum, S. W., and Tatum, E. L. (1953). *J. Biol. Chem.* **204,** 43–59.
Kinoshita, K. (1931). *Bot. Mag.* **45,** 45–61.
Kinoshita, K. (1937). *Acta Phytochim.* **9,** 129.
Kluyver, A. J., and Perquin, L. H. C. (1933). *Biochem. Z.* **266,** 668–681.
Kobayashi, T. (1967). *Process Biochem,* Sept., pp. 61–65.
La Nauze, J. M. (1966). *J. Gen. Microbiol.* **43,** 73–81.
Lockwood, L. B., and Batti, M. A. (1965). U.S. Patent 3,189,527.
Lockwood, L. B., and Schweiger, L. B. (1967). *In* "Microbial Technology" (H. J. Peppler, ed.), pp. 183–199. Van Nostrand-Reinhold, Princeton, New Jersey.
Lockwood, L. B., and Snell, R. L. (1976). "Development of Miles Laboratories Process for the Manufacture of Citric Acid," Miles Analecta, Vol. 3, No. 2, pp. 23–26.
Lockwood, L. B., and Stodola, F. H. (1946). *J. Biol. Chem.* **164,** 81–83.
Lockwood, L. B., Yoder, D. E., and Zienty, M. (1965). *Ann. N.Y. Acad. Sci.* **119** 844–867.
Mallea, O. (1950). *Ind. Quim.* **12,** 264–280.
May, O. E., Herrick, H. T., Moyer, A. J., and Wells, P. A. (1934). *Ind. Eng. Chem.* **26,** 575–578.

Meyrath, J. (1967). *Process Biochem.* Oct., pp. 25–28.
Mial, L. M. (1975). *In* "The Filamentous Fungi, Industrial Mycology" (J. E. Smith and D. R. Berry, eds.), Vol. I. Arnold, London.
Molliard, M. (1922). *C. R. Hebd. Seances Acad. Sci.* **174,** 881–883.
Moyer, A. J. (1953). *Appl. Microbiol.* **1,** 1–13.
Moyer, A. J., Umberger, E. J., and Stubbs, J. J. (1940). *Ind. Eng. Chem.* **32,** 1347.
Neubel, R. D., and Ratajak, E. J. (1962). U.S. Patent 3,044,941.
Peckham, G. T. (1944). *Chem. & Eng. News* **22,** 440–443.
Peckham, G. T., and Filachione, E. M. (1962). *Kirk-Othmer Encycl. Chem. Technol., 2nd Ed.,* Vol. 12, pp. 170–188.
Perquin, L. H. C. (1938). Dissertation Technische Hoogeschool, W. D. Mainema, Delft.
Pfeifer, V. F., Nelson, G. E. N., Vojnovich, C., and Lockwood, L. B. (1953). U.S. Patent 2,657,173.
Pfeifer, V. J., Vojnovich, C., and Heger, E. N. (1957). *Ind. Eng. Chem.* **44,** 2975–2980.
Scheele, W. (1789). *Crells Ann.* **2,** 1.
Schopmeyer, H. H. (1954). *In* "Industrial Fermentations" (L. A. Underkofler and R. J. Hickey, eds.), Vol. I, pp. 291–419. Chem. Publ. Co., New York.
Schweiger, L. B. (1957). U.S. Patent 2,970,084.
Scüzs, J. (1944). U.S. Patent 2,353,771.
Shu, P., and Johnson, M. J. (1947). *J. Bacteriol.* **54,** 161.
Shu, P., and Johnson, M. J. (1948a). *J. Bacteriol.* **56,** 577.
Shu, P., and Johnson, M. J. (1948b). *Ind. Eng. Chem.* **40,** 1202.
Steinberg, R. A. (1945). *Soil Sci.* **60,** 185–189.
Stubbs, J. J., Lockwood, L. B., Roe, E. T., Tabenkin, B., and Ward, G. E. (1940). *Ind. Eng. Chem.* **32,** 1626–1630.
Tsao, C. Y., Jr., and Su, S. C. (1964). *Sugar y Azucar* **59,** 36.
Tsao, C. Y., Jr., Huang, J. C., and Su, S. C. (1962). *Bull. Acad. Sci. Sinica* [N. S.] **3,** 26–38 and 48.
United States Tariff Commission (1974). *Chem. Mark. Rep.*
United States Tariff Commission (1976). *Chem. Mark. Rep.*
Ward, G. E. (1967). *In* "Microbial Technology" (H. J. Peppler, ed.), pp. 200–211. Van Nostrand-Reinhold, Princeton, New Jersey.
Wehmer, C. (1893). *Bull. Soc. Chim. Fr.* **9,** 728.
Woodward, J. C., Snell, R. L., and Nichols, R. S. (1949). U.S. Patent 2,492,673.
Yamada, K. (1965). "Science in Japan," p. 401. Am. Assoc. Adv. Sci., Washington, D.C.

Chapter 12

Plant Cell Suspension Cultures and Their Biosynthetic Potential*

W. G. W. KURZ
F. CONSTABEL

I. INTRODUCTION

The development of a plant from germination to flowering and the formation of a fruit is governed by an integrated system of internal and external factors, i.e., genotype, interrelationship of cells, tissues and organs, and habitat. An analysis of developmental processes, the ger-

*NRCC No. 15779.

MICROBIAL TECHNOLOGY, 2nd ed., VOL. I
Copyright © 1979 by Academic Press, Inc.
All rights of reproduction in any form reserved. ISBN 0-12-551501-4

mination of seeds, the formation of pigments, or the initiation of buds, for instance, confronts botanists with a complex set of factors which generally resists exact identification and computation. The very gradual change of plant morphology from a mere descriptive to an experimental discipline (Goebel, 1908) and the late emergence of morphogenesis as a unit of research focusing on the biochemistry of plant ontogeny (Wardlaw, 1968) reflect the difficulties of an analytical approach to plant development. Progress had to await the introduction of model systems which would display developmental steps in plant ontogeny in a simplified manner, not necessarily identical to *in situ* situations. The search for a model system resulted in organs, tissues, and cells excised from intact plants and grown under controlled environmental conditions *in vitro*. This cell culture system has successfully been employed in numerous morphogenetic analyses, mainly by varying (1) the origin of the cells and (2) the environment, i.e., physical and chemical external factors (Gautheret, 1959; Reinert and Bajaj, 1977; Steward, 1969; Street, 1973, 1974; White, 1963).

Results of morphogenetic investigations coupled with refinement of methods revealed a potential of cell cultures for application in various fields of plant sciences. As soon as cultured cells were found to be amenable to regeneration of entire plants, large-scale cloning was introduced to horticulture (Murashige, 1974). Cell culture, furthermore, offered a unique way to extend studies in plant pathology to causal relationships in tumor formation (Braun, 1968; Nester, 1977; Sacristan and Melchers, 1970), in host–parasite interaction (Hollings, 1965; Takebe, 1975), and in sanitation of pathogen-infected stock of crop plants (Kartha and Gamborg, 1975). Crop improvement through modification of somatic cells, cell hybridization, is entirely based on cell culture methods (Constabel, 1976; Gamborg *et al.,* 1974; Smith, 1974).

Plant cells when cultured as suspensions have a number of properties in common with cultures of microorganisms: (1) they grow in a sterile environment; (2) they are fairly homogeneous in size, rate of metabolism, and stage of development; (3) they have a doubling time which is longer than that of microorganisms but considerably shorter than of cells *in situ;* and (4) they can be grown on a large scale. In contrast to microorganisms, plant cells feature an immensely greater morphogenetic potential. This potential includes the synthesis and accumulation of plant products, terpenoids, glycosides (steroids, phenolics), and alkaloids. A number of these products are indispensable in the pharmaceutical industry. The idea of employing cell suspension cultures for industrial exploitation, therefore, has been widely considered (Becker, 1969; Furuya *et al.,* 1972; Jones, 1974; Puhan and Martin, 1970; Reinhard, 1967; Staba, 1969; Steck and Constabel, 1974; Zenk, 1976). The problem encountered would be to control environmental conditions of cell suspension cultures in such a manner that their morphogenetic potential is expressed and channeled into the production of desirable substances.

This chapter is intended to describe principles, experiments, and results of plant cell culture as they relate to product formation. For an introduction to detailed techniques which would guide students in establishing and maintaining plant cell cultures, excellent publications for laboratory use are available (Brown and Sommer, 1975; Gamborg and Wetter, 1975; Street, 1973; White, 1963).

II. PLANT CELL CULTURE TECHNIQUES

A. Plant Tissues

Seed plants, which occur in an infinite variety of forms as trees, herbs, or grasses, exhibit three basic morphological units or organs: root, stem, and leaf. The anatomy of these organs reveals coherent masses of cells, associated in tissues. Variation in form and function, differences in cell combination, and topographic distribution in the plant body determine the type of tissue. Parenchyma or ground tissue is the most versatile of all types of tissues.

Parenchyma cells are living nucleated cells, capable of growth and division. They vary in shape and are often isodiametric, but some maybe elongate. Their function is concerned with photosynthesis, storage, wound healing, origin of adventitious structures, and excretion. Parenchyma cells may occur in extensive continuous masses in the cortex and pith of stems and roots, in the mesophyll of leaves and fruits, or in the endosperm of seeds. They are also associated with other types of cells in morphologically heterogeneous tissues, such as phloem and xylem. Perpetually young parenchyma, primarily responsible for the formation of new cells, are the meristems. They are found in shoot and root tips as well as in a conical to cylindrical layer, cambium, in perennial stems and roots. In active meristems, part of the daughter cells remains meristematic, while others develop into the various tissue elements. These cells gradually change physiologically, chemically, and morphologically and assume more or less specialized functions through differentiation. The development of plant tissues is characterized by three types of growth: cell division, cell elongation, and cell differentiation.

B. Explants

Any tissue excised from a plant and transferred to nutrient medium *in vitro* is referred to as an explant.

While cell cultures may be obtained from any seed plant, some species can be manipulated more successfully than others. Variation in response to explantation is caused by the culture conditions employed but may also depend on the nature of the explants. Predominantly woody tissues will generally fail to grow; parenchyma rich in tannins may intoxicate the entire explant because of leakage from dead and wounded cells. In general, the selection of a species for establishing cell cultures will be dictated by the project under study and consequently will be directed to those species which feature useful characteristics.

Given a particular species, availability of parenchyma for explantation is a first consideration. Shoot tips appear most promising because during the growing season they harbor actively dividing cells, the shoot meristem. Shoot tip excision, however, requires microsurgery and isolation generally is limited to a low number of explants. Transferred to *in vitro* conditions shoot tips tend to retain their structure and regenerate a shoot and entire plant rather than yield a cell culture. Parenchyma from stems, rhizomes, tubers, and roots, by contrast, is easily accessible and will generally respond quickly to culture conditions *in vitro*. Preference for this material results also from the uniformity of the tissue. Explants from trees and shrubs can be obtained by excising phloem pieces. A method of explantation which applies to the widest range of species requires the germination of seeds and excision of segments from hypocotyls and young roots. It is estimated that the majority of cell strains maintained today originated from segments of seedlings. Recent advances in methodology permit the growth of haploid tissue and in a few cases the regeneration of haploid plants from cultured anthers.

Given the fact that all parenchyma cells of a plant have an identical morphogenetic potential, the origin of an explant, stem, root, or tuber, would be irrelevant. All tissue and cell cultures resulting from an explantation would have an equal potential for organogenesis and for synthesis and accumulation of natural products specific for the species in question. The selection of parenchyma for explantation, therefore, can largely be determined by practical reasons, i.e., accessibility, ease of excision and isolation, rapidity of response to *in vitro* conditions.

C. Culture Media

Successful plant cell culture depends on the choice of nutrient medium. In general, the medium is composed of five groups of ingredients: inorganic nutrients, carbon source, vitamins, growth regulators, and organic supplements.

Inorganic nutrients (mineral salts) are the same as required by intact plants, such as nitrate, sulfate, phosphate, potassium, calcium, magnesium, iron, and trace nutrients, and provide building stones for amino acids, chlorophyll, and vitamins, participate in oxidoreduction systems, and affect changes in plasma colloids. The concentration of mineral salts in media has to be well adjusted in order to prevent intoxication of cultured cells through ion antagonism. A concentration of 1–3 mM calcium, magnesium, and sulfate generally is adequate. For most purposes the nutrient should contain at least 25 mM each of potassium and nitrate. Replacement of nitrate by ammonium may result in growth inhibition (Gamborg, 1970). With respect to nitrate and ammonium it may be noted that depletion through uptake by the cells or tissues may change the pH of a medium. The required trace nutrients include I, B, Mn, Zn, Cu, Mo, and Co (Gamborg et al., 1976).

Plant cells cultured in vitro are heterotrophic with respect to carbohydrates, even when maintained in light. Growth therefore depends on an adequate supply of organic carbon. Sucrose at 2–4% is an efficient carbon source for most cell strains. The capacity of plant tissues to use a variety of carbon sources depends on the species in question. In cell cultures derived from carrots the following series of carbohydrates with decreasing efficiency in growth promotion was established; sucrose, glucose, maltose, raffinose, fructose, galactose, mannose, and starch (Koblitz, 1972).

Although most cell cultures are capable of synthesizing vitamins, the rate of synthesis may be too slow to support rapid growth. Myo-inositol may not be essential but is routinely added, since it has been shown to enhance callus growth. The supply of vitamins may become critical for the growth of cells cultured at low densities (Kao and Michayluk, 1975).

As a rule cell cultures are autotrophic with respect to amino acids. They are perfectly able to metabolize nitrate and ammonium; therefore, amino acids generally are not an essential ingredient of culture media. In case inorganic nutrients do not seem to be adequate, the best approach is to add 0.05–0.1 gm/liter casein hydrolysate to initiate and establish a culture. The amino acid mixture may later be replaced by L-glutamine (2–10 mM) or the organic nitrogen can be omitted from the medium completely (Gamborg et al., 1976).

Plant cells cultured in vitro commonly require media supplemented with growth hormones, auxins, and cytokinins, and sometimes gibberellins, alone or in combination (Street, 1966). Hormones play a dominant role in intercellular regulation of morphogenetic processes. Like all hormones, plant hormones are usually effective in small amounts and sites of synthesis and action may be different. One hormone may have a broad action spectrum and be capable of influencing a number of different processes. The mechanism of action of hormones is not fully understood. In particular the sites of action are still being debated. Assumptions are that they (a) activate all potentially active genes; (b) activate selected genes which in turn will activate other genes; or (c) do not affect genes but regulate one particular metabolic reaction, thus modifying a set of metabolic reactions which may result in an activation of one or several genes. The action spectrum of auxins is extremely wide. They induce cell division in explants, promote cell elongation, stimulate root formation, interact with the development of shoots, and affect the accumulation of secondary metabolites. A natural auxin is indole-3-acetic acid (IAA). Naphthalene-1-acetic acid (NAA) and the very active 2,4-dichlorophenoxyacetic acid (2,4-D) are synthetic products. Cytokinins are known to stimulate cell division and promote shoot formation. The most commonly used cytokinins are kinetin (6-furfurylaminopurine = 6-furfuryladenine) and 6-benzyladenine (6-benzylaminopurine). Cell cultures from monocotyledons may require an exogenous supply of zeatin (γ-methyl-γ-oxymethylallylaminopurine). Among many naturally occurring gibberellins only gibberellic acid (GA_3) has been employed in tissue culture to some extent. GA_3 appears to stimulate differentiation including the accumulation of secondary metabolites (Constabel and Nassif-Makki, 1971).

At times a number of undefined natural products have assisted in establishing plant tissue cultures, the liquid endosperm of coconuts (coconut water), of corn, as well as yeast and malt extract. Even today, a supplement of 10–15% coconut water may be recommended in cases where explants resist to grow despite the presence of auxins and cytokinins (Shantz and Steward, 1952; van Staden and Drewes, 1974; Steward, 1958). The advantages of avoiding the addition of any undefined supplements to nutrient media are evident.

Most water sources for the preparation of media may require demineralization, glass distillation, or both. The chemicals should be of the highest grade available. Growth regulators may require recrystallization before use. The pH is adjusted to a specified value between 5.5 and 6.8 and the medium autoclaved at 120°C for 15–20 minutes (up to 1 liter) or 30–40 minutes (volumes to 10 liters). Most components in the media tolerate autoclaving, although some caution should be exercised (Bragt et al., 1971). Exceptions may be IAA and glutamine. The media can be stored at room temperature, but +10°C is preferable (Gamborg et al., 1976).

Media after Murashige and Skoog (1962), Eriksson (1965), and Gamborg et al. (1968) have found widest acceptance in cell culture laboratories (Table I). Murashige and Skoog's medium appears to satisfy the nutritional as well as physiological needs of many plant cell strains.

A distinguishing feature of this medium is its high content of nitrate, potassium, and ammonium relative to other nutrient media. Eriksson's medium is similar to that of Murashige and Skoog but contains twice the amount of phosphate and much lower concentrations of micronutrients. Gamborg's medium has been tested with a wide range of cell cultures.

TABLE I. Culture Media

Nutrients	MS[a] mg/liter		ER[a] mg/liter		B5[a] mg/liter	
		mM		mM		mM
NH_4NO_3	1650	20.6	1200	15.0		
KNO_3	1900	18.8	1900	18.8	2500	25
$CaCl_2 \cdot 2 H_2O$	440	3.0	440	3.0	150	1.0
$MgSO_4 \cdot 7 H_2O$	370	1.5	370	1.5	250	1.0
KH_2PO_4	170	1.25	340	2.5		
$(NH_4)_2SO_4$					134	1.0
$NaH_2PO_4 \cdot H_2O$					150	1.1
		µM		µM		µM
KI	0.83	5.0			0.75	4.5
H_3BO_3	6.2	100	0.63	10	3.0	50
$MnSO_4 \cdot 4 H_2O$	22.3	100	2.23	10		
$MnSO_4 \cdot H_2O$					10	60
$ZnSO_4 \cdot 7l H_2O$	8.6	30			2.0	7.0
Zn versenate			15	37		
$Na_2MoO_4 \cdot 2 H_2O$	0.25	1.0	0.025	0.1	0.25	1.0
$CuSO_4 \cdot 5 H_2O$	0.025	0.1	0.0025	0.01	0.025	0.1
$CoCl_2 \cdot 6 H_2O$	0.025	0.1	0.0025	0.01	0.025	0.1
$Na_2 \cdot EDTA$	37.3	100	37.3	100	37.3	100
$FeSO_4 \cdot 7 H_2O$	27.8	100	27.8	100	27.8	100
Sucrose (gm)	30		40		20	
Inositol	100				100	
Nicotinic Acid	0.5		0.5		1.0	
Pyridoxine·HCl	0.5		0.5		1.0	
Thiamine·HCl	0.1		0.5		10.0	
Glycine	2.0		2.0			
IAA[b]	1–30					
NAA[c]			1.0			
Kinetin	0.04–10		0.02		0.1	
2,4-D[d]					0.1–1.0	
pH	5.7		5.8		5.5	

[a] Media of: MS, Murashige and Skoog (1962); ER, Eriksson (1965); B5, Gamborg *et al.* (1968).

[b] Indoleacetic acid.

[c] Naphthaleneacetic acid.

[d] 2,4-Dichlorophenoxyacetic acid.

This medium contains relatively low amounts of ammonium, a nutrient that may repress growth in batch cultures (Gamborg et al., 1976).

The selection of media presented here should not be interpreted as implying that other media are not suitable. The media are suggested as a starting point. For a more complete discussion of nutrition, several reviews are available (Dougall, 1972; Murashige and Skoog, 1962; Street, 1966; White, 1963).

D. Establishment of Callus and Cell Cultures

1. Explant Culture

A typical explant culture is a 2–5 mm³ sterile cube or segment excised from a stem, tuber, or root transferred onto 30 ml of nutrient agar medium inside a petri dish or culture tube. The culture is maintained in the dark, in light, or in a regime of light and dark cycles at 20°–30°C, preferably 25°–28°C. Histological analysis reveals that within 2–5 days the explanted tissue responds to the excision and implantation onto a nutrient medium with a wound reaction. Noninjured living nucleated cells adjacent to the wounded cells redifferentiate and reenter mitotic cycles. Given a uniform explant and suitable culture conditions the first division can occur in an almost perfectly synchronous manner (Mitotic Index = 70–80%) (Yoeman, 1974). The presence of hormones, in particular auxins, in the medium will induce further divisions of the daughter cells. Soon the surface of the explant and primarily the part exposed to the air will be covered with a thin layer of newly formed callus. Resumption of cell division by parenchyma cells in an explant may not be restricted to cells near the surface but may also progress toward the interior. As a result the explant will grow and, depending on the intensity of cell proliferation, will burst and become obliterated. Explants which consist of several types of cells show the formation of young tissue only around parenchyma cells and yield irregularly shaped callus. Tanniferous parenchyma cells were observed to undergo divisions and give rise to a sequence of nontanniferous derivatives (Constabel, 1968). Cells which store anthocyanins, chloroplasts, or Ca oxalate crystals were found to lose their pigments or inclusions, when reentering mitotic cycles, a process referred to as dedifferentiation (Buvat, 1944, 1945).

2. Callus Culture

A callus culture begins with the isolation of the newly formed tissue from the explant and its transfer to fresh medium and maintenance under conditions as mentioned before. The transferred callus continues to grow but, in time, may lose its uniformity. Actively dividing cells form a layer of meristem and build a globular mass of nondividing parenchyma or form small meristem islands producing a nodulated irregular callus. Differentiation in the callus may proceed further and result in the formation of tracheids. The homology between a callus meristem and a cambium is evident. As the callus grows the synthesis and accumulation of natural products may be substantial. Tannin, anthocyanin, betalain, alkaloid, and carotenoid production in callus has been reported (Krikorian and Steward, 1969).

After several weeks the size of the callus may permit cutting it into smaller pieces for inoculation of fresh media and establishment of subcultures. Selection of those callus pieces which resulted from optimum growth will be the material of choice for subculturing. This selection may mean the inception of a cell strain characterized by relatively rapid growth. In time this strain may alter its properties further and lose its capacity to produce roots, shoots, or natural products, in other words, lose some of its morphogenetic potential.

Heterogeneity of callus cannot only be found in the appearance of various kinds of tissues and cells but is also reflected in the level of ploidy of the cells. Poly- and aneuploid cells frequently occur in older callus cultures (Kao et al., 1970).

3. Cell Culture

Rapidly growing and soft callus tissue is suitable for transfer to liquid medium and may easily render a cell suspension culture. Typically, an inoculum of about 500 mg fresh weight is suspended in 50 ml nutrient contained in 250-ml Delong flasks. The cultures are maintained on gyratory shakers (120–150 rpm) under light and temperature conditions as before. While growth of a callus culture is more or less linear, cells suspended in liquid medium pass through growth phases similar to those of microorganisms and documented as a sigmoid curve. Cell suspension cultures have several advantages over callus cultures besides growing faster: (1) The suspensions can be pipetted; (2) they are less heterogeneous and differentiation of cells is less pronounced; (3) they can be cultured in volumes up to 1500 liters; (4) they can be subjected to more stringent environmental controls; and (5) they can be manipulated in their production of natural products by feeding precursors.

4. Anther Culture

Anther culture is a method which permits the callusing of pollen and growth of haploid tissue. This material may then be (1) exposed to mutagens to yield mutant cell lines, (2) treated with colchicine to result in homozygous cells, or (3) transferred to conditions conducive to the regeneration of plants, haploids. Anther culture could become a major tool in improving product formation in cell cultures (Gresshoff and Doy, 1972; Kasha, 1974; Zenk, 1976). Success in growing haploid cells depends on recognizing the stage of mononucleated pollen in the development of stamen and on the choice of a nutrient medium (Nitsch, 1972).

Anthers are excised from superficially sterilized flower buds (the exact developmental stage has to be determined by means of microscopic analysis of spores) and transferred to agar medium. Anther culture media appear to require a supplement of amino acids (Nitsch, 1974).

5. Protoplast Culture

The observations that protoplasts, i.e., cells minus cell wall, are able to take up inert as well as biologically active particles and that they can be induced to fuse have led to project somatic cell genetics, including the regeneration of parasexually hybrid plants. The experimental basis for this approach is the isolation and culture of protoplasts (Cocking, 1972; Gamborg et al., 1974).

Excised pieces of parenchyma or cultured cells are incubated in a solution containing cell wall digesting enzymes and osmotic stabilizers. Isolated protoplasts appear as perfectly spherical, wall-less cells. After isolation they have to be washed free of the enzyme solution and can now be subjected to experimental conditions or incubated with nutrient medium. Formation of a new cell wall may begin 20 minutes after removal of enzymes or may take several days. Cell division in protoplasts has been observed as early as 24 hours after isolation. Wall formation and resumption of cell divisions depend as much on the physiological conditions of the protoplast source as on the choice of nutrient medium. Rapidly growing cell suspension cultures have been a particularly suitable protoplast source. In attempts to achieve greatest variation in protoplast species, the mesophyll of young, expanded leaves has been used most extensively as starting material (Gamborg and Wetter, 1975). Prolonged culture and growth of protoplasts proper, i.e., wall-less cells, has not yet been achieved.

6. Plating Cells

Cloning of cell lines requires plating techniques. A simple way would be the dispersal of a mass of cells on the surface of nutrient agar medium by means of a spatula. A more efficient technique calls for the mixture of a cell suspension with agar medium autoclaved and cooled to 45°–40°C, and dispensing the mixture as a thin layer in petri dishes (Bergmann, 1960). Growth of plated cells may depend on the density of the cell population. Low densities require the conditioning of media by either supplementing media of growing cultures or applying feeder layers of inactivated cells (Raveh et al., 1973). The chemical nature of a conditioned medium is still poorly understood. Substances leaked or excreted from cells are a factor in conditioning.

7. Deep-Freeze Preservation

A cell suspension culture derived from seed plants is similar to cultures of yeast and other microorganisms. Storage and maintenance of such cultures is a pressing problem, the resolution of which would mean freedom from the problem of genotypic changes of clones. In a few cases, deep-freeze preservation of cultured cells (carrot and sycamore) has been achieved. After thawing, the cells demonstrated their full morphogenetic potential—carrot cells regenerated entire plants (Nag and Street, 1973; Street, 1975).

E. Regeneration of Organs and Plants

The cell theory (Schleiden, 1838; Schwann, 1839) postulates that cells are the unit of function of any organism. This may be interpreted by saying that living, nucleated cells carry the entire genetic information of the organism from which they are derived. As a conclusion cells cultured *in vitro* would have the morphogenetic potential of the original plant. Expression of this potential would be seen as shoot, root, and embryo formation, as regeneration of a complete plant.

The cell theory was proved to be correct when single carrot cells suspended in liquid medium were found to react like zygotes, forming embryos. These could subsequently be grown to mature, seed-bearing plants (Reinert, 1959; Steward, 1958). Since then embryogenesis has been observed in cell and anther cultures of a number of species (Murashige, 1974). Predictions are that this expression of totipotency of cultured cells will become a general phenomenon, once conducive environmental factors have been clearly defined. As a rule, embryogenesis is triggered by transfer of cells to media without hormones. However, only a fraction of the cells in question may respond to this treatment, and it is still impossible to identify with certainty those cells which are about to undergo the process of embryogenesis (Street and Withers, 1974).

An alternate and much more common route to plant regeneration from cultured cells is via shoot and root formation. The primordia of these organs originate from single cells or small groups of cells. Shoots tend to arise from the periphery, while roots emerge from the interior of a callus. A classical example for the regulation of organogenesis is found in callus derived from tobacco stem pith (Skoog and Miller, 1957). When this tissue is isolated and grown *in vitro* cell division occurs only on a medium containing both an auxin and a cytokinin. A combination of 2 mg/liter IAA and 0.1 mg/liter kinetin results in mere callus growth. If, however, the kinetin concentration is lowered to 0.02 mg/liter roots will develop; 0.5 mg/liter kinetin, conversely, will result in shoot formation. Shoots may root and thus regenerate an entire plant.

The inception of organs and embryos occurs most frequently in recently isolated tissues and this ability decreases with increasing duration of a culture. It is suggested that the expiration of the capacity to express totipotency may be related to the instability of genomes in cultured cells, for instance, to changes in the level of ploidy (Torrey, 1959).

III. CELL SUSPENSION CULTURES

One of the prerequisites for the use of plant cells in the commercial production of natural substances is their adaptability to submerged culture conditions. These conditions not only facilitate mass propagation but allow also stringent control of the culture environment as well as direction toward a high-yield output of the desired metabolite. In the past 25 years the development of various culture techniques has increased our knowledge of the conditions of culture to the extent that it would be now possible to cultivate most plant cells in suspension on a large scale. A cell derived from seed plants differs in its requirements considerably from a microbe and it is therefore not surprising that culture techniques developed for the latter cannot be adapted as such. Cell suspension cultures are generally heterogeneous, with single cells and cell aggregates growing at different rates. The proportion between free cells and aggregates, as well as their size, depends on the species and age of the culture, the composition of the medium, and environmental conditions. It is therefore not surprising that attempts to achieve ideal conditions for cell division and separation to yield a uniform culture of single cells have been numerous but only to a certain degree sucessful (Ganapathy et al., 1964; Kurz, 1973; Street and Henshaw, 1963; Street et al., 1965). The fundamental aspects of plant suspension cultures on a much broader basis have recently been reviewed (Street, 1973).

A. Physical Parameters of Culture

1. Inoculum

The size of the inoculum is important for the initiation of growth in suspension culture. Below a critical ratio of inoculum to volume of medium, cell division ceases. This observation led to the assumption that the medium has to be "conditioned" by factors released from the cells of the inoculum. Growth only resumes after a critical concentration of these factors is attained in the medium (Street and Henshaw, 1963; Street, 1966, pp. 665–680). Ten to fifteen percent v/v of an actively growing suspension culture has been found a satisfactory inoculum for most species (Veliky and Genest, 1972; Verma and van Huystee, 1971; Wang and Staba, 1963). The ratio of free cells to cell aggregates influences the growth rate (Torrey et al., 1962) and in some investigations the inoculum was prefiltered to obtain a more homogeneous cell suspension (Staba and Lamba, 1963; Torrey et al., 1961). Larger inocula may reduce the "conditioning time" of the medium but may lower the relative growth rate of the cells (Caplin, 1963).

2. Temperature

The temperature range within which plant cells are able to grow is slightly higher than for the plant proper and lies for most cultures between 20° and 28°C. Tulecke and Nickell (1960) studied the temperature effect on the growth of five suspension cultures and found optimal ranges between 20° and 21°C for *Lolium perenne* and an extreme temperature of 31°–32°C for *Rosa* sp.

3. Aeration and Agitation

In most suspension cultures effective aeration is essential for optimal growth and a minimum ratio of 1 : 4 (air : medium) was recommended for cultures grown under forced aeration (Routien and Nickell, 1956). However, this should only be taken as a rule of thumb. Aeration should be viewed in the whole context of a particular process as it has been found that forced aeration in some instances (Lamport, 1964; Muir et al., 1958; Verma and van Huystee, 1971) reduces the growth rate. Agitation has been achieved by placing culture vessels on rotary or reciprocating shakers; by passing air through filtered tubes (Wang and Staba, 1963), open-ended tubing on pipes (Kato et al., 1976; Lamport, 1964; Tulecke and Nickell, 1960; Tulecke, 1966; Wilson, 1976), or fritted disks (Graebe and Novelli, 1966; Reinhard, 1967); by compressed air (Kurz, 1973), magnetic stirrers (Bragt et al., 1971; Constabel et al., 1971, 1974), in conventional stirred-jar fermentors (Kato et al., 1972; Veliky and Martin, 1970; Veliky and Genest, 1972; Wang and Staba, 1963); or by using roller devices (Lamport, 1964; Muir et al., 1958; Short et al., 1969).

4. Light

Plant suspension cultures are usually grown in diffuse fluorescent light with light intensities ranging from 15 to 315 μ Einstein m^{-2} sec^{-1} ($\mu\epsilon$/m^2/second) (Hood, 1964; Mitsuoka and Nishi, 1974; Vasil et al., 1964; Venketeswaran, 1964). However, light seems only to be necessary for cultures containing chlorophyll and has no effect on the growth rate (Torrey and Reinert, 1961). A difference in the consistency of Haplopappus gracilis suspension culture was reported, showing that cultures grown in light were firm, while those grown in the dark were friable (Blakely and Steward, 1961).

5. pH

The pH range for plant suspension cultures is rather wide, being between 5.2 and 6.8, and the optimum depends on the tissue cultivated (Veliky and Martin, 1970; Verma and van Huystee, 1971). As autoclaving of the medium can change the pH, which might effect the growth of cells in suspension, special consideration should be given to the buffering of media (Holsten et al., 1965).

B. Culture Systems

The culture systems mentioned here were developed solely for the cultivation of plant suspensions and were designed to avoid shearing effects and other mechanical features which would rupture the thin-walled plant cell. Nevertheless microbial fermentor systems may be successfully adapted for cultivation of plant suspension cultures. The purpose of all systems is an even distribution of cells within the culture, thorough mixing of medium and cells, and adequate gas exchange between the liquid and gas phases of the culture.

Early systems were slowly rotating roller-tube devices holding different types of culture vessels (Caplin and Steward, 1949; de Ropp, 1946; White, 1953). One of them, the Auxophyton (Steward et al., 1958), was designed for rapid proliferation of explants in nippled Florence flasks. Through the slow movement of the Auxophyton, part of the medium was distributed as a thin film within the nipples of the culture vessel, which facilitated a high rate of gas exchange. A somewhat similar apparatus was designed by Muir and co-workers (1958). Good growth has also been obtained with rolling and spinning cultures, where glass bottles carrying the culture were rotated at different speeds at 45° to horizontal position (Lamport, 1964; Short et al., 1969). Reciprocal or rotary shakers are often used to grow plant cell suspensions in various types and sizes of flasks and bottles. The approximate speed for rotary shakers is up to 150 rpm and for reciprocal shakers up to 100 oscillations per minute. Large batch-type suspension culture devices have been described by several authors, where agitation and aeration is achieved either solely by forced aeration (Graebe and Novelli, 1966; Kurz, 1973; Tulecke and Nickell, 1960; Tulecke, 1966; Wilson, 1976) or by aeration and a mechanical stirrer (Melchers and Engelmann, 1955; Miller et al., 1968; Veliky and Martin, 1970; Wilson et al., 1971). These devices are more versatile compared with the one described earlier because of the ease with which growth conditions can be controlled through regulation of such parameters as temperature, agitation, aeration, and pH.

Because of the need of more uniform cell material for metabolic and physiological studies, grown under defined, nonlimiting culture conditions in a state of balanced growth, several continuous culture systems for cell suspension cultures have been developed in the past few years.

The system by Kurz (1973) (Fig. 1) employs compressed air for both agitation and aeration by feeding air pulses at regular intervals into the culture vessel, a flat-bottomed glass cylinder. As the compressed air (5–10 psi) enters the cylinder via a central pipe at the base, it expands into a large bubble having the same diameter as the culture vessel. As the bubble slowly moves upward through the vessel the entire culture passes as a thin layer between the fermentor wall and the surface of the air bubble, thus being aerated and stirred effectively. The reduction of air pressure from 5–10 psi to atmospheric pressure and expansion of the bubble causes a mild vibration in the culture, which probably is the main factor in producing a culture consisting of predominantly single cells, in continuous cultivation for long periods of time. This fermentor system was also successfully employed in the induction of partial synchrony of cell division in continuously grown plant cell suspension cultures (Constabel et al., 1974, 1977).

A conical glass V-fermentor was developed by Veliky and Martin (1970) (Fig. 2). In this system the culture is agitated by a Teflon-coated double bar magnet stirrer supported on a short glass rod at the bottom of the fermentor. The air is supplied through a 17-gauge hypodermic needle. This reduces the number of devices inserted into the culture to a minimum and avoids the buildup of cell clusters on them. The V-fermentor is mainly used for batch operations and in semicontinuous culture through intermittent renewal of medium and harvesting of culture.

The Phytostat described by Miller and co-workers (1968) (Fig. 3), as

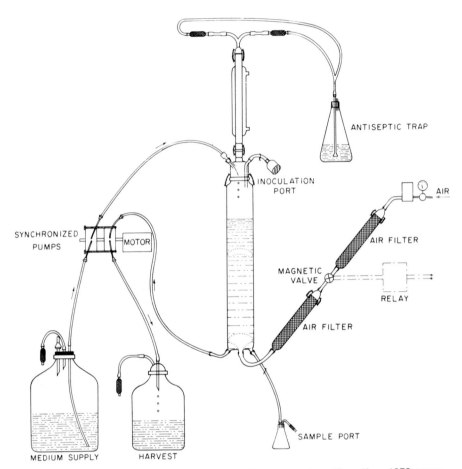

FIGURE 1. Schematic drawing of the Kurz fermentor system. (From Kurz, 1973; reproduced by permission of Academic Press.)

well as a similar device by Wilson and co-workers (1971), may be used for continuous or batch culture. They are both equipped with needle valves for the automatic collection of samples. However, because of tubes, stirrer, thermometer, and similar devices protruding into the culture, the buildup of cell masses on those at the interphase between culture and gas phase poses a problem.

C. Methods of Cultivation

The ideal state for a cell suspension culture is that of morphological, biochemical, and genetic homogeneity grown in a fully controllable

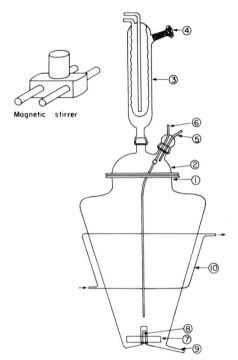

Magnetic stirrer

FIGURE 2. Schematic drawing of the V-fermentor: (1) Large-diameter flat-flange joint; (2) multisocket lid; (3) water-cooled condenser; (4) air exhaust with a sterilizing filter; (5) inlet for medium; (6) air inlet; (7) Teflon-coated double-bar magnetic stirrer; (8) short glass rod; (9) sampling outlet; (10) water jacket. (From Veliky and Martin, 1970; reproduced by permission of the National Research Council of Canada.)

environment. Within plant cell research, however, this state is far from being attained. It is not within the scope of this chapter to discuss different growth patterns in cell cultures in detail. This has been done in an excellent review (Street, 1973). Instead we shall try to evaluate different methods of cultivation with respect to production of natural substances.

1. Batch Culture

The majority of investigations involving plant cell suspension cultures have been performed under batch culture conditions. In batch culture, growth takes place in a "closed system"; the cells grow in a limited amount of medium and multiply to form a population in which succeeding generations progressively modify their environment. This produces a sequence of changes in the culture that is often mistakenly referred to as

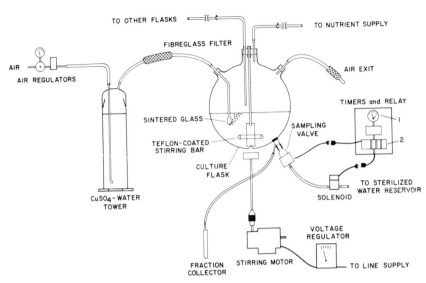

FIGURE 3. Schematic drawing of the Phytostat. (From Miller *et al.*, 1968; reproduced by permission of the authors and *Science*.)

the growth cycle. The growth rate of the cells changes during the period of cultivation so that the time required for the population to double usually increases as the culture progresses. Under these unbalanced conditions the metabolism and composition of the cells are also altered with successive generations and this makes it extremely difficult, if not impossible, to determine any factors responsible for these fluctuations. This notwithstanding, the batch culture technique in plant cell research has its place in the production of biomass and for exploratory research into synthesis of metabolites.

2. *Continuous Culture*

In continuous culture (chemostat), growth takes place in an "open system" of culture. In this method a continuous feed of medium flowing into a constant volume of growing culture produces a steady-state condition for growth. The environment imposes a constant growth rate on the cells so that the doubling time and overall metabolism of the cells remain constant and thus characteristic of the steady state. It is possible to change the growth rate by altering the flow through the system; in this way the growth of cells may be examined at different growth rates by analysis of the appropriate steady states and metabolic regulation in the production of secondary metabolites is possible. This significant advantage over the batch culture technique in connection with specifically designed cultural environments, such as the controlled feeding of precursors, will lead in the not too distant future to the commercial use of plant cell cultures in the production of secondary metabolites.

3. Continuous Synchronous Culture

In batch and continuous culture the cells divide randomly. In batch culture the doubling time of the successive generations is likely to be changing, but in the chemostat this remains constant. In continuous synchronous growth, an "open system" ideally, all the cells are at the same stage of development in the cell cycle. Thus, instead of an average condition as is obtained in the steady state of a chemostat from the randomly dividing population of cells, there is in the continuous synchronous culture a pattern of change which coincides with the cell cycle and repeats itself with each successive doubling of the cell population. The cell, thereby amplified by the size of the synchronous population, may be examined at any stage of the cell cycle and at any desired growth rate. This has the advantage that enzymes or metabolites occurring only at certain stages of the cell cycle can be obtained at maximal yields, a most useful criterion for consideration in a possible industrial production of such compounds.

The method of continuous synchronous cultivation of plant cells in suspension culture has been developed recently by imposing flushes of nitrogen or ethylene gas at regular intervals upon chemostat cultures at steady state (Constabel et al., 1974, 1977). On plant cells tested so far, this method gave up to 25% (mitotic index) of synchronization of the population over a period of several weeks. Earlier attempts of synchronization in plant suspension cultures (Eriksson, 1966, 1967; King et al., 1973) were based on batch culture conditions and synchrony only persisted for up to six generations.

IV. BIOSYNTHESIS AND ACCUMULATION OF NATURAL PRODUCTS IN PLANT CELL CULTURES

The metabolism of inorganic and organic compounds in plants results in a great number of products which are catabolized as soon as they are formed; these are referred to as intermediary or primary metabolites. At particular stages in the development of plants primary metabolites may be transformed into storage products, fats, starch, and proteins. Secondary metabolites are products which result from biosynthetic transformations of primary metabolites. They occur as terpenoids, glycosides (steroids, phenolics), and alkaloids. Although detected in various groups of animals, secondary metabolites typically are plant products. The diversity and specificity of secondary metabolites has been used as an argument in the classification of plants.

Secondary metabolites appear to originate in cell organelles and may be sequestered into the cytoplasm. As they seem not to be essential for the life of the cell (plant), they are either deposited in vacuoles or excreted from gland cells. Genotype, physiological condition, and state of development as much as location within a given plant determine the formation of metabolites. A meristematic cell, which divides frequently, will rarely exhibit terpenoids, phenolics, and alkaloids; a mature parenchyma cell, however, may be loaded with a whole spectrum of products. This observation can be explained reasonably well on the basis of differential gene activity.

The occurrence of secondary plant products in cell cultures has been well documented (Becker, 1969; Jones, 1974; Krikorian and Steward, 1969; Petiard et al., 1972; Reinhard, 1967; Zenk, 1976). In general the yield is low. It appears that commonly used media favor growth by cell division and not synthesis and accumulation of secondary metabolites. Relatively few cell suspension cultures have shown appreciable amounts of products (Table II). Two kinds of behavior may account for this exceptional result: Cell division and accumulation of products in individual cells do not exclude each other but occur simultaneously [anthocyanin production in *Haplopappus gracilis* cells (Constabel et al., 1971), anthraquinone production in *Morinda citrifolia* cells (Zenk et al., 1975)], or rapidly growing suspension cultures continuously sequester a substantial fraction of the cell population which matures and accumulates metabolites.

Temperature and photoperiod as well as nutrition and changes in the level of hormones may modify, induce, or repress the biosynthetic activity of cultured cells. As these factors can be strictly controlled they offer a chance to manipulate metabolite production.

The effect of various components of nutrient media on the synthesis and accumulation of secondary metabolites has been assessed in only a few cases (Matsumoto et al., 1971, 1973). In cell suspension cultures of soybean, exhaustion of nitrate was thought to cause a switch in cellular metabolism to reactions forming nitrogen-free products, phenylpropanoids for example (Hahlbrock et al., 1974). In *Morinda* cell cultures, nitrate did not affect the biosynthesis of anthraquinone. However, anthraquinone accumulation showed a drastic response to changes in the level of phosphate, 5 gm/liter KH_2PO_4 being the optimum concentration. Iron and calcium ions were necessary for growth and metabolite produc-

TABLE II. Natural Substances Produced by Plant Cell Suspension Cultures.

Products	Species	References
Tobacco alkaloids	Nicotiana tabacum	Solt (1957)
		Solt, et al. (1960);
	Nicotiana glauca	Speake, et al. (1964)
Tropane alkaloids	Datura stramonium	Chan and Staba (1965);
		Stohs (1969)
Indole alkaloids	Vinca minor	Petiard and Guinebault (1976)
		Boder et al. (1964);
	Catharanthus roseus	Patterson and Carew (1969)
		Stöckigt et al. (1976);
	Phaseolus vulgaris	Velicky (1972)
Purine alkaloids	Coffea arabica	Keller et al. (1972)
Quinoline alkaloids	Ruta graveolens	Boulanger et al. (1973)
Betalains	Phytolacca americana	Misawa et al. (1973)
Betaines	Medicago sativa	Sehti and Carew (1974)
Saponines	Panax ginseng	Furuya and Ishii (1973)
	Glycyrrhiza glabra	Tamaki et al. (1975)
	Stevia rebaudiana	Komatsu et al. (1976)
Cardiac glycosides	Digitalis purpurea	Staba and Lamba (1963)
	Apocynum cannabinum	Lee et al. (1972)
Glucosinolates	Reseda lutea	
	Trapaeolum majus	Kirkland et al. (1971)
Furanocoumarins	Ruta graveolens	Steck et al. (1971)
Anthocyanins	Populus nigra	Matsumoto et al. (1970)
		Fritsch et al. (1971);
	Haplopappus gracilis	Constabel et al. (1971)
Quinones	Nicotiana tabacum	Ikeda et al. (1974)
Anthraquinones	Morinda citrifolia	Zenk et al. (1975)
Carotenoids	Daucus carota	Sugano et al. (1971)
Peptides	Scopolia japonica	Misawa et al. (1973); (1975c)
Antibiotics	Phytolacca americana	Misawa et al. (1974b)
Plant virus inhibitors	Phytolacca americana	Misawa et al. (1974c); (1976)
	Agrostemma githago	Misawa and Sakato (1976)
		Misawa et al. (1975a)
Coccidiostatic substances	Catharanthus roseus	Misawa et al. (1975c)
Phosphodiesterase	Catharanthus roseus	Furuya et al. (1973)
	Phytolacca americana	Ukita et al. (1973)
	Nicotiana tabacum	Kubo (1974a)
	Daucus carota	Kubo (1974b)
	Scopolia japonica	
	Zea mays	
Amylases	Saccharum officinarum	Maretzki et al. (1971)
Glucanases	Triticum aestivum	
	Hordeum vulgare	Gamborg and Eveleigh (1968)

tion, while KI appeared to be dispensable. Glucose at 7.5% was superior to all other carbohydrates for anthraquinone production. *Morinda* cells had a definite requirement for organically-bound reduced nitrogen. This requirement was satisfied by supplementing the medium with casein hydrolysate. A dose–response curve showed that 0.2% casein hydrolysate was optimal for growth but suboptimal for anthraquinone production, which in turn was optimal at 0.3% (Zenk *et al.*, 1975).

Synthesis of secondary metabolites in cell cultures can be regulated by the kind and concentration of hormones in the medium in just the same manner as has been demonstrated for the differentiation of roots and shoots. The ability of 2,4-D to suppress and of NAA to enhance anthraquinone accumulation in *Morinda* cell suspension cultures (Zenk *et al.*, 1975), or of increased NAA concentrations to inhibit anthocyanin production in *Haplopappus* cells (Constabel *et al.*, 1971), or of increased 2,4-D concentrations to inhibit the accumulation of polyphenols in rose (Davies, 1972) and *Cassia* cultures (Shah *et al.*, 1976), and of nicotine in tobacco (Tabata *et al.*, 1971) as well as indole alkaloids in *Vinca minor* cultures (Petiard and Guinebault, 1976) are striking examples. Since kinetin is known to stimulate cell division it is reasonable to believe that it might be influencing metabolite production by enhancing cell division in a particular hormonal environment which stimulates the daughter cells to increased synthetic activity.

Morinda cells cultured in media supplemented with NAA not only proved that the biosynthetic potential of this material can exceed that of the original plant, but its biosynthetic capacity may also be taken as an excellent example for the expression of totipotency. Explants isolated from plant parts which normally do not synthesize and accumulate anthraquinones (leaves, stems, fruits) formed callus, and this material gave rise to suspension cultures which showed anthraquinone production (Zenk *et al.*, 1975).

The synthesis of flavonoids in parsley cultures as influenced by light can be traced back to the molecular level. As long as these cultures are kept in darkness, they multiply but do not form flavonoids. Once they are exposed to light the glucoside apiin can be detected in appreciable amounts. This reaction is due to the photocontrol of the synthesis of a key enzyme in flavonoid synthesis, phenylalanine-ammonia lyase (PAL), and of UDP-apiose synthetase (Hahlbrock and Wellmann, 1971). The photocontrol of PAL activity has been subjected to further analysis. In *Haplopappus* cell cultures blue light always leads to an increase of PAL synthesis and subsequent anthocyanin production. Irradiation with red and far red light prior to or after incubation with blue light had no effect on the enzyme synthesis (Gregor and Reinert, 1972). These results illustrate how cells in a culture undergo biochemical modifications in a controllable manner.

Metabolite accumulation cannot succeed when precursors are absent. An example is demonstrated by *Ruta graveolens* cultures producing the alkaloid dictamnine. Supplementing the medium with 4-hydroxy-2-quinolone stimulated the synthesis and accumulation of dictamnine from less than $10^{-4}\%$ to $0.1-0.2\%$ of dry weight (Steck *et al.*, 1973). *o*-succinylbenzoic acid (OSB) is an intermediary product in anthraquinone biosynthesis. Media supplemented with OSB stimulated the accumulation of anthraquinones in *Morinda* cell suspension cultures significantly. After 20 days of culture metabolite levels had increased by 100% (Zenk *et al.*, 1975). The production of alkaloids in cell cultures of *Datura stramonium* was considerably enhanced by feeding DL-tropic acid or tropine (Konoshima *et al.*, 1970); alkaloid production in tobacco cells was affected by methylputrescine (Neumann and Müller, 1971). The supplementation of nutrients with precursors, artificial or natural, therefore, can be a limiting factor in exploiting cultured cells for the production of secondary metabolites.

Aromatic secondary metabolites may not be as stable as their configuration, stage of oxidation and polymerization, and position in biosynthetic pathways would suggest. For many years diurnal or seasonal fluctuations in phenolics and terpenoids have been observed. It is only recently that through sterile systems as cell cultures the degradation of secondary metabolites has been proved (Berlin *et al.*, 1971; Ellis and Towers, 1970). A high rate of O_2 uptake accompanying the more active degradation indicates that it is the rapidly growing cultures which are most effective in degrading secondary metabolites (Constabel and Nassif-Makki, 1971).

Transformations of organic compounds by microorganisms have long been known. The most important are transformations of steroids. Advances in culture methods have permitted the studies of biotransformations of steroids in plant cells. Suspension cultures of *Digitalis lanata* performed a variety of transformations with digitalis glycosides, i.e., glycosylation, acetylation, hydroxylation, and reverse reactions (Reinhard, 1974). Biotransformations of progesterone and pregnenolone were obtained with callus cultures of *Nicotiana tabacum* and *Sophora angustifolia* (Furuya *et al.*, 1972). Tobacco cells which do not contain any compounds of morphinan structure were shown to perform the conversion (demethylation) of thebaine to codeine and morphine (Mothes, 1966).

Apart from secondary metabolites, plant cell cultures have proved to be a source of a variety of proteins, which in the future may become of economic importance, as well as enzymes, peptides, plant virus inhibitors, and antibiotics (Mitsuoka and Nishi, 1974) (Table II). It thus appears that plant cell cultures have infinite possibilities for producing useful natural substances in industry. Again, the remaining difficulty is to control the induction of product synthesis.

At present, attention is focused on employing mutants for high-yield product formation. Mutant cell lines would be characterized by increased enzymatic activities for the biosynthesis of a particular product, or by a defect in enzymes which degrade an accumulated product, or by a defect in enzymes located at a branching point of a metabolic pathway resulting in increased product formation. The culture of haploid cells derived from pollen (anther culture) affords the starting material. Exposure to X-rays or chemicals (ethyl-methanesulfonate, 8-azaguanidine) has proved to result in a variety of mutant cell lines. While the stability of the mutants may be a problem, progress is still hampered by a lack of adequate selection systems for cloning cells with higher than wild-type production of specific metabolites (Zenk, 1974).

Table II lists a variety of products which have been detected in cell suspension cultures. Reports on the occurrence of metabolites in callus cultures have not been included, because this kind of material appears less amenable to projected large-scale industrial operations. It becomes quite obvious from studying the table that substances which are known to exert physiological activities have attracted particular interest. *Vinca* and *Catharanthus* cultures which may produce cancerostatic alkaloids have received most attention. The range of products of cell culture origin would substantiate the feasibility of exploitation of plant cell cultures.

REFERENCES

Becker, H. (1969). *Mitt. Dtsch. Pharm. Ges.* **39**, 273–279.
Bergmann, L. (1960). *J. Gen. Physiol.* **43**, 841–851.
Berlin, J., Barz, W., Harms, H., and Haider, H. (1971). *FEBS Lett.* **16**, 141–146.
Blakley, L. M., and Steward, F. C. (1961). *Am. J. Bot.* **48**, 351–358.
Boder, G. B., Gorman, M., Johnson, I. S., and Simpson, P. J. (1964). *Lloydia* **27**, 328–333.
Boulanger, D., Bailey, B. K., and Steck, W. (1973). *Phytochemistry* **12**, 2399–2405.
Bragt, J. V., Mossel, D. A. A., Pierik, R. L. M., and Veldstra, H. (1971). Effects of sterilization on components in nutrient media. Misc. Pap. No. 9. Wageningen, Netherlands.
Braun, A. C. (1968). *Results Prob. Cell Differ.* **1**, 128–135.
Brown, C. L., and Sommer, H. S. (1975). "An Atlas of Gymnosperms Cultured in Vitro." Georgia Forest Research Council, Macon.
Buvat, R. (1944). *Ann. Sci. Nat., Bot. Biol. Veg.* **5**, 1–130.

Buvat, R. (1945). *Ann. Sci. Nat., Bot. Biol. Veg.* **6,** 1–119.
Caplin, S. M. (1963). *Am. J. Bot.* **50,** 91–94.
Caplin, S. M., and Steward, F. C. (1949). *Nature (London)* **163,** 920–921.
Chan, W. N., and Staba, E. J. (1965). *Lloydia* **28,** 1–26.
Cocking, E. C. (1972). *Annu. Rev. Plant Physiol.* **23,** 29–50.
Constabel, F. (1968). *Planta Med.* **16,** 241–247.
Constabel, F. (1976). *In Vitro* **12,** 743–747.
Constabel, F., and Nassif-Makki, H. (1971). *Ber. Dtsch. Bot. Ges.* **84,** 629–636.
Constabel, F., Shyluk, J. P., and Gamborg, O. L. (1971). *Planta* **96,** 306–316.
Constabel, F., Kurz, W. G. W., Chatson, B., and Gamborg, O. L. (1974). *Exp. Cell Res.* **85,** 105–110.
Constabel, F., Kurz, W. G. W., Chatson, B., and Kirkpatrick, J. W. (1977). *Exp. Cell Res.* **105,** 263–268.
Davies, M. E. (1972). *Planta* **104,** 50–65.
de Ropp, R. S. (1946). *Science* **104,** 371–373.
Dougall, D. K. (1972). *In* "Growth, Nutrition, and Metabolism of Cells in Culture" (G. H. Rothblat and V. J. Cristofalo, eds.), Vol. 2, pp. 372–406. Academic Press, New York.
Ellis, B. E., and Towers, G. H. N. (1970). *Phytochemistry* **9,** 1457–1461.
Engvild, K. C. (1974). *Physiol. Plant.* **32,** 390–393.
Eriksson, T. (1965). *Physiol. Plant.* **18,** 976–993.
Eriksson, T. (1966). *Physiol. Plant.* **19,** 900–910.
Eriksson, T. (1967). *Physiol. Plant.* **20,** 348–354.
Fritsch, H., Hahlbrock, K., and Grisebach, H. (1971). *Z. Naturforsch. Teil B* **26,** 581–585.
Furuya, T., and Ishii, T. (1973). Japanese Patent (Kokai) 73/31917.
Furuya, T., Syona, U., Kojima, H., Hirotani, M., Ikuta, A., Hikichi, M., Kawaguchi, K., and Matsumoto, K. (1972). *Ferment. Technol. Today, Proc. Int. Ferment. Symp. 4th, 1972,* pp. 705–709.
Furuya, A., Ukita, M., Kotani, Y., Misawa, M., and Tanaka, H. (1973). Japanese Patent (Kokai) 73/33093.
Gamborg, O. L. (1970). *Plant Physiol.* **45,** 372–375.
Gamborg, O. L., and Eveleigh, D. E. (1968). *Can. J. Biochem.* **46,** 417–421.
Gamborg, O. L., and Wetter, L. R. (1975). "Plant Tissue and Culture Methods." National Research Council of Canada, Ottawa.
Gamborg, O. L., Miller, R. A., and Ojima, K. (1968). *Exp. Cell Res.* **50,** 151–158.
Gamborg, O. L., Constabel, F., Fowke, L. C., Kao, K. N., Ohyama, K., Kartha, K. K., and Pelcher, L. E. (1974). *Can. J. Genet. Cytol.* **16,** 737–750.
Gamborg, O. L., Murashige, T., Thorpe, T. A., and Vasil, I. K. (1976). *In Vitro* **12,** 473–478.
Ganapathy, P. S., Hildebrandt, A. C., and Riker, A. J. (1964). *Am. J. Bot.* **51,** 669.
Gautheret, R. J. (1959). "La culture des tissus végétaux." Masson, Paris.
Goebel, K. (1908). "Einleitung in die experimentelle Morphologie der Pflanzen." Teubner, Leipzig.
Graebe, J. E., and Novelli, G. D. (1966). *Exp. Cell Res.* **41,** 509–520.
Gregor, H. D., and Reinert, J. (1972). *Protoplasma* **74,** 307–319.
Gresshoff, P. M., and Doy, C. H. (1972). *Aust. J. Biol. Sci.* **25,** 259–264.
Hahlbrock, K., and Wellmann, E. (1971). *Planta* **94,** 236–239.
Hahlbrock, K., Ebel, J., Oaks, A., Auden, J., and Liersch, M. (1974). *Planta* **118,** 75–84.
Hollings, M. (1965). *Annu. Rev. Phytopathol.* **3,** 367–396.
Holsten, R. D., Sugii, M., and Steward, F. C. (1965). *Nature (London)* **208,** 850.
Hood, K. J. (1964). *Plant Physiol.* **39,** Suppl. 11.

Ikeda, T., Matsumoto, T., Kato, K., and Noguchi, M. (1974). *Agric. Biol. Chem.* **38,** 2297–2298.

Jones, L. H. (1974). *In* "Industrial Aspects of Biochemistry" (B. Spencer, ed.), FEBS, Vol. 30, Part 2, pp. 813–833. Elsevier, Amsterdam.

Kao, K. N., and Michayluk, M. R. (1975). *Planta* **126,** 105–110.

Kao, K. N., Miller, R. A., Gamborg, O. L., and Harvey, B. L. (1970). *Can. J. Genet. Cytol.* **12,** 297–301.

Kartha, K. K., and Gamborg, O. L. (1975). *Phytopathology* **65,** 826–828.

Kasha, K. T., ed. (1974). "Haploids in Higher Plants." University of Guelph, Ontario.

Kato, K., Shiozawa, Y., Yamada, A., Nishida, K., and Noguchi, M. (1972). *Agric. Biol. Chem.* **36,** 889–902.

Kato, A., Kawazoe, S., Ijima, M., and Shinizu, Y. (1976). *J. Ferment. Technol.* **54,** 82–87.

Keller, H., Wanner, H., and Baumann, T. W. (1972). *Planta* **108,** 339–350.

King, P. J., Mansfield, K. J., and Street, H. E. (1973). *Can. J. Bot.* **51,** 1807–1823.

Kirkland, D. F., Matsuo, M., and Underhill, E. W. (1971). *Lloydia* **34,** 195–198.

Koblitz, H. (1972). "Gewebekulturen." Fischer, Jena.

Komatsu, K., Nozaki, W., Takemura, M., and Nakaminami, M. (1976). Japanese Patent (Kokai) 76/19169.

Konoshima, M., Tabata, M., Yamamoto, H., and Hiraoka, N. (1970). *Yakugaku Zasshi* **90,** 370–377.

Krikorian, A. D., and Steward, F. C. (1969). *In* "Plant Physiology" (F. C. Steward, ed.), Vol. 5B pp. 227–326. Academic Press, New York.

Kubo, Y. (1974a). Japanese Patent (Kokai) 74/85286.

Kubo, Y. (1974b). Japanese Patent (Kokai) 74/86589.

Kurz, W. G. W. (1973). *In* "Tissue Culture: Methods and Applications" (P. F. Kruse, Jr. and M. K. Patterson, eds.), p. 359. Academic Press, New York.

Lamport, D. T. A. (1964). *Exp. Cell Res.* **33,** 195–206.

Lee, P. K., Carew, D. P., and Rosazza, J. (1972). *Lloydia* **35,** 150–155.

Maretzki, A., Delacruz, A., and Nickell, L. G. (1971). *Plant Physiol.* **49,** 521–525.

Matsumoto, T., Nishida, K., Noguchi, M., and Tamaki, E. (1970). *Agric. Biol. Chem.* **34,** 1110–1114.

Matsumoto, T., Okunishi, K., Nishida, K., Noguchi, M., and Tamaki, E. (1971). *Agric. Biol. Chem.* **35,** 543–551.

Matsumoto, T., Nishida, K., Noguchi, M., and Tamaki, E. (1973). *Agric. Biol. Chem.* **37,** 561–567.

Melchers, G., and Engelmann, U. (1955). *Naturwissenschaften* **42,** 564–565.

Miller, R. A., Shyluk, J. P., Gamborg, O. L., and Kirkpatrick, J. W. (1968). *Science* **159,** 540–542.

Misawa, M. (1977). *In* "Plant Tissue Culture and Its Bio-Technological Application" (W. Barz, E. Reinhard, M. H. Zenk, eds.), pp. 17–26. Springer, Berlin, Heidelberg, New York

Misawa, M., and Sakato, K. (1976). Japanese Patent Appl. 76/113203.

Misawa, M., Hayashi, M., Nagano, Y., and Kawamoto, T. (1973). Japanese Patent (Kokai) 73/6153.

Misawa, M., Tanaka, H., Chiyo, O., and Mukai, N. (1974a). Japanese Patent (Kokai) 74/7491.

Misawa, M., Sakato, K., and Hayashi, M. (1974b). Japanese Patent (Kokai) 74/126894.

Misawa, M., Hayashi, M., Tanaka, H., Ko, K., and Misato, T. (1974c). *Biotechnol. Bioeng.* **17,** 1335–1347.

Misawa, M., Sakato, K., Hayashi, M., Takayama, S., Tanaka, H., Misato, T., and Ko, K. (1975a). Japanese Patent Appl. 75/9828.

Misawa, M., Sakato, K., and Tanaka, H. (1975b) Japanese Patent (Kokai) 75/5891.

Misawa, M., Hayashi, M., Shimada, K., and Omotani, Y. (1975c). Japanese Patent (Kokai) 75/101510.

Misawa, M., Sakato, K., Tanaka, H., Misato, T., and Ko, K. (1976). Japanese Patent (Kokai) 76/70889.

Mitsuoka, S., and Nishi, T. (1974). Japanese Patent (Kokai) 74/94897.

Mothes, K. (1966). *Naturwissenschaften* **53,** 317–323.

Muir, W. H., Hildebrandt, A. C., and Riker, A. J. (1958). *Am. J. Bot.* **45,** 589–597.

Murashige, T. (1974). *Annu. Rev. Plant Physiol.* **25,** 135–166.

Murashige, T., and Skoog, F. (1962). *Physiol. Plant.* **15,** 473–497.

Nag, K. K., and Street, H. E. (1973). *Nature (London)* **245,** 270–272.

Nester, E. (1977). "Crown Gall". Raven, New York.

Neumann, D., and Müller, E. (1971). *Biochem. Physiol. Pflanz.* **162,** 503–513.

Nitsch, C. (1974). *C. R. Hebd. Seances Acad. Sci.* **278,** 1031–1034.

Nitsch, J. P. (1972). *Z. Pflanzenzuecht.* **67,** 3–18.

Patterson, B. D., and Carew, D. P. (1969). *Lloydia* **32,** 131–140.

Petiard, V., and Guinebault, P. R. (1976). German Federal Republic Patentagency Offenlegungschrift 2,603,588.

Petiard, V., Demarly, Y., and Paris, R. R. (1972). *Plant. Med. Phytother.* **6,** 41–49.

Puhan, Z., and Martin, S. M. (1970). *Prog. Ind. Microbiol.* **9,** 13–39.

Raveh, D., Hubermann, E., and Galun, E. (1973). *In Vitro* **9,** 216–222.

Reinert, J. (1959). *Planta* **53,** 318–333.

Reinert, J., and Bajaj, Y. S. P., eds. (1977). "Applied and Fundamental Aspects of Plant Tissue Culture." Springer, Berlin and New York.

Reinhard, E. (1967). *Dtsch. Apoth.-Ztg.* **107,** 1201–1207.

Reinhard, E. (1974). *In* "Tissue Culture and Plant Science" (H. E. Street, ed.), pp. 433–459. Academic Press, New York.

Routien, J. B., and Nickell, L. G. (1956). U.S. Patent 2,747,334.

Sacristan, M. D., and Melchers, G. (1970). "Die Kulturpflanze," Suppl. 6. Akademie-Verlag, Berlin.

Schleiden, M. J. (1838). *Arch. Anat. Physiol. Physiol. Abt.* **137.**

Schwann, T. (1839). "Mikroskopische Untersuchungen über die Übereinstimmung in der Struktur und dem Wachstum der Tiere und der Pflanzen." Berlin.

Sehti, J. K., and Carew, D. P. (1974). *Phytochemistry* **13,** 321–324.

Shah, R. R., Subbaiah, K. V., and Mehta, A. R. (1976). *Can. J. Bot.* **54,** 1240–1245.

Shantz, E. M., and Steward, F. C. (1952). *J. Am. Chem. Soc.* **74,** 6133–6135.

Short, K. C., Brown, E. G., and Street, H. E. (1969). *J. Exp. Bot.* **20,** 572–578.

Skoog, F., and Miller, C. O. (1957). *Symp. Soc. Exp. Biol.* **11,** 118–130.

Smith, H. H. (1974). *BioScience* **24,** 269–276.

Solt, M. L. (1957). *Plant Physiol.* **32,** 480–484.

Solt, M. L., Dawson, R. F., and Christman, D. R. (1960). *Plant Physiol.* **35,** 887–894.

Speake, T., McCloskey, P., Smith, W. K., Scott, T. A., and Hussey, H. (1964). *Nature (London)* **201,** 614–615.

Staba, E. J. (1969). *Recent Adv. Phytochem.* **2,** 75–106.

Staba, E. J., and Lamba, S. S. (1963). *Lloydia* **26,** 29–35.

Steck, W., and Constabel, F. (1974). *Lloydia* **37,** 185–191.

Steck, W., Bailey, B. K., Shyluk, J. P., and Gamborg, O. L. (1971). *Phytochemistry* **10,** 191–195.

Steck, W., Gamborg, O. L., and Bailey, B. K. (1973). *Lloydia,* **36,** 93–95.

Steward, F. C. (1958). *Am. J. Bot.* **45,** 709–713.

Steward, F. C., ed. (1969). "Plant Physiology," Vol. 5B. Academic Press, New York.

Steward, F. C., Caplin, S. M., and Millar, F. K. (1952). *Am. J. Bot.* **16,** 58–77.

Stöckigt, J., Treimer, J., and Zenk, M. H. (1976). *FEBS Lett.* **70;** 267–270.

Stohs, S. J. (1969). *J. Pharm. Sci.* **58,** 703–705.

Street, H. E. (1966). *In* "Cells and Tissues in Cultures" (E. N. Wilmer, ed.), Vol. 3, pp. 534–629. Academic Press, New York.

Street, H. E. (1973). "Plant Tissue and Cell Culture," Bot. Monogr. Vol. 11. Univ. of California Press, Berkeley and Los Angeles.

Street, H. E., ed. (1974). "Tissue Culture and Plant Science." Academic Press, New York.

Street, H. E. (1975). *IAPTC Newsl.* **15,** 2–4.

Street, H. E., und Henshaw, G. G. (1963). *Symp. Soc. Exp. Biol.* **17,** 234–256.

Street, H. E., and Withers, L. A. (1974). *In* "Tissue Culture and Plant Science" (H. E. Street, ed.), pp. 71–100. Academic Press, New York.

Street, H. E., Henshaw, G. G., and Buiatti, M. C. (1965). *Chem. Ind. (London)* **1,** 27–33.

Sugano, N., Miya, S., and Nishi, A. (1971). *Plant Cell Physiol.* **12,** 525–531.

Tabata, M., Yamamoto, H., Hiraoka, N., Marumoto, Y., and Konoshima, K. (1971). *Phytochemistry* **10,** 723–729.

Takebe, I. (1975). *Annu. Rev. Plant Pathol.* **13,** 105–125.

Tamaki, E., Morishita, K., Nishida, K., Kato, K., and Matsumoto, T. (1975). Japanese Patent 75/16440.

Torrey, J. G. (1959). *In* "Cell, Organism and Milieu" (D. Rudnick, ed.), pp. 189–222. Ronald, New York.

Torrey, J. G., and Reinert, J. (1961). *Plant Physiol.* **36,** 483–491.

Torrey, J. G., Reinert, J., and Merkel, N. (1962). *Am. J. Bot.* **49,** 420–425.

Tulecke, W. (1966). *Ann. N.Y. Acad. Sci.* **139,** 162–175.

Tulecke, W., and Nickell, L. G. (1960). *Trans. N.Y. Acad. Sci.* [2] **22,** 196–206.

Ukita, M., Furuya, A., Tanaka, H., and Misawa, M. (1973). *Agric. Biol. Chem.* **37,** 2849–2854.

van Staden, I., and Drewes, S. E. (1974). *Physiol. Plant.* **32,** 347–352.

Vasil, I. K., Hildebrandt, A. C., and Riker, A. J. (1964). *Am. J. Bot.* **51,** 677.

Veliky, I. A. (1972). *Phytochemistry* **11,** 1405–1406.

Veliky, I. A., and Genest, K. (1972). *Lloydia* **35,** 450–456.

Veliky, I. A., and Martin, S. M. (1970). *Can. J. Microbiol.* **16,** 223–226.

Venketeswaran, S. (1964). *Am. J. Bot.* **51,** 669.

Verma, D. P. S., and van Huystee, R. B. (1971). *Exp. Cell Res.* **69,** 402–408.

Wang, C. J., and Staba, E. J. (1963). *J. Pharm. Sci.* **52,** 1058–1062.

Wardlaw, C. W. (1968). "Morphogenesis in Plants" Methuen, London.

White, P. R. (1953). *Am. J. Bot.* **40,** 517–524.

White, P. R. (1963). "The Cultivation of Animal and Plant Cells," 2nd Ed. Ronald, New York.

Wilson, G. (1976). *Ann. Bot. (London)* [N.S.] **40,** 919–932.

Wilson, S. B., King, P. J., and Street, H. E. (1971). *J. Exp. Bot.* **21,** 177–207.

Yoeman, M. M. (1974). *In* "Tissue Culture and Plant Science" (H. E. Street, ed.), pp. 1–18. Academic Press, New York.

Zenk, M. H. (1974). *In* "Haploids in Higher Plants" (K. J. Kasha, ed.), pp. 339–353. University of Guelph, Ontario.

Zenk, M. H. (1976). "Das physiologische Potential pflanzlicher Zelkulturen." Westdeutscher-Verlag, Düsseldorf.

Zenk, M. H., El-Shagi, H., and Schulte, U. (1975). *Planta Med. Suppl.* pp. 79–101.

Chapter 13

Polysaccharides

K. S. KANG
I. W. COTTRELL

MICROBIAL TECHNOLOGY, 2nd ed., VOL. I
Copyright © 1979 by Academic Press, Inc.
All rights of reproduction in any form reserved. ISBN 0-12-551501-4

I. INTRODUCTION

The polysaccharides derived from plants and seaweeds have been in use for thousands of years. Gum arabic was a component of embalming fluids used by the ancient Egyptians, and red seaweeds were cooked by the Chinese during the time of Confucius to produce a gel which was then flavored and sweetened.

Large-scale utilization of plant gums and seaweed gums, such as algin, and development of economical extraction processes for the seaweed gums began in the 1920s. Since that time other natural sources have been developed to yield gums such as guar and locust bean. Also, chemical modifications of starch and cellulose have been successful in producing water-soluble gums.

In the continuing search for novel, natural, water-soluble polysaccharides, particular attention has been directed in recent years to the production of extracellular polysaccharides of microorganisms. The first microbial polysaccharide to be commercialized was dextran; it was used as a blood plasma extender and in other applications. This polysaccharide has had only limited success as a commercial water-soluble gum, although it has been the subject of a large amount of research and many attempts at commercialization.

Some other bacteria that have been shown to produce polysaccharides with novel properties are *Alcaligenes faecalis* var. *myxogenes, Azotobacter vinelandii,* and *Beijerinckia indica.* These organisms produce curdlan, bioalgin, and PS-7, respectively. The yeast *Aureobasidium pullulans* produces pullulan, a polysaccharide with unique properties; scleroglucan, a fungal polysaccharide, is produced by a species of *Sclerotium.* However, none of these microbial polysaccharides has been produced on a large scale commercially.

At this time the only microbial polysaccharide that has reached and maintained large-scale commercial production is xanthan gum, the polysaccharide produced by the bacterium *Xanthomonas campestris.*

II. NATURE OF MICROBIAL POLYSACCHARIDES

A. Relation to Cell Structure

Microorganisms produce polysaccharides of three distinct types: extracellular, structural, and intracellular storage forms. The extracellular polysaccharides, the main subject of this chapter, can be further classified into two forms: (a) capsules that are integral with the cell wall, as well as structurally demonstrable microcapsules, and (b) loose slime components that accumulate in large quantities outside of the cell wall then diffuse into the culture medium. During growth these extracellular polysaccharides contribute a gummy texture to bacterial colonies on a solid medium or cause increased viscosity in a liquid medium.

The capsular components can be separated readily from the amorphous loose slime by centrifugation. The slime formers may produce slime in large quantities and, in some cases, the viscosity becomes so great that a liquid culture remains in place when the culture flask is inverted. The polysaccharides produced by slime formers have the greatest industrial potential, since their polysaccharides can be recovered in large quantities from the culture fluids.

B. Natural Functions of Polysaccharides

Extracellular polysaccharides probably protect microorganisms against various adverse environmental conditions (Wilkinson, 1958). For example, the remarkably high moisture-holding capacity of the polysaccharide enables bacteria to maintain at least a minimum of moisture in their immediate environment, even after prolonged exposure to low humidity. Also, the thick capsule associated with extracellular polysaccharide production by bacteria provides a definite protection against bacteriophages, simply by serving as a physical barrier against the agent. This fact was substantiated by Maxted (1952) for group A streptococci. The capsules also provide a partial protection against phagocytosis and amoebic attack (Wilkinson, 1958).

A strain of *B. indica* (Lopez and Becking, 1968) was reported to metabolize its own polysaccharide. However, slime-producing bacteria normally are not capable of catabolizing their own extracellular polysaccharides (Wilkinson, 1958). There are indications, however, that fungi degrade and utilize their own capsular polysaccharides (Szaniszlo *et al.*, 1968), and the relatively large amount of starch stored by *Polytomella coeca* during growth serves as an energy reserve (Bourne *et al.*, 1958).

The development of microbial populations in natural environments, such as soil, seawater, lakes, and ponds, occurs most abundantly on the surface of solid particles or films. Capsular polysaccharides and slime may serve to cement the cells and their microcolonies to solid substrata (Corpe, 1970).

C. Environmental Influences

Many polysaccharide producers respond to environmental factors directly. For some microorganisms the carbon source determines both the quantity of polysaccharide formation and the quality of the product synthesized. For example, several strains of *Bacillus polymyxa,* including USDA strain 354, produce a heteropolysaccharide consisting of glucose, mannose, and fructose when grown on a sucrose agar medium; whereas on a medium containing a monosaccharide, such as fructose, glucose, galactose, or arabinose, the polysaccharide formed consists of glucose, mannose, and uronic acid (Forsyth and Webley, 1949). In contrast, another strain of *B. polymyxa* produces a heteropolysaccharide when the organism is cultured in liquid medium with glucose as a carbon source: however, a homopolysaccharide that contains only fructose accumulates when sucrose is used as the carbon source (Ninomiya, 1967). Conversely, the extracellular heteropolysaccharides produced by many microorganisms, such as *X. campestris,* do not vary significantly in composition or molecular weight as a result of changes in the substrate.

Ordinarily the carbon source concentration affects the efficiency of carbon source conversion into polysaccharide. For example, it has been reported that the conversion efficiency of glucose to polymer by *X. campestris* decreased markedly with an increase in glucose concentration (Rogovin et al., 1961).

Although a nitrogen source is necessary for both cell growth and enzyme synthesis for polysaccharide formation, an excess of nitrogen, in general, reduces conversion of carbohydrate substrate to extracellular polysaccharide. Usually, the effects of aeration can be related to the relative efficiencies of energy production during aerobic and anaerobic culturing. Under anaerobic conditions, the "Pasteur effect" takes place due to a low efficiency of energy production. All polysaccharide production of commercial significance requires an aerobic process. Also, in all polysaccharide fermentation, the product yield parallels an increase in the viscosity of the fermentation liquor. Thus, as the liquor thickens, a marked decrease in aeration efficiency occurs, which can be attributed to a much lower mass transfer rate between the air bubbles and the liquor.

Temperature often is critical in polysaccharide synthesis. In general, the optimal temperature for cell growth also is the optimum for product formation. All polysaccharide producers of any commercial significance, such as *X. campestris,* are mesophilic organisms. In order to combat the inefficient mass transfer caused by the high viscosity, large amounts of air that usually are hot and high degrees of agitation are applied during fermentation.

Another important factor affecting polysaccharide production is pH. During the latter stages of acidic polysaccharide fermentations, such as those involving *Xanthomonas* (Moraine and Rogovin, 1966) or *Azotobacter* gums (Cohen and Johnstone, 1963), the pH decreases because polymers and certain organic and inorganic acids form. Nevertheless, the bacterial polysaccharides of possible commercial significance appear to have an optimal pH for synthesis between 6.0 and 7.5. For fungal gum production, the optimal pH lies between 4.0 and 5.5.

Many microorganisms have a strict requirement for certain mineral elements. Among these elements are K, P, Mg, Mn, and Ca. Other elements, such as Mo, Fe, Cu, and Zn, may also be required. However, certain of the minerals that might be added can inhibit product formation. As a result, the mineral requirements for polysaccharide synthesis vary from species to species.

III. MECHANISM OF SYNTHESIS

A. Homopolysaccharides

The mechanism for microbial synthesis of polysaccharides, from monosaccharides as well as from disaccharide donors, has been reviewed by a number of authors (Wilkinson, 1958; Hestrin, 1962; Anderson, 1963; Sutherland, 1972). In general, the synthesis of most polysaccharides involves the addition of a single monomeric unit to the nonreducing end of an existing primer polysaccharide molecule. The general reaction for homopolysaccharide synthesis may be represented by Eq. (1):

$$G\text{-}O\text{-}X + [(G\text{-}O)_n - G] \xrightarrow{\ E\ } [(G\text{-}O)_{n+1} - G] + X \tag{1}$$

In this equation G-O-X designates a donor system in which G is the glycoside and X is a nonpolymeric product, such as pyrophosphate, uracil diphosphate (UDP), or sugar. The symbol E denotes the transglycosidation enzyme, and n is the molar quantity. In mammalian glycogen synthesis the symbol X varies in different systems. X in the equation is UDP. The X also represents UDP in bacterial cellulose synthesis by *Acetobacter xylinum* (Glaser, 1958) and glucan synthesis by *Rhizobium japonicum* (Dedonder and Hassid, 1964).

In the case of starchlike polysaccharides synthesized by a strain of *Neisseria perflava* (Hehre et al., 1947; Hehre and Hamilton, 1948), by *P. coeca* (Lwoff et al., 1950), and by certain species of *Corynebacterium* and streptococci (Carlson and Hehre, 1949), the X in the above equation is phosphate. However, most *Neisseria* strains, except *N. perflava,* have an amylosucrase that catalyzes the polymerization of the glucose moiety of sucrose to yield a starchlike polysaccharide. In this reaction, the intermediary formation of glucose 1-phosphate (G-1-P) is not involved

(Hehre *et al.*, 1947). The production of a starchlike polysaccharide by an amylomaltase obtained from mutants of *Escherichia coli* (Doudoroff *et al.*, 1949; Monod and Torriani, 1950) is another example of transglycosylation. In these cases, maltose serves as the donor substrate.

For dextran and levan synthesized from disaccharide donors, the X in the equation is a single monomeric sugar unit—fructose for the former and glucose for the latter. Ebert and Patat (1962) propose a two-step mechanism for dextran synthesis as shown in Eq. (2): (2a) is the propagation reaction and (2b) is the acceptor reaction.

$$(EP)_n + S \rightarrow ESP_n \rightarrow \left[(EP)_{n+1}\right] + F \qquad (2a)$$
$$+$$
$$A \longrightarrow AESP_n \rightarrow ES + APN \qquad (2b)$$

Where E is the enzyme, S is sucrose, P_n is growing dextran chain, F is fructose, and ESP_n is a complex. The dissociation of ESP_n controls the overall reaction that results in growth of the polymer chain by one glucose unit with the liberation of one molecule of fructose. It was found that dextran sucrase has no detectable catalytic action on glucose 1-phosphate (G-1-P) (Hehre, 1943), or glucose 6-phosphate (G-6-P), fructose 6-phosphate (F-6-P), fructose 1,6-diphosphate (F-1,6-diP), and adenosine triphosphate (ATP) (Carlson *et al.*, 1953). Likewise, the levansucrases produced by *Aerobacter levanicum* (Hestrin *et al.*, 1955), *B. polymyxa,* and *Bacillus subtilis* (Hestrin, 1944; Hestrin and Shapiro, 1944) are not affected by inorganic phosphate, F-6-P, or F-1,6-diP.

Thus, the evidence indicates that dextransucrase, levansucrase, and amylosucrase directly transfer a single monomeric sugar unit, which originated from the disaccharide donors, Eq. (3):

$$n \text{ Sucrose} \xrightarrow{\substack{\text{dextransucrase and} \\ \text{amylosucrase}}} (\text{glucose})_n + n \text{ fructose}$$

$$n \text{ Sucrose} \xrightarrow{\text{levansucrase}} (\text{fructose})_n + n \text{ glucose} \qquad (3)$$

$$n \text{ Maltose} \xrightarrow{\text{amylomaltase}} (\text{glucose})_n + n \text{ glucose}$$

The branching of glycogen- and starchlike substances, as well as levan and dextran, results from the action of additional enzymes in the reaction mixture. The Q enzyme, which is capable of converting linear amylose to branched amylopectin, has been found in *N. perflava* (Hehre *et al.*, 1947) and *P. coeca* (Bebbington *et al.*, 1952). A similar branching enzyme operates in dextran synthesis (Bovey, 1959). Ebert and Brosche (1967) showed that branches in dextran molecules are produced by already formed dextran acting as acceptors. When tritium-labeled dextran is added to the reaction mixture for dextran synthesis, the branched dextran formed has unlabeled side chains and a labeled main chain.

B. Heteropolysaccharides

Exocellular heteropolysaccharide synthesis is catalyzed by multienzyme systems, and each enzyme assumes a high degree of specificity. Therefore, heteropolysaccharide synthesis is much more complex than homopolysaccharide synthesis. Information on the mechanism of synthesis of exocellular heteropolysaccharides of commercial significance is still very scant. Most of the information has been obtained from polysaccharide syntheses of immunologically and clinically important organisms.

Sutherland (1977) divided the process of exocellular polysaccharide synthesis into four major sequences: (1) substrate uptake, (2) intermediary metabolism, (3) formation of exopolysaccharide, and (4) modification and extrusion. These sequences are logical and useful in explaining the polysaccharide biosynthesis. Briefly, the substrate enters the cell, usually by active transport and group translocation involving substrate phosphorylation. After entry, the substrate is subjected to either catabolic pathways or pathways leading to the polysaccharide synthesis. The synthetic pathways involve substrate activation by formation of various sugar nucleotides, sugar phosphates, and the interconversion of sugar.

Monosaccharides from sugar nucleotides are added sequentially to growing polysaccharide chains in correct linkages. Isoprenoid alcohol phosphate serves as a carrier.

The role of the isoprenoid lipid, as monosaccharide carrier in polysaccharide synthesis, is well documented in *Streptococcus faecalis* (Umbreit et al., 1972), *Aerobacter aerogenes* (Troy et al., 1971), *Micrococcus lysodeikticus* (Scher et al., 1968), and various other bacteria (Sutherland, 1972). Finally, intermediate or incomplete polymer molecules are modified and extruded from the cell surface to form exocellular polysaccharides.

Bacterial algin exemplifies Sutherland's scheme of exocellular polysaccharide synthesis. This algin (which is a copolymer of D-mannuronic acid and L-guluronic acid) is produced by a strain of *A. vinelandii*. It is partially acetylated (Gorin and Spencer, 1966). Pindar and Bucke (1975) describe an anabolic pathway leading to the formation of alginic acid. More recently, Deavin et al. (1977) summarized algin synthesis by *A. vinelandii* as shown in Fig. 1.

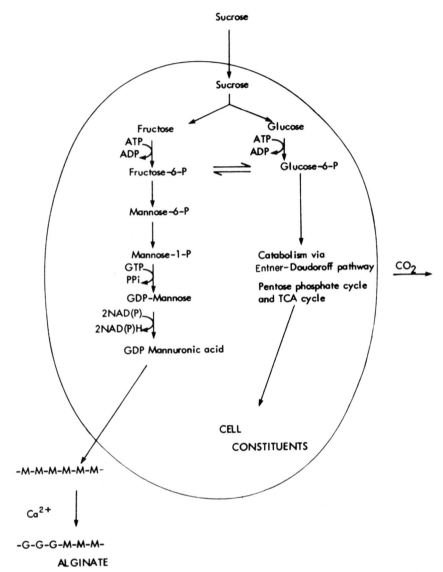

FIGURE 1. Metabolism of *Azotobacter vinelandii* in relation to alginate synthesis.

Sucrose, the carbon source, is transported into the cell and acted on by invertase, and G-6-P and F-6-P are formed by their respective kinases. Fructose 6-phosphate enters the alginate biosynthetic pathway in the sequence shown in Eq (4):

Mannose 6-phosphate (M-6-P) $\longrightarrow$ mannose 6-phosphate (M-1-P)

$$\downarrow \tag{4}$$

GDP–mannuronic acid $\longleftarrow$ glucose diphosphate (GDP)–mannose

As mannuronic acid residues accumulate, they then polymerize to form polymannuronate and are subsequently extruded from the cell. As Haug and Larsen (1971) demonstrated earlier, a polymannuronic acid C-5-epimerase partially converts D-mannuronic acid residues to L-guluronic acid residues extracellularly to form alginate. A part of the carbon source is converted via the Entner–Duodoroff pathway, pentose phosphate cycle, and tricarboxylic acid cycle for energy and biomass production.

Recently, Sutherland (1977) constructed a biosynthetic pathway for xanthan gum produced by *X. campestris* (Fig. 2) based on information obtained from other bacteria. Although only limited information is avail-

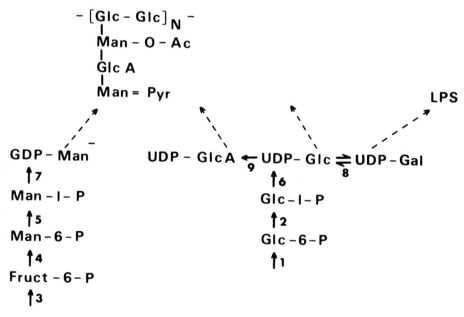

FIGURE 2. Biosynthesis of *Xanthomonas* polysaccharides. (1) Hexokinase, (2) phosphoglucomutase, (3) phosphoglucose isomerase, (4) phosphomannose isomerase, (5) phosphomannomutase, (6) UDP-Glc pyrophosphorylase, (7) GDP-Man pyrophosphorylase, (8) UDP-Gal epimerase, (9) UDP-Glc dehydrogenase.

able, enzyme specificity and the cytoplasmic membrane structure appear to be the factors that determine and control the polysaccharide synthesis.

IV. BACTERIAL POLYSACCHARIDES

A. Azotobacteraceae

1. Azotobacter Polysaccharides

All four genera listed in the family Azotobacteraceae (Breed et al., 1957) contain exocellular gum formers. The better known slime formers are A. vinelandii, B. indica, Azotomonas agilis, and Derxia gummosa. The nomenclature of microorganisms for this section follows that described in the 8th edition of Bergey's Manual (Buchanan and Gibbons, 1974) in which the species previously classified as Azotobacter indicus is described as B. indica.

Cohen and Johnstone (1964) isolated extracellular polysaccharides from three strains of A. vinelandii. Their analysis revealed that all three polymers contained galacturonic acid, D-glucose, and rhamnose in an approximate ratio of 43 : 2 : 1. The polymers probably contained mannuronolactone as well. The slimes from strains 155 and 102 were acetylated, whereas slime from strain 3A was not. The major slime components did not change whether the energy source supplied to the cells was sucrose, glucose, fructose, or ethanol.

For seven strains of A. vinelandii these workers (Cohen and Johnstone, 1963) also showed a relationship between pH and slime formation when these strains were cultured in Burk's broth. In general, they found that all slime-forming strains lowered the pH of the medium, whereas nonslime formers did not.

Gorin and Spencer (1966) discovered that certain strains of A. vinelandii cultured in Burk's medium produced extracellular polysaccharides consisting of D-mannuronic acid and a small amount of L-guluronic acid units. This polysaccharide resembled the algal polysaccharide, alginic acid. The only known difference between bacterial and algal alginates was the presence of O-acetyl in the former polysaccharide.

A structural study by Larsen and Haug (1971) that used three strains of A. vinelandii revealed that the sequences of uronic acid monomers were distributed in a blockwise manner as previously demonstrated in algal alginate. It was also shown that the D-mannuronic acid-L-guluronic acid (M/G) ratio could be controlled by adjusting the calcium concentration in the fermentation medium; the M/G ratio of the polymer was increased by decreasing calcium ion concentration. A subsequent study (Haug and Larsen, 1971) demonstrated that the polymannuronic acid C-5-epimerase, which catalyzes the transformation of the mannuronic acid residue to the G residue, was operating and the enzyme activity was influenced by Ca^{2+} (Deavin et al., 1977). Couperwhite and McCallum (1974) observed that EDTA in Burk's medium affected the M/G ratio. In the presence of EDTA the M/G ratio of the polymer increased markedly. This increase resulted from alteration of the epimerase, probably occurring when EDTA chelated Ca^{2+}.

Davidson and co-workers (1977) showed by using polyguluronate lyase and polymannuronate lyase that O-acetyl groups in the algin elaborated by A. vinelandii do not occur regularly in the polymer. They were associated with only one portion of the polymer. From this the authors theorized that the acyl groups protected certain portions of the mannuronic acid residues from epimerization during the formation of bacterial alginate.

Deavin and co-workers (1977) reported a continuous fermentation under phosphate-limited and oxygen-controlled conditions. Under these conditions the maximum yield of alginate (~45% of the sucrose utilized) occurred at the low respiration rates. At higher respiration rates the yield decreased dramatically, because a greater proportion of the substrate was being oxidized to carbon dioxide. The alginate yield in a typical batch culture was only 25% of the sucrose utilized.

2. Beijerinckia Polysaccharides

Quinnell and co-workers (1957) reported that the polysaccharide produced by one strain of B. indica consisted of glucose, glucuronic acid, and aldoheptose in the ratio of 3 : 2 : 1. On the basis of extensive structural work on the extracellular polysaccharide of another strain of B. indica, Parikh and Jones (1963) proposed a linear molecular structure composed of repeating units of D-glucuronic acid, D-glucose, and D-glycero-D-mannoheptose. Lopez and Becking (1968) reported that the polymer of B. indica strain Hawai-2 consisted of glucose, galactose, mannose, glucuronic acid, and galacturonic acid, but no heptose.

3. Polysaccharide PS-7

Kang and McNeely (1976a) developed a heteropolysaccharide, PS-7, by using a strain of B. indica that they isolated from soil. PS-7 is composed of glucose, rhamnose, and glucuronic acid in an approximate weight ratio of 6.6:1.5:1. PS-7 has an acetyl content of about 9%. Solutions of PS-7 gum are characterized by a high viscosity, a high degree of pseudoplasticity, and an excellent solubility in seawater and even in brine containing 25% salt. PS-7 also exhibits good pH and temperature stability. Figure 3 indicates the high viscosity of this polysaccharide.

Probably the most outstanding property of PS-7 is its exceptionally good suspending ability. Carico (1976a) compared the suspending ability of PS-7 to that of various other gums by means of a standard American Petroleum Institute sand-content tube. Figure 4 is a plot of the settling rates of sand in different polymer fluids. The dotted portion of the lines represents extrapolation to 100%, the time required for complete settling of sand to occur. PS-7 was the most efficient polymer for suspending the sand particles.

The high pseudoplasticity is depicted in Fig. 5 from a viscogram obtained using a Fann viscometer, Model 35. Note that the concentration of PS-7 is only one-half the concentration of xanthan gum or one-quarter of other polymer concentrations. These properties indicate that PS-7

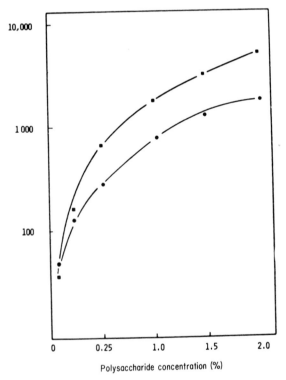

FIGURE 3. The viscosity of PS-7 (■) and xanthan gum (●).

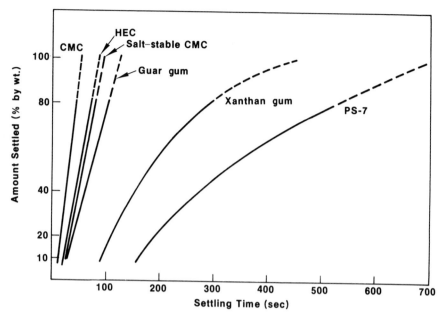

FIGURE 4. The settling rates of sand in different polymer fluids. CMC, carboxy-methylcellulose; HEC, hydroxyethylcellulose.

gum will find a variety of applications particularly in oil well drilling (Kang and McNeely, 1976b) and pseudoplastic water-based paint (Kang and McNeely, 1975).

Possible food applications have been also suggested (Kang and Kovacs, 1975). Recently, Kang and McNeely (1977) described production, properties, and possible applications of PS-7.

B. *Alcaligenes* Polysaccharides

Harada (1965) isolated *A. faecalis* var. *myxogenes* 10C3 from soil and found that it produced a unique extracellular polysaccharide when D-glucose or ethyleneglycol was used as a carbon source. The polymer was named succinoglucan 10C3, as it comprised approximately 10% succinic acid, 70–80% D-glucose, and small amounts of galactose and mannose. Methylation and periodate oxidation showed that this gum consisted of β-D-glucose linked through the (1→3), (1→4), and (1→6) positions and a small proportion of β-D-galactose linked through the (1→3) positions (Harada *et al.*, 1969). A high-yielding mutant (strain 22) produced 1.2 gm of the polysaccharide from 4 gm of glucose in 3 days

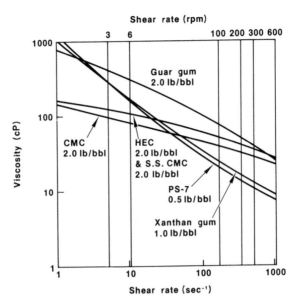

FIGURE 5. The viscosities of various polymers.

(Amemura and Harada, 1971). In 1% aqueous dispersions the free acid and the calcium salt exhibited high viscosities of 1560 and 2500 cP, respectively, as measured by a Cannon-Fenke viscometer at 30°C. It is interesting to note that the viscosity of the sodium salt was extremely low and was increased greatly by addition of many inorganic salts, especially ferric and aluminum ions (Harada and Yoshimura, 1965).

Another mutant derived from *A. faecalis* var. *myxogenes* 10C3 (mutant K) produced a homopolysaccharide known as curdlan (Harada *et al.*, 1968). This polymer consisted of a nonbranched linear chain of D-glucopyranosyl units joined entirely by β-D-(1→3) glycosidic linkages. In a defined or in a semisynthetic medium, the strain converted 50% of the added D-glucose into curdlan after 5 days' incubation at 32°C (Harada *et al.*, 1967). Thus, this polysaccharide, known to be neutral and soluble neither in water nor in acidic solution, dissolves in alkaline solutions, such as 0.5 N potassium hydroxide. Curdlan forms a resilient gel when a water suspension is heated above 54°C. The gel strength increases with the increase in temperature. The gel strength is almost constant between pH 2.5 and 10; strength decreases rapidly at pH slightly above 10 and it does not gel at about pH 12. The curdlan gel has properties intermediate between the brittleness of agar gel and the elasticity of gelatin, as shown in Fig. 6 (Harada, 1977).

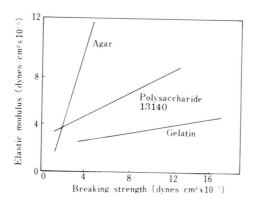

FIGURE 6. Relationship between the breaking strength and elastic modulus of various gels. Samples of gels were sliced into cylindrically shaped pieces 23-mm thick and then measured with Autograph model IM-100.

The curdlan-producing organism was further mutated to obtain strain NTK-u (Kimura *et al.*, 1963). This strain produces a thermogelable polysaccharide 1314) similar to curdlan. Various food applications of this polysaccharide, such as in processing boiled noodles, hamburger, and sausage, have been suggested by Harada (1977).

C. *Arthrobacter* Polysaccharides

Two strains of *Arthrobacter viscosus*, NRRL B-1797 and NRRL B-1973, are known to be extracellular polysaccharide producers (Gasdorf *et al.*, 1965; Knutson *et al.*, 1971). Polysaccharide B-1973 contains D-glucose, D-galactose, and D-mannuronate in the molar ratio of 1 : 1 : 1. The sugar composition of B-1797 polysaccharide is similar to that of B-1973, except that the former gum contains glucuronic acid. In addition, B-1973 gum is highly acetylated (25%), whereas B-1797 polysaccharide is only slightly acetylated. The polysaccharide B-1973 has been subjected to more extensive investigation (Jeanes *et al.*, 1965; Anonymous, 1962; Cadmus *et al.*, 1963; Sloneker *et al.*, 1968). The fermentation medium used for these investigations contained 2–4% glucose, 0.3% hydrolyzed casein, dipotassium acid phosphate adjusted to a pH of 7.0, and traces of magnesium and manganese salts. After clarification the native polysaccharide B-1973 was recovered by precipitation with methanol. Yields of 40–45%, based on glucose, have been obtained in a 4-day aerobic fermentation at 25°C.

These studies indicate that a high proportion of acetylated groups in the native polysaccharide interferes with hydration and molecular

alignment to the extent that poor films are formed when aqueous solutions are dried. However, flexible films are formed after deacetylation.
Also, solutions of the deacetylated polysaccharide show pseudoplastic
properties and greater shear stability than the native polysaccharide. At
0.1% concentration the deacetylated polysaccharide B-1973 shows a
more pronounced and extended increase in viscosity in the presence of
KCl than does the native polysaccharide (Jeanes *et al.*, 1965).

Arthrobacter stabilis NRRL B-3225 produces a heteropolysaccharide
that differs in chemical composition and properties from the *A. viscosus*
gums (Gill, 1972; Knutson *et al.*, 1971). Polysaccharide B-3225 consists
of glucose, galactose, and acetate in the molar ratio of 4 : 2 : 1 (Jeanes,
1974). This polysaccharide has equimolar proportions of pyruvate and
succinate linked as the half ester. Treatment with dilute alkali removes
acetyl and succinyl ester substituents, leaving the pyruvic acid ketal as
the only acidic group. This polysaccharide is characterized by a good
heat stability of its solution viscosity.

Arthrobacter polysaccharides have not been produced commercially.
The potential market appears to be small, probably due to the greater
salt sensitivity of the polysaccharide, its cross-linking tendencies on
heating, as well as a lack of toxicity studies.

D. *Methylomonas* Polysaccharides

Methylomonas mucosa produces a heteropolysaccharide in liquid
medium containing methanol as a carbon source. Acid hydrolysis followed by gas chromatographic analysis indicates that this polysaccharide is composed of glucose, galactose, mannose, and pyruvic acid
in a molar ratio of (1.5–2.0) : 1 : 1 : (1.5–2.0). The solution viscosity is
quite stable to a wide range of NaCl concentrations. However, the gum
is sensitive to pH in that the viscosity is at a maximum at pH 4.0 and is
slightly lower at pH 7.0. Above pH 7 or below pH 4 the viscosity decreases.

The *Methylomonas* polysaccharide is known to have drag-reducing
properties and to be pseudoplastic. On the basis of these properties
various industrial applications for this product, such as thickening and
suspending agents in foods, cosmetics, paints, oil well drilling, and oil
well flooding (Tannahill and Finn, 1975), have been suggested. The
crude polymer is negatively charged and has a 1% viscosity of approximately 350 cP as measured by a Brookfield viscometer at 30 rpm and
25°C. However, it has been claimed that the product viscosity can be
drastically increased and the fermentation time shortened by the controlled addition of methanol, the carbon source, to the fermentation

process near the end of the logarithmic phase of the bacterial growth. Recent studies by Tam and Finn (1977) on the growth kinetics for polymer production indicate that a semicontinuous fermentation is feasible for better reproducibility and a more economic fermentation than a batch process.

E. Miscellaneous Bacterial Polysaccharides

1. Zoogloea

It is well known that species of *Zoogloea* are one of the predominant bacterial groups in activated sludge (Dias and Bhat, 1964; Friedman *et al.*, 1968) and that polysaccharides are a major factor contributing to the physical characteristics of the sludge (Tenney and Stumm, 1965; Unz and Farrah, 1974). Friedman and his co-workers (1968) isolated a bacterium from sewage and identified it as *Zoogloea ramigera*.

Wallen and Davis (1972) reported that a water-soluble heteropolysaccharide was formed by *Z. ramigera* NRRL B-3669M when glucose was used as a carbon source. The polymer consisted of glucose, mannose, and galactose units that were linked through the (1→3) and (1→6) positions. This gum had a viscosity of about 2000 cP at 0.75% polymer concentration in water at 25°C, and the gum solution was quite pseudoplastic. Studies in our laboratory (unpublished results) indicate that this gum has fairly good stability against heat, pH, and mechanical shear.

2. Corynebacterium

Recently, Nagahama *et al.* (1977) isolated *Corynebacterium* species from effluents of the sweet potato starch-manufacturing plant. These isolates produce viscous anionic polysaccharides that contain 44–57% glucuronic acid and 9–12% pyruvic acid, the remaining constituents being mannose, galactose, and glucose in an approximately equimolar basis. These strains convert up to 66% of 3% "mashed potato" into polysaccharide in 4 days at 34°C. Although no detailed physical property study has been made, these polysaccharides appear to have a relatively high viscosity.

Corynebacterium equi. var. *mucilaginosus* ATCC 21521 produced a polysaccharide with *n*-paraffin mixture of $C_{12}-C_{17}$ (Yamatodani and Kanamaru, 1972). This polymer consisted of glucose and mannose, succinic acid, lactic acid, and acetic acid. The polymer yield in a typical fermentation with 7% *n*-paraffin mixture (v/v) was 1.7 gm/100 ml fermentation liquor. This gum was claimed to have a strong adhesive property at low concentrations. *Corynebacterium viscosus* also produced an acidic polysaccharide with an *n*-paraffin mixture of C_{13}–C_{16} (Yamada and Furukawa, 1973).

3. Rhizobium

The extrapolysaccharides that have been elaborated by *Rhizobium leguminosarum, Rhizobium phaseoli,* and *Rhizobium trifdii* contained D-glucose, D-glucuronic acid, D-galactose, pyruvic acid, and O-acetyl groups in the approximate proportions 5 : 2 : 1 : 2 : 3 (Zenvenhuizen, 1973). These polymers contained residues of (1→3)-linked D-glucose substituted at C-4 and C-6 by pyruvate (13%), (1→4)-linked D-glucose (32%), and (1→4)-linked D-glucuronic acid (20%); (1→4, 1→6)-linked branching residues of D-galactose and/or D-glucose (13%), and terminal D-glucose and/or D-galactose residues substituted at C-4 and C-6 by pyruvate (13%). *Rhizobium meliloti* gum did not contain glucuronate.

4. Bacillus

Cadmus *et al.* (1967) reported that *B. polymyxa* could utilize 3% glucose within 72–96 hours with 30–35% conversion. A different strain of *B. polymyxa* was isolated by Ninomiya and Kizaki (1969), which converted 5% glucose into polysaccharide with an efficiency of 36% after a 62-hour fermentation. In our laboratory, fermentation conditions were devised for a strain of *B. polymyxa* in which 3% glucose could be completely utilized within 20–24 hours with approximately 45% conversion efficiency (unpublished data). Therefore, under these rapid fermentation conditions, microbial contamination is not a major problem, since the native bacteria outgrow most contaminants.

Bacillus polymyxa slime is a heteropolysaccharide composed of D-glucose, D-mannose, D-galactose, and D-glucuronic acid in a ratio of 3 : 3 : 1 : 2. The gum exhibits relatively high viscosity and high pseudoplasticity. The viscosity increases sharply in aqueous solution when various salts are added, but it decreases drastically with a small increase in temperature.

Morita and Murao (1974) reported that *B. subtilis* FT-3 produced an acidic polysaccharide consisting of glucose, galactose, fucose, glucuronic acid, and O-acetyl in the molar ratio of 2 : 2 : 1 : 1 : (1.4). The mean molecular weight of this polymer was estimated as 10^4–10^6. Since these organisms are spore formers, the polysaccharide should be further processed in most cases either to remove or to kill the spores. This fact has discouraged commercial development of the *Bacillus* polysaccharides.

5. Levan

Levans are polymers of anhydro-β-D-fructose that contain mainly β-2,6-glycosidic linkages. Levan is produced by a number of microorganisms, including species of *Bacillus, Pseudomonas, Acetobacter,* and *Corynebacterium.* Biosynthesis of levan also can be readily accomplished when cell-free extracts of *B. subtilis* are used with the addition of sucrose.

Like dextran, levan is a branched-chain homopolysaccharide having molecular weights in the range of $1-100 \times 10^6$ (Hestrin, 1962). Despite the high molecular weight, the viscosities of aqueous solutions of levan are considerably lower than those of linear polysaccharides such as algin. Thus, levans appear to be unlikely prospects for development as commercial water-soluble gums.

6. Chromobacterium

Corpe (1960) described the formation of an extracellular polysaccharide by *Chromobacterium violaceum.* In peptone broth this organism converted glucose into a polysaccharide comprised of glucose, uronic acid, and amino sugar in a ratio of 5 : 1 : 1. Later, it was found (Corpe, 1964) that growth and polysaccharide synthesis in a defined medium were markedly enhanced when amino acids were substituted for ammonia as a nitrogen source, and the best polysaccharide yield occurred when the ratio of carbohydrate to nitrogen was 10 : 1. Studies conducted by Corpe (1960) indicate that the polysaccharide exhibits an unusual resistance to the microbial degradation by soil microorganisms and has some interesting rheological properties.

7. Starch- and Celluloselike Polymers

Many strains of *Neisseria* produce starchlike extracellular polysaccharides from sucrose (Carlson and Hehre, 1949). The product is a mixture of a linear α-1,4-linked amylose fraction and a branched-chain portion resembling amylopectin or glycogen. In these biosyntheses the enzyme amylosucrase transfers the D-glucopyranosyl units from the sucrose molecule to the end of the growing polymer chain.

As early as 1886 Brown reported that *Acetobacter xylinium* elaborated tough membranes of cellulose when grown in a suitable nutrient solution. Much later Tarr and Hibbert (1931a) reported cellulose formation by *Acetobacter pasteurianus, Acetobacter rancens, Sarcina ventriculi,* and *Bacterium xyloides.* The authors established the preferred substrate for *A. xylinium* as 5–10% of a hexose sugar, 0.1% asparagine, 0.5% potassium dihydrogen phosphate, 0.1% sodium chloride, and 0.5% ethanol. Aerobic fermentation of *A. xylinium* for 10 days at 30°C resulted in the formation of extracellular cellulose as a meshwork of intertwined microfibrils (Muhlethaler, 1949), a product identical to plant cellulose (Hestrin and Schramm, 1954).

Recently, Kjosbakken and Colvin (1975) demonstrated the existence of a soluble polymer of glucose in a cell-free particulate enzyme system from *A. xylinium.* This polymer was subsequently shown by King and Colvin (1976) to be an intermediate in bacterial cellulose synthesis. More recently, Colvin and his co-workers (1977) characterized this polymer as being a linear chain of β-(1→4) glucose with single glucose residues branched at position 2 of every third glucose, on the average.

A highly specialized use for bacterial cellulose in membranes for osmometry was reported (Masson *et al.*, 1946). Since starch and cellulose are available from plant sources at low prices and in very large amounts, it is unlikely that the microbial products could develop any significant commercial production.

V. FUNGAL POLYSACCHARIDES

A. Pullulans

Pullulan is the generic name applied to a type of α-D-glucan produced by *Aureobasidium (Pullularia) pullulans.* The polysaccharide was isolated and partially characterized by Bernier (1958) and further characterized by Bender *et al.* (1959).

Typical pullulans are linear polymers consisting of predominant repeating units of maltotriose attached via α-1,6-linkages. Small numbers of α-maltotetrose units are frequently present within the polymer chain (Catley and Whelan, 1971; Taguchi *et al.*, 1973). In addition, the presence of varying amounts of 1,3-linked glucosyl residues in pullulans elaborated by different strains is also reported (Sowa *et al.*, 1963). Pullulanase depolymerizes pullulan by hydrolyzing α-1,6-linkages (Drummond *et al.*, 1969; Taguchi *et al.*, 1973). The enzyme also attacks α-1,6 bonds in starch (Hathaway, 1971).

The molecular weight of a pullulan varies considerably depending on the fermentation conditions and strains used (Zajic and LeDuy, 1973). To meet specific applications Hayashibara Laboratories in Japan is said to be producing pullulan with molecular weights ranging between 1×10^4 and 4×10^5 (Yuen, 1974).

Catley and Whelan (1971) investigated the efficiency of several mono-

and disaccharides, glycerol, and acetate as carbon sources for pullulan elaboration and found that sucrose resulted in the best conversion to pullulan. However, it was reported that product yield over 70% could be obtained by replacing the conventional carbon sources with less expensive starch syrups (Yuen, 1974). The optimal dextrose equivalent of starch hydrolysate for pullulan production was 50–60%.

Yuen (1974) described pullulan as having characteristics similar to those of polyvinyl alcohol, and pullulan-made articles closely resemble those of styrene in transparency, gloss, and hardness, but are much more elastic. Pullulan forms very low oxygen-permeable films (Table I) and high-tensile fibers, the strengths of which are described as comparable to those of nylon.

A number of applications of pullulan and its derivatives have been proposed and patented primarily by Japanese workers. The applications include use as a flocculator of clay slimes in hydrometallurgical processes (Zajic and LeDuy, 1973), use in coating and packing material for foodstuffs and pharmaceuticals, in noncaloric food, in adhesives, and in the manufacture of special fibers and fabrics (Yuen, 1974).

B. *Rhinocladiella* Polysaccharides

The black yeastlike fungus *Rhinocladiella mansonii* NRRL Y-6272 elaborates an extracellular polysaccharide composed of residues of N-acetyl-D-glucosamine and N-acetyl-D-glucosaminuronic acid in a molar ratio of approximately 2 : 1 (Jeanes *et al.*, 1971). The organism converted 5% glucose to polysaccharide with a conversion efficiency of 13% in 4 days at 25°C. The medium used contained L-asparagine or urea

TABLE I. Oxygen Permeability Comparison of Various Films

Sample[a]	Oxygen permeability (ml/m², 24 hours, atm, 25°C)
PS	2.50
PM	2.01
PO	2.15
P_1	1.30
P_2	0.60
Cellophane	4.70
Moisture-proof cellophane	8.58
Polypropylene	1100.00

[a] PS, PM, PO, P_1 and P_2 represent pullulan films of different origins.

and several metal salts. The viscosity of the resulting cell-free polysaccharide in 1% aqueous solution was 9400 cP as measured by a Brookfield viscometer, Model LVF at 30 rpm (Burton *et al.*, 1976).

Another black yeastlike fungus, *Rhinocladiella elatior* Mangenot NRRL YB-4163, isolated from paper mill effluent, also produces a polysaccharide (Watson *et al.*, 1976). This polysaccharide is composed of 2-acetamido-2-deoxy-D-glucuronic acid residues. It is superior to YB-6272 in stabilization of oil–water emulsions and shows a high elasticity; when solutions (1%) of YP-4163 are poured or stirred, the entire solution tends to cling together.

C. Miscellaneous Fungal Polysaccharides

Bouveng *et al.* (1967) reported a water-soluble polysaccharide production when *Armillaria mellae* was grown in a broth containing D-glucitol as carbon source. They were able to fractionate this gum into four fractions by DEAE-cellulose chromatography. The two fractions consisted of a glycogen-type glucan and a β-D-(1→3) glucan, respectively. The third fraction was found to consist of a glucan and a xylomannan, and the last was comprised of D-galactose, D-mannose, fucose, and 3-O-methyl-D-galactose in a weight percentage ratio of 42 : 30 : 13 : 15.

Szaniszlo *et al.* (1968) described a capsular polysaccharide derived from *Leptosphaeria albopunctata,* a marine fungus. The gum is composed of "large amounts of glucose and minute amounts of mannose." Polysaccharide synthesis by this organism requires concentrations of salts similar to those in seawater. Furthermore, the synthesis is enhanced by NaCl concentrations above those of artificial seawater. Organisms of this type that produce sufficient gum to have industrial significance have not been reported.

Archer and his co-workers (1977) described an extracellular branched D-glucan produced by *Monilinia fructigena* in glucose–malt medium. More than 90% of this gum is glucose which has 1→4 main linkages, with a lower proportion of 1→2 linkages. The degree of branching varied between every fifth and every tenth residue. The gum contained small amounts of mannose and galactose.

A large number of fungal species are known, and new polysaccharides are frequently discovered. The authors recommend an excellent review article by Gorin and Spencer (1968) for the structural chemistry of yeast and fungal polysaccharides.

Antitumor Activities

Many fungal polysaccharides of the glucan type have been reported to have antitumor activities. A large amount of literature has been published in Japan on this subject. Whistler and his co-workers (1976) reviewed these and other microbial polysaccharides with bacterial, yeast, and plant origins. Several examples of fungal glucans having antitumor activities are given in Table II.

From their studies the authors concluded that the polysaccharides examined so far do not seem eligible for clinical trial in human cancer therapy, although they are quite active in suppressing transplanted tumors. Thus, they suggest further search for new polysaccharides having stronger antitumor activities than those so far discovered.

VI. YEAST POLYSACCHARIDES

A. Phosphomannan

Yeasts of the genus *Hansenula* and related genera produce extracellular phosphomannans from glucose (Slodki *et al.*, 1961). These polysaccharides yield on hydrolysis only D-mannose and D-mannose 6-phosphate. Developmental work was carried out on the phosphomannan produced by *Hansenula holstii* NRRL Y-2448 (Jeanes *et al.*, 1961a) at the Northern Utilization Research and Development Division of the USDA. This phosphomannan was produced in an aerobic 4-day fermentation at 28°C in a medium containing 6% glucose, organic nitrogen, potassium dihydrogen phosphate, and trace elements, with adjustment of pH to 5.0 (Anderson *et al.*, 1960). The product was isolated by alcohol precipitation from the fermentation beer. Yields were 40–55% based on utilization of the glucose substrate. *Hansenula capsulata* Y-1842 also produces a phosphomannan (Slodki, 1963).

Phosphomannan from *H. holstii* Y-2448 contains mannose and phosphorus in a molar ratio of 5 : 1 (Jeanes *et al.*, 1961a). The molecular weight by light scattering is $\sim 16 \times 10^6$. Jeanes and Watson (1962) proposed that the structure of this polymer consists of a repeating unit of 10 mannose residues which are distributed on an average of five between a phosphodiester group that is linked between C-6 of one mannose unit and C-1 of another. The proposed structure of phosphomannan is illustrated in Fig. 7.

A later study by Bretthauer *et al.* (1973) involving a mild acid hydrolysis of the polymer showed that at least 65% of the mannose and phosphate in the native polymer occur as homogeneous monophosphomannanpentose units.

Soft, gellike solutions are formed by phosphomannan at concentrations as low as 0.25% in water. The viscosity versus concentration curve is unusual for a water-soluble gum in that the viscosity reaches a

TABLE II. Antitumor Activity of Some Fungal Glucans against Subcutaneously Implanted Sarcoma 180 in Mice.[e]

Glucan	Source	Linkages	Dose (mg/kg × number)	Route of injection	Complete regression	Inhibition ratio (%)
Lentinan[a]	Lentinus edodes	(1→3)-β-D	25 × 10	i.p.	2/9	73.0
			5 × 10	i.p.	7/10	97.5
			1 × 10	i.p.	6/10	95.1
			0.2 × 10	i.p.	6/10	78.1
Schizophyllan[b]	Schizophyllum commune	(1→3)-β-D, (1→6)-β-D	5 × 10	i.p.	4/10	89
			1 × 10	i.p.	7/10	81
			0.5 × 10	i.p.	7/10	82
			5 × 4	i.v.	5/10	100
			1 × 4	i.v.	4/10	96
			10 × 10	s.c.	4/10	82
			1 × 10	s.c.	0/10	11
Pachymaran[c]	Pachyman from Poria cocos	(1→3)-β-D	5 × 10	i.p.	4/9	96
Scleroglucan[d]	Sclerotium glucanicum	(1→3)-β-D, (1→6)-β-D	50 × 10	i.p.	2/10	41.2
			5 × 10	i.p.	5/10	88.2
			0.5 × 10	i.p.	7/10	91.3
Fraction LC-1[a]	Lentinus edodes	(1→4)-β-D, (1→6)-β-D	30 × 10	i.p.	8/9	96.5
			15 × 10	i.p.	10/10	100
			5 × 10	i.p.	8/10	99.0

[a] From Chihara, G. et al. (1969).
[b] From Komatsu, N. et al. (1969).
[c] From Chihara, G. et al. (1970).
[d] From Singh, P. P. (1974).
[e] Treatment with glucans was started 24 hr after tumor implantation, and the results were recorded after 5 weeks.

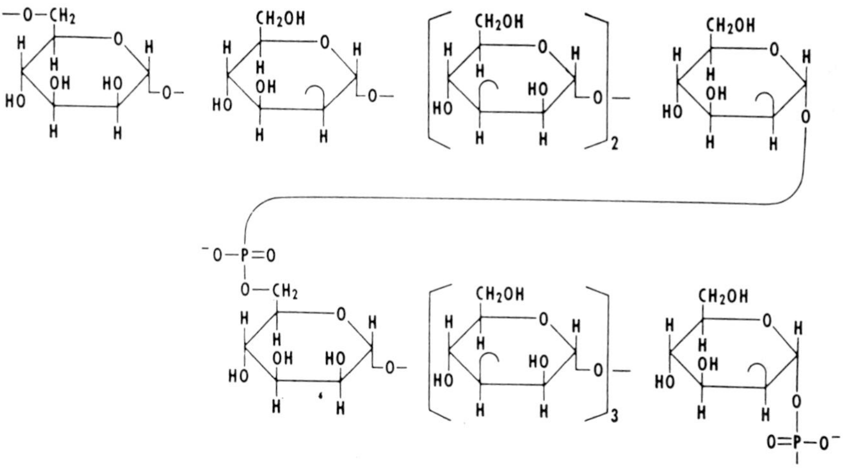

FIGURE 7. Structural types and relative proportions of constituent mannose units in phosphomannan Y-2448. Absence of branching and some details of sequence have not been established.

maximum of about 2500 cP at a 1.5% polysaccharide concentration and a minimum of 1700 cP at 3%, as measured by a Brookfield viscometer. Solutions of approximately 0.65% concentration have the same viscosity as those of 3% concentration. Solutions are thixotropic and are highly sensitive to salts. At 0.5% KCl concentration, solutions having 1.5, 2, and 3% polymer concentrations lose 94, 87, and 63%, respectively, of their viscosity is distilled water. The polymer is also sensitive to vigorous mechanical shearing and heat (Jeanes et al., 1961a). Weak acids split the diester cross-link, causing a loss of viscosity (Anonymous, 1958).

Development of phosphomannan as a commercial water-soluble gum appears to be difficult. The applications are limited primarily because of the sensitivity to salts, shear, and heat, as well as its instability in acid solutions.

B. Polysaccharide Y-1401

This polysaccharide was produced by *Cryptococcus laurentii* var. *flavescens* NRRL Y-1401 in a 6-day aerobic fermentation at 25°C with 5% glucose substrate in the presence of autolyzed brewers' yeast and inorganic salts (Cadmus et al., 1962). The conversion efficiency was 30–35% based on the glucose consumed. Polysaccharide Y-1401 contains D-mannose, D-xylose, D-glucuronic acid, and O-acetyl in the molar ratio of 4 : 1 : 1 : (1.5).

Certain *Tremella* heteropolysaccharides (Slodki *et al.*, 1966), including *Tremella mesenterica* Fries NRRL Y-6151 (Slodki, 1966), contain the same components found in the *Cryptococcus* polymers. Solutions of *Tremella* polysaccharides have much lower viscosities than those of the Y-401 polymer (Jeanes *et al.*, 1964). These polysaccharides have not been produced commercially.

VII. COMMERCIALLY PRODUCED POLYSACCHARIDES

A. Xanthan Gum

1. History

A large number of *Xanthomonas* species have been documented (Stolp and Starr, 1964) and 158 "nomen species" and 44 doubtful species are listed in the 8th Edition of Bergey's Manual (Buchanan and Gibbons, 1974). However, only a few species, such as *Xanthomonas campestris*, *X. phaseoli*, *X. malvacearum*, and *X. carotae*, were reported as efficient gum formers (Lilly *et al.*, 1958). Recently, a strain of *X. manihotis* was reported to be an efficient polysaccharide producer (Chen and Tsuo, 1976).

The Northern Utilization Research and Development Division of the USDA conducted an extensive screening program for gum producers stored in their large microbial culture collection. Of the various biosynthetic polysaccharides produced in their laboratory polysaccharide B-1459 (xanthan gum), produced by *X. campestris* NRRL B-1459, was found to have characteristics that rendered it very promising as a commercial product (Jeanes *et al.*, 1961b). The Kelco Division of Merck & Co., Inc., San Diego, California, carried out the pilot plant feasibility studies for xanthan gum in 1960, semicommercial production of Kelzan xanthan gum in 1961, and substantial commercial production in early 1964.

2. The Safety and Regulatory Status of Xanthan Gum

With respect to safety properties xanthan gum is one of the most extensively investigated polysaccharides (McNeely and Kovacs, 1975). Short-term acute feeding tests on rats and dogs at the Pharmacology Laboratory of the Western Regional Research Laboratory of the USDA have indicated that xanthan gum causes no acute toxicity or growth-inhibiting activity (Booth *et al.*, 1963). Xanthan gum was also found to be a nonsensitizer (Robbins *et al.*, 1964; Woodard Research Corp., unpublished report, 1973), and it caused no eye or skin irritation (Woodard Research, Corp. unpublished report, 1971).

In short-term subacute toxicological studies of xanthan gum it was not possible to determine acute oral toxicity (LD_{-50}) to rats and dogs. The test animals fed over a 24-hour period—with doses of as much as 45 gm/kg to rats (Woodard Research Corp., unpublished report, 1968a) and 20 gm/kg to dogs (Woodard Research Corp., unpublished report, 1968b) —had no fatalities and showed no signs of toxicity or changes in their internal organs.

The long-term feeding studies on xanthan gum consisted of 2-year studies of rats and dogs and a three-generation reproduction study of the rat (Woodard et al., 1973). No significant effects on growth rate, survival, hematological values, or organ weights were found, and no tumors were detected.

On the basis of the above feeding studies the Food and Drug Administration (FDA) issued a food additive order on March 19, 1969 permitting the use of xanthan gum in food products without any specific quantity limitations (Anonymous, 1969a). In the United States the FDA regulations permit addition of xanthan gum to many standardized foods, such as cheeses, cheese products, milk and cream products, mellorine (imitation ice cream), food dressings, table syrups, and vegetables in butter sauce (Anonymous, 1977a). In the standards for these products xanthan gum is either mentioned by name or with the phrase "safe and suitable stabilizers" in the listing of approved ingredients. The FDA has also approved xanthan gum as a component of paper and paperboard that is intended for food contact.

USDA regulations also permit the use of xanthan gum in sauces, gravies, and breadings used with meat and poultry products (Anonymous, 1977b). The United States Environmental Protection Agency (EPA) has exempted xanthan gum from tolerance requirements when it is used as an inert ingredient in pesticide formulations (Anonymous, 1976).

The Canadian Governor-in-Council has also formally approved the general use of xanthan gum in foods (Anonymous, 1971). Xanthan gum is on Annex II of the European Economic Community emulsifier/stabilizer list (Anonymous, 1974). The Joint Expert Committee of the Food and Agriculture Organization/World Health Organization of the United Nations (FAO/WHO) has issued an acceptable daily intake (ADI) for xanthan gum (Anonymous, 1975). In addition, many other countries have approved xanthan gum for various food uses (Kovacs and Kang, 1977).

3. Production

Xanthomonas campestris is a bacterium originally isolated from the rutabaga plant; it produces viscid or gummy colonies on agar media (Breed et al., 1957). Lilly and co-workers (1958) reported that X. campestris produced substantial quantities of extracellular polysaccharides during the fermentation of a carbohydrate substrate. This fermentation had a close similarity to that of X. *campestris* growth with cabbage extracts as the sole substrate.

Sutton and Williams (1970) determined that the chemical components of xanthan gum are the same whether produced by an industrial-type fermentation or on living cabbage tissues under natural conditions. Furthermore, xanthan gum from both sources produced the same precipitin bands when reacted with an antiserum specific for the polysaccharide. Additional tests demonstrated that the metal ion analysis profile and the rheological properties of both the commercially available food-grade xanthan gum and the xanthan gum produced on a pure cabbage extract are similar (Kang, 1972). These data indicate that the commercially produced xanthan gum is identical to the naturally occurring xanthan gum.

The nutritional requirements for minimum growth of 30 different species of the genus *Xanthomonas* were reported by Starr (1946). A mixture of glucose, ammonium chloride, a phosphate buffer, magnesium sulfate, and certain trace minerals was found to be sufficient for minimum growth. During the fermentation the pH of the medium decreases due to the formation of metabolic acids and xanthan gum, which also contains acidic functions. If the pH reaches a critical point, such as 5.0, the gum production drastically decreases. A nearly neutral pH allows the gum synthesis to continue until all the carbohydrate substrate is exhausted, at which point fermentation is complete. In the commercial process, with maintenance of optimal conditions for gum production, including a high degree of aeration and a constant temperature, the fermentation proceeds at a much faster rate than on the cabbage plant.

In aerobic fermentation at around 28°C, 1–5% glucose concentrations have been found to provide the best xanthan gum yields (Lilly *et al.*, 1958). At higher glucose concentrations an actual decrease in conversion efficiency is observed. Glucose, sucrose, and starch have been found to be efficient in polysaccharide production. Recently, Charles and Radjai (1977) reported the use of acid whey as a carbon source. Cottage cheese whey was hydrolyzed by immobilized lactase. The authors reported that glucose in the whey hydrolysate was used more rapidly than galactose, but both sugars were almost completely consumed without diauxy.

After the completion of fermentation the liquor is pasteurized, and the polysaccharide is recovered by precipitation with isopropyl alcohol. This is followed by drying, milling, testing, and packaging. Meticulous precautions must be taken to meet all microbiological, chemical, and physical specifications. Detailed procedures for culture maintenance, polysaccharide production, purification, and analysis of xanthan gum were recently described by Jeanes *et al.* (1976). A flow sheet diagram for a typical xanthan gum manufacturing process is shown in Fig. 8.

A number of alternate methods have been suggested for the recovery of xanthan gum from the fermentation liquor. Each process starts with the pasteurization of the liquor to kill the bacterial cells. Since xanthomonads are nonspore formers and the vegetative cells are sensitive to elevated temperatures, complete pasteurization can be readily accomplished.

Drum- or spray-drying of the fermentation beer yields a crude polysaccharide product (Rogovin *et al.*, 1965). Long-chain quaternary ammonium salts have been proposed as precipitants (Rogovin and Albrecht, 1964). However, it is doubtful that these methods would be acceptable for the manufacture of food-grade products. Xanthan gum can be recovered also by precipitation with calcium ions followed by washing of the insoluble calcium complex with acid or salt (McNeely and O'Connell, 1966).

A number of continuous fermentation methods have been reported (Silman and Rogovin, 1972; Patton and Lindblom, 1962). More recently, Ellwood and Evans (1977) described a single-stage continuous fermentation of *Xanthomonas juglandis* and *X. campestris.* However, the problem of maintaining sterility even in the batch fermentors indicates that the installation of a continuous fermentation system on a commercial scale would significantly increase the contamination risk.

A clear product can be produced by diluting the fermentation liquor and clarifying by filtration. A process whereby xanthan gum is treated

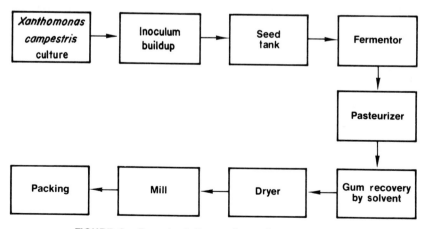

FIGURE 8. Flow sheet diagram for xanthan gum process.

with hypochlorite to produce a clear solution was patented by Colegrove (1970). Another method to clarify xanthan gum utilized high pH at high temperatures (Patton, 1973).

4. Composition, Structure, and Conformation

Xanthan gum is a high molecular weight, natural polysaccharide containing D-glucose (2.8 moles), D-mannose (3.0 moles), D-glucuronic acid (2.0 moles), acetic acid (~4.7%), and pyruvic acid (~3%) (Sloneker and Jeanes, 1962). A primary structure for this polysaccharide was proposed by Sloneker et al. (1964) and is shown in Fig. 9. This structure has a 15-sugar repeat that implies a biosynthetic complexity hitherto unknown in the production of extracellular bacterial polysaccharides. The primary structure of xanthan gum was reinvestigated by Jansson et al. (1975), using an improved degradative technique (Lindberg et al., 1973) and improved methylation analysis (Bjorndal et al., 1970), and was shown to be composed of pentasaccharide repeating units containing two glucose units, two mannose units, and one glucuronic acid unit (Fig. 10). The main chain of xanthan gum contains β-D-glucose units linked through the 1- and 4-positions; i.e., the chemical structure of the xanthan gum backbone is identical to that of cellulose. The side chain contains the two mannose units and the glucuronic acid unit. The terminal β-D-mannose unit is linked glycosidically to the 4-position of β-D-glucuronic acid, which in turn is glycosidically linked to the 2-position of α-D-mannose. This three-sugar side chain is linked to the 3-position of

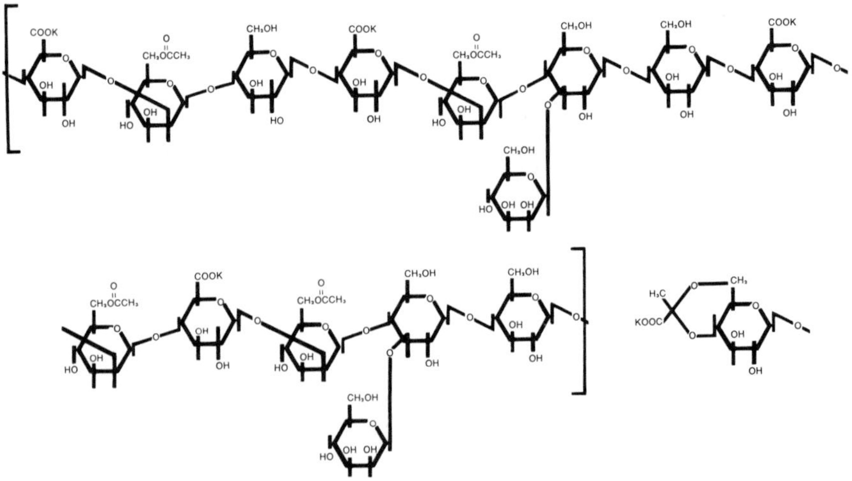

FIGURE 9. Previously accepted structure of xanthan gum.

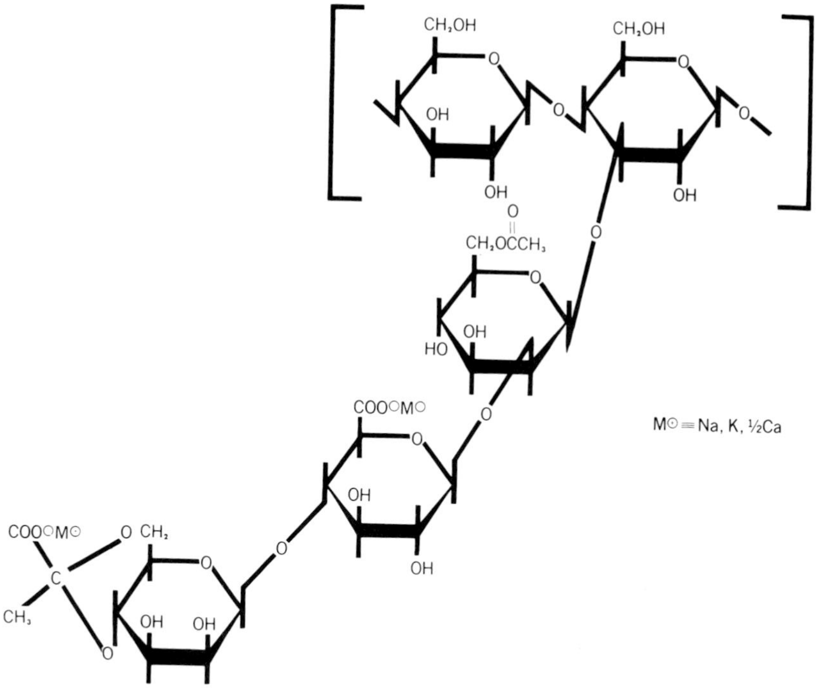

FIGURE 10. Structure of xanthan gum.

every other glucose unit, on an average, in the main chain. The distribution of the side chains has not been determined. The D-mannose unit adjacent to the main chain contains an acetyl group at position 6, and about half of the terminal D-mannose units are ketalically linked through the 4- and 6- positions to pyruvic acid. The distribution of the pyruvate groups has not been elucidated. The molecular weight has been reported to range between 2 and 50×10^6 (or 5×10^7) daltons (Dintzis *et al.*, 1970). This wide range is probably due to an association phenomenon between polymer chains that results in aggregates. The covalent or chemical structure of xanthan gum detailed here only partially describes the xanthan gum molecule. A complete description of the xanthan gum molecule is necessary to determine the shape or conformation in solution. Although most polysaccharides occur as random coils in solution, certain polysaccharides, such as carrageenan, have a special shape in solution (Rees, 1970). The unusual properties of xanthan gum in solution indicate that xanthan gum has a specific conformation in solution. Considerable research has been carried out recently to elucidate the secondary and tertiary structure of xanthan gum. X-Ray diffraction studies

(Moorhouse *et al.*, 1977a,b) and electron microscopic studies (Holzworth and Prestridge, 1977) of oriented fibers indicate that xanthan gum is either a single-stranded or a multistranded helix. Indications are that it is a double helix, but firm data have yet to be obtained.

The solution conformation of xanthan gum has been studied by means of optical rotation, circular dichroism, nuclear magnetic resonance, viscosity, and potentiometric titration measurements (Holzworth, 1976; Morris *et al.*, 1977). These studies have suggested that the xanthan gum molecule exists in either a rodlike, ordered conformation or as a wormlike chain with a low degree of flexibility. The nature of the association between chains and the role of the side chain in the overall conformation have yet to be determined. Elucidation of the complex conformational relationship between xanthan gum molecules will enable molecular interpretations of the unique properties of xanthan gum to be made.

5. Chemical Derivatives and Graft Copolymers

Several chemical modifications of xanthan gum have been reported. These include derivatives, such as carboxymethyl xanthan gum (Schweiger, 1966a), diethylaminoethyl xanthan gum (Schweiger, 1966b), the propylene glycol ester of xanthan gum (Schweiger, 1966c), xanthan gum sulfate (Schweiger, 1969), and formaldehyde cross-linked xanthan gum (Patton, 1962). Also, deacetylated xanthan gum has been prepared (Jeanes and Sloneker, 1961).

Graft copolymers of xanthan gum have been prepared by grafting acrylic monomers onto xanthan gum in the fermentation broth using ceric ion initiation (Pettitt, 1973). Specifically, acrylamide and/or acrylic acid have been grafted onto xanthan gum to produce xanthan gum-g-poly(acrylamide), xanthan gum-g-poly(acrylic acid), or xanthan gum-g-poly(acrylamide-coacrylic acid), respectively. Also, acrylamide and 2-acrylamido-2-methylpropane sulfonic acid have been grafted onto xanthan gum to produce xanthan gum-g-poly(acrylamide-co-2-acrylamido-2-methylpropane sulfonic acid) (Cottrell, 1977; Cottrell *et al.*, 1978). A sulfonic acid-containing graft copolymer has also been prepared by derivatization of xanthan gum-g-poly(acrylamide) using formaldehyde and sodium metabisulfite (Cottrell *et al.*, 1978). Although many derivatives of xanthan gum have been prepared, none has been put into commercial production at this time.

6. Properties

a. Rheological Properties. One of the most important properties of xanthan gum is its ability to control the rheological properties of fluids.

Xanthan gum will dissolve with stirring in either hot or cold water to produce a high-viscosity solution even at low gum concentrations. The relationship between solution viscosity and the concentration of xanthan gum is shown in Fig. 11, as measured using a Brookfield Model LVF viscometer at 60 rpm and 25°C.

Figure 12 shows the effect of salts, such as sodium chloride, on the viscosity of xanthan gum solutions. At low concentrations of xanthan gum, 0.1% or lower, addition of sodium chloride causes a slight reduc-

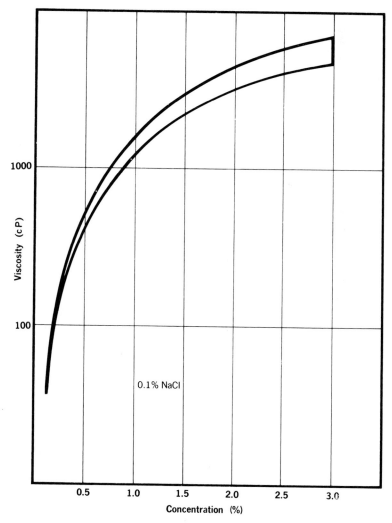

FIGURE 11. Effect of xanthan gum concentration on viscosity.

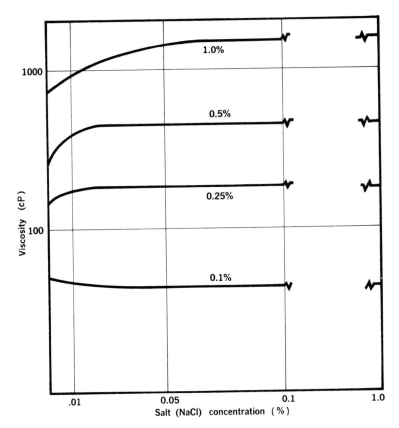

FIGURE 12. Effect of salt concentration on xanthan gum solution viscosity.

tion in viscosity. At concentrations of xanthan gum of 0.25% or greater, addition of sodium chloride causes an increase in viscosity. The peak viscosity is reached at a sodium chloride concentration of 0.02–0.07%, depending upon the xanthan gum concentration; the higher the concentration of xanthan gum, the higher the concentration of salt necessary to achieve maximum viscosity.

Aqueous solutions of xanthan gum are highly pseudoplastic. The relationship between viscosity and shear rate for xanthan gum solutions ranging in concentration from 0.5 to 2.5% is shown in Fig. 13. The viscosity decreases rapidly as the rate of shear is increased. This relationship is instantaneous and reversible.

Aqueous solutions of xanthan gum with a concentration of 0.75% or greater have a rheological yield point. This is demonstrated in Fig. 14. The working yield value is defined as the shear stress required to

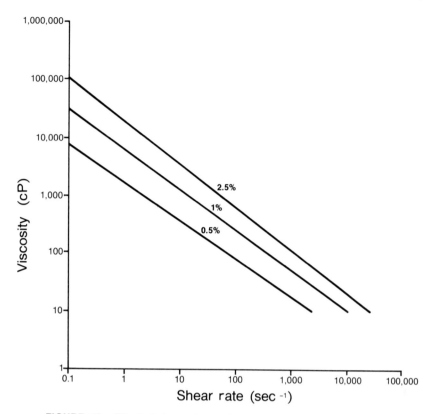

FIGURE 13.   Effect of shear rate on viscosity of xanthan gum solutions.

produce a shear rate of 0.01/second. For xanthan gum the working yield value is 15 dynes/cm² in the absence of salt and 52 dynes/cm² in the presence of potassium chloride (1%). The novel rheological properties of xanthan gum solution are related to end use in Sections VII, A, 7–10.

b. Effect of Heat on the Properties of Xanthan Gum Solutions.   The viscosity of solutions of most polysaccharides generally decreases when they are heated. Solutions of xanthan gum in a salt-free system, however, increase in viscosity after the initial decrease (Jeanes *et al.*, 1961b). This behavior is consistent with the unwinding of an ordered conformation, such as a double helix, changing into a random coil conformation with a resulting increase in effective hydrodynamic volume of the molecule and, therefore, a resulting increase in viscosity.

In the presence of a small amount of salt, 0.1% sodium chloride, the viscosity of a xanthan gum solution is virtually unaffected by tempera-

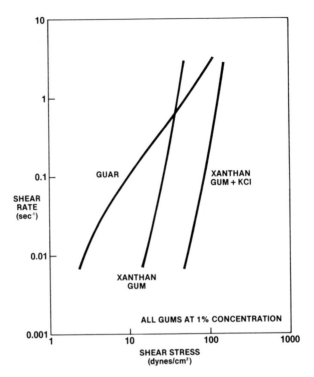

FIGURE 14. The rheological yield points of guar gum, xanthan gum, and xanthan gum with KCl in aqueous solutions.

tures from 25°F (−4°C) to 200°F (93°C) (Fig. 15). Xanthan gum solutions have excellent thermal stability.

c. Effect of pH on the Viscosity of Xanthan Gum Solutions. The viscosity of xanthan gum solutions in the presence of a low level of salt, 0.1% sodium chloride, is independent of pH over the pH range 1.5–13 (Fig. 16).

d. Compatibility of Xanthan Gum Solutions. Xanthan gum will dissolve directly in many acid solutions, such as 5% sulfuric acid, 5% nitric acid, 5% acetic acid, 10% hydrochloric acid, and 25% phosphoric acid. These solutions will be reasonably stable at ambient temperature (~25°C) for several months.

Xanthan gum will also dissolve directly in 5% sodium hydroxide solutions to produce thickened solutions. However, thickened 10–15% sodium hydroxide solutions must be prepared by adding 50% sodium

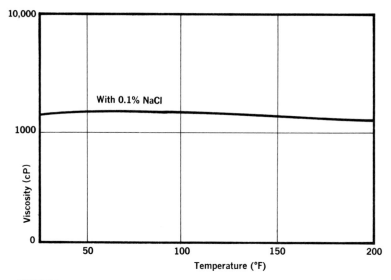

FIGURE 15. Effect of temperature on 1% xanthan gum solution viscosity.

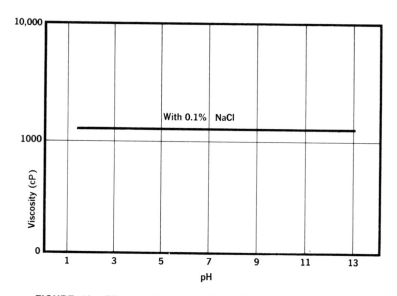

FIGURE 16. Effect of pH on viscosity of 1% xanthan gum solution.

hydroxide solutions to aqueous xanthan gum solutions. These highly alkaline, viscous solutions have excellent stability at ambient temperature (~25°C). Most salts are compatible with xanthan gum solutions, except that polyvalent metal ions cause gelation or precipitation of xanthan gum at high pH.

The viscosity of xanthan gum solutions in the presence of several acids, bases, and salts together with the effect of long-term storage (3 months) is shown in Table III.

TABLE III.   Compatibilities of Xanthan Gum

Material	Percentage of material	Percentage of xanthan gum[a]	Viscosity (cP)[b] Initial	After 90 days
Organic and inorganic acids				
Acetic acid	10.0	2.0	4200	4050
	20.0	2.0	4300	4400
Citric acid	10.0	1.0	1200	900
	20.0	1.0	1400	1030
Hydrochloric acid	5.0	2.0	3400	2750
	10.0	2.0	3500	1250
Phosphoric acid	30.0	2.0	4850	5250
	40.0	2.0	5600	7000
Sulfuric acid	5.0	2.0	3610	3150
	10.0	2.0	3800	3100
Tartaric acid	10.0	1.0	1210	900
	20.0	1.0	1430	1010
Organic and inorganic salts and bases				
Aluminum sulfate	0.5	1.0	2860	gel
	5.0	1.0	1460	1330
Ammonium chloride	10.0	1.0	1120	980
	30.0	1.0	1380	1050
Diammonium phosphate	5.0	1.0	1100	1100
	10.0	1.0	1190	1230
Barium chloride	5.0	1.0	1240	1180
	15.0	1.0	1330	1130
Calcium chloride	10.0	1.0	1260	1300
	20.0	1.0	1580	1610
Chrome alum	5.0	1.0	1430	gel
Cobalt chloride	5.0	1.0	1280	1110
	15.0	1.0	1600	1160
Cobalt sulfate	5.0	1.0	1300	1120
	15.0	1.0	1170	1170

(continued)

TABLE III (*Continued*)

Material	Percentage of material	Percentage of xanthan gum[a]	Viscosity (cP)[b]	
			Initial	After 90 days
Cupric chloride	5.0	1.0	1180	970
	15.0	1.0	1500	2000
Ferrous sulfate	5.0	1.0	1410	300
	15.0	1.0	1600	gel
Magnesium chloride	5.0	1.0	1130	1070
	15.0	1.0	1210	1180
Potassium chloride	5.0	1.0	1240	1200
	15.0	1.0	1200	1115
Sodium bisulfite	5.0	1.0	1210	1160
	15.0	1.0	1540	1485
Sodium tetraborate	5.0	1.0	1140	1140
	15.0	1.0	1630	1520
Sodium carbonate	5.0	1.0	1130	1050
Sodium chloride	5.0	1.0	1110	1090
	15.0	1.0	1330	1200
Sodium chromate	5.0	1.0	1620	gel
Sodium citrate	5.0	1.0	1240	1170
	15.0	1.0	1390	1240
Sodium hydroxide	5.0	1.0	1360	810
	10.0	1.0	1390	115
Disodium hydrogen phosphate	5.0	1.0	1310	1100
	10.0	1.0	1190	1230
Sodium sulfate	5.0	1.0	1070	1100
	15.0	1.0	1460	1550
Zinc chloride	5.0	1.0	1200	1090
	15.0	1.0	1460	1320

[a] 0.5% xanthan gum (preserved with 0.1% formaldehyde): initial viscosity, 290 cP; after 90 days, 285 cP; 1.0% xanthan gum (preserved with 0.1% formaldehyde): initial viscosity, 860 cP; after 90 days, 990 cP; 2.0% xanthan gum (preserved with 0.1% formaldehyde): initial viscosity, 2350 cP; after 90 days, 2500 cP.

[b] Brookfield Model LVF Viscometer at 60 rpm, appropriate spindle.

Solutions of xanthan gum are compatible with methanol, ethanol, isopropanol, and acetone up to a final concentration of 50–60%. Concentrations greater than this range cause gelation or precipitation of the xanthan gum. Xanthan gum will not dissolve in most organic solvents, but it is soluble in formamide at 25°C and in glycerol and ethylene glycol at 65°C.

Enzymes such as protease, cellulase, hemicellulase, pectinase, and amylase will not degrade xanthan gum in solution. As with all polysaccharides or gums, xanthan gum is degraded by strong, oxidizing agents, such as peroxides, persulfates, and hypochlorites. This degradation is accelerated at elevated temperatures.

Most of the commercially available thickeners, such as sodium alginate and starch, have excellent compatibility with xanthan gum. Galactomannans, such as locust bean gum and guar, exhibit synergistic viscosity increases with xanthan gum. This "useful incompatibility" (Kovacs, 1973a) is described in the following section.

e. Interaction with Galactomannans. One of the more novel properties of xanthan gum is its ability to react with galactomannans, such as guar (Rocks, 1971) and locust bean gum (Schuppner, 1971). In the case of guar the interaction is made manifest by a synergistic increase in viscosity; i.e., the viscosity of the combination is greater than would be expected from the viscosities of solutions of xanthan gum and guar gum alone. The viscosity increase resulting from the synergistic interaction is shown in Fig. 17. In the case of locust bean gum the interaction is made manifest by a large synergistic viscosity increase at low gum concentrations (Fig. 18), but as the concentration of the gum is increased, a heat-reversible gel is formed (Fig. 19). The utility of the xanthan gum–galactomannan combinations is discussed in Sections VII, A, 7 and 8.

Considerable research has been carried out in recent years to determine the specific interactions between polysaccharides and galactomannans that result in synergistic viscosity increases or gelation (Dea et al., 1972, 1977). The studies have suggested that the interaction of agarose and carrageenan with galactomannans results from the association between the carrageenan and agarose double helices, as well as sequences of unsubstituted mannose residues in the galactomannan. A schematic representation of the galactomannan conformation is shown in Fig. 20, and the double-helical conformation of polysaccharides was discussed in section VII, A, 4. A possible model for the interaction between xanthan gum and locust bean gum is shown in Fig. 21. This model is similar to that suggested by Morris et al.. (1977) to explain the interaction between xanthan gum and galactomannans which results in a three-dimensional network of polysaccharide molecules that contain water molecules in the "open" spaces within the network. Hence, the state of the water in the system is modified with either a resultant increase in viscosity or gelation (Rees, 1969).

7. Industrial Applications

Because of its unique properties xanthan gum has utility in a wide range of industrial applications.

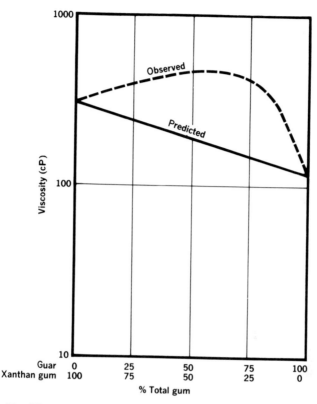

FIGURE 17. Effect of colloid ratio on xanthan gum/guar gum solutions (0.5% total gum).

a. Petroleum. *(1) Oil Well Drilling Fluids.* Xanthan gum is used at low levels, 0.5–1.5 lb/bbl, to viscosify and control the rheological properties of water-based drilling fluids. Specifically, the pseudoplasticity or shear-thinning property provides low viscosity at the drill bit, where the amount of shear is very high, and high viscosity in the annulus, where the amount of shear is low. Therefore, xanthan gum serves a double purpose in allowing faster penetration at the bit because of low viscosity; at the same time its high viscosity suspends the cuttings in the annulus (Carico, 1976b). The excellent compatibility of xanthan gum with salts, such as sodium chloride and calcium chloride, and the resistance to thermal degradation are additional properties that make xanthan gum the ideal drilling fluid additive.

(2) Completion and Workovers. The shear thinning and suspending properties of xanthan gum are utilized to prepare completion and work-

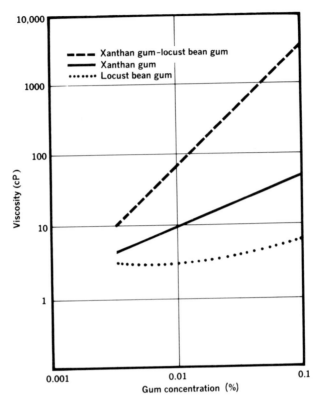

FIGURE 18. Viscosity as a function of colloid concentration; xanthan gum–locust bean gum combinations exhibit a synergistic increase in viscosity. Solutions were prepared by diluting at 140°F from 1% concentration.

over fluids with the desired rheological properties (Lipton and Burnett, 1976). After completion xanthan gum can be completely degraded with "breakers" such as hypochlorite or hypochloric acid. Because of the chemical nature of xanthan gum these fluids cause minimum formation damage. Weighted completion fluids containing high levels of calcium chloride or calcium bromide can also be formulated with xanthan gum because of its excellent salt compatibility.

(3) *Fracturing Fluids*. The combination of unique properties allows xanthan gum to be used to formulate water-based hydraulic fracturing fluids. Xanthan gum imparts high viscosity at low concentrations, and the shear-thinning property provides ease of pumping under the high-shear conditions, yet suspension of proppants (such as sand) under low-shear

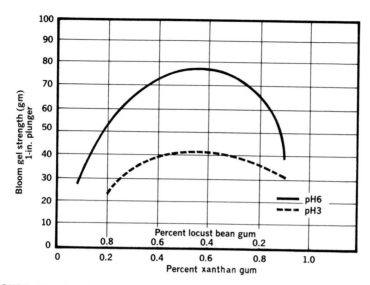

FIGURE 19. Gel strength of 1% xanthan gum–locust bean gum gel as a function of colloid ratio.

conditions. Since xanthan gum contains a very small percentage of insoluble material, formation damage (blockage of the rock material by insoluble material) is kept to a minimum. Xanthan gum can be removed from the formation by adding a controlled amount of hypochlorite or hydrochloric acid to the fluid before injection into the well. The excellent thermal stability, shear stability, and salt compatibility are additional properties that enable xanthan gum to find utility in fracturing fluids.

(4) Enhanced Oil Recovery. Xanthan gum produces viscous pseudo-plastic aqueous solutions that provide excellent mobility control for increased oil displacement efficiency in secondary or tertiary flooding processes. As with the other petroleum applications described the thermal stability, shear stability, and salt compatibility are important properties of xanthan gum, because they provide additional functionality to viscous water-flooding fluids. The development of xanthan gum-based water-flooding fluids, which can be injected into formations having low permeability, has been the subject of much research (Lipton, 1974; Burnett, 1975; Colegrove, 1976).

FIGURE 20. Schematic representation of galactomannan conformation.

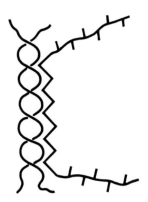

FIGURE 21. Possible model for the interaction between xanthan gum and locust bean galactomannan, resulting in gel formation.

b. Textile Printing and Dyeing.   Thickeners are used in textile printing pastes to control the rheological properties of the paste during application and, also, to prevent migration of the dye and thereby produce sharp, clean patterns. Xanthan gum is an excellent thickener for this application because of its high viscosity at low concentrations and its unique rheological properties.

Xanthan gum in combination with guar gum imparts excellent thermal stability and desirable rheological properties to carpet printing pastes. Xanthan gum is compatible with most components of the printing pastes, and the wash-out properties of xanthan gum are excellent (Racciato, 1976).

Xanthan gum is used to thicken dye solutions used in space dyeing. This technique is used to print yarns that are primarily used in multicolor carpet manufacture. The xanthan gum prevents flushing and bleeding of the dyes on the yarn prior to drying.

c. Ceramic Glazes.   The unique rheological properties of xanthan gum at low concentration enable the insoluble components of the glaze to be suspended for extended periods of time. Xanthan gum is compatible with the components of the glaze and it also prevents agglomeration of the glaze components during grinding, thereby allowing a reduction in grinding time. Also, the drying time is controlled and the level of imperfections is reduced (McNeely et al., 1966). Ceramic processes based on extrusion are improved by the addition of xanthan gum since its pseudoplastic nature imparts plasticity to the mix.

d. Cleaners. Acid and alkaline cleaners with excellent functionality can be formulated with xanthan gum because it has good viscous stability in strong acids and alkalies. The pseudoplasticity of xanthan gum produces cleaners that have very high viscosity at a very low shear rate; this allows the cleaner to cling to inclined surfaces for extended periods of time and maximizes contact time. Abrasive cleaners can be formulated with xanthan gum so that the abrasive component is uniformly suspended and the flow properties of the cleaner are not impaired.

e. Slurry Explosives. Xanthan gum is used to thicken slurry explosive formulations containing 60–80% sodium, ammonium, and calcium nitrate because of the excellent compatibility of xanthan gum with high levels of salts. The thickened system is reacted with a cross-linking agent to form a gel that is either pumped into fissures in the rock for immediate use or is packed into small-diameter casings. The cross-linked xanthan gum prevents the highly soluble nitrate salts from being leached out of the gel when it is immersed in water (Colegrove, 1967). Also, xanthan gum is used in combination with guar to provide better heat stability.

f. Other Applications. Xanthan gum has also found utility in ink, paint, paper, and wallpaper adhesives and suspension polymerization (Hartnek *et al.*, 1974). A process to prepare a concentrated pumpable suspension of xanthan gum has been developed for use in industrial applications that require metered or automated addition of xanthan gum (Colegrove, 1975).

8. Food Applications

Xanthan gum has found widespread utility in the food industry as a result of its unique properties. These unique properties are summarized as follows: (a) solubility in hot or cold water; (b) high viscosity at low concentrations; (c) little variation in viscosity with changing temperatures; (d) excellent solubility and stability in acid systems; (e) unique rheological properties that provide high viscosity under low shear and low viscosity under high shear; (f) excellent suspending properties due to high yield value; (g) excellent compatibility with a wide range of salts; (h) excellent thermal stability; (i) good emulsifying properties; and (j) freeze-thaw stability.

Some of the application areas where xanthan gum has utility have been reviewed (Rocks, 1971; McNeely and Kang, 1973; Kovacs, 1973b; Kovacs and Kang, 1977; Andrew, 1977; Cottrell and Kang, 1978). Several application areas that illustrate the utility of xanthan gum in foods follow.

a. Pastry Fillings. A typical bakery jelly used to fill pastries or donuts contains a high concentration of sugar, a low level of starch, and small amounts of colors and flavors. Addition of 0.2% xanthan gum to the formulation provides a filling that has a low level of syneresis and, as a result, the filling is not absorbed by the pastry. The pseudoplasticity of xanthan gum, i.e., very high viscosity at very low shear, is responsible for this performance. The pseudoplasticity imparted by the xanthan gum also allows the fillings to be easily pumped during the filling operation because the viscosity is low under high shear.

b. Sauces and Gravies. Spaghetti sauce contains high levels of tomato solids and cheese, together with small amounts of spices and flavors. Addition of xanthan gum to the formulation provides a sauce that has excellent heat stability for extended periods of time and that maintains constant viscosity over wide ranges of temperature. The high viscosity at very low shear rate allows the sauce to cling to the spaghetti during consumption. The sauce also has excellent flavor release and mouthfeel because of the low viscosity under the high-shear conditions in the mouth. These important properties of the sauce result from the pseudoplasticity imparted by the xanthan gum. Cheese sauces and barbecue sauces can also be formulated using xanthan gum. The same properties that are important for sauces are also important for gravies. Addition of a small amount of xanthan gum to the gravy mix provides these properties.

c. Pourable Salad Dressings. Salad dressings are vegetable oil–water emulsions containing vinegar, sugar, egg yolk, and flavors. Addition of 0.25% xanthan gum provides excellent stability to the emulsion for periods up to 1 year. The dressings can be easily pumped during the filling operation and they also flow easily from the bottle during use (O'Connell, 1962). These unique properties are a result of the rheological properties of xanthan gum, specifically the high working yield value and the high degree of pseudoplasticity.

d. Dairy Products. Xanthan gum in combination with locust bean gum and guar gum stabilizes cottage cheese creaming emulsions. In addition to providing stability the xanthan gum blend also emulsifies the fat, prevents whey-off, and improves the cling of the dressing to the curd (Kovacs and Titlow, 1976). A blend of xanthan gum, locust bean gum, and guar gum improves the physical and organoleptic properties of pasteurized processed cheese spread (Kovacs and Igoe, 1976). Blends of xanthan gum, locust bean gum, and guar gum are effective stabilizers in ice cream, ice milk, and milk shakes. At relatively low concentrations in ice cream and ice milk, the water-absorbing characteristics provide excellent mix viscosity control. The frozen products exhibit smooth texture, creamy body, and exceptional tolerance to heat shock. A low-pH dessert gel can be prepared by combining a highly acid fruit juice with milk using xanthan gum in combination with locust bean gum and carboxy methyl cellulose (CMC). The resulting dessert gel has a smooth, creamy consistency and relatively long refrigerated shelf-life. Certain milk-based beverages can be stabilized with xanthan gum and guar gum (Igoe, 1977).

e. Other Applications. Xanthan gum has found utility in canned foods, dry mixes, frozen foods, juice drinks, relishes, spoonable dressings, and syrups. These are only a few examples of the foods that benefit from xanthan gum, and it is to be expected that more applications that use the unique properties of xanthan gum will emerge in the future.

9. Agricultural Applications

a. Animal Feed. Liquid feed supplements often contain ingredients which tend to stratify, settle out, or alter their physical stability with time, such as ground limestone, magnesium oxide, trace minerals, salts, or fat-soluble vitamins. Addition of 1–2 lb xanthan gum per ton of supplement minimizes any stratification or separation of these ingredients during shipping or extended storage (Andrew, 1977; Petrowski, 1978).

Calf milk replacers have traditionally used skimmed milk powder as the primary protein source. It is advantageous, at times, to use vegetable or fermentation proteins, but these proteins are partially or totally insoluble in water and settle out of the milk replacer in a very short time. Addition of 0.03–0.04% xanthan gum suspends the insoluble proteins to give a product having uniform quality (Andrew, 1977).

b. Agricultural Chemicals. Xanthan gum alone or in combination with locust bean gum improves the efficiency of flowable fungicides, herbicides, and insecticides by uniformly suspending the solid components of the formulation in an aqueous system or by stabilizing emulsions and multiphase liquid systems. The unique rheological properties of xanthan gum also improve sprayability, reduce drift, and increase pesticide cling and permanence (Schuppner, 1972; Gibsen and Saddington, 1973).

10. Pharmaceutical and Cosmetic Applications

a. Pharmaceutical Applications. Xanthan gum can be used to stabilize emulsion cream formulations containing pharmaceuticals. The rheological properties of xanthan gum account for the stabilization and ease of application (Felty, 1975). The aluminum salt of xanthan gum may be used to suspend barium sulfate in a radiographic opacifier. Xanthan gum will also suspend pharmaceuticals, such as antibiotics, to provide formulations of uniform dosage.

b. Cosmetic Applications. Denture cleaning compositions containing dilute solutions of phosphoric acid can be formulated with xanthan gum. The gum suspends the diatomaceous earth in the formulation in the presence of the acid (Regan and Regan, 1975). Toothpaste can be formulated with xanthan gum. The unique rheological properties of xanthan gum enable the paste to cling to the brush yet spread over the teeth during brushing. Xanthan gum provides body to creams and lotions but allows ease of application.

B. Dextrans

1. History

Structurally, dextrans are defined today as a large class of α-D-glucans in which α-1,6 linkages are predominant (Jeanes, 1977). Dextrans are the oldest microbial water-soluble gums known to man. In 1822 a report was made that cane sugar juice, after storage, had changed to a thick mucilage and was unfit for use (Brown, 1906). Pasteur, in 1861, noted that the viscous fermentations of sucrose were caused by a microbial action. Scheibler (1869) named the gum "dextran" because of its similarity to dextrin. He found that it yielded only D-glucose on acid hydrolysis. In 1930 Hucker and Pederson classified three distinct species of *Leuconostoc*—*L. mesenteroides*, *L. dextranicum*, and *L. citrovorum*—as dextran producers. Tarr and Hibbert (1913b) reported optimal conditions for dextran production from sucrose.

Jeanes (1952) of the Northern Utilization Research and Development Division of the USDA published a bibliography of 410 references related to the clinical use of dextran. The same author updated the bibliography, which includes dextran information from 1861 through mid-1976 (Jeanes, 1978). Comprehensive reviews were made by Murphy and Whistler (1973) and by Jeanes (1977).

2. Production

a. Whole-Culture Fermentation. A large number of bacterial strains is capable of producing dextran (Jeanes, 1952; Jeanes *et al.*, 1954). Commercial production of dextran has been carried out with only two of these strains, *L. mesenteroides* and *L. dextranicum*. Jeanes (1965) described methods for the preparation of three types of dextrans from strains NRRL B-512F, B-1146, and B-523. For B-512F dextran, Jeanes recommended a medium for both inoculum buildup and production that contained 10% sucrose, 0.5% dipotassium acid phosphate, 0.25% Difco yeast extract, and 0.02% magnesium sulfate heptahydrate. Dextran production by *Streptococcus bovis* (Bailey and Oxford, 1958) and *Streptococcus* NRRL B-1351 (Rogovin *et al.*, 1960) were also reported.

Many publications and patents disclose media composition and the use of adjuncts for the whole-culture fermentation of dextran and the process used for the commercial production of dextran (Murphy and Whistler, 1973; Jeanes, 1968). In the commercial whole-culture process the inoculum is increased in several stages in a medium similar to that of the final fermentation at 25°C in an agitated fermentor. After the fermentation, dextran is recovered by precipitation with alcohol and purified to make it suitable for clinical use. The use of molasses as an inexpensive source of substrate to produce technical grade dextrans was also reported (Behrens and Wuensche, 1969).

b. Enzymatic Synthesis. This method offers tremendous advantages over the whole-culture method for simpler product purification and greater product uniformity. Hehre (1941) accomplished the enzymatic synthesis of dextran from sucrose in cell-free extracts of *L. mesenteroides*. Hehre (1941) and later Tsuchiya *et al.* (1955) investigated the properties of the transglycosidase enzyme, dextransucrase. Optimal pH values for enzyme production and for dextran synthesis were 6.7 and 5.1, respectively.

Jeanes (1965) described a method of enzymatic dextran synthesis with *L. mesenteroides* NRRL B-512F. The medium for the initial stage of inoculum buildup contained 2% sucrose, 0.5% dipotassium acid phosphate, 0.5% Difco yeast extract, and 0.25% Difco Tryptone. The media for successive stages include small amounts of metal salts and exclude Tryptone. Bacterial cells are removed by centrifugation. Dextran is produced enzymatically by adding sucrose to the cell-free culture liquid at pH 5.0–5.2 and holding the temperature at 25°–30°C. Since dextransucrase incorporates only the glucose portion of sucrose into dextran, approximately 50% of the original weight of sucrose remains in solution as D-fructose after removal of the dextran. Methods for the recovery of this by-product (Koepsel et al., 1956) have been used commercially to produce fructose with dextran as a by-product. A semicontinuous process (Behrens et al., 1965) and fully continuous fermentation were described by Ogino (1973).

3. Derivatives

Like cellulose and starch, dextran contains no charged groups and is readily substituted. A large number of chemically modified dextrans have been reported, including acetates, nitrates, sulfates, phosphates, and various ethers (Murphy and Whistler, 1973). The cross-linked insoluble dextrans and the iron–dextran complex are of major commercial interest.

Products for laboratory use in gel filtrations and as molecular sieves are produced commercially from dextran cross-linked and insolubilized with epichlorohydrin (Anonymous, 1960a,b, 1964; Smiley, 1966). These materials have proved useful for rapid molecular weight determinations on small quantities of polymers. Epichlorohydrin-insolubilized dextran also serves as a base for ion-exchange resins. The diethylaminoethyl, carboxymethyl, and sulfoethyl derivatives of cross-linked dextran are available (Anonymous, 1965). They are effectively used in the purification, separation, and isolation of enzymes and hormones (Kagadel and Akerstroem, 1971).

4. Properties

The molecular weight of dextran ranges from approximately 1.5×10^4 to 2×10^7 or higher, depending on the method of preparation. Solutions of dextran in water are quite low in viscosity for such a high molecular weight polymer. A 2% solution of native dextran has a viscosity of approximately 150 cP. Dextran has relatively good compatibilities with salts, acids, and bases. It is also resistant to degradation at high temperatures.

5. Applications

a. Blood Plasma Extender. A considerable research effort was expended in the period from 1948 to 1955 to develop dextran as a blood plasma extender (Grönwall, 1957). Sterile, pyrogen-free dextran solutions (~6%) having a molecular weight between 50,000 and 100,000 were stockpiled in pint bottles. In emergency situations these solutions could be used to restore blood volume in patients suffering from shock due to blood loss. The United States' military stockpiling program for dextran as a blood plasma extender was discontinued in 1955, and the producing plants were shut down. Since 1955 production of dextran for use as a blood plasma extender has been limited.

b. Oil Well Drilling. Dextran was used as an oil well drilling fluid additive in the 1950s (Owen, 1950). Several patents (Owen, 1959; Sparks, 1962; Cypert and Patton, 1963) have been issued. Because of the low viscosity of dextran in aqueous solutions it was used in relatively high concentrations in drilling fluids to control fluid loss. As a result, dextran was uneconomical, and its use in drilling fluids was discontinued. Recently, a limited number of further studies are being performed (Jeanes, 1977). However, the authors are not aware of a major drilling operation where dextrans or modified dextrans are being used.

c. Iron–Dextran Complex. The complex of iron with dextran has attained limited use as a source of nutritional iron. The iron in this complex was reported to be completely available to form the hemoglobin complex. This source of iron was used in combating anemia (Martin et al., 1955; Cox et al., 1965). The solution contains 5% iron and 20% dextran of molecular weight of 5000.

d. Miscellaneous Uses. Dextran was tested extensively in thickened water flooding for the secondary recovery of oil. It proved to be uneconomical because of its low viscosity in solution. Various applications in foods, metallurgy, photographic emulsions, drug encapsulations, and other applications have been reviewed (Murphy and Whistler, 1973; Jeanes, 1977). However, most of these applications have not reached the practical stage primarily because other, better performing, competitive products are available.

C. *Sclerotium* Polysaccharides

Many species of *Sclerotium* produce glucans (Johnson et al., 1963; Halleck, 1967) in a medium containing suitable carbon sources, such as sucrose and glucose, in the presence of autolyzed yeast and certain salts. Similar polysaccharides are produced by other fungi, such as *Claviceps purpurea* (Perlin and Taber, 1963; Buck et al., 1968), *Stereum sanguinolentum* (Axelsson et al., 1968), *Plectania occidentalis* NRRL 3137 (Wallen et al., 1965; Davis et al., 1966), and *Helotium* sp. NRRL 3129 (Davis et al., 1966). The polysaccharide scleroglucan has been developed by the Pillsbury Company and commercialized under the trade name of Polytran. This polysaccharide was described by Rodgers (1973).

1. Structure

The typical scleroglucan produced by *Sclerotium glucanicum* has a structure consisting of a main chain of β-D-(1→3)-glucopyranose units to which glucopyranose units are joined by β-D-(1→6) linkages to every third unit of the main chain (Johnson *et al.*, 1963). The number of single-unit side chains varies, depending upon the organisms producing glu-cans: 25% in a gum produced by *C. purpurea* (Perlin and Taber, 1963) and 50% in a glucan elaborated by *P. occidentalis* NRRL 3137 (Axelsson *et al.*, 1968). The degree of polymerization (DP) of the scleroglucan from *S. glucanicum* varies widely between ~110 and 1600. However, the commercial scleroglucan has an average DP of ~800 (Rodgers, 1973).

2. Production

Scleroglucan is produced commercially by aerobic submerged fer-mentation in a medium containing D-glucose, corn steep liquor, nitrate as the major source of nitrogen, and mineral salts (Rodgers, 1973). Synthesis of scleroglucan proceeds simultaneously with mycelial growth and, as a result of the polysaccharide production, the medium develops a gellike consistency. The pH of the fermentation, initially 4.5, drops to about 2 as a result of the accumulation of oxalic acid.

After fermentation the fermentation liquor is heated to inactivate glucanase enzymes. This is followed by homogenization to detach the glucan from the mycelial tissue. The homogenized liquor is diluted to decrease viscosity and filtered to eliminate mycelia and other particulate matter. The gum in the clear filtrate is alcohol precipitated, dried, and milled. For some industrial applications the homogenized fermentation liquor is spray-dried without purification. Scleroglucan content in this grade is about 30%.

3. Properties

Scleroglucan disperses readily in water upon mechanical stirring to give high-viscosity solutions (Fig. 22). This polysaccharide produces a self-supporting sliceable gel at 25°C when dispersed in water at a concentration of 1.2–1.5%. However, at temperatures below 10°C even very dilute solutions form diffusely structured gels. Such gels tend to shrink and undergo syneresis when left undisturbed for long periods of time. However, these gels disperse quickly when a mild agitation is applied. It forms a stable pH-irreversible gel by slowly complexing with borate in alkaline solutions.

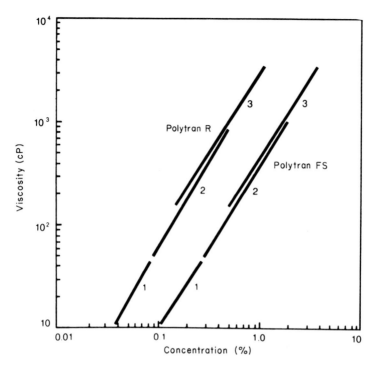

FIGURE 22. Relation of viscosity of solutions of refined (Polytran R) and commercial (Polytran FS) scleroglucans to concentration. Brookfield LVT viscometer; spindles 1, 2, and 3; 30 rpm.

The viscosity of scleroglucan solutions is relatively stable at temperatures between 10° and 90°C, as shown in Fig. 23. The effect of pH on stability of 0.5% solutions of commercial scleroglucan heated at 121°C for 0.25 and 20 hours is given in Fig. 24. The viscosity is stable at the autoclaving temperature (121°C) from initial pH 3 to 11 for 20 hours. The zero-hour curve indicates the viscosity of a solution aged 24 hours, heated to 90°C, cooled, and adjusted to varied pH values. This solution exhibits good stability between pH 1 and 10.

Scleroglucan has good compatibility with various salts. However, aluminum sulfate and magnesium sulfate precipitate the gum at 10% salt concentration. Probably the most outstanding property of scleroglucan is its synergistic viscosity increase when combined with bentonite. The apparent viscosities of 0.15% purified scleroglucan and 5.0% bentonite are near 200 and 300 cP, respectively. A combined viscosity is ~4000 cP as measured by the Brookfield viscometer LVT at 30 rpm with No. 3 spindle. However, the stability of this synergistic viscosity in the application situation, such as in drilling muds, is questionable.

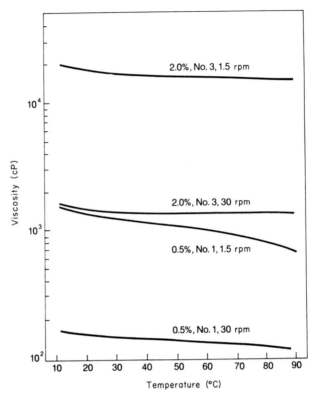

FIGURE 23. Relation of viscosity of solutions of commercial scleroglucan to temperature at varied concentration and shear rate. Brookfield LVT viscometer, spindles 1 and 3.

4. Applications

Scleroglucan has been suggested for use in many applications: inks and coatings; ceramics (Halleck, 1969); drilling muds (Anonymous, 1969a); pharmaceuticals (Sheth and Lachman, 1969); and cosmetics (Rodgers, 1973). Despite these suggested applications scleroglucan has not captured any significant portion of industrial applications with the possible exception of ceramics.

D. Zanflo Polysaccharides

Zanflo is the trademark of a heteropolysaccharide produced by an extensively mutated strain of a bacterium that was isolated from soil. This mutant grows at 30°C only in the presence of added iron. The organism will not grow at 37°C whether or not iron is present in the medium.

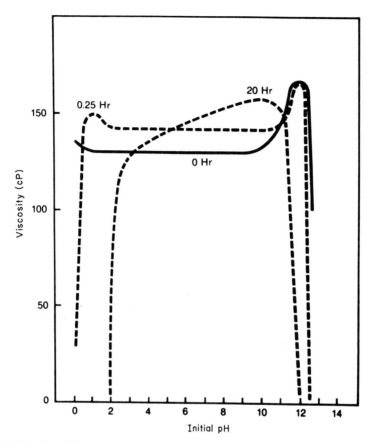

FIGURE 24. Effect of pH on stability of 0.5% solutions of commercial scleroglucan heated at 121°C for 0.25 and 20 hours. Brookfield LVT viscometer, spindle 1, 30 rpm.

Zanflo gum is produced by aerobic submerged fermentation in a medium containing phosphate as a buffering agent, ammonium nitrate and a soy protein product as nitrogen sources, magnesium sulfate, trace minerals, and a carbon source. This organism requires specific carbon sources for optimal polysaccharide production. Lactose and hydrolyzed starch are preferred over glucose, sucrose, or maltose. This polysaccharide contains 97% carbohydrate and 3% protein. The carbohydrate portion was found to contain glucose, galactose, glucuronic acid, and fucose in the molar ratio of 3 : 2 : (1.5) : 1. Uronic acid accounts for approximately 20% of the polysaccharide on a weight basis. It is noteworthy that fucose is not commonly found as a structural constituent of exocellular bacterial heteropolysaccharides.

This gum is a high-viscosity polysaccharide, as shown by the viscosity concentration curve in Fig. 25. The viscosity is considerably higher than that of xanthan gum and becomes more outstanding at higher concentrations. At a 1.5% gum concentration xanthan gum has a viscosity of 2500 cP, whereas Zanflo polysaccharide has a viscosity of 5000 cP.

The relationship between viscosity and temperature, as shown in Fig. 26, is almost linear. From this graph one can calculate a decrease in viscosity of 25 cP/°C as the temperature is increased. The effect of pH on Zanflo polysaccharide and xanthan gum is indicated in Fig. 27. The viscosity of this gum remains stable from pH 5 to 10 but decreases on either side of this range.

One of the most striking properties of this polysaccharide is its compatibility with cationic dyes. Anionic gums react with cationic dyes, such as methylene blue chloride, to form a fibrous precipitate; this precipitate

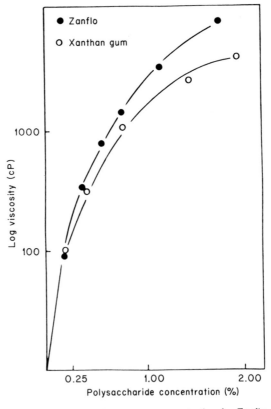

FIGURE 25. Viscosity versus concentration for Zanflo.

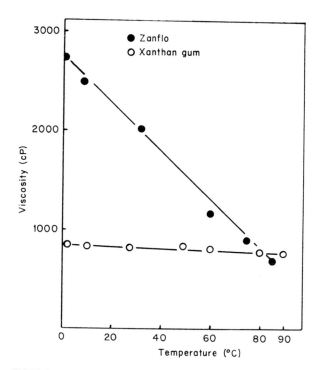

FIGURE 26. Effect of temperature on Zanflo viscosity (1%).

limits their industrial applications. However, Zanflo polysaccharide, even though it contains a substantial amount of uronic acid, does not precipitate with methylene blue chloride at any pH. Zanflo gum has a high resistance to enzymatic attack and has excellent flow and leveling properties. Because of these properties, as well as those described earlier, Zanflo polysaccharide has already found applications in the paint industry.

In our laboratories we developed another extensively mutated strain (tTR-45) of a bacterium that was isolated from a rhizosphere soil sample. This mutant requires thymine or thymidine for growth at 30°C but will not grow at 37°C. This strain produces a heteropolysaccharide consisting of 33% mannose, 29% glucose, 21% galactose, and 17% glucuronic acid. The polysaccharide also contain ~5.7% acetyl and ~4.9% pyruvate. The viscosity of this polymer is even higher than that of Zanflo gum. For example, a 1% solution of this polymer has a viscosity of ~3000 cP as compared to ~2000 for Zanflo gum. The flow and leveling properties of this polymer are as good as those of Zanflo gum.

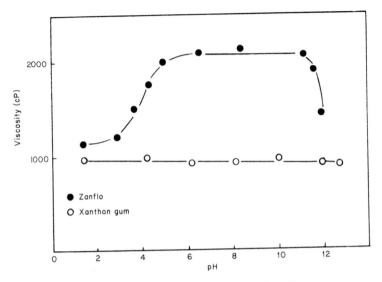

FIGURE 27. Effect of pH on Zanflo viscosity.

VIII. OUTLOOK AND FUTURE DEVELOPMENT

Because world labor rates will continue to climb, it can be expected that the price of the traditional plant gums, such as gum arabic, will increase. Also, because of the lack of uniformity and problems with purity it can be expected that these gums will be replaced in industrial usage.

The number of microbial species that could be polysaccharide producers is large. In view of the encouraging results obtained during the last two decades, it is to be expected that several of the microbial polysaccharides described in this chapter will achieve commercial production on a substantial scale in the future. It is also expected that additional new microbial polysaccharides having novel properties will be discovered. However, the new microbial polysaccharides must have a superior cost/performance compared with existing gums, and they must also comply with all safety and regulatory requirements. Therefore, it must be emphasized that commercialization of new microbial polysaccharides will be extremely arduous and costly.

Further elucidation of metabolic pathways leading to the polysaccharide synthesis should be made to establish control mechanisms of microbial polysaccharides. This sort of information is still lacking, even though it is fundamental for a thorough understanding of polysaccharide production and quality differentiation of microbial polysaccharides through metabolic manipulation and mutation.

REFERENCES

Amemura, A., and Harada, T. (1971). *Hakko Kogaku Zasshi* **49,** pp. 559–564.

Anderson, R. F. (1963). *In* "Biochemistry of Industrial Micro-organisms" (C. Rainbow and A. H. Rose, eds.), pp. 300–316. Academic Press, New York.

Anderson, R. F., Benedict, R. U., Cadmus M. C., and Slodki, M. E. (1960). *Arch. Biochem. Biophys.* **89,** 289.

Andrew, T. R. (1977). *In* "Extracellular Microbial Polysaccharides" (P. A. Sandford and A. Laskin, eds.), ACS Symp. Ser. No. 45, pp. 231–241. Am. Chem. Soc., Washington, D.C.

Anonymous (1958). "Information on Phosphomannan Y-2448," CA-N-7. USDA North. Util. Res. Dev. Div., Peoria, Illinois.

Anonymous (1960a). "Sephadex in Gel Filtration." Pharmacia, Uppsala.

Anonymous (1960b). British Patent 854,715 (Aktiebolaget Pharmacia). (Nov. 23) ibid., 974,054 (Nov. 4, 1964).

Anonymous (1962). "Microbial Polysaccharide B-1973," CA-N-21. USDA Agric. Res. Serv., Peoria, Illinois.

Anonymous (1964). British Patent 974,054 (Aktiebolaget Pharmacia).

Anonymous (1965). "Sephadex Ion Exchangers." Pharmacia, Uppsala.

Anonymous (1969a). 21 CFR 172.695.

Anonymous (1969b). Federal Food and Drug Administration (FDA).

Anonymous (1971). Canadian Governor-in-Council. *Can. Gazette Part II* **105.**

Anonymous (1974). European Economic Community Emulsifier/Stabilizer List, Annex II.

Anonymous (1975). Food and Agriculture Organization/World Health Organization (FAO/WHO) Joint Expert Committee.

Anonymous (1976). U.S. Environmental Protection Agency.

Anonymous (1977a). 21 CFR, Parts 131, 133, 155 and 168.180.

Anonymous (1977b). 9 CFR, 318.7 (c) (4) and 381.147 (f) (3).

Archer, S. A., Clamp, J. R., and Migliore, D. (1977). *J. Gen. Microbiol.* **102,** 157–167.

Axelsson, K., Bjorndal, H., and Ericksson, K. E. (1968). *Acta Chem. Scand.* **22,** 1363–1364.

Bailey, R. W., and Oxford, A. E. (1958). *Nature (London)* **182,** 185.

Bebbington, A., Bourne, E. J., and Wilkinson, J. A. (1952). *J. Chem. Soc.* pp. 246–253.

Behrens, U., and Wuensche, L. (1969). *Cana Azucar* **3,** 39; *Chem. Abstr.* **73,** 5743Y (1970).

Behrens, U., Ringpfeil, M., Bagert, A., and Krueger, K. (1965). East German Patent 35,367; *Chem. Abstr.* **63,** 7624 (1965).

Bender, H., Lehmann, J., and Wallenfels, K. (1959). *Biochim. Biophys. Acta* **36,** 309–316.

Bernier, B. (1958). *Can. J. Microbiol.* **4,** 195–204.

Bjorndal, H., Hellerquist, C. G., Lindberg, B., and Svensson, S. (1970). *Angew. Chem. Int. Ed. Engl.* **9,** 610–619.

Booth, A. N., Deeds, F., and Hendrickson, A. P. (1963). *Toxicol. Appl. Pharmacol.* **5,** 478–484.

Bourne, E. J., Stacey, M., and Wilkinson, I. A., (1950). *J. Chem. Soc.* 2694–2698.

Bouveng, H. O., Fraser, R. L., and Lindberg, B. (1967). *Carbohydr. Res.* **4,** 20–31.

Bovey, F. A. (1959). *J. Polym. Sci.* **35,** 183–190.

Breed, R. S., Murray, E. G. D., and Smith, N. R., Eds. (1957). "Bergey's Manual of Determinative Bacteriology," 7th ed. Williams & Wilkins, Baltimore, Maryland.

Bretthauer, R. K., Kaczorowski, G. J., and Weise, M. J. (1973). *Biochemistry* **12,** 1251–1256.

Brown, A. J. (1886). *J. Chem. Soc.* **49,** 172 and 432.

Brown, C. A., Jr. (1906). *J. Am. Chem. Soc.* **28,** 453.

Buchanan, R. E., and Gibbons, N. E., eds. (1974). "Bergey's Manual of Determinative Bacteriology," 8th Ed. Williams & Wilkins, Baltimore, Maryland.

Buck, K. W., Chain, E. B., Chen, A. W., and Dickerson, A. G. (1968). *J. Gen. Microbiol.* **51,** 377–385.

Burnett, D. B. (1975). *Soc. Pet. Eng. AIME* **SPE 5372.**

Burton, K. A., Cadmus, M. C., Lagoda, A. A., Sandford, P. A., and Watson, P. R. (1976). *Biotechnol. Bioeng.* **18,** 1669–1677.

Cadmus, M. C., Anderson, R. F., and Lagoda, A. A. (1962). *Appl. Microbiol.* **10,** 153.

Cadmus, M. C., Gasdorf, H., Lagoda, A. A., Anderson, R. F., and Jackson, R. W. (1963). *Appl. Microbiol.* **11,** 488–492.

Cadmus, M. C., Burton, K. A., Lagoda, A. A., and Smiley, K. L. (1967). *Bacteriol. Proc.* A102.

Carico, R. D. (1976a). *Oil and Gas J.* **74**(27), 81–84.

Carico, R. D. (1976b). *Soc. Pet. Eng. AIME* **SPE 5870.**

Carlson, A. S., and Hehre, E. J. (1949). *J. Biol. Chem.* **177,** 281.

Carlson, W. W., Rosano, C. L., and Whiteside-Carlson, V. (1953). *J. Bacteriol.* **54,** 136.

Catley, B. J., and Whelan, W. J. (1971). *Arch. Biochem. Biophys.* **143,** 138–142.

Charles, M., and Radjai, M. K. (1977). *In* "Extracellular Microbial Polysaccharides" (P. A. Sandford and A. Laskin, eds.), ACS Symp. Ser. No. 45, pp. 27–39. Am. Chem. Soc., Washington, D.C.

Chen, W. P., and Tsuo, C. H. (1976). *Taiwan Sugar* **23,** 14–16.

Chihara, G. *et al.* (1976). *Nature (London)* **222,** 687–688.

Chihara, G. *et al.* (1970). *Nature (London)* **225,** 943–944.

Cohen, G. H., and Johnstone, D. B. (1963). *Nature (London)* **198,** 211.

Cohen, G. H., and Johnstone, D. B. (1964). *J. Bacteriol.* **88,** 329–338.

Colegrove, G. T. (1967). U.S. Patent 3,326,733.

Colegrove, G. T. (1970). U.S. Patent 3,516,983.

Colegrove, G. T. (1975). U.S. Patent 3,894,879.

Colegrove, G. T. (1976). U.S. Patent 3,966,618.

Colvin, J. R., Chéné, L., Sowden, L. C., and Takai, M. (1977). *Can. J. Biochem.* **55,** 1057–1063.

Corpe, W. A. (1960). *Can. J. Microbiol.* **6,** 153–163.

Corpe, W. A. (1964). *J. Bacteriol.* **88,** 1433–1441.

Corpe, W. A. (1970). *In* "Adhesion in Biological System" (R. S. Manly, ed.), pp. 73–87. Academic Press, New York.

Cottrell, I. W. (1977). German Offen. 2,643,140.

Cottrell, I. W., and Kang, K. S. (1978). *Dev. Ind. Microbiol.* **19,** pp. 117–131.

Cottrell, I. W., Shim, J. L., Best, G. H., and Empey, R. H. (1978). *In* "Carbohydrate Sulfates" (R. G. Schweiger, J. L. Shin, and G. H. Best, eds.), ACS Symp. Ser. No. 77, pp. 193–202. Am. Chem. Soc., Washington, D.C.

Couperwhite, I., and McCallum, M. F. (1974). *Arch. Microbiol.* **97,** 73–80.

Cox, J. S., King, R. E., and Reynolds, G. F. (1965). *Nature (London)* **207,** 1202–1203.

Cypert, J. D., and Patton, J. T. (1963). U.S. Patent 3,084,122; *Chem. Abstr.* **59,** 1422 (1963).

Davidson, I. W., Sutherland, I. W., and Lawson, C. J. (1977). *J. Gen. Microbiol.* **98,** 603–606.

Davis, E. N., Rhodes, R. A., and Schulk, H. R. (1966). *Appl. Microbiol.* **13,** 267–271.

Dea, I. C. M., McKinnon, A. A., and Rees, D. A. (1972). *J. Mol. Biol.* **68,** 153–172.

Dea, I. C. M., Morris, E. R., Rees, D. A., Welsh, E. J., Barnes, H. A., and Price, J. (1977). *Carbohydr. Res.* **57,** 249–272.

Deavin, L., Jarman, T. R., Lawson, C. J., Righaleto, R. C., and Slocombe, S. (1977). In "Extracellular Microbial Polysaccharides" (P. A. Sandford and A. Laskin, eds.), ACS Symp. Ser. No. 45, pp. 14–26. Am. Chem. Soc., Washington, D.C.

Dedoner, R. A., and Hassid, W. Z. (1964). Biochim. Biophys. Acta 90, 239.

Dias, F., and Bhat, J. (1964). Appl. Microbiol. 12, 412–417.

Dintzis, F. R., Babcock, G. E., and Tobin, R. (1970). Carbohydr. Res. 13, 257–267.

Doudoroff, M., Hassid, W. A., Putman, E. W., Potter, A. L., and Lederberg, J. (1949). J. Biol. Chem. 179, 921.

Drummond, G. S., Smith, E. E., Whelan, W. G., and Jai, H. (1969). FEBS Lett. 5, 85.

Ebert, K. H., and Brosche, M. (1967). Biopolymers 5, 423–430.

Ebert, K. H., and Patat, F. (1962). Z. Naturforsch. Teil B 17, 738.

Ellwood, D. C., and Evans, G. T. C. (1977). Netherlands Patent 7612-448.

Felty, L. G. (1975). U.S. Patent 3,906,108.

Forsyth, W. G. C., and Webley, D. M. (1949). Biochemistry 44, 455.

Friedman, B., Dugan, P., Pfister, R., and Ramsen, C. (1968). J. Bacteriol. 96, 2144–2153.

Gadsdorf, H. J., Benedict, R. G., Cadmus, M. C., Anderson, R. F., and Jackson, R. W. (1965). J. Bacteriol. 90, 147–150.

Gibsen, K. F., and Saddington, A. W. (1973). U.S. Patent 3,717,452.

Gill,.J. W. (1972). U.S. Patent 3,632,570.

Glaser, L. (1958). J. Biol. Chem. 232, 627.

Gorin, P. A. G., and Spencer, J. F. T. (1968). Adv. Carbohydr. Chem. 23, 367.

Gorin, P. A. T., and Spencer, J. F. T. (1966). Can. J. Chem. 44, 993–998.

Grönwall, A. (1957). In "Dextran and Its Use in Colloidal Infusion Solutions." Academic Press, New York.

Halleck, F. E. (1967). U.S. Patent 3,301,848.

Halleck, F. E. (1969). U.S. Patent 3,447,940.

Harada, T. (1977). In "Extracellular Microbial Polysaccharides" (P. A. Sandford and A. Laskin, eds.), ACS Symp. Ser. No. 45, pp. 265–283. Am. Chem. Soc., Washington, D.C.

Harada, T. (1965). Arch. Biochem. Biophys. 112, 65–69.

Harada, T., and Yoshimura, T. (1965). Agric. Biol. Chem. 29, 1027.

Harada, T., Maeda, I., Hiroshi, S., Masada, M., and Misaki, A. (1967). Agric. Biol. Chem. 31, 1184–1188.

Harada, T., Misaki, A., and Saito, H. (1968). Arch. Biochem. Biophys. 124, 292–298.

Harada, T., Misaki, A., Hiroshi, S., and Toshiko, I. (1969). Biochemistry 8, 4645–4650.

Hartnek, H. G., Empey, R. A., Pettitt, D. J., and Van Winkle, T. L. (1974). U.S. Patent 3,852,257.

Hathaway, R. J. (1971). U.S. Patent 3,556,942.

Haug, A., and Larsen, B. (1971). Carbohydr. Res. 17, 297–308.

Hehre, E. J. (1941). Science 93, 237.

Hehre, E. J. (1943). Proc. Soc. Exp. Biol. Med. 54, 240.

Hehre, E. J., and Hamilton, D. M. (1948). J. Bacteriol. 55, 197.

Hehre, E. J., Carlson, A. E., and Neill, V. M. (1947). Science 106, 523.

Hestrin, S. (1944). Nature (London) 154, 581.

Hestrin, S. (1962). In "The Bacteria" (I. C. Gunsalus and R. Y. Stanier, eds.), Vol. 3, pp. 373–388. Academic Press, New York.

Hestrin, S., and Shapiro, S. (1944). Biochem. J. 38, 2.

Hestrin, S., and Shramm, N. (1954). Biochem. J. 58, 345.

Hestrin, S., Feingold, D. S., and Avigad, G. (1955). (J. Am. Chem. Soc. 77, 6710.

Holzworth, G. (1976). Biochemistry 15, 4333–4339.

Holzworth, G., and Prestridge, E. B. (1977). *Science* **197,** 757–759.
Hucker, G. J., and Pederson, C. S. (1930). *N.Y., Agric. Exp. Stn., Geneva, Tech. Bull.* **167,** 3.
Igoe, R. S. (1977). U.S. Patent 4,046,925.
Jansson, P. E., Kenne, L., and Lindberg, B. (1975). *Carbohydr. Res.* **45,** 275–282.
Jeanes, A. (1952). *In* "Dextran, A Selected Bibliography," Bur. Agric. Ind. Chem. AIC-288. USDA, Peoria, Illinois.
Jeanes, A. (1965). *Methods Carbohydr. Chem.* **5,** 118–127.
Jeanes, A. (1968). *Encyl. Poly. Sci. Techno.* (N. M. Bikales, ed.) **4,** 693–711.
Jeanes, A. (1974). *Food Technol. (Chicago)* **28**(5), 34–40.
Jeanes, A. (1978). Dextran Bibliography, *US, Dep. Agric., Agric, Res. Serv.,* Misc. Publ. #1355.
Jeanes, A. (1977). *In* "Extracellular Microbial Polysaccharides" (P. A. Sandford and A. Laskin, eds.), ACS Symp. Ser. No. 45, pp. 284–298. Am. Chem. Soc., Washington, D.C.
Jeanes, A., and Sloneker, J. H. (1961). U.S. Patent 3,000,790.
Jeanes A., and Watson, P. R. (1962). *Can. J. Chem.* **40,** 1318–1325.
Jeanes, A., Haynes, W. C., Wilham, C. A., Rankin, J. C., Melvin, E. H., Austin, M., Cluskey, J. E., Tsuchiya, H. M., and Rist, C. E. (1954). *J. Am. Chem. Soc.* **76,** 5041–5052.
Jeanes, A., Dimler, R. J., Pittsley, J. E., and Watson, P. R. (1961a). *Arch. Biochem. Biophys.* **92,** 343.
Jeanes, A., Pittsley, J. E., and Santi, F. R. (1961b). *J. Appl. Polym. Sci.* **5,** 519–526.
Jeanes, A., Pittsley, J. E., and Watson, P. R. (1964). *J. Appl. Polym. Sci.* **8,** 2775–2787.
Jeanes, A., Knutson, C. A., Pittsley, J. E., and Watson, P. R. (1965). *J. Appl. Polym. Sci.* **9,** 627–638.
Jeanes, A., Burton, K. A., Cadmus, M. C., Knutson, C. A., Rowin, G. L., and Sandford, P. A. (1971). *Nature (London), New Biol.* **233,** 259–260.
Jeanes, A., Cadmus, M. C., Knutson, C. A., Rogovin, P., and Silman, R. W. (1976). *U.S., Agric. Res. Serv., North Cent. Reg.* [*Rep.*] **ARS-NC-51.**
Johnson, J. J., Jr., Kirkwood, S., Misaki, A., Nelson, T. E., Scaletti, J. V., and Smith, F. (1963). *Chem. Ind. (London),* p. 820.
Kagadel, L., and Akerstroem, S. (1971). *Acta Chem. Scand.* **25,** 1855–1859.
Kang, K. S. (1972). Kelco Div. of Merck & Co., Inc., San Diego, California (unpublished data).
Kang, K. S., and Kovacs, P. (1975). *Int. Food Sci. Technol., Proc. Int. Congr., 4th, 1974,* pp. 33–34.
Kang, K. S., and McNeely, W. H. (1975). U.S. Patent 3,894,976.
Kang, K. S., and McNeely, W. H. (1976a). U.S. Patent 3,960,832.
Kang, K. S., and McNeely, W. H. (1976b). U.S. Patent 3,979,303.
Kang, K. S., and McNeely, W. H. (1977). *In* "Extracellular Microbial Polysaccharides" (P. A. Sandford and A. Laskin, eds.), ACS Symp. Ser. No. 45, pp. 220–230. Am. Chem. Soc., Washington, D.C.
Kimura, H., Moritaka, S., and Misaki, M. (1973). *J. Food Sci.* **38,** 668–670.
King, G. G. S., and Colvin, J. R. (1976). *Appl. Polym. Symp.* **28,** 623–636.
Kjosbakken, J., and Colvin, J. R. (1975). *Can. J. Microbiol.* **21,** 111-120.
Knutson, C. A., Pittsley, J. E., and Jeanes, A. (1971). ACS *Abstr. Pap., 161st Meet., Am. Chem. Soc.* CARB. 28.
Koepsell, H. J., Jackson, R. W., and Hoffman, C. A. (1956). U.S. Patent 2,729,587.
Komatsu, N., Okubo, S., Kikumoto, S., Kimura, K., Saito, G., and Sakai, S. (1969). *Gann* **60,** 137–144.
Kovacs, P. (1973a). *Food Technol.* **27,** 26–30.
Kovacs, P. (1973b). *Food Trade Rev.* **11,** 17–22.

Kovacs, P., and Igoe, R. S. (1976). *Food Prod. Dev.* **10,** 32, 34, 36, and 38.

Kovacs, P., and Kang, K. S. (1977). *In* "Food Colloids" (H. D. Graham, ed.), pp. 500–522. Avi Publ. Co., Westport, Connecticut.

Kovacs, P., and Titlow, B. D. (1976). *Am. Dairy Rev.* **4,** 34J–34N.

Larsen, B., and Haug, A. (1971). *Carbohydr. Res.* **17,** 287–296.

Lilly, V. G., Leach, J. G. and Wilson, H. A. (1958). *Appl. Microbiol.* **6,** 105–108.

Lindberg, B. Lörngren, J., and Thompson, J. L. (1973). *Carbohydr. Res.* **28,** 351–357.

Lipton, D. (1974). *Soc. Pet. Eng. AIME* **SPE 5099.**

Lipton, D., and Burnett, D. B. (1976). *Soc. Pet. Eng. AIME* **SPE 5872.**

Lopez, R., and Becking, J. H. (1968). *Microbiol. Expan.* **21,** 1–23.

Lwoff, A., Ionesco, H., and Gutman, A. (1950). *Biochim. Biophys. Acta* **4,** 270.

McNeely, W. H., and Kang, K. S. (1973). *In* "Industrial Gums" (R. L. Whistler and J. N. Be Miller, eds.), 2nd ed., pp. 473–498. Academic Press, New York.

McNeely, W. H., and Kovacs, P. (1975). *In* "Physiological Effects of Food Carbohydrates" (A. Jeanes and J. Hodge, eds.), ACS Symp. Ser. No. 15, pp. 269–281. Am. Chem. Soc., Washington, D.C.

NcNeely, W. H., and O'Connell, J. J. (1966). U.S. Patent 3,232,929.

McNeely, W. H., Fairchild, W. P., and Hunter, A. R. (1966). Canadian Patent 727,071.

Martin, L. E., Bates, C. M., Beresford, C. R., Donaldson, J. D., McDonald, F. F., Dunlop, D., Sheard, P., London, E., and Twigg, G. D. (1955). *Br. J. Pharmacol. Chemother.* **10,** 375.

Masson, C. R., Menzies, R. F., and Cruikshank, L. (1946). *Nature (London)* **157,** 74.

Maxted, W. R. (1952). *Nature (London)* **170,** 1020–1021.

Monod, J., and Torriani, A. M. (1950). *Ann. Inst. Pasteur, Paris* **78,** 65.

Moorhouse, R., Walkinshaw, M. D., and Arnott, S. (1977a). *In* "Extracellular Microbial Polysaccharides" (P. A. Sandford and A. Laskin, eds.), ACS Symp. Ser. No. 45, pp. 90–102. Am. Chem. Soc., Washington, D.C.

Moorhouse, R., Walkinshaw, M. D., Winter, W. T., and Arnott, S. (1977b). *In* "Cellulose Chemistry and Technology" (J. C. Arthur, ed.), ACS Symp. Ser. No. 48, pp. 133–152. Am. Chem. Soc., Washington, D.C.

Moraine, R. A., and Rogovin, S. P. (1966). *Biotechnol. Bioeng.* **8,** 511–524.

Morita, N., and Murao, S. (1974). *J. Ferment. Technol.* **52,** 438–444.

Morris, E. R., Rees, D. A., Young, G., Walkinshaw, M. D., and Darke, A. (1977). *J. Microbiol.* **110,** 1–16.

Muhlethaler, K. (1949). *Biochim. Biophys. Acta* **3,** 527.

Murphy, P. T., and Whistler, R. L. (1973). *In* "Industrial Gums" (R. L. Whistler and J. N. Be Miller, Eds.), 2nd ed., pp. 513–542. Academic Press, New York.

Nagahama, T., Fujimoto, S., and Kanie, M. (1977). *Agric. Biol. Chem.* **41,** 9–16.

Ninomiya, E. (1967). Japanese Patent 42-7600.

Ninomiya, E., and Kizaki, T. (1969). *Angew. Makromol. Chem.* **6,** 179.

O'Connell, J. J. (1961). U.S. Patent 3,067,038.

Ogino, S. (1973). Japanese Patent 48-2800.

Owen, W. E. (1950). *Sugar* **45,** 42.

Owen, W. I. (1959). U.S. Patent 2,868,725.

Parikh, V. M., and Jones, J. K. N. (1963). *Can. J. Chem.* **41,** 2826–2835.

Pasteur, L. (1861). *Bull. Soc. Chim. Fr.,* pp. 30–31.

Patton, J. T. (1962). U.S. Patent 3,020,207.

Patton, J. T. (1973). U.S. Patent 3,729,460.

Patton, J. T., and Lindblom, G. P. (1962). U.S. Patent 3,020,206.

Perlin, A. S., and Taber, W. A. (1963) *Can. J. Chem.* **41,** 2278–2282.

Petrowski, G. E. (1977). *Proc., AFMA Symp.* pp. 57–59.

Pettitt, D. J. (1973). U.S. Patent 3,708,446.

Pindar, D. F., and Bucke, C. (1975). *Biochem. J.* **152,** 617–622.

Quinnell, C. M., Knight, S. G., and Wilson, P. W. (1957). *Can. J. Microbiol.* **3,** 277–288.

Racciato, J. S. (1976). *Am. Dyest. Rep.* **65**(11), 51 and 69.

Rees, D. A. (1969). *Adv. Carbohydr. Chem. Biochem.* **24,** 209.

Rees, D. A. (1970). *Sci. J.* **6**(12), 47–51.

Regan, B. F., and Regan, G. B. (1975). U.S. Patent 3,899,437.

Robbins, D. J., Booth, A. N., and Moulton, J. E. (1964). *Food Cosmet. Toxicol.* **2,** 545–550.

Rocks, J. K. (1971). *Food Technol. (Chicago)* **25,** 22–31.

Rodgers, N. E. (1973). *In* "Industrial Gums" (R. L. Whistler and J. N. Be Miller, eds.), 2nd ed., pp. 499–511. Academic Press, New York.

Rogovin, S. P., and Albrecht, W. J. (1964). U.S. Patent 3,119,812.

Rogovin, S. P. Senti, F. R., Benedict, R. G., Tsuchiya, H. M., Watson, P. R., Tobin, R., Sohns, V. E., and Slodki, M. E. (1960). *J. Biochem. Microbiol. Technol. Eng.* **2,** 381.

Rogovin, S. P., Anderson, R. F., and Cadmus, M. C. (1961). *J. Biochem. Microbiol. Technol. Eng.* **3,** 51.

Rogovin, S. P., Albrecht, W. J., and Sohns, V. (1965). *Biotechnol. Bioeng.* **7,** 161–169.

Scheibler, Z. (1869). *Verh. Dtsch. Zucker-Ind.* **19,** 472.

Scher, M., Lennarz, W. J., and Sweeley, C. C. (1968). *Proc. Natl. Acad. Sci. U.S.A.* **59,** 1313–1320.

Schuppner, H. R. (1971). U.S. Patent 3,557,016.

Schuppner, H. R. (1972). U.S. Patent 3,659,026.

Schweiger, R. G. (1966a). U.S. Patent 3,236,831.

Schweiger, R. G. (1966b). U.S. Patent 3,244,695.

Schweiger, R. G. (1966c). U.S. Patent 3,256,271.

Schweiger, R. G. (1969). U.S. Patent 3,446,796.

Sheth, P., and Lachman, L. (1969). U.S. Patent 3,421,920.

Silman, R. W., and Rogovin, S. P. (1972). *Biotechnol. Bioeng.* **14,** 23–31.

Singh, P. P. (1974). *Carbohydr. Res.* **37,** 245–247.

Slodki, M. E. (1963). *Biochim. Biophys. Acta* **69,** 96.

Slodki, M. E. (1966). *Can. J. Microbiol.* **12,** 495–499.

Slodki, M. E., Cadmus, M. C., and Wickerham, L. J. (1961). *J. Bacteriol.* **82,** 269.

Slodki, M. E., Bandoni, R. J., and Wickerham, L. J. (1966). *Can. J. Microbiol.* **12,** 489–494.

Sloneker, J. H., and Jeanes, A. (1962). *Can. J. Chem.* **40,** 2066–2071.

Sloneker, J. H., Orentas, D. G., and Jeanes, A. (1964). *Can. J. Chem.* **42,** 1261–1269.

Sloneker, J. H., Orentas, D. G., Knutson, C. A., Watson, P. R., and Jeanes, A. (1968). *Can. J. Chem.* **46,** 3353–3361.

Smiley, K. L. (1966). *Food Technol.* **20,** 1206–1210.

Sowa, W., Blackwood, A. C., and Adams, G. A. (1963). *Can. J. Chem.* **41,** 2314–2319.

Sparks, W. J. (1962). U.S. Patent 3,053,765; *Chem. Abstr.* **58,** 8838 (1963).

Starr, M. P. (1946). *J. Bacteriol.* **51,** 131–143.

Stolp, H., and Starr, M. P. (1964). *Phytopathol. Z.* **51,** 442–478.

Sutherland, I. W. (1972). *Adv. Microb. Physiol.* **8,** 143–208.

Sutherland, I. W. (1977). *In* "Extracellular Microbial Polysaccharides" (P. A. Sandford and A. Laskin, eds.), ACS Symp. Ser. No. 45, pp. 40–57. Am. Chem. Soc., Washington, D.C.

Sutton, J. C., and Williams, P. H. (1970). *Can. J. Bot.* **48,** 645–651.

Szaniszlo, P. J., Wirsen, C., Jr., and Mitchell, R. (1968). *J. Bacteriol.* **96,** 1474–1483.

Taguchi, R., Kikuchi, Y., Sakuno, Y., and Kobashi, T. (1973). *Agric. Biol. Chem.* **37,** 1583–1588.

Tam, K. T., and Finn, R. K. (1977). *In* "Extracellular Microbial Polysaccharides" (P. A. Sandford and A. Laskin, eds.), ACS Symp. Ser. No. 45, pp. 58–80. Am. Chem. Soc., Washington, D.C.

Tannahill, A. T., and Finn, R. K. (1975). U.S. Patent 3,878,045.

Tarr, H. L. A., and Hibbert, H. (1931a). *Can J. Res.* **4,** 372.

Tarr, H. L. A., and Hibbert, H. (1931b). *Can. J. Res.* **5,** 414.

Tenney, M., and Stumm, W. (1965). *J. Water Pollut. Control Fed.* **37,** 1370–1388.

Troy, F. A., Frerman, F. E., and Heath, E. C. (1971). *J. Biol. Chem.* **246,** 118–133.

Tsuchiya, H. M., Hellman, N. N., Koepsell, H. J., Corman, J., Stringer, C. S., Rogovin, S. P., Bogard, M. O., Bryant, G., Feger, V. H., Hoffman, C. A., Senti, F. R., and Jackson, R. W. (1955). *J. Am. Chem. Soc.* **77,** 2412–2419.

Umbreit, J. N., Stone, K. J., and Strominger, J. L. (1972). *J. Bacteriol.* **112,** 1302–1305.

Unz, R. F., and Farrah, S. R. (1974). *Envion. Prot. Technol. Ser.* **EPA-670/2-74-018.**

Wallen, L. L., and Davis, E. N. (1972). *Environ. Sci. Technol.* **6,** 161–164.

Wallen, L. L., Rhodes, R. A., and Shulke, H. R. (1965). *Appl. Microbiol.* **13,** 272–278.

Watson, P. R., Sandford, P. A., Burton, K. A., Cadmus, M. C. and Jeanes, A. (1976). *Carbohydr. Res.* **46,** 259–265.

Whistler, R. L., Bushway, A. A., Nakamura, W., Singh, P. P., and Tokuzen, R. (1976). *Adv. Carbohydr. Chem. Biochem.* **32,** 235–275.

Wilkinson, J. F. (1958). *Bacteriol. Rev.* **22,** 46.

Woodard, G., Woodard, M. W., McNeeley, W. H., Kovacs, P., and Cronin, M. T. I. (1973). *Toxicol. Appl. Pharmacol.* **24,** 30–36.

Woodard Research Corp. (1973). Unpublished report. Herndon, VA.

Woodard Research Corp. (1971). Unpublished report. Herndon, VA.

Woodard Research Corp. (1968a). Unpublished report. Herndon, VA.

Woodard Research Corp. (1968b). Unpublished report. Herndon, VA.

Yamada, K., and Furukawa, T. (1973). Japanese Patent 48-33396.

Yamatodani, T., and Kanamaru, T. (1972). U.S. Patent 3,674,642.

Yuen, S. (1974). *Process Biochem.* **9,** 7–9.

Zajic, J. E., and LeDuy, A. (1973). *Appl. Microbiol.* **25,** 628–635.

Zevenhuizen, L. P. T. M. (1973). *Carbohydr. Res.* **26,** 409–419.

Chapter 14

Microbial Transformation
of Steroids and Sterols

O. K. SEBEK
D. PERLMAN

I. INTRODUCTION

Naturally occurring steroids possess remarkable hormonal properties which are of critical importance to human well-being. They include hormones of the adrenal cortex (cortisone, cortisol, corticosterone), the progestational hormone (progesterone), the androgens or male hormones (testosterone, dihydrotestosterone), and the female sex hormones (estradiol, estrone) (see Scheme 1).

They are all derived from a tetracyclic hydrocarbon (perhydrocyclopentanophenanthrene) and differ primarily in the number, type, and location of the substituent functional groups and in the number and position of the double bonds (Scheme 2).

In the last 30 years, thousands of derivatives of these compounds have been synthesized and tested for their endocrine and other activities. Many of them were found to be active as antiinflammatory and progestational agents and also as sedatives and anabolic and antitumor substances. Some are effective in allergic, dermatologic, and ocular diseases, in cardiovascular therapy, and in veterinary products, and some

483

MICROBIAL TECHNOLOGY, 2nd ed., VOL. I
Copyright © 1979 by Academic Press, Inc.
All rights of reproduction in any form reserved. ISBN 0-12-551501-4

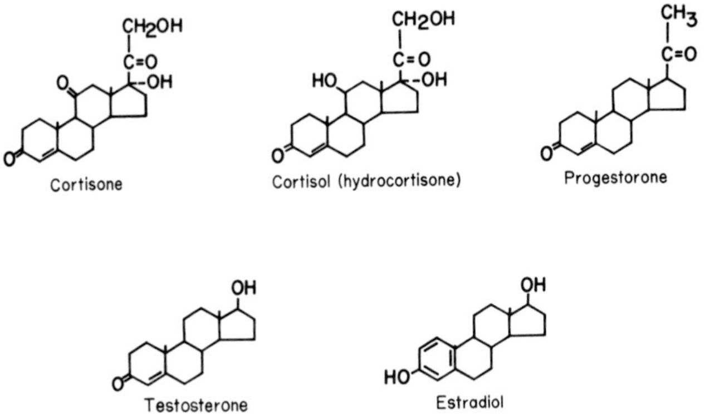

Cortisone

Cortisol (hydrocortisone)

Progestorone

Testosterone

Estradiol

SCHEME 1

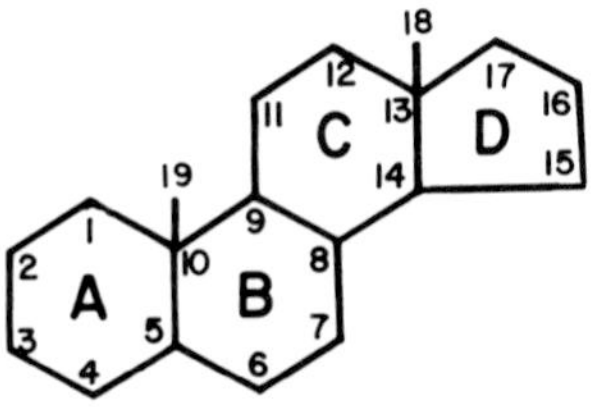

SCHEME 2

have been outstandingly successful as oral contraceptives (Applezweig, 1962, 1978).

The demonstration of such extraordinary activities had its origin in the discovery by Hench and his associates in 1949 of the remarkable antiinflammatory property of cortisone in rheumatoid arthritis (Hench *et al.*, 1949, 1950).

In the wake of this discovery, the dramatic surge in the demand for cortisone to treat millions of arthritic sufferers went unmet since no adequate method for its manufacture existed. The chemical synthesis from deoxycholic acid (Sarett, 1946) developed at Merck and Company, was workable but complicated and uneconomical: 31 steps were needed to obtain 1 kg of cortisone acetate from 615 kg of deoxycholic acid. The oxygen shift from C-12 to C-11 alone, which is essential for the activity of cortisone (and of all adrenal corticosteroid hormones), required nine steps* and is shown in Scheme 3.

Economic synthesis of cortisone, cortisol, and other corticosteroids eventually developed from three lines of information: (1) diosgenin from

---

*The efficiency of this shift was subsequently improved and deoxycholic acid, obtained from beef bile, is at present a substrate in some manufacturing processes.

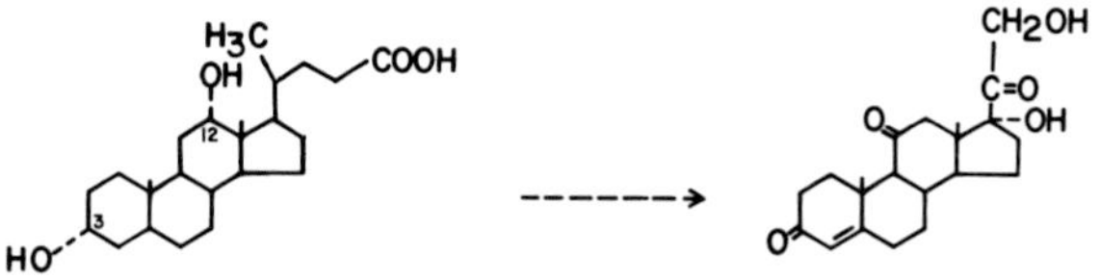

Deoxycholic acid   Cortisone

SCHEME 3

the Mexican barbasco plant (*Dioscorea composita*) and stigmasterol from soybeans (*Glycine max*) were found abundant and inexpensive raw materials for chemical conversion to 16-dehydropregnenolone and pregnenolone, respectively; (2) deoxycorticosterone was hydroxylated to corticosterone by perfusion with isolated beef adrenal glands* (Hechter *et al.*, 1949) (see Scheme 4); (3) yeasts and bacteria were reported to

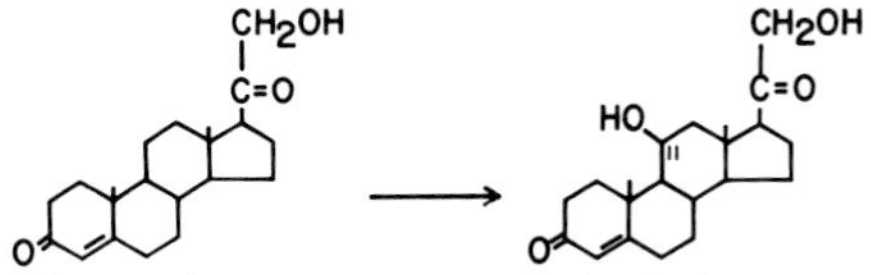

Deoxycorticosterone   Corticosterone

SCHEME 4

carry out nuclear reduction, oxidation, isomerization, and hydrolysis of steroids (Mamoli and Vercellone, 1937a,b).

By applying this information, Peterson and Murray succeeded in 1952 in introducing the essential oxygen at C-11 by oxidizing progesterone to 11α-hydroxyprogesterone by means of first *Rhizopus arrhizus* and then by *Rhizopus nigricans* in practically quantitative yields (>85%) (Peterson and Murray, 1952; see also Murray and Peterson, 1953) (Scheme 5).

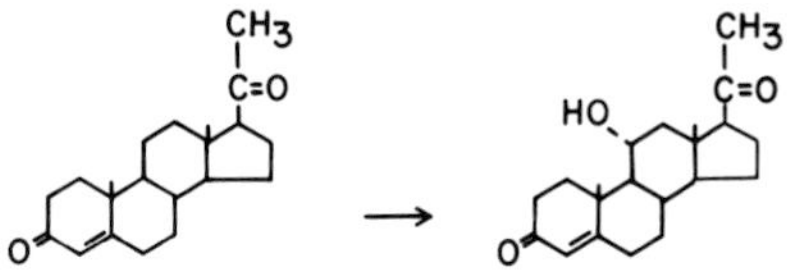

Progesterone   11α-Hydroxyprogesterone

SCHEME 5

Almost simultaneously, *Streptomyces argenteolus* was reported to oxidize progesterone to 16α-hydroxyprogesterone (Perlman *et al.*, 1952) (Scheme 6).

---

*This process was employed at one time by G. D. Searle and Company; it produced the first reasonably large quantities of cortisol for clinical testing.

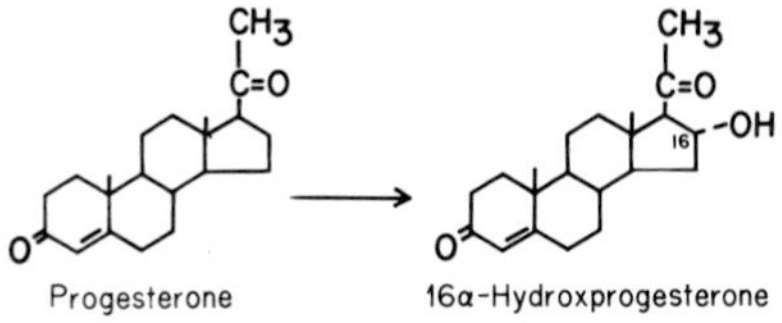

SCHEME 6

This reaction became of considerable importance 5 years later in the synthesis of 16α-hydroxy-9α-fluoroprednisolone (triamcinolone) which has greatly improved antiinflammatory activity.

Following these reports, the disclosures of other hydroxylations were made in a rapid sequence: *Cunninghamella blakesleeana* (Hanson *et al.*, 1953) and *Curvularia lunata* (Shull *et al.*, 1953; Shull and Kita, 1955) were found to carry out 11β-hydroxylation at C-11 in good yields (about 60%) and the latter fungus is used in the manufacture of cortisol from compound S (Scheme 7).

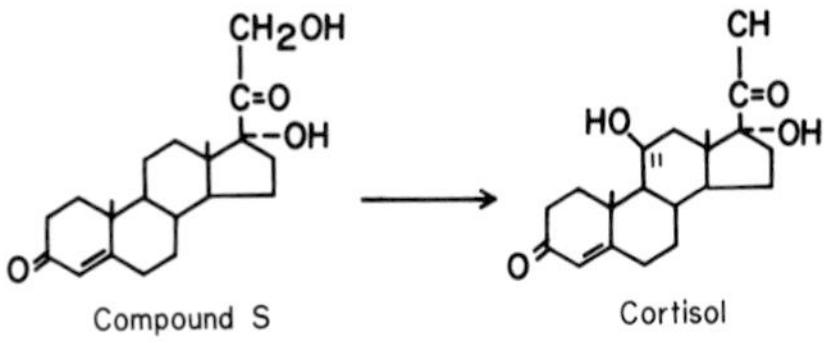

SCHEME 7

These observations with their highly practical applications stimulated an intensive search for additional microbial modifications of these compounds. In time, all available carbon atoms were found to be hydroxylated by different organisms and by now the number of microbial hydroxylations exceeds the number of hydroxylations carried out by mammalian systems (Table I). In many cases, polyhydroxylations in different combinations were also noted with dihydroxylations (such as 6β, 11α; 7α, 14α; 9α, 14α; or 11β,21) prevailing. They are not, however, of practical importance. The oxygen atom involved in these reactions is derived from gaseous oxygen and not from water, and the hydroxyl group thus formed retains the same configuration as the hydrogen that has been replaced (Hayano *et al.*, 1958a,b).

The therapeutic properties of cortisone and cortisol were further improved by microbial introduction of a 1,2-double bond whereby prednisone (1-dehydrocortisone) and prednisolone (1-dehydrocortisol) were formed, respectively (for example, see Scheme 8).

This 1-dehydrogenation was first described to be carried out by *Cylin-*

TABLE I. Comparison of Hydroxylation of Steroids by Microbial and Mammalian Systems

Microbial systems	Mammalian systems
1α	1α
1β	2α
2α	2β
2β	6α
3β	6β
$5\alpha^a$	7α
$5\beta^b$	11β
6β	12α
7α	15α
7β	16α
9α	17α
$10\beta^c$	18
11α	19
11β	21
12α	
12β	
14α	
15α	
15β	
16α	
16β	
17α	
18	
19	
21	

^{a} In A-norprogesterone series.
^{b} In cardiac glycosides.
^{c} In 19-nortestosterone.

drocarpon radicicola and Streptomyces lavendulae (Fried et al., 1953), by Fusarium solani and Fusarium caucasicum (Vischer and Wettstein, 1953), and by Septomyxa affinis, which is also used in the preparation of 6α-methylprednisolone (Medrol), a valuable synthetic corticoid which does not cause significant sodium retention (Sebek and Spero, 1959; Spero et al., 1956). These organisms also degrade the side chain,

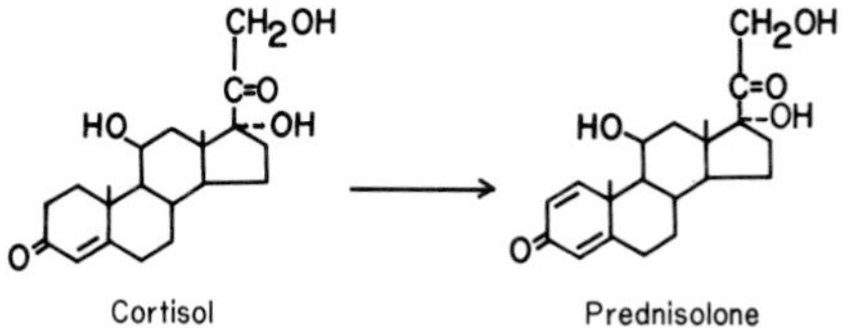

Cortisol   Prednisolone

SCHEME 8

oxidize the hydroxyl group at C-17, and bring about expansion of ring D of some steroids (such as progesterone and deoxycorticosterone) (see Scheme 9).

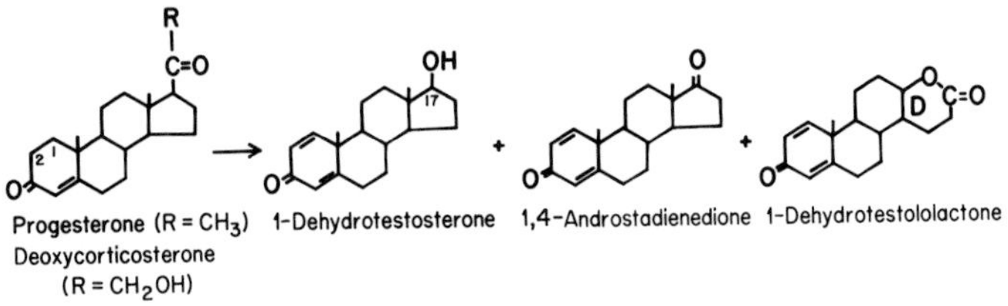

Progesterone (R = CH₃)   1-Dehydrotestosterone   1,4-Androstadienedione   1-Dehydrotestololactone
Deoxycorticosterone
(R = CH₂OH)

SCHEME 9

In contrast, *Arthrobacter (Corynebacterium) simplex* which was the first organism used commercially to 1-dehydrogenate cortisol to prednisolone (Nobile *et al.*, 1955) is devoid of these attendant reactions. The same is true of two other 1-dehydrogenating bacteria (*Bacillus sphaericus* and *Bacterium cyclooxydans*) which have also been used in industrial practice.

A process called "pseudo-crystallofermentation" was also described which was reported to give high yields of the 1-dehydrogenated product. According to this procedure, *A. simplex* transforms finely powdered solid cortisol in 1–50% concentrations to crystalline prednisolone in high (>93%) yields directly in the fermentation beers within 5 days (Kondō and Masuo, 1961).

II. COMMERCIAL DEVELOPMENT

As these considerations indicate, microbial modifications were directly responsible for the generation of new analogs, which in turn served in many cases as substrates for a still larger number of steroid products. They had also a significant effect on the cost of the desired hormones. Thus the 11α-hydroxylation of progesterone brought the 1949 price of cortisone in a very short time from $200 to $6 per gram and, through further improvements, to less than $1 per gram at present.

To carry out these conversions on a large scale, the culture is grown in fermentation tanks with aeration and agitation. The steroid is dissolved in a suitable solvent and added at different growth stages, preferably toward the end of the growth phase, and the transformation is allowed to proceed under the same aerobic conditions to a maximum. Since steroids are essentially insoluble or sparingly soluble in water, it is of considerable importance that they react in aqueous suspensions and that the reactions are completed in most cases within a reasonable time.

Thus 11α-hydroxylation of 2 gm of progesterone per liter by *R. nigricans* proceeds satisfactorily in 48 hours. The product is released into the medium from which it is recovered by extraction with methylene chloride. With further development (use of surfactants, improved methods of steroid addition) the substrate levels have been increased with the concomitant yield increases of the product. Solid substrate addition also improved the efficiency of 11α-hydroxylation. Thus 2 g of finely ground aqueous suspension of progesterone with 0.01% wetting agent (Tween 80) per 100 ml was converted by *Aspergillus ochraceus* to 11α-hydroxyprogesterone in 90% yields in 3 days with a minimum of a side-product (6β, 11α-dihydroxyprogesterone, Weaver *et al.*, 1960). A study of this hydroxylation indicated that 14% of the substrate progesterone remained unconverted regardless of the concentrations at which it had been added. This was due to the formation of a crystalline structure consisting of 6 parts of the product and 1 part of progesterone, which thus became unavailable for the transformation. Continuous feeding of the substrate overcame this drawback. A similar substrate–product interaction was observed during 1-dehydrogenation by *S. affinis* of 11β,21-dihydroxy-4, 17(20)-pregnadien-3-one (dienediol) to 11β,21-dihydroxy-1,4,17(20)-pregnatrien-3-one (trienediol), an intermediary step in the synthesis of 6α-methylprednisolone (see above). This difficulty was overcome by multistage feeding of the substrate (Maxon *et al.*, 1966).

For illustration, some of the steroid transformations of commercial importance are listed in Table II. The fermentation conditions used in laboratory studies of these and related transformations are summarized in Table III (also see Perlman, 1976).

The complexity of industrial processes may be illustrated by the information in Fig. 1, which outlines the conversion of stigmasterol to six steroids of clinical importance.

Twenty-one chemical and microbial steps are involved in the conversion of stigmasterol to oxylone, and 10 steps to prednisolone and to cortisone. By continuous refinement of the process operations as high as 10 mole-% conversions of the starting material to final product have been realized.

III. TYPES OF MICROBIAL TRANSFORMATIONS

In addition to the six reactions of commercial importance (see Table II), many other microbial transformations have been discovered. They are listed in Table IV and include introduction of nuclear and side-chain hydroxy groups; cleavage of carbon-to-carbon linkages in the side chain as well as in ring D; introduction of a 1,2-double bond in ring A; oxidation

TABLE II. Some Steroid Transformations of Commercial Importance

Reaction	Substrate → product	Microorganism	Some industrial producers
11α-Hydroxylation	Progesterone → 11α-hydroxyprogesterone	*Rhizopus nigricans*	The Upjohn Company
11β-Hydroxylation	Compound S → cortisol	*Curvularia lunata*	Pfizer, Inc.; Gist-Brocades
16α-Hydroxylation	9α-Fluorocortisol → 9α-fluoro-16α-hydroxycortisol	*Streptomyces roseochromogenus*	E.R. Squibb and Sons; Lederle Laboratories
1-Dehydrogenation	Cortisol → prednisolone	*Arthrobacter simplex*	Schering Corporation
	Dienediol[a] → trienediol[a]	*Septomyxa affinis*	The Upjohn Company
1-Dehydrogenation, side-chain cleavage, and ring D expansion	Progesterone → 1-dehydrotestololactone	*Cylindrocarpon radicicola*	E.R. Squibb and Sons
Side-chain cleavage	β-Sitosterol → androstadienedione and/or androstenedione	*Mycobacterium* spp.	G.D. Searle and Company

[a] See text for chemical names.

TABLE III. Media and Fermentation Conditions Used in Microbial Transformations of Steroids

Microorganism	Steroid substrate	Steroid product (approx. yields wt. %)	Composition of medium[a]	Length of incubation, temperature, aeration	Reference
Alcaligenes faecalis	Cholic acid	Ketocholic acids (90–100%)	A	2 days (monoketo acid); 4 days (diketo acid); 6 days (triketo acid); 37-39°; surface culture	Schmidt and Hughes (1944)
Corynebacterium mediolanum	21-Acetoxy-3β-hydroxy-5-pregnen-20-one	21-Hydroxy-4-pregnene-3,20-dione (30%)	B	6 days, 36-37°C, pure oxygen with agitation	Mamoli, (1944)
Cunninghamella blakesleeana (H334)	Compound S	Cortisone (19%), cortisol (65%)	C	3 days, 28°C, rotary shaker (250 rpm)	O'Connell, et al. (1955)
Cylindrocarpon radicicola (ATCC 11011)	Progesterone	1-Dehydrotestolo-lactone (50%)	D	3 days, 25°C, recipro-cating shaker (120 spm)	Fried and Thoma, (1956)
Fusarium solani	Progesterone	1,4-Androstadiene-3,17-dione (85%)	E	4 days 25°C, rotary shaker (100 rpm)	Vischer and Wettstein (1953)
Rhizopus arrhizus ATCC 11145	4-Androstene-3,17-dione	11α-Hydroxy-4-androstene-3,17-dione (25%)	F	4 days, 28°C small aerated tank (6-7 mM O_2/liter/min)	Murray and Peterson (1953)
Streptomyces albus	Estradiol	Estrone (90-95%)	G	6 hours of substrate oxidation with resting cells, 30°C	Heusghem and Welch (1948)
Streptomyces aureus	Progesterone	15α-Hydroxy-4-pregnene-3,20-dione (11%)	H	3 days, 25°C, rotary shaker (280 rpm)	Fried et al. (1956)

[a] Composition of the media: (A) 4.7 gm $(NH_4)_2SO_4$, 0.5 gm asparagine, 5 gm NaCl, 0.65 gm NaOH, 2 gm glycerol, inorganic salts, 5 gm cholic acid, distilled water to 1 liter; (B) 60 ml yeast water, 10 ml of 0.2 M Na_2HPO_4, 1 gm KH_2PO_4, 0.2 gm 21-acetoxy-3-hydroxypregnan-20-one; (C) 3 gm $NaNO_3$, 0.5 gm KCl, 30 gm dextrin, 0.01 gm $FeSO_4 \cdot 7H_2O$, 10 mg Tween 80, 0.5 gm $MgSO_4 \cdot 7H_2O$, 1,3 gm $K_2HPO_4 \cdot 3H_2O$, distilled water to 1 liter, pH 7.2; (D) 3 g corn steep solids, 3 gm $NH_4H_2PO_4$, 2.5 gm $CaCO_3$, 2.2 gm soybean oil, 0.5 gm progesterone, distilled water to 1 liter, pH 7.0; (E) 15 gm peptone, 6 ml corn steep liquor, 50 gm glucose, distilled water to 1 liter, pH 6.0, progesterone (0.25 gm) added after 2 days of incubation; (F) 20 gm lactalbumin digest, 5 ml corn steep liquor, 50 gm glucose, tap water to 1 liter, pH 5.5–5.9; androstenedione (0.25 gm) added after 27 hours of incubation; (G) nutrient broth (to grow cells), phosphate buffer (pH 7.0) (to oxidize estradiol); and, (H) 2.2 gm soybean oil, 15 gm soybean meal, 10 gm glucose, 2.5 gm $CaCO_3$, 0.25 gm progesterone, water to 1 liter

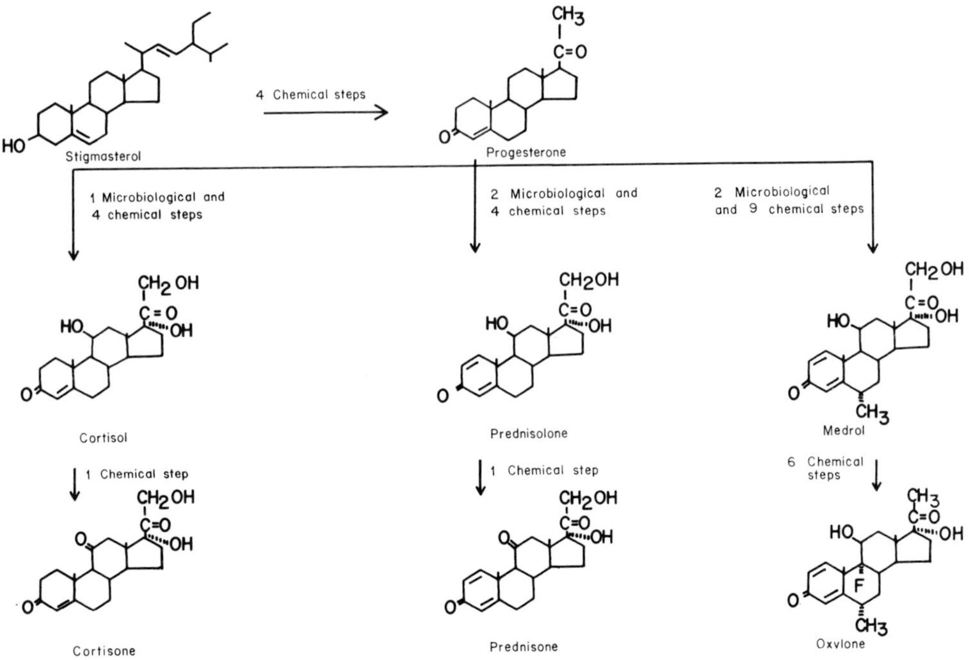

FIGURE 1. Conversion of stigmasterol to six clinically important steroids.

of hydroxyl groups to ketones; reduction of carbonyl groups; and hydrolysis and formation of esters.

The methods generally used in screening microorganisms for ability to transform steroids are: (1) the microorganism is first grown in suitable media for 1 or 2 days; (2) the steroid, dissolved in a water-soluble solvent (e.g., dimethylformamide, acetone, propylene glycol), is then added; (3) incubation is continued for 1 or 5 days; (4) the transformation products are obtained by extraction of the beer with methylene chloride or nonpolar solvents (chloroform, ethyl acetate) and by purification by column chromatography and other methods; (5) the structure of the transformation products is determined by classical methods of organic chemistry. Most of the fungi and streptomycetes examined have been found to transform some steroids, and sufficient information has accumulated to predict the type of transformation that might be expected if the genus and species of the culture is known or, conversely, to predict the identity of the microorganism knowing the transformation (Čapek et al., 1966, 1976; Charney and Herzog, 1968, 1980; Vézina and Rakhit, 1974).

The transformations can be performed by cells at different stages of growth. Thus the steroid may be added to: (1) the growing culture (either at the time of inoculation or toward the end of the growth phase); (2) the

TABLE IV. Types of Microbial Transformation of Steroids

A. Oxidation
    1. Conversion of secondary alcohol to ketone
    2. Introduction of primary hydroxyl on steroid side chain
    3. Introduction of secondary hydroxyl on steroid nucleus
    4. Introduction of tertiary hydroxyl on steroid nucleus
    5. Dehydrogenations of ring A of steroid nucleus in positions 1(2) and 4(5)
    6. Aromatization of ring A of the steroid nucleus
    7. Oxidation of the methylene group to ketone group
    8. Cleavage of side chain of pregnane at C-17 to form ketone
    9. Cleavage of side chain of pregnane at C-17 and opening of D ring to form testololactone
   10. Cleavage of side chain of steroids to form carboxyl group
   11. Cleavage of side chain of pregnane steroids at C-17 to form secondary alcohol
   12. Formation of epoxides
   13. Decarboxylation of acids

B. Reduction
    1. Reduction of ketone to secondary alcohol
    2. Reduction of aldehyde to primary alcohol
    3. Hydrogenation of double bond at position 1(2) of ring A
    4. Hydrogenation of double bond at positions 4(5) of ring A and at 5(6) of ring B
    5. Elimination of secondary alcohol
    6. Formation of homosteroids of the androstane series from pregnane derivatives

C. Hydrolysis
    1. Saponification of steroid esters
    1. Acetylation

D. Esterification

nonproliferating ("resting") cell suspensions (or cell-free extracts); (3) microbial systems which carry out simultaneously other reactions as the major fermentation operations; (4) the spore suspensions; or (5) the immobilized whole cells (and cell-free enzymes). The last two methods are receiving considerable attention (Abbott, 1976; Vézina et al., 1968) but have not been used on a commercial scale.

IV. SIDE-CHAIN DEGRADATION OF STEROLS

In the last 15 years, the degradation of the side chain of cholesterol and of related plant sterols by microorganisms has been investigated extensively, and the results of these studies have become of considerable economic importance (Martin, 1977). The first data of this kind by Turfitt in 1948 reported on the conversion of cholestenone and bile acids to small amounts of 3-keto-4-androstene-17-carboxylic acid (Turfitt, 1948) and by Whitmarsh in 1964 on the degradation of cholesterol into a mixture of 3-ketobisnor-4-cholenic acid, 3-ketobisnor-1,4-choladienic acid, 4-androstene-3,17-dione, and 1,4-androstadiene-3,17-dione in the presence of small amounts of 8-hydroxyquinoline (Whitmarsh, 1964). By this approach, other compounds were found which prevent the degradation of the steroid nucleus while allowing the degradation of the cholesterol side chain (Nagasawa et al., 1969, 1970): iron- or copper-chelating agents (such as quinoline-2-carboxylic acid, 2,2′-bipyridine, 1,10-phenanthroline); divalent ions that replace iron or block -SH functions (Ni^{2+}, Co^{2+}, Pb^{2+}, SeO_3^{2-}, AsO_2); or redox dyes. Sih and his collaborators showed that the degradation of the sterol nucleus can be blocked also by the chemical modification of the substrate cholesterol at C-19 (19-hydroxy-, 19-nor-, 6β, 19-oxidosterols). The same workers also established that the cholesterol side chain is degraded to the C-17 ketosteroid by a stepwise β-oxidation via the intermediary C-24 and C-22 carboxylic acids (Sih et al., 1968) as shown in Scheme 10.

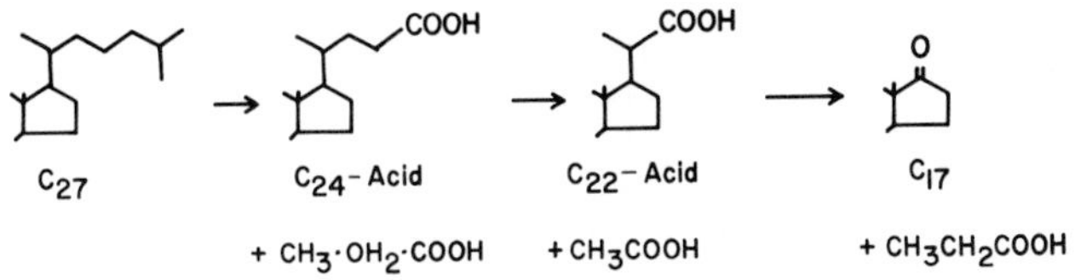

SCHEME 10

When these inhibitors are not present or when the substrate cholesterol is not chemically modified, most of the cholesterol-utilizing microorganisms degrade the nucleus as well. Mycobacteria (and pseudomonads) metabolize it through the sequence shown in Scheme 11.

With this information, the problem of accumulating the desired prod-

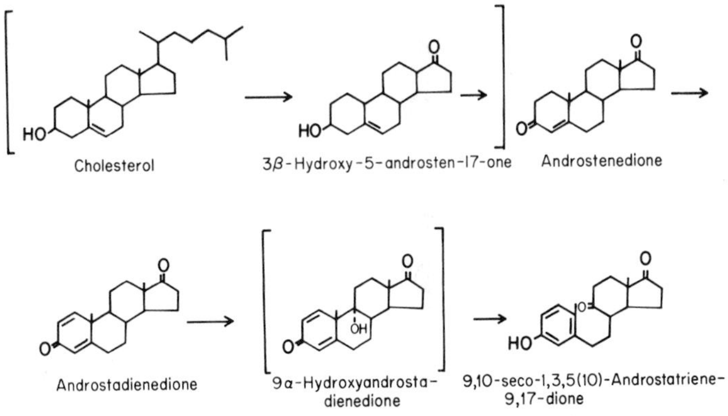

SCHEME 11

ucts 4-androstene-3,17-dione (androstenedione) and 1,4-androstadiene-3,17-dione (androstadienedione) from cholesterol and related plant sterols in the absence of inhibitors or chemically modified substrates has been solved. Marsheck and his collaborators (1972) prepared a mutant of a soil mycobacterium which converted cholesterol, stigmasterol, and sitosterols to androstadienedione. By additional uv irradiation, a new mutant was selected which accumulated androstenedione as the main product. Both of these compounds (androstadienedione and androstenedione) are convenient substrates for the synthesis of other steroids. This discovery is of signal practical importance. It has made soy sterols (especially β-sitosterol and campesterol) commercially attractive new steroid substrates, since they are readily available in large quantities from soybeans and tall oil (waste product of the pulpwood kraft process; Conner et al., 1976).

REFERENCES

Abbott, B. J. (1976). Adv. Appl. Microbiol. **20,** 203.

Applezweig, N. (1962). "Steroid Drugs." McGraw-Hill, New York.

Applezweig, N. (1978). In "Crop Resources" (D. S. Seigler, ed.), p. 149. Academic Press, New York.

Čapek, A., Hanč, O., and Tadra, A. (1966). "Microbial Transformations of Steroids." Academia, Prague.

Čapek, A., Fassatiová, O., and Hanč, O. (1976). Folia Microbiol. (Prague) **21,** 70.

Charney, W., and Herzog, H. (1968). "Microbiological Transformations of Steroids," 1st ed. Academic Press, New York.

Charney, W., and Herzog, H. (1980). "Microbiological Transformations of Steroids," 2nd ed. Academic Press, New York.

Conner, A. H., Nagaoka, M., Rowe, J. W., and Perlman, D. (1976). Appl. Microbiol. **32,** 301.

Fried, J., Thoma, R. W., and Klingsberg, A. (1953). J. Am. Chem. Soc. **75,** 5764.

Fried, J., and Thoma, R. W.. (1956). U.S. Patent 2,744,120.

Hanson, F. R., Mann, K. M., Nielson, E. D., Anderson, H. V., Brunner, M. P., Karnemaat, J. N., Colingsworth, D. R., and Haines, W. J. (1953). J. Am. Chem. Soc. **75,** 5369.

Hayano, M., Gut, M., Dorfman, R. I., Sebek, O. K., and Peterson, D. H. (1958a). J. Am. Chem. Soc. **89,** 2336.

Hayano, M., Saito, A., Stone, D., and Dorfman, R. I. (1958b). Biochim. Biophys. Acta **21,** 380.

Hechter, O., Jacobsen, R. P., Jeanloz, R., Levy, H., Marshall, C. W., Pincus, G., and Schenker, V. (1949). J. Am. Chem. Soc. **71,** 3261.

Hench, P. S., Kendall, E. C., Slocumb, C. H., and Polley, H. F. (1949). Mayo Clin. Proc. **24,** 181.

Hench, P. S., Kendall, E. C., Slocumb, C. H., and Polley, H. F. (1950). Arch. Intern. Med. **85,** 545.

Heusghem, C., and Welsch, M. (1948). Bull. Soc. Chim. Biol. **31,** 282.

Kondō, E., and Masuo, E. (1961). J. Gen. Appl. Microbiol. **7,** 113.

Mamoli, L. (1944). U.S. Patent 2,341,110.

Mamoli, L., and Vercellone, A. (1937a). *Ber. Dtsch. Chem. Ges.* **70,** 470 and 2079.
Mamoli, L., and Vercellone, A. (1937b). *Hoppe-Seyler's Z. Physiol. Chem.* **245,** 93.
Marsheck, W. J., Kraychy, S., and Muir, R. D. (1972). *Appl. Microbiol.* **23,** 72.
Martin, C. K. A. (1977). *Adv. Appl. Microbiol.* **22,** 29.
Maxon, W. D., Chen, J. W., and Hanson, F. R. (1966). *Ind. Eng. Chem., Process Des. Dev.* **4,** 421.
Murray, H. C., and Peterson, D. H. (1953). U.S. Patent 2,646,370.
Nagasawa, M., Bae, M., Tamura, G., and Arima, K. (1969). *Agric. Biol. Chem.* **33,** 1644.
Nagasawa, M., Watanabe, N., Hashiba, H., Murakami, M., Bae, M., Tamura, G., and Arima K. (1970). *Agric. Biol. Chem.* **34,** 838.
Nobile, A., Charney, W., Perlman, P. L., Herzog, H. L., Payne, C. C., Tully, M. E., Jevnik, M. A., and Herschberg, E. B. (1955). *J. Am. Chem. Soc.* **77,** 4184.
O'Connell, P. W., Mann, K. M., Nielson, E. D., and Hanson, F. R. (1955). *Appl. Microbiol.* **3,** 16.
Perlman, D. (1976). *In* "Applications of Biochemical Systems in Organic Chemistry" (J. B. Jones, C. J. Sih, and D. Perlman, eds.), Wiley, New York.
Perlman, D., Titus, E., and Fried, J. (1952). *J. Am. Chem. Soc.* **74,** 2126.
Peterson, D. H., and Murray, H. C. (1952). *J. Am. Chem. Soc.* **74,** 1871.
Sarett, L. (1946). *J. Biol. Chem.* **162,** 591.
Schmidt, L. H., and Hughes, H. B. (1944). U.S. Patent 2,360,447.
Sebek, O. K., and Spero, G. B. (1959). U.S. Patent 2,897,218.
Shull, G. M., and Kita, D. A. (1955). *J. Am. Chem. Soc.* **77,** 763.
Shull, G. M., Kita, D. A., and Davisson, J. W. (1953). U.S. Patent 2,658,023.
Sih, C. J., Tai, H. H., Tsong, Y. Y., Lee, S. S., and Coombe, R. G. (1968). *Biochemistry* **7,** 808.
Spero, G. B., Thompson, J. L., Magerlein, B. J., Hanze, A. R., Murray, H. C., Sebek, O. K., and Hogg, J. A. (1956). *J. Am. Chem. Soc.* **78,** 6213.
Turfitt, G. E. (1948). *Biochem. J.* **42,** 376.
Vézina, C., and Rakhit, S. (1974). *In* "CRC Handbook of Microbiology" (A. I. Laskin and H. A. Lechevalier, eds.), Vol. IV, p. 117. CRC Press, Cleveland, Ohio.
Vézina, C., Sehgal, S. N., and Singh, K. (1968). *Adv. Appl. Microbiol.* **10,** 221.
Vischer, E., and Wettstein, A. (1953). *Experientia* **9,** 271.
Weaver, E. A., Kenney, H. E., and Wall, M. E. (1960). *Appl. Microbiol.* **8,** 345.
Whitmarsh, J. M. (1964). *Biochem. J.* **90,** 23P.

Chapter 15

Vitamin B$_{12}$

J. FLORENT
L. NINET

I. INTRODUCTION

A. Historical Background

Vitamin B$_{12}$, or cyanocobalamin, is an important biological compound active as an hematopoietic factor in mammals and as a growth factor for many microbial and animal species. Although its existence was suspected in 1925 by Wipple and Robscheit-Robbins when studying the therapeutic effect of beef liver in pernicious anemia, vitamin B$_{12}$ was isolated in 1948 as pure crystals independently by Rickes et al. and by Smith. In 1955 Hodgkin et al. unraveled its structure by means of X-ray crystallography. Ten years were spent from the first efforts of Woodward and Eschenmoser until a full chemical synthesis was achieved (Krieger,

MICROBIAL TECHNOLOGY, 2nd ed., VOL. I
Copyright © 1979 by Academic Press, Inc.
All rights of reproduction in any form reserved. ISBN 0-12-551501-4

1973; Maugh, 1973). With 70 steps required, this difficult synthesis is, for industrial purposes, of little value and today the vitamin B_{12} group of compounds is obtained by fermentation processes.

B. Structure of Vitamin B_{12}

Vitamin B_{12} belongs to the large family of cobaltocorrinoids exhibiting the general formula presented in Fig. 1.

The X radical for the cobalamins, which are the most interesting cobaltocorrinoids from the medical point of view, is 5,6-dimethylbenzimidazole. The group includes cyanocobalamin, or true vitamin B_{12}; hydroxocobalamin, or vitamin B_{12a}; methylcobalamin, or mecobalamin; and 5′-deoxyadenosylcobalamin, or cobamamide (or coenzyme B_{12} of Barker), where Y is a cyano, hydroxyl, methyl, or 5′-deoxyadenosyl radical, respectively.

Other cobaltocorrinoids, with different heterocyclic bases (substituted benzimidazoles, purines) as the X radical, are also produced by various microorganisms either spontaneously or after the introduction of the corresponding base in the fermentation medium. Pseudovitamin B_{12} and factor III, in which X is adenine and 5-hydroxybenzimidazole, respectively, are the most frequently encountered. Natural analogs designated as "incomplete" are also known, for instance, etiocobalamin, or factor B, which is devoid of the X–ribose–phosphate moiety.

These "complete" or "incomplete" analogs have been reviewed by Mervyn and Smith (1964). Generally, they have low activity as growth factors for vertebrates but are very potent for the microorganisms. The cobalamins, in biological tests, are always the most active.

C. Sources of Vitamin B_{12}

Although vitamin B_{12} is present in small amounts in almost every animal tissue, e.g., 1 mg/kg in beef liver, it originates from microorganisms. Depending on the nature of their nutritional habits and digestive physiology, animals obtain the vitamin from their own intestinal flora or from other animals through their meat diet. An exogenous supply is mandatory for man.

Vitamin B_{12} derived from cultures of microorganisms soon supplanted beef liver as a practical source of the vitamin for therapeutic purposes. Around 1950, materials rich in biomasses, such as activated sludges or broths of antibiotic-producing streptomyces, were used for isolating vitamin B_{12} either in a crude form for animal feeding or in à pure state for a medical use (Noyes, 1969). Later, high-producing bacterial strains were specially selected for commercial production.

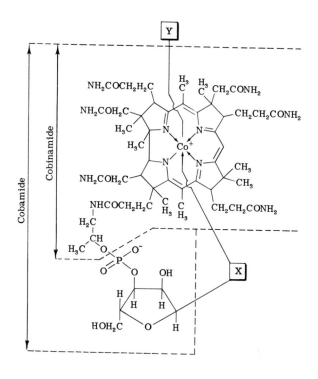

FIGURE 1. General structure of cobaltocorrinoids with references to cobalamins.

Whatever the strain and the culture conditions, a mixture of cobala-mins and some analogs is always obtained. Industrially the isolation of each distinct substance would be almost impractical. The extraction is generally performed after simple transformation of all cobalamins into cyanocobalamin by treatment with cyanide. The latter is not the most abundant natural substance but it is one of the most stable and active derivatives. Despite this transformation, vitamin isolation remains tedious (see Section II, A, 3).

D. Assay Methods

The accurate assay of cobalamins, as a group or individually, in culture broths or crude products is not simple. Numerous methods have been developed but none is entirely satisfactory (Rosenthal, 1968). For this reason, it is often simpler to use cyanide transformation with simul-taneous concentration of the preparation before assaying. In practice, the method of isotopic dilution, extraction, and spectrophotometry is one of the best, although lengthy and expensive (Bacher et al., 1954). There-fore, microbiological assays with cobalamin-dependent strains as Lac-tobacillus leichmanii ATCC 4797 and 7830, Escherichia coli 113-3, Euglena gracilis, and Ochromonas malhamensis, although nonspecific, are often chosen for routine work.

In practice the purer the preparation, the easier the analysis. Hence it is advisable to carry out preliminary separations by thin-layer or column chromatography (Bilkus and Mervyn, 1970; Tortolani et al., 1970a,b; Vogelmann and Wagner, 1973; Tortolani and Mantovani, 1974; Mat-sumoto et al., 1974; Lotti, 1975) or by electrophoresis (Kato and Shimizu, 1962; Tortolani and Ferri, 1974). Afterward, several classical methods are available for assaying the isolated fractions or pure products: spec-trophotometry in the visible spectrum (Farmaceutici Italia, 1971b; Cel-letti et al., 1976) or in the IR spectrum (Goldstein et al., 1974), atomic absorption (Whitlock et al., 1976), potentiometry (Goldstein and Duca, 1976), and enzymatic analysis (Hermann and Mueller, 1976; Sheehan and Hercules, 1977). Cyano- and hydroxocobalamins are described in several national pharmacopoeias.

II. PRODUCTION OF VITAMIN B_{12}

A. General

During the last 20 years several accounts on the vitamin B_{12} production processes have appeared (Perlman, 1959, 1967; Prescott and Dunn,

1959; Mervyn and Smith, 1964; Friedmann and Cagen, 1970). The main lines of this research have been: (1) to obtain a better knowledge of the biosynthetic pathway in order to sustain biochemical and genetic studies; (2) to improve the better known strains (propionibacteria and *Pseudomonas*) by mutation; (3) to replace traditional sugars by cheaper nutrients in the media; and (4) to improve extraction and purification techniques.

1. Biogenesis of Vitamin B$_{12}$

Since the reviews by Wagner (1966) and Friedmann and Cagen (1970), a great deal of attention has been devoted to biosynthesis with *Streptomyces olivaceus* (Sato *et al.*, 1968, 1971) and *Propionibacterium shermanii* (Lowe and Turner, 1970; Mueller *et al.*, 1970, 1971; Renz and Bauer-David, 1972; Russel, 1974; Scott, 1975, 1976; Scott *et al.*, 1976; Battersby and McDonald, 1976; Battersby *et al.*, 1976, 1977a,b; Bykhovsky and Zaitseva, 1976; Ford and Friedmann, 1976; Imfeld *et al.*, 1976).

With few exceptions all the authors agree with the general pathway shown in Fig. 2. According to this pathway, several heme derivatives (cytochromes and catalase) which are vital elements for microbial life can also repress the first steps of vitamin B$_{12}$ biosynthesis (Vorobeva *et al.*, 1967; Simon, 1968; Bykhovsky *et al.*, 1969, 1976; Brazenas and Kanopkaite, 1970a; Miyazaki *et al.*, 1972). Likewise, oxidation–reduction potentials and metal concentrations can play a leading part in the course of vitamin B$_{12}$ fermentations.

2. Producing Strains, Culture Media, and Yields

Numerous microorganisms have been considered as potential sources of vitamin B$_{12}$. Because of their rapid growth and their high productivity, two propionibacteria and one *Pseudomonas* cultivated on carbohydrates were selected rather than the *Streptomyces* and other genera for industrial purposes as shown in Table I. Other species have been recently examined mainly in relation to the need for organisms which can grow on cheaper nutrients than carbohydrates (food industries wastes, see Section II, B; hydrocarbons and alcohols, see Section II, C). In this respect, methanogenic bacteria grown on methanol might become serious challengers in the next decade.

The best strains are spontaneous or induced mutants, screened for their resistance to different agents: cobalt or manganese ions, antibiotics, etc. (Chinoin-Gyogyszer, 1965; Aries, 1974). Induced mutants are obtained with treatments by uv or X-rays or with such chemical agents as *N*-methyl-*N*′-nitro-*N*-nitrosoguanidine, nitrosoethylurea, ethylenimine,

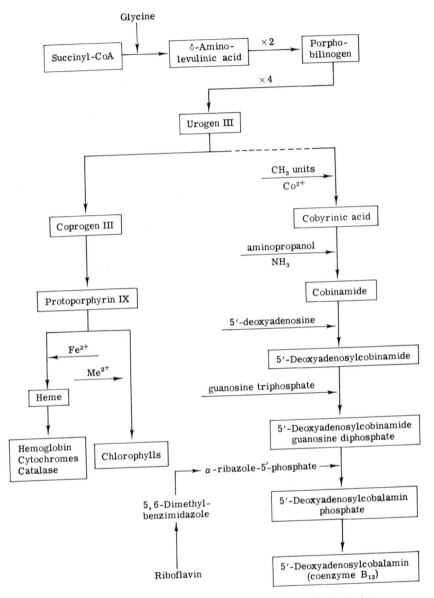

FIGURE 2. General pathway for the biosynthesis of cobalamins.

TABLE I. Vitamin B$_{12}$ Production by Different Strains

Strain	Carbon source	Yield (mg/liter)	References
Micromonospora sp.	Glucose	11.5	Wagman et al. (1969)
Nocardia rugosa	Glucose–cane molasses	14	Farmaceutici Italia (1971a)
Propionibacterium freudenreichii	Glucose	25	Uclaf (1960)
Propionibacterium shermanii	Glucose	23	Speedie and Hull (1960)
Propionibacterium shermanii	Glucose	28	Chinoin-Gyogyszer (1965)
Propionibacterium shermanii	Glucose	39	Pierrel S.p.A. (1965b)
Propionibacterium vannielli	Glucose	25	Pliva T.F.K.P. (1964)
Pseudomonas denitrificans	Beet molasses	59	Merck and Co., Inc., (1971)
Streptomyces olivaceus	Glucose–lactose	8.5	Terada et al. (1959)
Mixed methanogenic bacteria	Methanol	35	Richter Gedeon V.G. (1975a,b)
Bacterium FM-02T	Methanol	2.6	Toraya et al. (1975)
Methanobacillus omelianskii	Methanol	8.8	Pantskhava and Bykhovsky (1966)
Protoaminobacter ruber	Methanol	2.5	Kojima et al. (1976)
Corynebacterium and Rhodopseudomonas	n-Paraffins	2.3	Nakao et al. (1974)
Nocardia gardneri	Hexadecane	4.5	Kyowa Hakko Kogyo Co., Ltd. (1970)

diethylsulfate, and mustard gas (Merck and Co., Inc., 1971; Gruzina *et al.*, 1974; Aries, 1974; Georgopapadakou *et al.*, 1976). These mutants exhibit new morphological or biochemical characteristics (size and color of colonies, antibiotic resistance, substrate utilization).

Whatever the strains and the culture conditions may be, it is necessary to add to the medium certain essential elements for vitamin biosynthesis: cobalt ions in all cases, and often 5,6-dimethylbenzimidazole. Supplementation with other potential precursors of corrinoids, such as glycine, threonine, δ-aminolevulinic acid, and aminopropanol, also can be beneficial.

3. Isolation of Vitamin B_{12}

During the last 20 years the extraction processes have been improved many times but usually they include the following main stages: (1) solubilization of cobalamins and conversion to cyanocobalamin and (2) isolation of a crude product, 80% pure (utilizable directly for animal feeding), followed by purification to a 95–98% level (for medical use).

In order to extract the vitamin B_{12}, the whole broth, or an aqueous suspension of harvested cells, is heated at 80°–120°C for 10–30 minutes at pH 6.5–8.5. The conversion to cyanocobalamin is obtained by treating the heated broth or cell suspension with cyanide or thiocyanate, often in the presence of sodium nitrite or chloramin B. If the cyanide treatment is postponed to a later stage in order to reduce tedious handling procedures, a conversion to the more stable sulfato- or sulfitocobalamin is advisable (Aschaffenburger Zellstoffwerke A.G., 1966; Aries, 1974).

The extraction operation units are classical (Robinson, 1955), and they are combined to achieve an efficient and inexpensive process. For instance, the following methods are used:

1. Adsorption on such different supports as Amberlite IRC 50, alumina, carbon, and elution by hydroalcoholic or hydrophenolic mixtures. Many new supports have been tested recently (Merck and Co., Inc., 1964; Olin Mathieson Chem. Corp., 1965; Aschaffenburger Zellstoffwerke A.G., 1966; Kamikubo, 1967; Rohm and Haas Co., 1968; Sato *et al.*, 1968, 1977; Zodrow *et al.*, 1968; Duric, 1969; Seisan Kaihatsu Kagaku Kenkyusho, 1970; Tortolani *et al.*, 1970a; Iashi, 1973; Vogelmann and Wagner, 1973; Lotti, 1975).

2. Extraction from aqueous solutions by phenol or cresol alone or in mixture with benzene, butanol, carbon tetrachloride, or chloroform (Stefaniak *et al.*, 1968; Seisan Kaihatsu Kagaku Kenkyusho, 1970; Merck and Co., Inc., 1971; Aries, 1974).

3. Precipitation or crystallization from various solutions by evapora-

tion, dilution with appropriate solvents or addition of such reagents as tannic acid or p-cresol (Olin Mathieson Chem. Corp., 1965; Pierrel S.p.A., 1965a). Some impurities can also be removed by common precipitating agents, such as calcium or zinc hydroxide.

An extraction process presently used is briefly outlined in Fig. 3.

B. Fermentation Processes from Carbohydrates

As previously indicated, propionibacteria and *Pseudomonas* are used for industrial vitamin B$_{12}$ production.

1. Production by Propionibacteria

Several microaerophilic propionibacteria produce cobaltocorrinoids in conventional carbohydrate media correctly supplemented with cobalt, without aeration. However, the production of cobalamins, mainly adenosylcobalamin, depends on an internal or external supply of 5,6-dimethylbenzimidazole (DBI). For this process *Propionibacterium freudenreichii* ATCC 6207, *P. shermanii* ATCC 13673, and their substrains and mutants, which can synthesize their own DBI, are the most appropriate. Since aeration favors DBI formation, it is advisable to use a two-stage culture to obtain the highest yields. In the first stage a practically anaerobic culture is run to almost total depletion of sugar, in order to promote the growth of the bacteria and etiocobalamin biosynthesis. Then, in the second stage, an aeration shift leads to DBI biosynthesis and conversion of etiocobalamin to deoxyadenosylcobalamin. The two stages can be conducted batchwise in the same tank or continuously in two connected fermentors (Speedie and Hull, 1960). Anaerobically grown cells can also be collected by centrifugation and incubated, as a heavy suspension, with air and, if required, with DBI and cyanide. DBI supplementation is only required during the second stage, if selected propionibacteria are not able to synthesize their own, but, due to the inhibitory effect of DBI on the corrinoid biosynthesis, addition of DBI during the first stage must be precluded (Friedmann and Cagen, 1970).

Many improvements of this process have been attempted in different ways and several of them are described in detail by Noyes (1969). Little information is available on the strain preservation and culture maintenance conditions.

The fermentation media consist mainly of glucose or inverted molasses (10–100 gm/liter) with small amounts of ferrous, manganous, and magnesium salts in addition to cobalt salts (10–100 mg/liter), buffering or neutralizing agents, and nitrogenous compounds. Among these, yeast

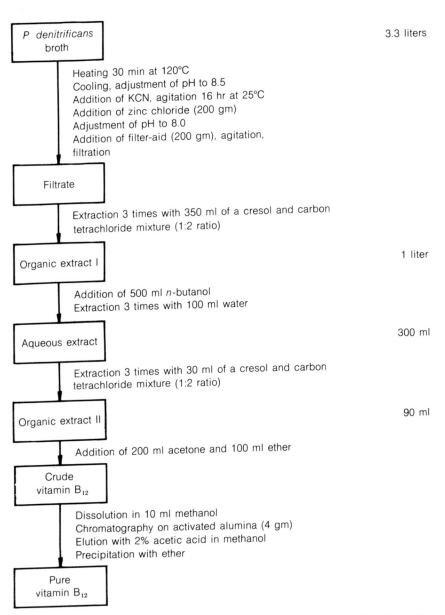

FIGURE 3. Crystalline vitamin B_{12} isolation process. (Adapted from Merck and Co., Inc., 1971.)

preparations, casein hydrolysates, and other traditional sources are currently used but corn steep liquor (30–70 gm/liter) is often preferred since it supplies some lactic and pantothenic acids, which enhance the growth of the bacteria. The latter acid, which stimulates corrinoids biosynthesis, may also be added as a supplement (Zodrow *et al.*, 1963). Generally the temperature of the culture is 30°C and the pH is controlled at around 6.5–7.0.

Recent studies have been conducted in order to increase yields by supplementing the above media with lactose (Danilova and Bankina, 1974), a large range of trace salts (Kotala and Luba, 1965), and some porphobilinogen precursors, such as glycine, and δ-aminolevulinic acid (Hercules Inc., 1968; Bykhovsky, 1969). Substituting the carbon and nitrogen sources by skim milk or lactoserum is also possible (Beskhov *et al.*, 1969, 1973; Popova and Ilieva, 1970; Popova and Bakchevanska, 1972; Giec *et al.*, 1976; Janicka *et al.*, 1976). These modifications may even be beneficial.

Reports indicate that propionibacteria produce from 25 to 40 mg of vitamin B$_{12}$ per liter (see Table I). Even though the values presently reached at an industrial scale are higher than these, the 216 mg/liter claimed by Aries patent (1974) seems beyond any expectation.

Figure 4 gives an example of a current process using *P. shermanii;* the main culture is successively conducted with two modes of aeration.

2. Production by Pseudomonas

Several *Pseudomonas* species are claimed to be vitamin B$_{12}$ producers but currently the most used is a *Pseudomonas denitrificans* strain (Miller and Rosenblum, 1960). In contrast with the propionibacteria fermentation, the growth of the *Pseudomonas* keeps pace with cobalamin biosynthesis under aerobic conditions throughout the culture. Accordingly, the fermentation is conducted with aeration and agitation in a single vat, batchwise or continuously. This strain only requires traditional components, such as sucrose, yeast extract and several metallic salts, in the growth medium but from 10 to 25 mg of DBI per liter and from 40 to 200 mg of cobaltous nitrate per liter must be supplemented at the starting of the culture in order to enhance vitamin production (McDaniel, 1961; Daniels, 1970). Likewise, betaine and, to some extent, choline have favorable effects in activating some biosynthetic steps or altering the membrane permeability (Demain *et al.*, 1968; Daniels, 1970; Demain and White, 1971; White and Demain, 1971; Kaplan and Birnbaum, 1973). Glutamic acid stimulates the growth of bacteria (Daniels, 1966; Koike and Hattori, 1975). Owing to their cheapness and high betaine and glutamic acid contents, beet molasses, a multivalent nutrient of choice,

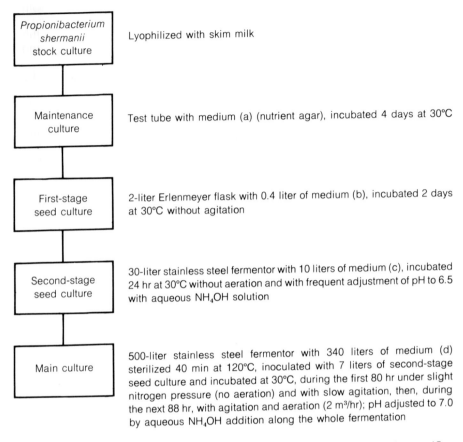

Propionibacterium shermanii stock culture	Lyophilized with skim milk
Maintenance culture	Test tube with medium (a) (nutrient agar), incubated 4 days at 30°C
First-stage seed culture	2-liter Erlenmeyer flask with 0.4 liter of medium (b), incubated 2 days at 30°C without agitation
Second-stage seed culture	30-liter stainless steel fermentor with 10 liters of medium (c), incubated 24 hr at 30°C without aeration and with frequent adjustment of pH to 6.5 with aqueous NH_4OH solution
Main culture	500-liter stainless steel fermentor with 340 liters of medium (d) sterilized 40 min at 120°C, inoculated with 7 liters of second-stage seed culture and incubated at 30°C, during the first 80 hr under slight nitrogen pressure (no aeration) and with slow agitation, then, during the next 88 hr, with agitation and aeration (2 m^3/hr); pH adjusted to 7.0 by aqueous NH_4OH addition along the whole fermentation

Medium (a): Tryptone, 10 gm; yeast extract, 10 gm; filtered tomato juice, 200 ml; agar, 15 gm; with tap water to 1 liter and pH adjusted to 7.2.

Medium (b): Medium (a) without agar.

Medium (c): Corn-steep liquor, 20 gm; dextrose, 90 gm; with tap water to 1 liter and pH adjusted to 6.5.

Medium (d): Corn-steep liquor, 40 gm; dextrose (sterilized separately), 100 gm; cobalt chloride, 20 mg; with tap water to 1 liter and pH adjusted to 7.0.

FIGURE 4. Vitamin B_{12} from *Propionibacterium shermanii*. Semipilot plant scale fermentation process. (Adapted from Speedie and Hull, 1960; and Pierrel S.p.A., 1965b.)

is preferentially used in industrial fermentations at a 60–120 gm/liter concentration in conjunction with from 2 to 5 gm of ammonium phosphate per liter and some oligoelements. Temperature optimum is 28°C, with a pH of about 7.0. By strain selection and mutation, the vitamin B_{12} production has jumped from 0.6 to 60 mg/liter within 12 years (Long, 1962; Lago and Demain, 1969; Merck and Co., Inc., 1971).

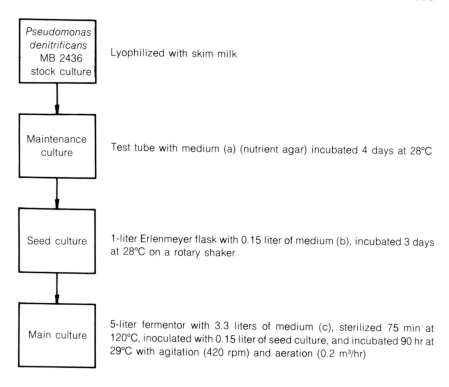

Medium (a): Beet molasses, 60 gm; brewers' yeast, 1 gm; N-Z-amine, 1 gm; diammonium
phosphate, 2 gm; magnesium sulfate, 1 gm; manganese sulfate, 200 mg; zinc
sulfate, 20 mg; molybdate sulfate, 5 mg; agar, 25 gm; with tap water to 1 liter
and pH adjusted to 7.4 (White *et al.*, 1973).
Medium (b): Medium (a) without agar.
Medium (c): Beet molasses, 100 gm; yeast, 2 gm; diammonium phosphate, 5 gm; magnesium sulfate, 3 gm; manganese sulfate, 200 mg; cobalt nitrate, 188 mg;
5,6-dimethylbenzimidazole, 25 mg; zinc sulfate, 20 mg; sodium molybdate, 5
mg; with tap water to 1 liter and pH adjusted to 7.4.

FIGURE 5. Vitamin B$_{12}$ from *Pseudomonas denitrificans.* Laboratory-scale fermentation process. (Adapted from Merck and Co., Inc., 1971.)

Such a process is outlined in Fig. 5. Using the extraction process
outlined in Fig. 3, 98% pure cyanocobalamin is obtained with a 75%
yield from a broth containing 59 mg/liter.

C. Fermentation Processes from Other Substrates

Continuing the general trend of replacing carbohydrates by cheaper
nutrients in industrial fermentations, several alcohols and hydrocarbons

have been studied in vitamin B_{12} production. Only a short account of the main approaches is given here since none of the designed processes is presently competitive.

1. Alcohols

Among the alcohols tested, methanol seems the most attractive and special attention was paid to methanogenic anaerobic bacteria in view of the cobalamin participation in their specific metabolism.

Some of these bacteria are mesophilic. They can be isolated as a mixture from sewage sludges and incubated together with a semicontinuous addition of methanol, ammonium and cobalt salts, and DBI, in order to achieve a 3–12 gm/liter methanol concentration. At harvest 35 mg of vitamin B_{12} and 10 mg of etiocobalamin per liter are present (Richter Gedeon V.G., 1963, 1967, 1975a,b; Szemler and Székely, 1969).

Thermophilic methanogenic bacteria, such as *Methanobacillus* and *Methanobacterium,* produce around 2 mg of vitamin B_{12} per liter from 8 gm of methanol per liter (Pantskhava and Pcjelkina, 1970; Bykhovsky *et al.,* 1972). By growing *Methanobacterium soehngenii* first in a distillers' solubles medium and then adding 8 gm of methanol per liter, an 8 mg/liter mixture of cobalamins and factor III is obtained (Pantskhava and Bykhovsky, 1966).

Aerobic mesophilic bacteria appear to be less useful. Both the FM-02T bacterium, cultivated according to the special "exponential-fed batch cultivation" technique, and *Protoaminobacter ruber* produce only 2.5 mg of vitamin B_{12} per liter from 12 gm of methanol per liter (Toraya *et al.,* 1975; Kojima *et al.,* 1976). With *Klebsiella* sp. 101, *Pseudomonas* AM1 and *Microcyclus eburneus,* yields do not exceed 0.2 mg/liter (Nishio *et al.,* 1975, 1976a; Kamikubo *et al.,* 1976).

However, the future of methanol utilization may be considered as promising in view of the mutational potentialities of the strains which have not been exploited.

Up to now, the interest of cultivating *Bacillus badius* or *Arthrobacter* species, or the methylotroph bacterium FM-02T, respectively, on ethanol, isopropanol, and propanediol has to be proved (Toraya *et al.,* 1975; Kojima *et al.,* 1976; Nishio *et al.,* 1976b).

2. Hydrocarbons

Despite the efforts of the past 10 years and the wide range of potentially useful strains, vitamin B_{12} production from hydrocarbons does not look attractive. Apart from yields of 4.5 mg/liter from 20 gm of hexadecane per liter by *Nocardia gardneri* (Kyowa Hakko Kogyo Co., Ltd., 1970) and 2.3 mg/liter from 100 gm of *n*-paraffins per liter by a mixed culture of *Corynebacterium* sp. and *Rhodopseudomonas spheroides* (Nakao *et al.*, 1974), the production from all other species does not exceed 0.5 mg/liter as, for example, with *Pseudomonas* (Morikawa and Kamikubo, 1969; Kanegafuchi Chem. Ind. Co., Ltd., 1972; Kamikubo *et al.*, 1976), mycobacteria (Fukui *et al.*, 1967; Mil'ko and Rabotnova, 1967), *Corynebacterium simplex* (Fujii *et al.*, 1966; Fujii and Fukui, 1970; Kyowa Hakko Kogyo Co., Ltd., 1973), *B. badius* (Kamikubo *et al.*, 1976; Nishio *et al.*, 1976b), and such phototrophic species as *R. spheroides* (Morikawa *et al.*, 1970) and *Rhodospirillum fulvum* (Uehisa, 1972).

III. PRODUCTION OF VITAMIN B$_{12}$ DERIVATIVES AND ANTAGONISTS

Although formerly isolated from beef liver and microbial cells as their cyano derivative, cobalamins are most frequently encountered in nature as a mixture of cyano-, hydroxo-, deoxyadenosyl-, and methylcobalamins. These four compounds, for instance, are present in human serum with distributions of 0–10, 8–15, 22–39, and 39–62%, respectively, and they are bound more or less to specific transport agents, the transcobalamins (McDonald *et al.*, 1977). Each of these derivatives controls several enzymatic reactions which could play a role in man in the genesis of pernicious anemia, neuropathy, and growth deficiency (Barker, 1972; Kamikubo *et al.*, 1972; Fenton *et al.*, 1976). In order to clarify the situation and to obtain more powerful therapeutic agents, an intensive effort to produce these different cobalamins has been made.

Recent studies have also showed the existence, in some microbial cultures, of vitamin B$_{12}$ antagonists which could well turn out to be useful tools in the study of the different aspects of vitamin B$_{12}$ activity.

A. Cobalamin Derivatives

The various cobalamins are directly extracted from the broth used for cyanocobalamin production but without the introduction of cyanide. Since the isolation and separation steps in the mixtures are especially tedious, it is nowadays considered advisable to prepare the pure substances by biochemical or chemical transformation of cyanocobalamin.

1. *Hydroxocobalamin*

This compound was formerly isolated from cultures of antibiotic-producing *Streptomyces,* then from the cells of numerous bacteria, such

as *P. shermanii* (Ilieva, 1971). During the extraction, successive transformations of native cobalamins in their sulfato, nitrato, and chloro derivatives are required before a final hydrolytic treatment by Amberlite IRA
400 (OH⁻) to generate hydroxocobalamin (Kaczka *et al.*, 1956; Pierrel
S.p.A., 1963a).

The chemical synthesis from cyanocobalamin is also carried out
through sulfito and nitrito derivatives (Pierrel S.p.A., 1963b; Smith,
1965).

2. 5'-Deoxyadenosylcobalamin (Coenzyme B_{12})

Coenzyme B_{12} was recently reviewed by Abeles and Dolphin (1976). It
is extracted directly from the broths of *P. shermanii* (Barker *et al.*, 1963;
Chinoin-Gyogyszer, 1971; Ilieva and Popova, 1974), *P. freudenreichii*
(Sifa, 1964), *Nocardia rugosa* (Farmaceutici Italia, 1971a), *N. gardneri*
(Kyowa Hakko Kogyo Co., Ltd., 1970), *Klebsiella* sp. 101 (Nishio *et al.*,
1976a), *Methanobacillus kuzneceovii* (Pcjelkina *et al.*, 1970), and *Protoaminobacter ruber* (Shimizu and Sato, 1976), where it can represent up
to 80% of the total native cobalamins.

Coenzyme B_{12} is also obtained by bioconversion of cyano- or hydroxocobalamin according to several processes using: (a) Whole broths
of propionibacteria (Kaken Kagaku K.K., 1965; Sifa, 1965, 1967) (for
instance, 42 mg of cyanocobalamin per liter are added to a 3-day-old
anaerobic culture of *P. freudenreichii;* by incubating 2 days more with
aeration, the coenzyme B_{12} content amounts to 52 mg/liter); (b) suspensions of *P. shermanii* cells in buffered glucose solutions (these are
collected by centrifugation, washed with water and occasionally dried
with acetone) (Fukui *et al.*, 1964; Pawelkiewicz, 1971); (c) cell-free extracts of *Clostridium tetanomorphum* (Weissbach *et al.*, 1961; Vitols *et
al.*, 1966), *Bacillus megaterium,* and *P. shermanii* (Brady *et al.*, 1962;
Brazenas and Kanopkaite, 1970b) in the presence of ATP or flavin
cofactor.

When the concentration of added compound does not exceed 50
mg/liter, the transformation yield is near 100% from hydroxocobalamin
and 50% from cyanocobalamin.

Coenzyme B_{12} is always endocellular; after harvest of the cells by
centrifugation, extraction is performed in the cold by an acetone–water
mixture, or at 80°–100°C during a short time by a 2% phenol aqueous
solution or an ethanol–water mixture. Purification of coenzyme B_{12} is
achieved by the methods described in Section II, A, 3. All operations
must be conducted in the cold, under subdued light, and avoiding any
drastic pH conditions and the presence of cyanide in order to reach a

global yield of 80–85% in the extraction process (Kaken Kaguku K.K., 1965; Chinoin-Gyogyszer, 1971).

Presently coenzyme B$_{12}$ is manufactured by condensing a 5'-ester of adenosine with vitamin B$_{12r}$ or B$_{12s}$, which both result from cyano- or hydroxocobalamin reduction (Hoffmann-La Roche, 1963; Glaxo Group, Ltd., 1964).

3. Mecobalamin (Methylcobalamin)

The presence of mecobalamin was detected in whole broths of *E. coli*, *Streptomyces rimosus* (Lindstrand, 1964, 1965), propionibacteria (Skupin et al., 1970), *M. kuzneceovii* (Pcjelkina et al., 1970), and *Klebsiella* sp. 101 (Kamibuko et al., 1975). In *S. olivaceus* and *P. ruber* cultures, mecobalamin present in 0.3 mg/liter amounts up to 50% of the total cobalamins during the growth period; this proportion is reduced to 5% when the culture is continued to reach the maximal yield of cobalamins (Sato et al., 1971, 1977).

Cell-free extracts of *Methanosarcina barkeri* (Blaylock and Stadtman, 1966), *Clostridium thermoaceticum* (Kuratomi et al., 1966; Ljungdahl and Wood, 1969), and *M. kuzneceovii* (Pantskhava et al., 1973) easily convert cyanocobalamin or vitamin B$_{12s}$ to mecobalamin.

Chemically mecobalamin is synthesized by reaction of methyl bromide with reduced forms of cyano- or hydroxocobalamin (Hoffmann-La Roche, 1963; Glaxo Group, Ltd., 1964). In aqueous solution mecobalamin is sensitive to light but does not react with cyanide (Barker, 1968; Lorenzi and Oddono, 1972).

B. Vitamin B$_{12}$ Antagonists

Substances which inhibit, in special conditions, the growth of vitamin B$_{12}$-dependent microorganisms are produced by several bacteria. These antagonists belong to two groups. The first is constituted by corrinoids deprived of cobalt and found among photosynthetic bacteria such as *Chromatium* (Perlman and Toohey, 1966; Toohey, 1971, 1974). The second includes peptides and uncommon amino acids, such as N-5-hydroxy-L-arginine, 4-oxo-L-isoleucine, and the compound 102804, which are distributed among anthracomorph bacteria (Perlman et al., 1974; Kageyama et al., 1977). Cultures of all these bacteria do not require special conditions; at the moment they are still in an experimental stage.

Other vitamin B$_{12}$ antagonists can also be formed by the chemical

transformation of natural corrinoids (Ford, 1959; Koppenhagen and Pfiffner, 1970, 1971; Koppenhagen *et al.*, 1974).

IV. MARKETING PROSPECTS FOR COBALAMINS

The world market for cyanocobalamin is presently estimated to be
9000–10,000 kg/year, a major part of which is directed to the synthesis of
other cobalamins. The pharmaceutical compositions take some 3500 kg
of cyanocobalamin, 2000 kg of hydroxocobalamin, 1000 kg of coenzyme
B_{12}, and a small amount of mecobalamin, the remaining part being used
in animal feed. Apparently there is at present no prospect for any marked
expansion of this market.

Some of the main pharmaceutical companies involved in the field are
Chinoin (Hungary), Farmitalia (Italy), Glaxo (United Kingdom), Merck
(United States), Rhône-Poulenc (France), Richter (Hungary), and Rous-
sel (France).

The bulk selling price of cyanocobalamin dropped from $500 to $15
per gm between 1951 and 1963. Since 1974 the price has fluctuated
between $6 and $1.80 per gm.

REFERENCES

Abeles, R. H., and Dolphin, D. (1976). *Acc. Chem. Res.* **9**, 114–120.
Aries, R. (1974). French Patent 2,209,842.
Aschaffenburger Zellstoffwerke A.G. (1966). French Patent 1,426,051.
Bacher, F. A., Boley, A. E., and Shonk, C. E. (1954). *Anal. Chem.* **26**, 1146–1149.
Barker, H. A. (1968). *In* "The Vitamins" (W. H. Sebrell, Jr. and R. S. Harris, eds.), 2nd ed.,
   Vol. 2, pp. 184–212. Academic Press, New York.
Barker, H. A. (1972). *Annu. Rev. Biochem.* **41**, 55–90.
Barker, H. A., Smyth, R. D., and Hogenkamp, H. P. (1963). *Biochem. Prep.* **10**, 27–35.
Battersby, A. R., and McDonald, E. (1976). *Philos. Trans. R. Soc. London, Ser. B* **273**,
   161–180.
Battersby, A. R., Hollenstein, R., McDonald, E., and Williams, D. C. (1976). *J. Chem. Soc.,
   Chem. Commun.* **14**, 543–544.
Battersby, A. R., Ihara, M., McDonald, E., Redfern, J. R., and Golding, B. T. (1977a). *J.
   Chem. Soc., Perkin Trans. 1*, **2**, 158–166.
Battersby, A. R., McDonald, E., Hollenstein, R., Ihara, M., Satoh, F., and Williams, D. C.
   (1977b). *J. Chem. Soc., Perkin Trans. 1*, **2**, 166–178.
Beskhov, M. N., Karova, E. A., and Murgov, I. D. (1969). *Nauch. Tr., Vissh Inst. Khranit.
   Vkusova Prom-st., Plovdiv.* **16**, 337–351; *Chem. Abstr.* **77**, 59987 (1972).
Beskhov, M. N., Keskinova, D., Khristozova, T., and Ploshchakova, M. (1973). *Nauch. Tr.,
   Vissh Inst. Khranit. Vkusova Prom-st., Plovdiv.* **20**, 211–216; *Chem. Abstr.* **83**, 129945
   (1975).

Bilkus, D. I., and Mervyn, L. (1971). *In* "Cobalamins" (H. R. V. Arnstein, ed.), pp. 17–21. Churchill, London.

Blaylock, B. A., and Stadtman, T. C. (1966). *Arch. Biochem. Biophys.* **116,** 138–152.

Brady, R. O., Castanera, E. G., and Barker, H. A. (1962). *J. Biol. Chem.* **237,** 2325–2332.

Brazenas, G., and Kanopkaite, S. (1970a). *Liet. TSR Mokslu Akad. Darb. Ser. C.* **3,** 157–164; *Chem. Abstr.* **75,** 127033 (1971).

Brazenas, G., and Kanopkaite, S. (1970b). *Liet. TSR Mokslu Akad. Darb. Ser. C.* **3,** 165–173; *Chem. Abstr.* **75,** 127034 (1971).

Bykhovsky, V. Y. (1969). Russian Patent 227,528.

Bykhovsky, V. Y., and Zaitseva, N. I. (1976). *Prikl. Biokhim. Mikrobiol.* **12,** 365–371.

Bykhovsky, V. Y., Zaitseva, N. I., and Bukin, V. N. (1969). *Dokl. Akad. Nauk SSSR* **185,** 459–461.

Bykhovsky, V. Y., Zaitseva, N. I., and Yavorskaya, A. N. (1976). *Prikl. Biokhim. Mikrobiol.* **12,** 491–494.

Bykhovsky, V. Y., Pantskhava, E. S., Zaitseva, N. I., Pcjelkina, V. V., Fedorov, S. A., Golubovsky, S. A., and Yarovenko, M. L. (1972). Russian Patent 359,264.

Celletti, P., Moretti, G. P., and Petrangeli, B. (1976). *Farmaco, Ed. Prat.* **31,** 413–419.

Chinoin-Gyogyszer (1965). Hungarian Patent 152,435.

Chinoin-Gyogyszer (1971). French Patent 2,062,367.

Daniels, H. J. (1966). *Can. J. Microbiol.* **12,** 1095–1098.

Daniels, H. J. (1970). *Can. J. Microbiol.* **16,** 809–815.

Danilova, N. P., and Bankina, L. P. (1974). Russian Patent 454,250.

Demain, A. L., and White, R. F. (1971). *J. Bacteriol.* **107,** 456–460.

Demain, A. L., Daniels, H. J., Schnable, L., and White, R. F. (1968). *Nature (London)* **220,** 1324–1325.

Duric, N. (1969). *Process Biochem.* **4**(12), 35–50.

Farmaceutici Italia (1971a). French Patent 2,078,641.

Farmaceutici Italia (1971b). French Patent 2,083,992.

Fenton, W. A., Ambani, L. M., and Rosenberg, L. E. (1976). *J. Biol. Chem.* **251,** 6616–6623.

Ford, J. E. (1959). *J. Gen. Microbiol.* **21,** 693–701.

Ford, S. H., and Friedmann, H. C. (1976). *Arch. Biochem. Biophys.* **175,** 121–130.

Friedmann, H. C., and Cagen, L. M. (1970). *Annu. Rev. Microbiol.* **24,** 159–208.

Fujii, K., and Fukui, S. (1970). *Eur. J. Biochem.* **17,** 552–560.

Fujii, K., Shimizu, S., and Fukui, S. (1966). *Hakko Kogaku Zasshi.* **44,** 185–191; *Chem. Abstr.* **68,** 28441 (1968).

Fukui, S., Tamao, Y., Kato, T., Takahashi, T., and Shimizu, S. (1964). *J. Vitaminol.* **10,** 202–210.

Fukui, S., Shimizu, S., and Fujii, K. (1967). *Hakko Kogaku Zasshi.* **45,** 530–540; *Chem. Abstr.* **69,** 18021 (1968).

Georgopapadakou, N. H., Petrillo, J., Scott, A. I., and Low, B. (1976). *Genet. Res.* **28,** 93–100.

Giec, A., Weker, H., and Skupin, J. (1976). *Bull. Acad. Pol. Sci., Ser. Sci. Biol.* **24,** 497–503.

Glaxo Group, Ltd. (1964). British Patent 963,373.

Goldstein, S., and Duca, A. (1976). *J. Pharm. Sci.* **65,** 1831–1833.

Goldstein, S., Ciupitoiu, A., Vasilescu, V., and Duca, A. (1974). *Can. J. Pharm. Sci.* **9,** 96–98.

Gruzina, V. D., Erokhina, L. I., and Ponomareva, G. M. (1974). *Genetika* **10,** 121–127.

Hercules Inc. (1968). U.S. Patent 3,411,991.

Hermann, R., and Mueller, O. (1976). *Hoppe-Seyler's Z. Physiol. Chem.* **357,** 1695–1698.

Hodgkin, D. C., Pickworth, J., Robertson, J. H., Trueblood, K. N., and Prosen, R. J. (1955). *Nature (London)* **176,** 325–328.

Hoffmann-La Roche (1963). Belgian Patent 631,589.

Iashi (1973). German Patent 2,210,973.

Ilieva, M. (1971). *Nauch. Tr., Vissh Inst. Khranit Vkusova Prom-st., Plovdiv.* **18,** 371–377; *Chem. Abstr.* **79,** 40871 (1973).

Ilieva, M., and Popova, Y. (1974). *Nauch. Tr., Vissh Inst. Khranit Vkusova Prom-st., Plovdiv.* **21,** 105–112; *Chem. Abstr.* **85,** 141264 (1976).

Imfeld, M., Townsend, C. A., and Arigoni, D. (1976). *J. Chem. Soc., Chem. Commun.* **14,** 541–542.

Janicka, I., Maliszewska, M., and Pedziwilk, F. (1976). *Acta Microbiol. Pol.* **25,** 205–210.

Kaczka, E. A., Wolf, D. E., and Folkers, K. (1956). U.S. Patent 2,738,302 (to Merck and Co., Inc.).

Kageyama, M., Burg, K. A., and Perlman, D. (1977). *J. Antibiot.* **30,** 283–288.

Kaken Kagaku, K.K. (1965). Japanese Patent 65/4,440.

Kamikubo, T. (1967). Japanese Patent 67/14,802.

Kamikubo, T., Takeda, Y., Hayashi, M., and Friedrich, W. (1972). *Agric. Biol. Chem.* **36,** 164–165.

Kamikubo, T., Nishio, N., and Yano, T. (1975). Japanese Patent 75/132,186 (to Kyowa Hakko Kogyo Co., Ltd.).

Kamikubo, T., Hayashi, M., and Nishio, N. (1976). *Abstr., Int. Ferment. Symp., 5th, 1976,* No. 21-15, p. 401.

Kanegafuchi Chem. Ind. Co., Ltd. (1972). Japanese Patent 72/15,748.

Kaplan, L., and Birnbaum, J. (1973). *Abstr. Annu. Meet. Am. Soc. Microbiol.* p.8 (E 45).

Kato, T., and Shimizu, S. (1962). *Vitamins* **26,** 470–472.

Koike, I., and Hattori, A. (1975). *J. Gen. Microbiol.* **88,** 1–10.

Kojima, I., Sato, H., Maruhashi, K., and Fujiwara, Y. (1976). *Abstr., Int. Ferment. Symp., 5th, 1976,* No. 7-04, p. 134.

Koppenhagen, V. B., and Pfiffner, J. J. (1970). *J. Biol. Chem.* **245,** 5865–5873.

Koppenhagen, V. B., and Pfiffner, J. J. (1971). *Fed. Proc., Fed. Am. Soc. Exp. Biol.* **30,** 1088 (Abstr. 209).

Koppenhagen, V. B., Elsenhans, B., Wagner, F., and Pfiffner, J. J. (1974). *J. Biol. Chem.* **249,** 6532–6540.

Kotala, L., and Luba, J. (1965). *Acta Pol. Pharm.* **22,** 419–422.

Krieger, J. H. (1973). *Chem. & Eng. News* **51** (11), 16–19, 25, 27, and 29.

Kuratomi, K., Poston, J. M., and Stadtman, E. R. (1966). *Biochem. Biophys. Res. Commun.* **23,** 691–695.

Kyowa Hakko Kogyo Co., Ltd. (1970). Japanese Patent 70/36,159.

Kyowa Hakko Kogyo Co., Ltd. (1973). Japanese Patent 73/38,880.

Lago, B. D., and Demain, A. L. (1969). *J. Bacteriol.* **99,** 347–349.

Lindstrand, K. (1964). *Nature (London)* **204,** 188–189.

Lindstrand, K. (1965). *Acta Chem. Scand.* **19,** 1762–1763.

Ljungdahl, L. G., and Wood, H. G. (1969). *Annu. Rev. Microbiol.* **23,** 515–538.

Long, R. A. (1962). U.S. Patent 3,018,225 (to Merck and Co., Inc.).

Lorenzi, E., and Oddono, F. (1972). *Farmaco, Ed. Prat.* **27,** 280–297.

Lotti, B. (1975). *Boll. Chim. Farm.* **114,** 416–420.

Lowe, D. A., and Turner, J. M. (1970). *J. Gen. Microbiol.* **64,** 119–122.

McDaniel, L. E. (1961). U.S. Patent 3,000,793 (to Merck and Co., Inc.).

McDonald, C. M., Farquharson, J., Bessent, R. G., and Adams, J. F. (1977). *Clin. Sci. Mol. Med.* **52,** 215–218.

Matsumoto, J., Takuma, J., Takahashi, K., and Murakami, M. (1974). *Yamanouchi Seiyaku Kenkyu Hokoku.* **2**, 34–39; *Chem. Abstr.* **84**, 65304 (1976).

Maugh, T. H. (1973). *Science* **179**, 266–267.

Merck and Co., Inc. (1964). U.S. Patent 3,163,637.

Merck and Co., Inc. (1971). French Patent 2,038,828.

Mervyn, L., and Smith, E. L. (1964). *Prog. Ind. Microbiol.* **5**, 151–201.

Mil'ko, E. S., and Rabotnova, I. L. (1967). *Prikl. Biokhim. Mikrobiol.* **3**, 277–282.

Miller, I. M., and Rosenblum, C. (1960). U.S. Patent 2,939,822 (to Merck and Co., Inc.).

Miyazaki, A., Hayashi, M., and Kamikubo, T. (1972). *Vitamins* **45**, 131–135.

Morikawa, H., and Kamikubo, T. (1969). *Hakko Kogaku Zasshi.* **47**, 470–477; *Chem. Abstr.* **73**, 64967 (1970).

Morikawa, H., Hayashi, M., and Kamikubo, T. (1970). *Asahi Garasu Kogyo Gijutsu Shoreikai Kenkyu Hokoku.* **16**, 95–102; *Chem. Abstr.* **75**, 18560 (1971).

Mueller, G., Dieterle, W., and Siebke, G. (1970). *Z. Naturforsch., Teil B* **75**, 307–309.

Mueller, G., Gross, R., and Siebke, G. (1971). *Hoppe-Seyler's Z. Physiol. Chem.* **352**, 1720–1722.

Nakao, Y., Hisano, K., Kanemaru, T., and Yamano, T. (1974). Japanese Patent 74/15,796 (to Takeda Chem. Ind., Ltd.).

Nishio, N., Yano, T., and Kamikubo, T. (1975). *Agric. Biol. Chem.* **39**, 21–27.

Nishio, N., Yano, T., Hayashi, M., and Kamikubo, T. (1976a). *Agric. Biol. Chem.* **40**, 1035–1037.

Nishio, N., Ueda, M., Omae, Y., Hayashi, M., and Kamikubo, T. (1976b). *Agric. Biol. Chem.* **40**, 2037–2043.

Noyes, R. (1969). "Chemical Process Review," No. 40. Noyes Dev. Corp., Park Ridge, New Jersey.

Olin Mathieson Chem. Corp. (1965). U.S. Patent 3,164,582.

Pantskhava, E. S., and Bykhovsky, V. Y. (1966). *Prod. Mikrobnogo Sin., Akad. Nauk Latv. SSR, Inst. Mikrobiol.* 81–94; *Chem. Abstr.* **67**, 10335 (1967).

Pantskhava, E. S., and Pcjelkina, V. V. (1970). Russian Patent 265,043.

Pantskhava, E. S., Pcjelkina, V. V., and Bukin, V. N. (1973). *Biokhimiya* **38**, 507–514.

Pawelkiewicz, J. (1971). *In* "Methods in Enzymology" (S. P. Colowick and N. O. Kaplan, eds.), Vol. 18, Part C, pp. 99–101. Academic Press, New York.

Pcjelkina, V. V., Pantskhava, E. S., and Bukin, V. N. (1970). *Biokhimiya* **35**, 1007–1013.

Perlman, D. (1959). *Adv. Appl. Microbiol.* **1**, 87–122.

Perlman, D. (1967). *In* "Microbial Technology" (H. J. Peppler, ed.), pp. 283–287. Van Nostrand-Reinhold, Princeton, New Jersey.

Perlman, D., and Toohey, J. I. (1966). *Nature (London)* **212**, 300–301.

Perlman, D., Vlietinck, A. J., Matthews, H. W., and Lo, F. F. (1974). *J. Antibiot.* **27**, 826–832.

Pierrel S.p.A. (1963a). French Patent 1,325,304.

Pierrel S.p.A. (1963b). French Patent 1,325,308.

Pierrel S.p.A. (1965a). British Patent 1,007,971.

Pierrel S.p.A. (1965b). British Patent 1,007,972.

Pliva T.F.K.P. (1964). Belgian Patent 651,153.

Popova, Y., and Bakchevanska, S. (1972). *Nauch. Tr., Vissh Inst. Khranit. Vkusova Prom-st., Plovdiv.* **19**, 237–244; *Chem. Abstr.* **83**, 145681 (1975).

Popova, Y., and Ilieva, M. (1970). *Nauch. Tr., Vissh Inst. Khranit. Vkusova Prom-st., Plovdiv.* **17**, 151–159; *Chem. Abstr.* **79**, 30449 (1973).

Prescott, S. C., and Dunn, C. G. (1959). *In* "Industrial Microbiology," 3rd ed., pp. 482–496. McGraw-Hill, New York.

Renz, P., and Bauer-David, A. J. (1972). *Z. Naturforsch., Teil B* **27**, 539–544.

Richter Gedeon V.G. (1963). Hungarian Patent 151,065.
Richter Gedeon V.G. (1967). French Patent 1,503,402.
Richter Gedeon V.G. (1975a). Belgian Patent 821,535.
Richter Gedeon V.G. (1975b). Belgian Patent 821,536.
Rickes, E. L., Brink, N. G., Koniuszy, F. R., Wood, T. R., and Folkers, K. (1948). *Science* **107,** 396–397.
Robinson, F. M. (1955). *Encycl. Chem. Technol.* **14,** 813–828.
Rohm and Haas Co. (1968). Belgian Patent 710,151.
Rosenthal, H. L. (1968). *In* "The Vitamins" (W. H. Sebrell, Jr. and R. S. Harris, eds.), 2nd ed., Vol. 2, pp. 145–170. Academic Press, New York.
Russel, C. S. (1974). *J. Theor. Biol.* **47,** 145–151.
Sato, K., Shimizu, S., and Fukui, S. (1968). *Agric. Biol. Chem.* **32,** 1–11.
Sato, K., Ohmori, H., Shimizu, S., and Fukui, S. (1971). *Agric. Biol. Chem.* **35,** 333–350.
Sato, K., Ueda, S., and Shimizu, S. (1977). *Appl. Environ. Microbiol.* **33,** 515–521.
Scott, A. I. (1975). *Ann. N.Y. Acad. Sci.* **244,** 356–370.
Scott, A. I. (1976). *Philos. Trans. R. Soc. London, Ser. B* **273,** 303–318.
Scott, A. I., Kajiwara, M., Takahashi, T., Armitage, I. M., Demou, P., and Petrocine, D. (1976). *J. Chem. Soc., Chem. Commun.* **16,** 544–546.
Seisan Kaihatsu Kagaku Kenkyusho (1970). Japanese Patent 70/13,434.
Sheehan, T. L., and Hercules, D. M. (1977). *Anal. Chem.* **49,** 446–450.
Shimizu, S., and Sato, K. (1976). *Abstr., Int. Ferment. Symp., 5th, 1976,* No. 21-16, p. 402.
Sifa (1964). French Patent 1,368,892.
Sifa (1965). French Patent 1,422,040.
Sifa (1967). French Patent Addition 90,362 (addition to French Patent 1,422,040).
Simon, A. (1968). *Zentralbl. Bakteriol., Parasitenkd., Infektionskr. Hyg., Abt. 2* **122,** 143–154.
Skupin, J., Pedziwilk, F., and Jaszewski, B. (1970). *Bull. Acad. Pol. Sci., Ser. Sci. Biol.* **18,** 511–515.
Smith, E. L. (1948). *Nature (London)* **161,** 638–639.
Smith, E. L. (1965). U.S. Patent 3,167,539 (to Glaxo Lab., Ltd.).
Speedie, J. D., and Hull, G. W. (1960). U.S. Patent 2,951,017 (to Distillers Co., Ltd.).
Stefaniak, O., Lubienska, B., and Janicka, M. (1968). *Acta Microbiol. Pol.* **17,** 83–90.
Szemler, L. L., and Székely, A. D. (1969). *Process Biochem.* **4** (12), 25–27.
Terada, O., Ohishi, K., and Kinoshita, S. (1959). *Vitamins* **18,** 16–33.
Toohey, J. I. (1971). *In* "Methods in Enzymology" (S. P. Colowick and N. O. Kaplan, eds.), Vol. 18, Part C, pp. 71–75. Academic Press, New York.
Toohey, J. I. (1974). U.S. Patent 3,846,237.
Toraya, T., Yongsmith, B., Tanaka, A., and Fukui, S. (1975). *Appl. Microbiol.* **30,** 477–479.
Tortolani, G., and Ferri, P. G. (1974). *J. Chromatogr.* **88,** 430–433.
Tortolani, G., and Mantovani, V. (1974). *J. Chromatogr.* **92,** 201–206.
Tortolani, G., Bianchini, P., and Mantovani, V. (1970a). *J. Chromatogr.* **53,** 577–579.
Tortolani, G., Bianchini, P., and Mantovani, V. (1970b). *Farmaco, Ed. Prat.* **25,** 772–775.
Uclaf (1960). French Patent 1,264,016.
Uehisa, Y. (1972). Japanese Patent 72/50,397.
Vitols, E., Walker, G. A., and Huennekens, F. M. (1966). *J. Biol. Chem.* **241,** 1455–1461.
Vogelmann, H., and Wagner, F. (1973). *J. Chromatogr.* **76,** 359–379.
Vorobeva, L. I., Konovalova, L. V., and Lisenkova, L. L. (1967). *Prikl. Biokhim. Mikrobiol.* **3,** 264–269.
Wagman, G. H., Gannon, R. D., and Weinstein, M. J. (1969). *Appl. Microbiol.* **17,** 648–649.
Wagner, F. (1966). *Annu. Rev. Biochem.* **35,** 405–434.

Weissbach, H., Redfield, B., and Peterkofsky, A. (1961). *J. Biol. Chem.* **236,** PC40 (Prelim. Commun.).

White, R. F., and Demain, A. L. (1971). *Biochim. Biophys. Acta* **237,** 112–119.

White, R. F., Kaplan, L., and Birnbaum, J. (1973). *J. Bacteriol.* **113,** 218–223.

Whitlock, L. L., Melton, J. R., and Billings, T. J. (1976). *J. Assoc. Off. Anal. Chem.* **59,** 580–581.

Wipple, G. H., and Robscheit-Robbins, F. S. (1925). *Am. J. Physiol.* **72,** 395–407.

Zodrow, K., Stefaniak, O., Chelkowski, J., and Szczepska, K. (1963). *Acta Microbiol. Pol.* **12,** 263–266.

Zodrow, K., Stefaniak, O., Lubienska, B., and Janicka, M. (1968). *Acta Microbiol. Pol.* **17,** 91–94.

Chapter 16

Microbial Process for Riboflavin Production

D. PERLMAN

I. INTRODUCTION

Microbiologically produced riboflavin has long been available in yeast and related preparations in association with many other vitamins of the B-complex. It is a unique vitamin in that it can be produced *de novo* to a high concentration rather rapidly by certain microorganisms. Fermentation processes have been described in which the riboflavin content of the fermented medium amounted to more than 7 gm/liter. The organisms involved in these processes are ascomycetes, namely, *Eremothecium ashbyii* and *Ashbya gossypii*.

Although riboflavin is produced by many microorganisms, including bacteria, yeasts, and molds, other microorganisms require riboflavin for growth. The organisms which have been found to produce sufficient riboflavin to be of interest from a commercial standpoint are listed in Table I, where the yields are mentioned along with the sensitivity of the microorganism to iron.

Commercial fermentation processes for production of riboflavin or riboflavin concentrates are relatively recent, having been developed in the past 40 years. Aside from food yeasts, the first organisms employed

521

Copyright © 1979 by Academic Press, Inc.
All rights of reproduction in any form reserved. ISBN 0-12-551501-4

TABLE I. Microorganisms Producing Considerable Amounts of Riboflavin and the Effects of Iron on Biosynthesis of the Vitamin

Microorganism	Riboflavin yield (mg/liter)	Optimum iron conc. (mg/liter)	References
Clostridium acetobutylicum	97	1–3	Meade et al. (1947)
Mycobacterium smegmatis	58	Not critical	Mayer and Rodbart (1946)
Mycocandida riboflavina	200	Not critical	McClary (1951)
Candida flareri	567	0.04–0.06	Levine et al. (1949)
Eremothecium ashbyii	2480	Not critical	Moss and Klein (1949)
Ashbya gossypii	6420	Not critical	Szczeniak et al. (1973)

primarily for riboflavin production was *Clostridium acetobutylicum* which, when grown in grain mashes or on whey low in iron, yielded residues containing 4–5 mg riboflavin per gram fermentation solids. These fermentations were succeeded in about 1940 by a process using *E. ashbyii* with yields of about 2 mg/ml. In 1946 processes using the ascomycete *A. gossypii* were started. In 1978 the only manufacturer using the microbiological process is Merck & Co., Inc. (United States). Companies manufacturing riboflavin by chemical synthesis include Hoffmann-La Roche Inc. (United States and Switzerland), Takeda Chemical Industries Ltd. (Japan), Pfizer, Inc. (U.S.A.), and E. Merck (West Germany).

II. THE FERMENTATION PROCESS

Development of the current processes used for riboflavin production by *A. gossypii* NRRL Y-1056 focused on three aspects: (1) preparation of the culture medium; (2) selection of mutant cultures and optimization of inoculum preparation; and (3) optimization of the fermentation conditions, e.g., incubation temperature, aeration level, and fermentor design. The most important increases in fermentation productivity have resulted from the changes in the culture media and the selection of high riboflavin-producing mutant cultures.

Initial studies (Tanner et al., 1949) showed that good growth of *A. gossypii* occurred in a medium containing glucose, corn steep liquor, and animal stick liquor, tankage, or meat scraps. It was also shown that a sterilization time of less than 30 minutes, a low concentration of "young inoculum," e.g., 2% v/v, and efficient aeration were important. Maximum yields of 700 mg/liter (10 days) were obtained when the medium was autoclaved for 15–30 minutes (as compared to 325 mg/liter when it was autoclaved for 90 minutes). Pilot-plant reproduction of these conditions resulted in yields of 500–850 mg/liter.

Further improvements resulted when enzymatically degraded collagen and lipids were used as energy sources together with accessory factors present in corn steep liquor. Distillers' solubles, and brewers' yeast were included in the medium with yields of the order of 4200 mg/liter. The lipid used as an energy source was proposed earlier for the *E. ashbyii* fermentation (Rudert,1945; Phelps,1949) and also worked well for the *A. gossypii* fermentation. Typical data on stimulation of riboflavin production as a result of including in the media various peptones are summarized in Table II, as is the effect of supplementation with glycine.

Following earlier studies, Pfeifer *et al.* (1950) suggested that some improvement in productivity was possible as a result of culture selection. Pridham and Raper (1952) reported on methods they used to obtain mutants yielding 200–300 mg of riboflavin per liter more than the parent. Since only 700 mutants were tested, the chance of obtaining an improved mutant was slim. In addition, they suggested several methods of handling the cultures, including: (1) incorporating a reducing agent,

TABLE II. Effect of Supplementing Growth Medium with Peptones and Glycine on Riboflavin Production by *Ashbya gossypii*

Nature of supplement	Riboflavin yield (mg/liter)
A. Peptones[a]	
Peptic hydrolysate of animal tissue	1520
Equal amounts of a peptic hydrolysate of animal tissue and pancreatic digest of casein	1280
Pancreatic digest of lactalbumin	1000
Pancreatic digests of casein	340
Pancreatic digest of gelatin	3620
Papaic digest of soybean meal	673
B. Glycine[b] (added in gm/liter)[c]	
0	3280
1	3640
2	3980
3	4200

[a] From Malzahn *et al.* (1959).

[b] From Malzahn *et al.* (1963).

[c] The basal medium contained (w/v): corn steep liquor solid, 2.25%; Wilson's peptone W-809, 3.5%; soybean oil, 4.5%.

e.g., sodium dithionate, into the stock culture media to decrease the production of low-yielding mutants; (2) reisolating superior substrains frequently; and (3) preserving proved strains by immediate lyophilization or storage under mineral oil. Malzahn *et al.* (1963) recommended a procedure in which the cultures were transferred at 4-day intervals on malt–yeast extract–glucose–agar slants. Cultures were frozen after 4 days' growth to maintain their flavinogenic capacity. They also noted that productivity was improved in mutants obtained by treatment of the cells with ultraviolet radiation, uranyl nitrate, nitrogen mustard, or ethyleneamine, with selection based on increased pigmentation of colonies when grown on the malt–yeast extract–agar medium. One isolate produced 3620 mg riboflavin per liter when soybean oil was the energy source, compared with 1420 mg/liter for the parent grown in the same medium.

The optimal fermentation conditions include aeration at about one-third volume of air per volume of liquid per minute, and agitation with three impellers at a rate of 1.0 hp/1000 liters of medium. When foaming was excessive, it was controlled by initial addition of emulsified silicone antifoam, and later by addition of soybean oil (which also acted as a nutrient). Sterilization of the medium was accomplished by heating at 121°C for 3 hours (which improved productivity), and the incubation temperature was maintained at 28°C for the 7-day incubation period (Malzahn. *et al.*, 1963).

III. RECOVERY OF RIBOFLAVIN FROM FERMENTED MEDIA

Upon completion of the fermentation, the solids were dried to a crude product for animal-feed supplementation or processed to a United States Pharmacopoeia-grade product. In either case, the pH value of the fermented medium was adjusted to pH 4.5. For the feed-grade product, the broth was concentrated to about 30% solids and dried on double-drum driers.

When a crystalline product was required, the fermented broth was heated for 1 hour at 121°C to solubilize the riboflavin. Insoluble matter was removed by centrifugation, and riboflavin recovered by conversion to the less soluble form (Michaels *et al.*, 1936). Both chemical (Hines, 1945a) and microbiological (Hines, 1945b) methods of conversion have been used. The precipitated riboflavin was then dissolved in water or polar solvents (Dale, 1947) or an alkaline solution (Morehouse, 1958), oxidized by aeration, and recovered by recrystallization from the aqueous or polar solvent solution or by acidification of the alkaline solution.

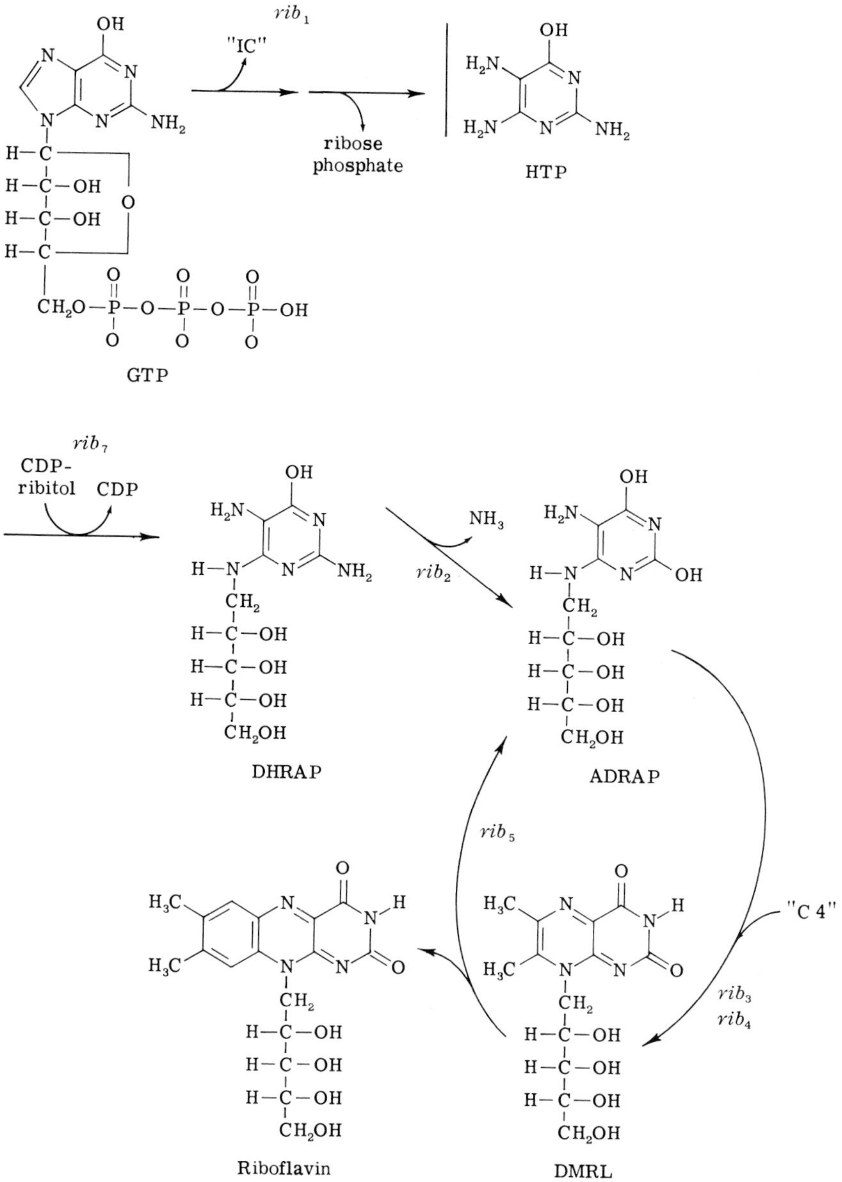

FIGURE 1. Probable pathway of riboflavin biosynthesis. The genes (rib_1 to rib_5, rib_7) in *Saccharomyces cerevisiae* which correspond to each reaction are shown. Abbreviations: HTP 6-hydroxy-2,4,5-triaminopyrimidine; DHRAP, 2,5-diamino-6-hydroxy-4-ribitylamino-pyrimidine; ADRAP 5-amino-2,6-dihydroxy-4-ribitylaminopyrimidine; DMRL, 6,7-di-methyl-8-ribityllumizine.

The crystals obtained from the alkaline solution were type A or B (Dale, 1952), and heating the mixture to boiling resulted in conversion to the more desirable type A (Dale, 1957).

IV. BIOSYNTHETIC STUDIES

Progress in determining the biosynthetic pathways (mainly in the yeasts and *A. gossypii*) were summarized by Goodwin (1959) and more recently by Demain (1972). The pathway shown in Fig. 1 is currently accepted as the most likely. Demain (1972) has categorized the "state of the art" as depending upon the degree of accumulation of the vitamin. He recognizes weak overproducers, moderate overproducers, and strong overproducers. Clostridia are representatives of the weak over-producers and are markedly sensitive to the presence of iron in the medium. The moderate overproducers include yeasts, and iron inhibits their overproduction. The strong overproducing group includes species of *Ashbya* and *Eremothecium,* and iron has no effect on production. He concludes that iron control of riboflavin overproduction must be exer-cised before the final step of the pathway, since production of DMRL (see Fig. 1) is inhibited by iron in clostridia. If an iron–flavoprotein is the repressor of riboflavin biosynthesis, how can one explain the insensitiv-ity to iron of riboflavin overproduction by ascomycetes? Perhaps these species are constitutive with respect to riboflavin-synthesizing enzymes, or the empirically developed conditions that favor riboflavin overproduc-tion inhibit formation of the repressor. One such condition could be the temperature of incubation.

REFERENCES

Dale, J. K. (1947). U.S. Patent 2,421,142.

Dale, J. K. (1952). U.S. Patent 2,603,633.

Dale, J. K. (1957). U.S. Patent 2,797,215.

Demain, A. L. (1972). *Annu. Rev. Microbiol.* **26,** 369.

Goodwin, T. W. (1959). *Prog. Ind. Microbiol.* **1,** 139.

Hines, G. E., Jr. (1945a). U.S. Patent 2,367,644.

Hines, G. E., Jr. (1945b). U.S. Patent 2,387,023.

Levine, H., Oyaas, J. E., Wasserman, L., Hoogerheide, J. C., and Stern, R. M. (1949). *Ind. Eng. Chem.* **41,** 1665.

McClary, J. E. (1951). U.S. Patent 2,537,148.

Malzahn, R. C., Phillips, R. F., and Hanson, A. M. (1959). U.S. Patent 2,876,169.

Malzahn, R. C., Phillips, R. F., and Hanson, A. M. (1963). *Abstr. 63rd Annu. Meet. Am. Soc. Microbiol.* p. 21.

Mayer, R. L., and Rodbart, R. (1946). *Arch. Biochem.* **11,** 49.

Meade, R. E., Pollard, H. L., and Rodgers, N. E. (1947). U.S. Patent 2,433,680.

Michaels, L., Schubert, M., and Symthe, C. V. (1936). *J. Biol. Chem.* **116,** 587.

Morehouse, A. L. (1958). U.S. Patent 2,822,361.

Moss, A. R., and Klein, R. (1949). British Patent 615,847.

Pfeifer, V. F., Tanner, F. W., Jr., Vojnovich, C., and Traufler, D. H. (1950). *Ind. Eng. Chem.* **42,** 1776.

Phelps, A. S. (1949). U.S. Patent 2,473,818.

Pridham, T. G., and Raper, K. B. (1952). *Mycologia* **44,** 452.

Rudert, F. J. (1945). U.S. Patent 2,374,503.

Szczeniak, T., Karabin, L., and Wituch, K. (1973). Polish Patent 66,611.

Tanner, F. W., Jr., Vojnovich, C., and Van Lanen, J. M. (1949). *J. Bacteriol.* **58,** 737.

Chapter 17

Carotenoids

L. NINET
J. RENAUT

I. INTRODUCTION

Carotenoids form one of the most important families of natural pigments. They are liposoluble tetraterpenes, usually yellow to red in color, which are formed by the condensation of isoprenyl units. The first carotenoids were discovered around 1800 and now more than 400 different compounds are known (Isler, 1977), the majority of which consist of hydrocarbons (carotenes) or their oxygenated derivatives (xanthophylls), both containing 40 carbon atoms. The chemical structures of the main carotenoids related to fermentation processes are shown in Fig. 1.

Some of the carotenoids are normally present in food and these have essential biological functions, particularly as precursors of vitamin A. In view of these biological activities, they are used as food supplements to

MICROBIAL TECHNOLOGY, 2nd ed., VOL. I
Copyright © 1979 by Academic Press, Inc.
All rights of reproduction in any form reserved. ISBN 0-12-551501-4

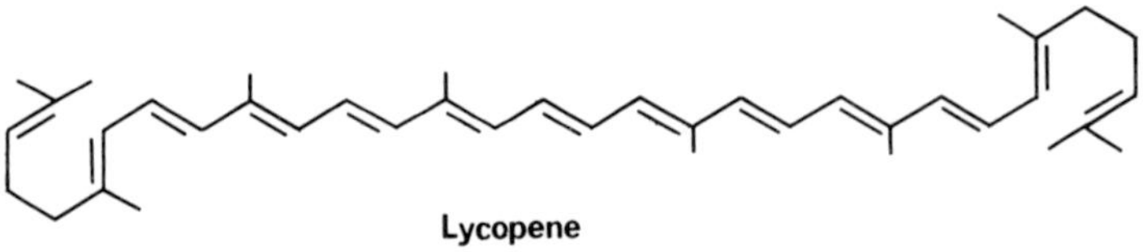

Lycopene

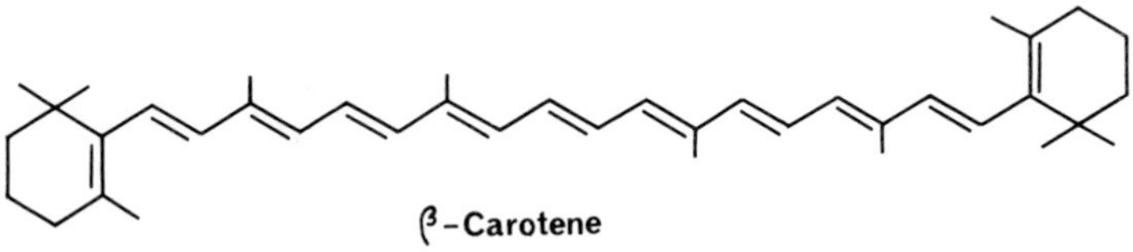

β-Carotene

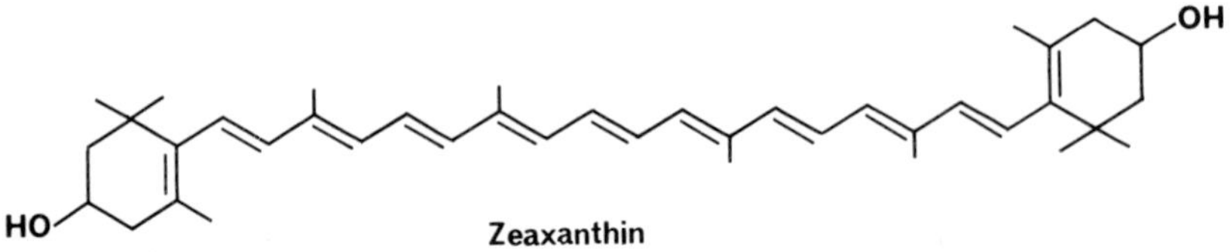

Zeaxanthin

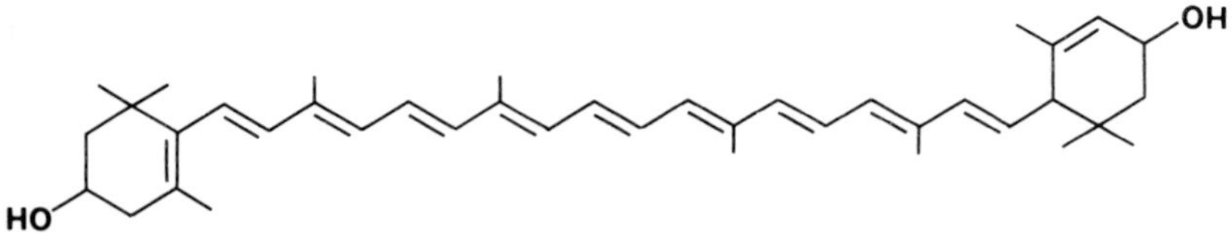

Lutein

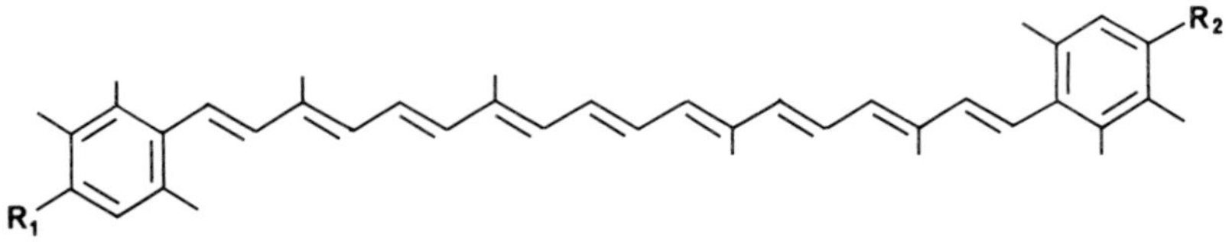

Other compounds

FIGURE 1. Chemical structures of principal carotenoids produced by microorganisms. R groups for "other compounds" are as follows:

R_1	R_2	Compound
H	H	Isorenieratene (leprotene)
OH	H	3-Hydroxyisorenieratene
OH	OH	3,3′-Dihydroxyisorenieratene

prevent or cure vitamin deficiencies. In addition other pigmented carotenoids are used both as food additives for intensifying or modifying the color in fat, oil, cheese, and beverages and also as animal feed supplement to enhance the color of such foods as egg yolks and chicken flesh. It is for these latter uses, as food and feed additives, that pure products derived from organic processes, rather than from plant sources, have played an ever-increasing role. However, since 1956, attempts have also been made to produce by fermentation either pure carotenoids as β-carotene, lycopene, and xanthophylls or crude colored biomass which may be added directly to the animal feed.

While the carotenoids are widely found in plants and animals, only microorganisms and plants have the necessary systems to synthesize a wide range of these substances. All of them arise from minor variations in the final steps of a common biosynthetic pathway. This pathway is summarized in Fig. 2 (Goodwin, 1972; Beytia and Porter, 1976). In practice β-carotene and related carotenes are mainly found in fungi and algae, while the xanthophylls occur almost exclusively in bacteria and algae (Ciegler, 1965a; Goodwin, 1972; Weedon, 1971). However the rates of synthesis vary considerably in different microorganisms so that the number of such organisms which can be used in industrial fermentation processes is limited (Ciegler, 1965a).

II. PRODUCTION OF β-CAROTENE

A. Earlier Developments

Many species of algae and fungi (e.g., *Neurospora crassa, Penicillium sclerotiorum, Phycomyces blakesleeanus*) and also yeasts (*Rhodotorula*) were considered for use in β-carotene production but were found to be unsuitable (Ceigler, 1965a).

Following the publication of the work by Barnett *et al.* (1956) on some particular fungi in the Mucorales group, the interest concentrated on the development of industrial fermentation with Choanephoraceae, mainly *Blakeslea trispora*. The use in the same culture of the two sexual forms of this species was found to lead to a remarkable increase in mycelial carotenoid content. The special conditions required for cultures in liquid media with mixed sexual forms of *B. trispora* have been shown by several authors, particularly from the Northern Regional Research Laboratory, Peoria, Illinois. First the medium should be viscous and rich in vegetable oils, kerosene, and surface-active agents. Second, α- or β-ionone (or another terpenoid) is added during the incubation period

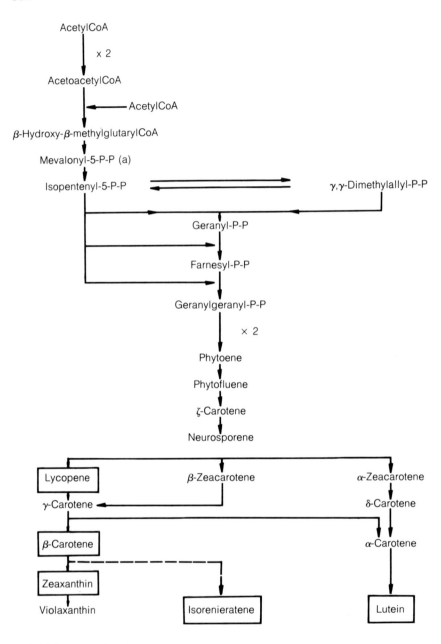

(a) P-P = pyrophosphate

FIGURE 2. Pathway of carotenoid biosynthesis.

according to a previous study on *Phycomyces* (Mackinney *et al.*, 1952), together with an antioxidant to prevent oxidation of the β-carotene. The medium is inoculated with equal volumes of mature cultures of each sexual form of *B. trispora*. However there is no loss of yield if the mixed culture is seeded with an excess of *B. trispora* (−) form (Zajic, 1964) or if the ionones are replaced by the mycelia of various fungi or yeasts, whether unwashed or washed with hexane (Ciegler, 1965b). The yield of β-carotene with this procedure amounts to 1 gm/liter in the medium, i.e., 20 mg/gm of dried mycelium. The work in this area was reviewed by Ciegler (1965a) and Hanson (1967).

B. Recent Developments

1. Fermentation

Much work has been devoted to ways of simplifying and improving the fermentation process with *B. trispora* and, as shown below, the main approaches have been to modify the composition of the medium or to suppress the interdependence of the two sexual forms.

a. Modification of the Medium Composition. The medium normally contains from 1 to 8% of vegetable or animal oils, and these can be partially or totally replaced by cheaper oils, such as linoleic and linolenic acid-rich oils (autoclave oils and soapstock), which are by-products of oil refining (Pazola *et al.*, 1969; Bohinski, 1970). The yield of β-carotene with this substitution remains 1 gm/liter (Gutcho, 1973d).

b. Substitution of the β-Ionone. As β-ionone is not incorporated into β-carotene (Reyes *et al.*, 1964), chemically related compounds have been examined as replacement products. The Krebs cycle keto acids are only partially efficient as substitutes (Bjork and Neujahr, 1969), but bicyclic terpenes of turpentine (Cederberg and Neujahr, 1969) as well as other natural terpenes (Kolot *et al.*, 1973) are suitable for both partial and total replacement of β-ionone. Other substances tested include isoprene dimers and trimers (Nakamura *et al.*, 1974) and cyclohexane, cyclohexanone, and their trimethyl derivatives (Kolot *et al.*, 1973; Ninet *et al.*, 1963a, 1964, 1969). Among these, 2,6,6-trimethyl-1-acetyl-cyclohexene added at the level of 1 gm/liter is the most efficient at providing the same β-carotene yield as β-ionone itself. Cheap citrus fruit peels, rich in citric acid, were also added in the medium instead of ionone (Ciegler, 1965a). When this material was treated with alkali, the β-carotene content amounted to 1.7 gm/liter (Upjohn Co., 1965b).

c. New Activators of Carotene Production. The observation that carotene production was somewhat enhanced by the presence of dimethylformamide used as a solvent for the antioxidant prompted a systematic survey of numerous amides and nitrogenous compounds for similar but more effective activity. α-Pyrrolidone and succinimide, added as 3–4 gm/liter, showed some activity but the best results so far have been obtained with isoniazid, iproniazid, and 4-formylpyridine at concentrations of 1, 1, and 2 gm/liter, respectively. Interestingly, these activators, as well as being active in their own right, show an enhanced effect when used in combination with β-ionone or with 2,6,6-trimethyl-1-acetylcyclohexene giving a yield of β-carotene of 3 gm/liter, i.e., 30 mg/gm of dried mycelium (see Table I) (Ninet *et al.*, 1963b, 1965b, 1966, 1969). It is not yet known whether these substances are only permeation agents for *B. trispora* or act as inducers or activators of some enzymes operating in the carotenoid pathway. The detailed methods for utilization of these activators were described by Hanson (1967) and by Gutcho (1973c).

d. Suppression of the Sexual Interdependence. The (−) sexual form of *B. trispora* in the absence of the (+) form is capable of synthesizing significant quantities of β-carotene if the hormonal substances (trisporic acids), produced in cultures of the mixed forms, are added to the medium (Prieto *et al.*, 1964; Sebek and Jäger, 1964). These hormones can be extracted, with chloroform, from the filtered broth of a mated culture, the main component being trisporic acid C (Caglioti *et al.*, 1964) (see Fig. 3). Studies on the structure, biosynthesis, and mode of action of

TABLE I. Improvements in β-Carotene Production with New Activators[a]

	β-Carotene (mg/liter) with:		
Added substances	None	Ionone (1 gm/liter)	TMACH[b] (1 gm/liter)
None	850	1550	1350
Succinimide	1350	2200	1850
Isoniazid	2200	3050	2950
Iproniazid	2200	3200	3200

[a] From Ninet *et al.* (1969).

[b] TMACH, 2,6,6-trimethyl-1-acetylcyclohexene. Strains : *Blakeslea trispora* NRRL 2456 (+) and NRRL 2457 (−). Basal culture medium: distillers' solubles, 60 gm; "Fox head" starch, 60 gm; soya bean oil, 35 gm; cotton seed oil, 35 gm; yeast extract, 1 gm; potassium phosphate, 0.5 gm; manganese sulfate, 0.2 gm; thiamin hydrochloride, 0.01 gm; kerosene, 20 gm; ethoxyquin, 0.5 gm; with tap water to 1 liter.

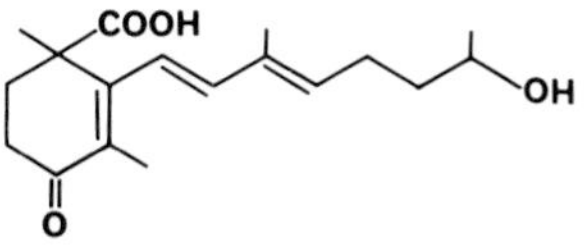

FIGURE 3. Structure of trisporic acid.

these hormones were summarized by Gooday (1974). They appear to arise from metabolism of β-carotene, or related substances, normally secreted by the fungus in mixed cultures (Feofilova et al., 1976; Feofilova and Bekhtereva, 1976). At the same time, they behave similarly to β-ionone, acting competitively with it in derepressing the enzyme converting mevalonylpyrophosphate to γ, γ'-dimethylallylpyrophosphate (Rao and Modi, 1977) and play a role in the formation and the fusion of zygophores to zygospores. These findings were utilized in the development of a two-stage process for β-carotene production by Farmaceutici Italia (1965) and Jäger (1966). The trisporic acids are produced in a mixed culture of B. trispora (+) NRRL 2456 and (−) NRRL 9159 in a medium containing distillers' solubles, starch, cotton seed oil, and kerosene which gives a yield of 3.5 gm/liter. They are then extracted from the broth and added to a culture of (−) NRRL 9159 in the same medium, supplemented with β-ionone at a level of 0.5 gm/liter. This gives a β-carotene yield of 1.2 gm/liter (Farmaceutici Italia, 1965).

The highest yielding process so far reported is summarized in Fig. 4. However, despite the increased productivity so far achieved with the B. trispora cultures, this process cannot compete with chemical synthesis. This might be reversed by improving the yields through strain mutations but the multinucleate nature of the spores makes this difficult and only rare attempts have been made (Skryabin, 1967).

A very active ^{14}C-labeled β-carotene was prepared for metabolism and stability studies by adding [1,2 ^{14}C]sodium acetate to a 48-hour-old mixed culture of B. trispora (Purcell and Walter, 1971; Gutcho, 1973f).

2. Extraction

In contrast with the considerable literature on fermentation processes, there are relatively few publications on the extraction of carotenes from broth. The extraction technology of β-carotene is almost identical, whether it be from plant or from microbial sources, as it is always endocellular. For animal feedstuffs, the mycelium can be directly used after drying, but as the stability of the β-carotene in these biomasses is poor, antioxidants (e.g., 0.5 gm/liter of ethoxyquin) are added to the broth or to the derived biomass to stabilize the product.

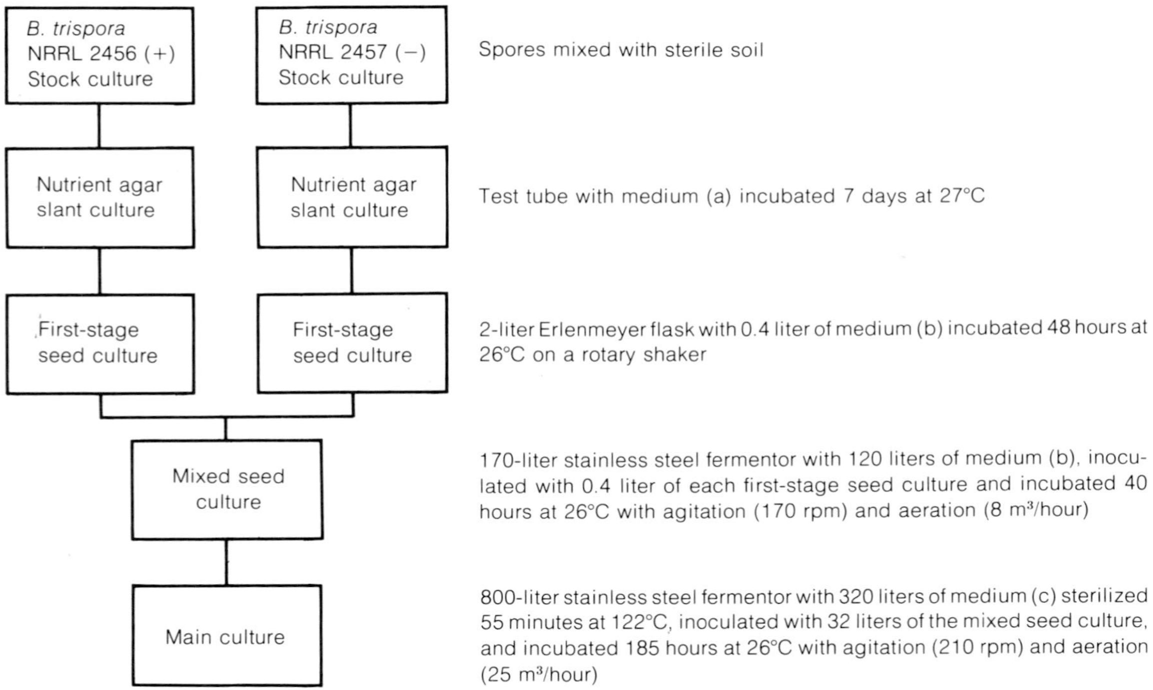
B. trispora
NRRL 2456 (+)
Stock culture
B. trispora
NRRL 2457 (−)
Stock culture
Spores mixed with sterile soil
Nutrient agar slant culture
Nutrient agar slant culture
Test tube with medium (a) incubated 7 days at 27°C
First-stage seed culture
First-stage seed culture
2-liter Erlenmeyer flask with 0.4 liter of medium (b) incubated 48 hours at 26°C on a rotary shaker
Mixed seed culture
170-liter stainless steel fermentor with 120 liters of medium (b), inoculated with 0.4 liter of each first-stage seed culture and incubated 40 hours at 26°C with agitation (170 rpm) and aeration (8 m³/hour)
Main culture
800-liter stainless steel fermentor with 320 liters of medium (c) sterilized 55 minutes at 122°C, inoculated with 32 liters of the mixed seed culture, and incubated 185 hours at 26°C with agitation (210 rpm) and aeration (25 m³/hour)

Medium (a): medium M_5 (from Hesseltine and Benjamin, 1959.)

Medium (b): Corn steep, 70 gm; corn starch, 50 gm; potassium dihydrogen phosphate, 0.5 gm; manganese sulfate, 0.1 gm; thiamin hydrochloride, 0.01 gm; with tap water to 1 liter.

Medium (c): Distillers' solubles, 70 gm; cornstarch, 60 gm; soya bean oil, 30 ml; cottonseed oil 30 ml; ethoxyquin, 0.35 gm; manganese sulfate, 0.2 gm; thiamin hydrochloride, 0.5 mg; isoniazid, 0.6 gm; kerosene 20 ml; with tap water to 1 liter and pH adjusted to 6.3. The last two substances are each sterilized separately by filtration. After 48 hours of growth 1 gm/liter of β-ionone and 5 ml/liter of kerosene are added aseptically; then a total amount of 42 gm/liter of glucose, as an aqueous 53% solution, is continuously injected up to the end of the culture.

FIGURE 4. A high-yielding β-carotene process.

More commonly the mycelium mass, after dehydration by vacuum drying or by treatment with low molecular weight aliphatic alcohols, is treated with hydrocarbon or chlorohydrocarbon solvents to extract the total carotenoids (of which 75–92% is β-carotene). This is followed by a series of precipitations and crystallizations with solvents mixtures to obtain crystalline β-carotene (Upjohn Co., 1965a). Such a process is illustrated in Fig. 5.

In some cases, the mycelium is extracted with refined vegetable oils to obtain an oily solution rich in carotenoids which is then treated by steam-stripping to remove odorous and volatile impurities (Pazola *et al.*, 1971).

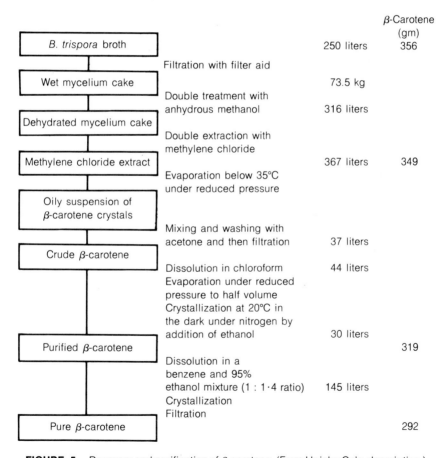

		β-Carotene (gm)
B. trispora broth	250 liters	356
Filtration with filter aid		
Wet mycelium cake	73.5 kg	
Double treatment with anhydrous methanol	316 liters	
Dehydrated mycelium cake		
Double extraction with methylene chloride		
Methylene chloride extract	367 liters	349
Evaporation below 35°C under reduced pressure		
Oily suspension of β-carotene crystals		
Mixing and washing with acetone and then filtration	37 liters	
Crude β-carotene		
Dissolution in chloroform	44 liters	
Evaporation under reduced pressure to half volume		
Crystallization at 20°C in the dark under nitrogen by addition of ethanol	30 liters	
Purified β-carotene		319
Dissolution in a benzene and 95% ethanol mixture (1 : 1·4 ratio)	145 liters	
Crystallization Filtration		
Pure β-carotene		292

FIGURE 5. Recovery and purification of β-carotene. (From Upjohn Co's. description.)

3. β-Carotene Assay

The mixed cultures of *B. trispora* produce mainly the trans isomer of β-carotene. To determine its concentration in the broth the wet mycelium is harvested and either mechanically disintegrated or thoroughly dried. In both cases the carotenoids are extracted with an appropriate solvent and the extract is chromatographed on magnesium oxide and celite mixture or alumina columns, and the optical density of the eluates is measured at 450 nm. This final stage can also be used directly for assaying crude or pure preparation of β-carotene (Ciegler, 1965a; Liaaen-Jensen, 1971).

III. PRODUCTION OF LYCOPENE

A. Earlier Developments

Lycopene is a deep red, tomato-like pigment which shows no provitamin A activity and which is used as a coloring agent for foodstuffs. Microorganisms are not rich sources of this material. However it can be obtained by the mixed culture of *B. trispora,* strains NRRL 9216 (+) and NRRL 9159 (−), in a medium including glucose, oleic acid, and fish stick liquor. The broth pH must be kept above 6.6 by adding sodium carbonate. The lycopene yield is about 150 mg/liter (Swarthout, 1963).

The methods of extraction and determination are similar to those used for β-carotene.

B. Recent Developments

Recent developments have permitted the production of higher yields of lycopene, using cultures of either *B. trispora* or *Streptomyces chrestomyceticus* var. *rubescens.*

1. Blakeslea trispora Cultures

Recent studies on *B. trispora* cultures to improve production of β-carotene have shown that a number of substances, particularly nitrogenous heterocyclic bases (such as imidazole, pyridine, quinoline, and some of their substituted derivatives), can block carotenoid biosynthesis at the level of lycopene (Ninet *et al.*, 1965a, 1969). The position of the substitution is critical, some of the substituted substances having no effect on lycopene production or even enhancing β-carotene production. The effective agents, added to the broth at concentrations of 0.1–0.3 gm/liter, enable the production of lycopene to reach 1 gm/liter. The culture process is virtually identical with that for β-carotene production with the exception that β-ionone is absent, the latter having no effect on lycopene production. The details of the process are given by Gutcho (1973b).

Studies on the pigmentation of fruit and vegetables led to the discovery of other substances that can enhance the production of lycopene by fungi or photosynthetic bacteria (Coggins et al., 1970; Hsu et al., 1972, 1974; Yokoyama, 1974). These substances are essentially tertiary amines or their derivatives; for example, triethylamine, tributylamine, hordenine, 2-(4-chlorophenylthio)triethylamine, α-diethylaminopropiophenone, 4-[β-(diethylamino)ethoxy]benzophenone, and chalcone derivative of triethylamine. They act by derepressing the enzymes involved in carotene synthesis in B. trispora and by inhibiting the carotenoid cyclization enzymes (Elahi et al., 1973). A 5-day mixed culture of B. trispora (+) NRRL 2895 and (−) NRRL 2896 in a medium based on cotton seed and corn flours and soya bean oil, supplemented at 48 hours with 1.5 gm/liter of triethylamine, will yield 0.4 gm lycopene per liter of broth, whereas in the presence of 1.25 gm/liter 4-(2-diethylaminoethyl)phenol (hordenine), mixed B. trispora (+) NRRL 9216 and (−) NRRL 9156 will provide 0.7 gm lycopene per liter (Ito et al., 1973; Ito and Yamaguchi, 1973). Nicotine, which favors lycopene biosynthesis by several microorganisms, was apparently not tested with Blakeslea species (Goodwin, 1972; McDermott et al., 1973; Elahi et al., 1973).

2. *Streptomyces chrestomyceticus* subsp. *rubescens Cultures*

This strain was obtained after repeated mutagenic treatments of Streptomyces chrestomyceticus. After a 6-day growth in a medium including starch, soya bean flour, and ammonium sulfate, the yield was 0.5 gm lycopene per liter (Bianchi et al., 1968; Gutcho, 1973b).

IV. PRODUCTION OF XANTHOPHYLLS AND OTHER PIGMENTS

A. Earlier Developments

Xanthophylls are mixtures of colored carotenoids which include lutein, canthaxanthin, cryptoxanthin, neoxanthin, violaxanthin, and zeaxanthin. These substances are coloring agents useful in poultry farming and are supplied by the ground maize kernels ordinarily included in the diet. However, it was also tempting to add to the feedstuffs the dried pigmented

biomasses of one of the numerous microorganisms which produce xanthophylls.

The most extensively studied microorganisms are:

1. Basidiomycetes of the Dacrymycetaceae family; an example is
Dacrymyces deliquescens which, when cultured over 5 days in liquid
cultures containing glucose, glycerol, and corn steep liquor and under
external illumination, provides 40 mg xanthophylls per liter (i.e., 4
mg/gm of dried mycelium) together with some β-carotene (Farrow and
Tabenkin, 1961).

2. Green algae, particularly *Spongiococcum excentricum* and *Chlorella pyrenoidosa*. The latter, in a 7-day liquid culture with continuous addition of glucose, urea, and phosphate and under external illumination, yields 650 mg xanthophylls per liter (i.e., 2.1 mg/gm of dried
biomass) (Kathrein, 1960, 1963, 1964; Farrow and Tabenkin, 1964a,b;
Thériault, 1965).

Antioxidants are required, as with β-carotene feedstuffs, to prevent
oxidation of the xanthophylls. For instance, ethoxyquin is added to the
broth or the biomass at a 2.5% level.

However, none of these processes has as yet been used industrially.

B. Recent Developments

1. Zeaxanthin

Recent studies devoted to zeaxanthin production have concentrated
on the use of marine flavobacteria. When those are grown in a medium
based on glucose and corn steep, the zeaxanthin content amounts to
10–40 mg/liter and supplementation with palmitic esters, methionine,
pyridoxine, and ferrous salts raises the zeaxanthin yields up to 190
mg/liter (i.e., 16.4 mg/gm of dried biomass) (Shocher and Wiss, 1972;
Shepherd and Dasek, 1974; Shepherd *et al.*, 1974). When the temperature is reduced and the nutrients are added continuously, a content of
335 mg/liter can be obtained with a mutant strain of *Flavobacterium*
(Dasek *et al.*, 1973). The methods of extraction and assay for zeaxanthin
are similar to those for the carotenoids.

2. Other Xanthophylls

Studies relating to the production of carotenoid mixture of undeter-
mined composition and which might be used as a feed additive in the
form of the fermentation biomass are briefly discussed under this head-
ing.

While the actinomycetes generally produce low yields of carotenoids (Krasil'nikov et al., 1965; Garkavenko et al., 1972), a *Streptomyces chrestomyceticus* var. *aurantioideus* mutant cultured over 5 days in a medium containing glucose and dried yeast gives 0.5 gm/liter of total carotenoids (Marnati et al., 1967; Gutcho, 1973e), whereas *Streptomyces mediolani* cultured over 5 days in a medium containing dextrin, casein, and corn steep liquor yields 0.9 gm/liter with a large proportion of isorenieratene and its oxidation derivatives (Franceschi and Grein, 1969; Gutcho, 1973a).

A strain of *Mycobacterium phlei,* when grown over 7 days on beet molasses and urea and under illumination, yields 0.3 gm/liter of oxycarotenoids (Noury and Van Der Lande, 1967), whereas mycobacteria or corynebacteria grown on paraffins produce very low amounts (Kyowa Hakko Kogyo Co., Ltd., 1972; Yamada and Furukawa, 1973).

With a number of yeasts grown on various sugars, ethanol, or paraffins, the content is less than 5 mg/liter (i.e., 0.5 mg/gm of dried biomass) (Nikolaev, 1966; Vallet and Brabant, 1971; Bekers et al., 1972; Protein Biosynthesis Institute, 1975).

Possibly a more extensive consideration of different activating agents and the development of improved mutant strains could lead to higher production in the future.

V. MARKETING PROSPECTS FOR CAROTENOIDS

The market for carotenoids is small at present, the worldwide sales of β-carotene amounting to some 27 tons/year, one-half of which is accounted for by the United States market. The carotenoids presently sold are those derived from synthetic and plant sources and consist mainly of β-carotene, canthaxanthin, citrinaxanthin, and two C 30 apocarotenoids. Most of the xanthophylls used are extracted from plant sources, such as maize kernel, alfalfa, xanthasouci, and algae. The fermentation derived products have not so far been exploited commercially, mainly due to the limited markets and relatively low prices of the competitive products. β-Carotene is mainly used as a coloring agent, first of all in the food industry and fairly less often in drugs and cosmetics manufacturing. As a provitamin A its market seems rather narrow. There are possibilities for its use in animal feedstuffs and as a source material in canthaxanthin synthesis. Lycopene is listed as a food coloring agent for some countries, but it is rarely used.

The other products are principally used in poultry farming as feedstuffs, the sales of citrinaxanthin for this use being 3–5 tons/year.

There is also some interest in the use of canthaxanthin as a "suntanning" agent taken orally.

As indicated above, fermentation production of carotenoids has not so far been of commercial value, and interest in studying new processes has been decreasing, except perhaps in Eastern Europe. However, any process producing a major improvement in yields or any factors markedly changing the price of the present raw materials could completely reverse this situation and make the fermentation derived carotenoids a viable proposition.

REFERENCES

Barnett, H. L., Lilly, V. G., and Krause, R. F. (1956). *Science* **123,** 141.

Bekers, M. J., Krause, I. J., Seile, M. K., Viesturs, U. E., Tschazkij, P. A., and Kluna, G. W. (1972). German Patent 2,059,387.

Beytia, E. D., and Porter, J. W. (1976). *Annu. Rev. Biochem.* **45,** 113–142.

Bianchi, M. L., Franceschi, G., Marnati, M. P., and Spalla, C. (1968). British Patent 1,128,440 (to Farmaceutici Italia).

Bjork, L., and Neujahr, H. Y. (1969). *Acta Chem. Scand.* **23,** 2908–2909.

Bohinski, R. C. (1970). U.S. Patent 3,492,202 (to Chas. Pfizer and Co., Inc.).

Caglioti, L., Cainelli, G., Camerino, B., Mondelli, R., Prieto, A., Quilico, A., Salvatori, T., and Selva, A. (1964). *Chim. Ind. (Milan)* **46,** 1–6.

Cederberg, E., and Neujahr, H. Y. (1969). *Acta Chem. Scand.* **23,** 957–961.

Ciegler, A. (1965a). *Adv. Appl. Microbiol.* **7,** 1–34.

Ciegler, A. (1965b). U.S. Patent 3,226,302 (to U.S.).

Coggins, C. W., Jr., Henning, G. L., and Yokoyama, H. (1970). *Science* **168,** 1589–1590.

Dasek, J., Shepherd, D., and Traelnes, K. R. (1973). Belgian Patent 790,289 (to Société des Produits Nestlé S.A.).

Elahi, M., Chichester, C. O., and Simpson, K. L. (1973). *Phytochemistry* **12,** 1627–1632.

Farmaceutici Italia (1965). French Patent 1,414,828.

Farrow, W. M., and Tabenkin, B. (1961). U.S. Patent 2,974,044 (to Hoffmann-La Roche and Co.).

Farrow, W. M., and Tabenkin, B. (1964a). French Patent 1,376,027 (to Hoffmann-La Roche and Co.).

Farrow, W. M., and Tabenkin, B. (1964b). Belgian Patent 639,531 (to Hoffmann-La Roche and Co.).

Feofilova, E. P., and Bekhtereva, M. N. (1976). *Mikrobiologiya* **45,** 557–558.

Feofilova, E. P., Fateeva, T. V., and Arbuzov, V. A. (1976). *Mikrobiologiya* **45,** 169–171.

Franceschi, G., and Grein, A. (1969). Belgian Patent 723,901 (to Farmaceutici Italia).

Garkavenko, A. I., Savchenko, L. F., and Razumovskii, P. N. (1972). Russian Patent 327,250.

Gooday, G. W. (1974). *Annu. Rev. Biochem.* **43,** 39–43.

Goodwin, T. W. (1972). *Prog. Ind. Microbiol.* **11,** 29–88.

Gutcho, S. J. (1973a). *Chem. Technol. Rev.* (Chemicals by Fermentation) **19,** 304.

Gutcho, S. J. (1973b). *Chem. Technol. Rev.* (Chemicals by Fermentation) **19,** 308–311.

Gutcho, S. J. (1973c). *Chem. Technol. Rev.* (Chemicals by Fermentation) **19,** 311–315.

Gutcho, S. J. (1973d). *Chem. Technol. Rev.* (Chemicals by Fermentation) **19,** 315–318.

Gutcho, S. J. (1973e). *Chem. Technol. Rev.* (Chemicals by Fermentation) **19**, 318–320.
Gutcho, S. J. (1973f). *Chem. Technol. Rev.* (Chemicals by Fermentation) **19**, 320–322.
Hanson, A. M. (1967). *In* "Microbial Technology" (H. J. Peppler, ed.), pp. 222–250. Van Nostrand-Reinhold, Princeton, New Jersey.
Hesseltine, C. W., and Benjamin, C. R. (1959). *Mycologia* **51**, 887–901.
Hsu, W. J., Yokoyama, H., and Coggins, C. W., Jr. (1972). *Phytochemistry* **11**, 2985–2990.
Hsu, W. J., Poling, S. M., and Yokoyama, H. (1974). *Phytochemistry* **13**, 415–419.
Isler, O. (1977). *Experientia* **33**, 555–573.
Ito, S., and Yamaguchi, Y. (1973). Japanese Patent 73/16,189 (to Takasago Perfumery Co., Ltd.).
Ito, S., Yamaguchi, Y., Yoshida, T., and Komatsu, A. (1973). Japanese Patent 73/16,190 (to Takasago Perfumery Co., Ltd.).
Jäger, H. K. (1966). French Patent 1,457,619 (to Upjohn Co.).
Kathrein, H. R. (1960). U.S. Patent 2,949,700 (to Grain Processing Corp.).
Kathrein, H. R. (1963). U.S. Patent 3,108,402 (to Grain Processing Corp.).
Kathrein, H. R. (1964). U.S. Patent 3,142,135 (to Grain Processing Corp.).
Kolot, F. B., Vakulova, L. A., and Samokhvalov, G. I. (1973). *In* "Vitaminy i Vitaminnye Preparaty" (V. A. Yakovlev, ed.), pp. 241–246. Izd. "Meditsina," Moscow.
Krasil'nikov, N. A., Korenyako, A. I., Gavrilova, O. A., Artamanova, O. I., Leonov, L. I., Nikitina, N. I., Dushenkova, L. I., Ulezlo, I. V., and Khokhlova, Y. M. (1965). Russian Patent 169,749.
Kyowa Hakko Kogyo Co., Ltd. (1972). Japanese Patent 72/04,503.
Liaaen-Jensen, S. (1971). *In* "Carotenoids" (O. Isler, ed.), pp. 61–73. Birkhäuser, Basel.
McDermott, J. C. B., Britton, G., and Goodwin, T. W. (1973). *J. Gen. Microbiol.* **77**, 161–171.
Mackinney, G., Nakayama, T., Buss, C. D., and Chichester, C. O. (1952). *J. Am. Chem. Soc.* **74**, 3456–3457.
Marnati, M. P., Prieto, A., and Spalla, C. (1967). U.S. Patent 3,330,737 (to Farmaceutici Italia).
Nakamura, T., Tsuji, M., Abe, H., Fujii, K., Makimoto, T., and Ootera, J. (1974). Japanese Patent 74/31,896 (to Kuraray Co., Ltd.).
Nikolaev, P. I. (1966). Russian Patent 178,449.
Ninet, L., Renaut, J., and Tissier, R. (1963a). French Patent 1,325,656 (to Rhône-Poulenc S. A.).
Ninet, L., Renaut, J., and Tissier, R. (1963b). French Patent 1,344, 264 (to Rhône-Poulenc S. A.).
Ninet, L., Renaut, J., and Tissier, R. (1964). French Patent 1,377,523 (to Rhône-Poulenc S. A.).
Ninet, L., Renaut, J., and Tissier, R. (1965a). French Patent 1,403,839 (to Rhône-Poulenc S. A.).
Ninet, L., Renaut, J., and Tissier, R. (1965b). French Patent 1,412,506 (to Rhône-Poulenc S. A.).
Ninet, L., Renaut, J., and Tissier, R. (1966). French Patent 1,449,879 (to Rhône-Poulenc S. A.).
Ninet, L., Renaut, J., and Tissier, R. (1969). *Biotechnol. Bioeng.* **11**, 1195–1210.
Noury and Van Der Lande N. V. Koninklijke Industrieele Maatschappij (1967). French Patent 1,467,317.
Pazola, Z., Switek, H., Janicki, J., and Michnikowska, W. (1969). U.S. Patent 3,476,646 (to Centralne Laboratorium Przemyslu Koncentratow Spozywczych-Poland).
Pazola, Z., Switek, H., and Janicki, J. (1971). East German Patent 76,661.
Prieto, A., Spalla, C., Bianchi, M., and Biffi, G. (1964). *Chem. Ind. (London)* **13**, 551.

Protein Biosynthesis Institute (1975). Russian Patent 448,219.

Purcell, A. E., and Walter, W. M., Jr. (1971). U.S. Patent 3,579,424 (to U.S.).

Rao, S., and Modi, V. V. (1977). *Experientia* **33,** 31–33.

Reyes, P., Chichester, C. O., and Nakayama, T. O. M. (1964). *Biochim. Biophys. Acta* **90,** 578–592.

Schocher, A. J., and Wiss, O. (1972). Belgian Patent 770,744 (to Hoffmann-La Roche and Co.).

Sebek, O. K., and Jäger, H. K. (1964). *Abstr. 148*th *Meet., Am. Chem. Soc.* pp. 9Q–10Q.

Shepherd, D., and Dasek, J. (1974). Belgian Patent 816,766 (to Société des Produits Nestlé S.A.).

Shepherd, D., Dasek, J., and Carels, M. (1974). Belgian Patent 816,767 (to Société des Produits Nestlé S.A.).

Skryabin, G. K. (1967). Russian Patent 185,316.

Swarthout, E. J. (1963). French Patent 1,333,942 (to Miles Laboratories, Inc.).

Thériault, R. J. (1965). *Appl. Microbiol.* **13,** 402–416.

Upjohn Co. (1965a). Dutch Patent 64/11,184.

Upjohn Co. (1965b). Dutch Patent 65/00,788.

Vallet, G., and Brabant, P. G. (1971). French Patent 2,057,482 (to Distillerie Brabant et Cie.).

Weedon, B. C. L. (1971). *In* "Carotenoids" (O. Isler, ed.), pp. 29–59. Birkhäuser, Basel.

Yamada, K., and Furukawa, T. (1973). Japanese Patent 73/21,515 (to Mitsui Petrochemical Industries, Ltd.).

Yokoyama, H. (1974). Cited in *Chem. Eng. News.* **52**(15), 19–21.

Zajic, J. E. (1964). U.S. Patent 3,128,236 (to Grain Processing Corp.).

Subject Index

A